MOLECULAR STRUCTURES AND VIBRATIONS

Molecular Structures and Vibrations

SELECTED CONTRIBUTIONS TO THEORETICAL AND EXPERIMENTAL STUDIES OF POLYATOMIC MOLECULES BY SPECTROSCOPIC METHODS AND GAS ELECTRON DIFFRACTION

EDITED BY

S.J. CYVIN

Dr. Techn.

Professor of Theoretical Chemistry
Institute of Physical Chemistry
Technical University of Norway
Trondheim

ELSEVIER PUBLISHING COMPANY

AMSTERDAM·LONDON·NEW YORK

ELSEVIER PUBLISHING COMPANY
335 JAN VAN GALENSTRAAT
P.O. BOX 330, AMSTERDAM, THE NETHERLANDS

AMERICAN ELSEVIER PUBLISHING COMPANY, INC.
52 VANDERBILT AVENUE
NEW YORK, NEW YORK 10017

Cover design by E. ASTRUP

Library of Congress Card Number: 72-87952

ISBN 0-444-41069-4

With 63 illustrations and 79 tables.

Printed in The Netherlands

Preface

OTTO BASTIANSEN

My first stumbling steps on the road of science were taken during World War II in an occupied country. As thousands of other students at that time most of my energy was spent outside the university world and on activities other than study or research. By the end of the war I discovered that I had never in my life met in person one single scientist from outside my own country, though I knew hundreds of them by name and by their work. I also discovered that during the years of scientific isolation I had made myself a picture of a great many of the persons behind those names, without knowing anything about them except their publications. I had even, - certainly subconciously, - made myself an image of their outer appearance. I was not even aware of this till I met in person my first literature acquaintance and discovered to my disappointment that he had no resemblance whatsoever with the fictitious image.

How different isn't it now? Today when you open a journal in your own field and study the list of authors, you find a list of friends. Among them you find dear friends whom you have known for many years. Often is friendship among scientific colleagues of a higher quality than usual friendship. On one hand the common devotion to science ties people together. On the other hand friendship among scientific colleagues is often won against national, linguistic and political barriers, sometimes even against a barrier of scientific antagonism. The origin of a friendship may well be a scientific controversy. I have several friends I won over an argument. The quality of the friendship seems not to depend upon whether you won the argument or suffered a defeat.

This interhuman contact has been made possible by modern technology: You dial yourself into the offices of colleagues over continents. A trans-Atlantic call is today easier to get through than a call to a neighbouring city was a few years ago. Letters bringing ideas, data and personal messages reach from anywhere to everywhere in a few days. Scientists meet each other more and more frequently in universities, in research centra, on meetings, and at airports. They also more often come to stay with each other for longer periods of time working together on problems of joint interest.

The ways new ideas and new data are spread over the world of science are many. Obviously the frequent intercollegial contacts contribute to accelerate the informal exchange of knowledge. Usually a paper has been subject to scrutinizing discussions among fellow workers at various locations before finding its final shape in a journal. Often an article has even lost its value as a novelty when it finally reaches the libraries. In spite of this, the flow of conventional scientific journals and books seems not to slow down. Even the contemporary economical setback of science does not seem to hamper severely the rate of publication. Most probably the classical scientific journal is going to continue to play the dominant part for a long time to come. Apparently scientists find the scientific journal the useful way of documenting their thoughts. The scientific journals also act to establish priorities, and in that way they continue to contribute building up the intercollegial status hierarchy.

One of the greatest scientists of our time, the late J. D. Bernal, was for many years interested in modernizing the means of spreading scientific information. Undoubtedly his ideas have already had valuable effects and may in the future be of still greater importance. It seems necessary to experiment on new types of publication, and perhaps the present volume may be considered as an example of such an experiment. The idea of inviting a series of scientists, representing one field or a group of related fields, to contribute articles to an isolated volume, is not a new idea. That is the way memorial publications are made, and such publications are not always well accepted by reviewers. However, there is one great difference I think I see between a memorial publication and the present volume. The guests invited to contribute to the former kind of publication are rather chosen on criteria of prominence and status than on the criterium of obtaining a scientifically exciting volume. I think that the editor of this volume has chosen his authors among people whom he knows have an original idea or some particularly interesting findings. The editor has for many years been a central person in his field. A large number of people have come to work with him in Trondheim, and he has co-workers spread over the whole world. He knows what kind of problems these people are dealing with and must have felt a need for seeing some of their problems presented together in one volume.

Perhaps I am overinterpreting the ideas and the aims of our friend Sven J. Cyvin when I am trying to find a deeper philosophy behind his initiative. Perhaps the driving force of his task is primarily the joy of seeing some of his friends' latest ideas being put together in one volume. But whatever Cyvin's motivation, he made us sit down and contribute to a real international enterprise. And the editor has not been saving himself. He is not playing the part of the big administrator putting other people to work. He has had his hands full of all kinds of work, ranging from the chief editor's work to the typist's work. Every single word of this volume has been typed for the offset edition by Cyvin himself!

Scientists have for good reasons claimed that their activity promotes international understanding. I like to see this volume as a genuine contribution to the best form of internationalism. Internationalism has been put into practice in many ways. One of the classical approaches is the missionary approach: Some people find out that their own idea or creed is so good that it ought to be spread all over the world with the hope that their ideology eventually is going to prevail and at the end be the only one surviving. Another approach is to work towards a world unity based on an average of all cultures, like the melting pot culture average of Ibsen's "Peer Gynt". To me a genuine internationalism must be based upon coexistence of all the cultures of the world. It is not possible for all of us to learn to understand the multitude of cultures and languages of the world, but it is necessary to learn to respect them. The present book contains shorter abstracts in languages that are not commonly used for scientific communication. I take these contributions as a symbol for the respect and the appreciation of cultures that are unknown to many of us.

When estimating the value of science for international understanding the role of scientists as culture mediators is hardly the most important. However, the fact that so many people with their roots in so many different cultures are linked together in an international friendship-union must be of considerable importance. If one puts points on a globe marking out all university and research centra in the world, and then draws one line between such points for each intercollegial friendship relation, one would end up with a multitudinous network of lines covering the whole globe. It is our hope that this friendship network system may help acting as a safety net for an endangered world. It is my hope that Cyvin's activity as a scientist and his efforts with this volume may have a little effect also to this end.

List of Contributors

Ingrid ALFHEIM; Siv.ing.[1]
G. ARULDHAS; M.Sc., Ph.D.[2]
K. BABU JOSEPH; M.Sc., Ph.D.[2]
Enrique J. BARAN; Professor, Dr.Chem.[3]
Otto BASTIANSEN; Professor, dr.phil.[4]
Jon BRUNVOLL; Siv.ing., lic.techn.[1]
D.W.J. CRUICKSHANK; Professor, Sc.D.[5]
Bjørg N. CYVIN; Siv.ing.[1]
Sven J. CYVIN; Professor, dr.techn.[1]
Giovanna DELLEPIANE; Professor[6]
Nina M. EGOROVA; B.Math.[7]
Irene ELVEBREDD; Siv.ing.[1]
Mariangela GUSSONI; Dr.[6]
Georg HAGEN; Siv.ing., lic.techn.[1]
István HARGITTAI; Dr., C.Sc.[8]
Kenneth HEDBERG; Professor, Ph.D.[9]
G. KOVÁCS; Dr.[10]
Kozo KUCHITSU; Professor, D.Sc.[11]
Vladimir S. MASTRYUKOV; Dr.phil.[12]
Tove MOTZFELDT; Cand.real.[4]
Achim MÜLLER; Professor, Dr.rer.nat.[13]
Valeria D. OPPENHEIM; Dr.phil.[12]
G. Stuart PAWLEY; Ph.D.[14]
Øystein RA; Siv.ing., dr.techn.[1]
Nikolay G. RAMBIDI; Professor, Sc.D.[7]
Lothar SCHÄFER; Assoc.Professor, Dr.rer.nat.[15]
Karlheinz SCHMIDT; Dipl.Chem.[13]
Hans Martin SEIP; Siv.ing., dr.phil.[4]
Alan SNELSON; Ph.D.[16]
Georg Ole SØRENSEN; Lektor, cand.polyt.[17]
P.SOHÁR; C.Sc., Ph.D.[18]

Vladimír ŠPIRKO; Ing., C.Sc.[19]
Reidar STØLEVIK; Cand.real.[4]
Natalia A. TARASENKO; Dr.phil.[12]
F. TÖRÖK; Professor, D.Sc.[20]
V. UNNIKRISHNAN NAYAR; M.Sc., Ph.D.[21]
Gy. VARSÁNYI; Professor, D.Sc.[22]
Lev V. VILKOV; Dr.sc.[12]
B. VIZI; Senior lecturer, dr.phil.[23]
Giuseppe ZERBI; Professor[6]
Z. ZUBOVICS; M.Sc.[18]

ADDRESSES

1 Institute of Physical Chemistry, The Technical University of Norway, N-7034 Trondheim-NTH, Norway.

2 Department of Physics, University of Cochin, Cochin-22, India.

3 Cátedra de Química Inorgánica, Facultad de Ciencias Exactas, Universidad Nacional de la Plata, La Plata, Argentina.

4 Department of Chemistry, University of Oslo, Oslo 3, Norway.

5 Department of Chemistry, University of Manchester Institute of Science and Technology, Manchester M60 1QD, England.

6 Consiglio Nazionale delle Ricerche, Istituto di Chimica delle Macromolecole, Via A. Corti 12, 20133 Milano, Italy.

7 Institute for High Temperatures, Korowinskoye rd. 127412 Moscow, USSR.

8 Center for Studies on Chemical Structures, Hungarian Academy of Sciences, Budapest, VIII., Puskin u. 11-13, Hungary.

9 Department of Chemistry, Oregon State University, Corvallis, Oregon 97331 USA.

10 Central Research Institute for Chemistry, Hungarian Academy of Sciences, Budapest, II., Pusztaszeri u. 57-69, Hungary.

11 Department of Chemistry, Faculty of Science, The University of Tokyo, Bunkyo-ku, Tokyo, Japan.

12 Laboratory of Electron Diffraction, Department of Chemistry, Moscow State University, Moscow 117234, USSR.

13 Sektion für Spektroskopie und Molekülstruktur, Institut für Chemie, Universität Dortmund, 46 Dortmund, West Germany.

14 Department of Physics, University of Edinburgh, Edinburgh EH9 3JZ, Scotland.

15 Department of Chemistry, University of Arkansas, Fayetteville, Arkansas 72701 USA.

16 IIT Research Institute, 10 West 35 Street, Chicago, Illinois 60616 USA.

17 University of Copenhagen, Chemical Laboratory V, H.C.Ørsted Institute, DK-2100 Copenhagen, Denmark.

18 Research Institute for Pharmaceutical Chemistry, Budapest, IV., Szabadságharcosok u. 47-49, Hungary.

19 Institute of Physical Chemistry, Czechoslovak Academy of Sciences, Prague 6, Flemingovo n. 2, Czechoslovakia.

20 Eötvös Lóránd University, Department of General and Inorganic Chemistry, Budapest, VIII., Múzeum krt. 6-8, Hungary.

21 Department of Physics, St. Thomas College, Kozhencherry, Kerala, India.

22 Department of Physical Chemistry, The Technical University, Budapest, XI., Budofoki u. 8, Hungary.

23 Institute of General and Inorganic Chemistry, University for Chemical Industries, Veszprém, Hungary.

Contents

CHAPTER 1

Molecular Vibrations, Coriolis Coupling, and Centrifugal Distortion: Basic Concepts

S. J. CYVIN

1.1. Introduction

The theory of molecular vibrations was presented in the book of Wilson, Decius, and Cross[1], which is recognized as a standard reference, along with the Russian counterpart[2] published some years earlier. The famous **GF** matrix method due to Wilson[3,4] involves a formulation of the important secular equation in the theory of molecular vibrations, which was simultaneously developed and formulated by Eliashevich.[5] As to the theory of Coriolis coupling, the formulation of Meal and Polo[6,7] involving ζ^{α} and $\mathbf{C}^{\alpha}$ matrices has been generally adopted; cf., e.g., Allen and Cross,[8] and Cyvin.[9] The latter monograph is mainly devoted to the studies of mean square amplitudes and related quantities, and includes a treatise of the $\boldsymbol{\Sigma}$ matrix.[10] The theory of centrifugal distortion, as advanced by Kivelson and Wilson,[11,12] is also treated in the books by Allen and Cross[8] and by Wollrab.[13] It has in part been reformulated by several authors.[14-16] The $\mathbf{T}_{\varsigma}$ matrices appear in the modification of this theory due to Cyvin et al.[17,18]

The purpose of this chapter is to recall some of the basic definitions and relations from the aforementioned topics, in order to facilitate the references in the subsequent chapters, and to summarize an essential part of the applied notation. The material is presented in a condensed form without detailed explanations of the applied symbols, for which the above cited works should be consulted. In particular it is adherred to the usage of symbols in Ref. 9, but in contrast to the formulation therein, the below relations are formulated for real matrices throughout. Furthermore the present notation distinguishes between vectors and matrices.

1.2. Matrices of Geometrical Nature

1. The equilibrium positions of atoms in a molecular model may be given in terms of the (equilibrium) position vectors,

$$\vec{R}_a = \{X_a^{\,e},\ Y_a^{\,e},\ Z_a^{\,e}\}\,; \tag{1.1}$$

$a = 1,2,\ldots,N$. $\mathbf{R}^e$ is used to denote the $3N \times 1$ column matrix of the vector components from (1), i.e.

$$\mathbf{R}^e = \{X_1^{\,e}, Y_1^{\,e}, Z_1^{\,e}, \ldots\ldots, X_N^{\,e}, Y_N^{\,e}, Z_N^{\,e}\}. \tag{1.2}$$

The elements of the inertia tensor* in the equilibrium position are given in matrix notation by[20]

$$I_{\alpha\beta}^{\ \ e} = (\mathbf{R}^e)'\,\mathbf{I}^\alpha \mathbf{m}\,(\mathbf{I}^\beta)'\mathbf{R}^e\,; \tag{1.3}$$

$\alpha,\ \beta = x, y$ or z. The relation holds whenever α and β are equal or different. $\mathbf{m}$ is the $3N \times 3N$ diagonal matrix of atomic masses, and the $\mathbf{I}^\alpha$ matrices consist of N 3×3 diagonal blocks as given in the following.

---

* The products of inertia are defined here as

$$I_{xy} = -\sum_a m_a X_a Y_a\,,\ I_{yz} = -\sum_a m_a Y_a Z_a\,,\ I_{zx} = -\sum_a m_a Z_a X_a\,.$$

This conforms with the definition of Kivelson and Wilson[11,12] and other works based on their papers. Unfortunately in the pioneering work of H.H. Nielsen[19] the other alternative without the minus signs has been used. The latter definition (opposite to ours) was also adopted by Wilson, Decius, and Cross.[1]

$$I^x_a = \begin{bmatrix} 0 & 0 & 0 \\ 0 & 0 & 1 \\ 0 & -1 & 0 \end{bmatrix}, \quad I^y_a = \begin{bmatrix} 0 & 0 & -1 \\ 0 & 0 & 0 \\ 1 & 0 & 0 \end{bmatrix}, \quad I^z_a = \begin{bmatrix} 0 & 1 & 0 \\ -1 & 0 & 0 \\ 0 & 0 & 0 \end{bmatrix}. \qquad (1.4)$$

2. The $\mathbf{B}$ matrix determines the transformation between Cartesian displacement coordinates ($\mathbf{X}$) and a set of internal vibrational coordinates, $\mathbf{S}$, which incidentally may be identified with symmetry coordinates;

$$\mathbf{S} = \mathbf{B}\mathbf{x}. \qquad (1.5)$$

The elements of the $\mathbf{B}$ matrix are purely geometrical quantities. They may be determined from the position vectors (1) for the model in question.

The components of $\vec{s}$ vectors (cf. Section 3.3.1 of Ref.9) just constitute the $\mathbf{B}$ matrix elements.

3. The $\mathbf{G}$ matrix (connected with vibrational kinetic energy) is

$$\mathbf{G} = \mathbf{B}\mathbf{m}^{-1}\mathbf{B}'. \qquad (1.6)$$

It is clearly determined by the geometry of the molecule and the atomic masses. The elements of $\mathbf{G}$ are given in terms of $\vec{s}$ vectors by

$$G_{ij} = \sum_a \mu_a (\vec{s}_{ia} \cdot \vec{s}_{ja}), \qquad (1.7)$$

where $\mu_a = 1/m_a$.

4. For the Coriolis $\mathbf{C}^\alpha$ matrices one has similarly to Eq. (6):

$$\mathbf{C}^\alpha = \mathbf{B}\mathbf{I}_\mu^\alpha \mathbf{B}', \qquad (1.8)$$

where $\mathbf{I}_\mu^\alpha = \mathbf{m}^{-1}\mathbf{I}^\alpha$. In terms of $\vec{s}$ vectors:

$$C_{ij}^\alpha = \sum_a \mu_a (\vec{s}_{ia} \times \vec{s}_{ja}) \cdot \vec{e}_\alpha \; ; \qquad (1.9)$$

$\vec{e}_x = \{1, 0, 0\}$, $\vec{e}_y = \{0, 1, 0\}$, and $\vec{e}_z = \{0, 0, 1\}$.

5. The $\mathbf{A}$ matrix defines the opposite transformation of Eq. (5);

$$\mathbf{x} = \mathbf{A}\mathbf{S}. \tag{1.10}$$

Notice that the (truncated) **B** and **A** matrices have the dimensions of $s \times 3N$ (where $s = 3N - 6$ or $3N - 5$) and $3N \times s$, respectively. Hence these two matrices are not simply inverse of each other, but at least $\mathbf{BA} = \mathbf{E}$ holds, where **E** has the dimension of $s \times s$. The matrix may be determined by the Crawford-Fletcher formula[21]

$$\mathbf{A} = \mathbf{m}^{-1}\mathbf{B}'\mathbf{G}^{-1}, \tag{1.11}$$

which automatically takes care of the Eckart conditions through the inverse **G** matrix.

The components of $\vec{t}^{\,o}$ vectors (cf. Section 3.3.2 of Ref.9) constitute the **A** matrix elements.

6. Let **M** be the inverse **G** matrix; then

$$\mathbf{M} = \mathbf{A}'\mathbf{m}\mathbf{A}. \tag{1.12}$$

In terms of $\vec{t}^{\,o}$ vectors:

$$M_{ij} = \sum_a m_a (\vec{t}^{\,o}_{ia} \cdot \vec{t}^{\,o}_{ja}). \tag{1.13}$$

7. For the $\mathbf{MC}^{\alpha}$ matrix product one has:

$$\mathbf{MC}^{\alpha} = \mathbf{A}'\mathbf{I}^{\alpha}\mathbf{B}', \tag{1.14}$$

$$(MC^{\alpha})_{ij} = \sum_a (\vec{t}^{\,o}_{ia} \times \vec{s}_{ja}) \cdot \vec{e}_{\alpha}. \tag{1.15}$$

8. For $\bar{\mathbf{C}}^{\alpha}$, another matrix of relevance to Coriolis coupling, and its elements, one has:

$$\bar{\mathbf{C}}^{\alpha} = \mathbf{MC}^{\alpha}\mathbf{M}, \tag{1.16}$$

$$\bar{\mathbf{C}}^{\alpha} = \mathbf{A}'\mathbf{I}_m^{\alpha}\mathbf{A}, \tag{1.17}$$

$$\bar{C}_{ij}^{\alpha} = \sum_a m_a (\vec{t}^{\,o}_{ia} \times \vec{t}^{\,o}_{ja}) \cdot \vec{e}_{\alpha}. \tag{1.18}$$

In Eq. (17) $I_m^\alpha = mI^\alpha$.

9. The $T_{\alpha\beta,S}$ matrices, which pertain to the centrifugal distortion, are given by[20]

$$T_{\alpha\beta,S} = 2BI^\alpha(I^\beta)'R^e = 2BI^\beta(I^\alpha)'R^e. \quad (1.19)$$

These matrices are invariant with respect to a translation of the Cartesian coordinate axes.

The partial derivatives at equilibrium of the inertia tensor components with respect to the S_i coordinates are given in matrix notation by

$$J_{\alpha\beta,S} = G^{-1}T_{\alpha\beta,S}. \quad (1.20)$$

10. The U matrix connects valence coordinates and symmetry coordinates;

$$S = UR, \quad (1.21)$$

and is orthogonal; $UU' = E$.

All the transformation matrices and matrices of vibrational-rotational constants summarized in paragraphs 2 through 10 above may be determined entirely from the geometry of the molecular model in question and the atomic masses.

1.3. Matrices of Physical Nature

1. The normal-coordinate transformation matrix (L) indicates the form of the normal vibrations;

$$S = LQ, \quad (1.22)$$

where Q represents the normal coordinates.

2. If ω_k is the frequency (in wave numbers) associated with a given normal coordinate, Q_k, we may form several kinds of frequency parameters.

$$\lambda_k = 4\pi^2c^2\omega_k^2, \quad \sigma_k = 1/\lambda_k = 1/(4\pi^2c^2\omega_k^2), \quad (1.23)$$

$$\delta_k = \langle Q_k^2 \rangle = \frac{h}{8\pi^2 c\omega_k} \coth \frac{hc\omega_k}{2kT} . \tag{1.24}$$

$\boldsymbol{\lambda}$, $\boldsymbol{\sigma}$, and $\boldsymbol{\delta}$ [cf. Eqs.(3.57), (4.20), and (7.9) of Ref. 9] are diagonal matrices with the λ_k, σ_k, and δ_k parameters, respectively, along their main diagonals.

3. Under this paragraph a number of relations containing the $\mathbf{L}$ matrix or its inverse ($\mathbf{K} = \mathbf{L}^{-1}$) are summarized.

$$\mathbf{G} = \mathbf{L}\mathbf{L}', \quad \mathbf{M} = \mathbf{K}'\mathbf{K}, \tag{1.25}$$

$$\mathbf{N} = \mathbf{L}\boldsymbol{\sigma}\mathbf{L}', \quad \mathbf{F} = \mathbf{K}'\boldsymbol{\lambda}\mathbf{K}, \tag{1.26}$$

$$\mathbf{C}^{\alpha} = \mathbf{L}\boldsymbol{\zeta}^{\alpha}\mathbf{L}', \quad \mathbf{M}\mathbf{C}^{\alpha} = \mathbf{K}'\boldsymbol{\zeta}^{\alpha}\mathbf{L}', \quad \bar{\mathbf{C}}^{\alpha} = \mathbf{K}'\boldsymbol{\zeta}^{\alpha}\mathbf{K}, \tag{1.27}$$

$$\boldsymbol{\Sigma} = \mathbf{L}\boldsymbol{\delta}\mathbf{L}', \tag{1.28}$$

$$\mathbf{J}_{\alpha\beta,Q} = \mathbf{L}'\mathbf{J}_{\alpha\beta,S} = \mathbf{K}\mathbf{T}_{\alpha\beta,S} . \tag{1.29}$$

Here $\mathbf{F}$ is the force-constant matrix connected with the potential energy of the molecule. $\mathbf{N}$ is the compliance matrix. For a complete set of internal coordinates without redundancies $\mathbf{N}$ simply is the inverse of $\mathbf{F}$. $\boldsymbol{\zeta}^{\alpha}$ is the matrix of Coriolis constants ζ^{α}. $\boldsymbol{\Sigma}$ is the mean-square amplitude matrix with elements $\Sigma_{ii} = \langle S_i^2 \rangle$, $\Sigma_{ij} = \langle S_i S_j \rangle$. $\mathbf{J}_{\alpha\beta,Q}$ is the matrix of partial derivatives at equilibrium of the inertia tensor components with respect to the normal coordinates. The elements of a 6 x 6 tensor pertaining to the centrifugal distortion are[17]

$$t_{\alpha\beta\gamma\delta} = \mathbf{J}'_{\alpha\beta,Q}\boldsymbol{\sigma}\mathbf{J}_{\gamma\delta,Q} = \mathbf{J}'_{\alpha\beta,S}\mathbf{N}\mathbf{J}_{\gamma\delta,S} \tag{1.30}$$

or in terms of the $\mathbf{T}_{\alpha\beta,S}$ matrices:

$$t_{\alpha\beta\gamma\delta} = \mathbf{T}'_{\alpha\beta,S}\mathbf{M}\mathbf{N}\mathbf{M}\mathbf{T}_{\gamma\delta,S} . \tag{1.31}$$

1.4. Calculation of Physical Constants

1.4.1. Introduction

Assume the matrices of geometrical nature (cf. Section

1.2) to be known. The vibrational analysis is completed when (a) **L** and all ω_k have been determined, or (b) **F** (or equivalent information) has been established. It is well known that a complete set of vibrational frequencies alone is in general not sufficient to determine the force field of the molecule.

(a) With the knowledge of **L** the Coriolis constants (ζ^α) may be calculated. If in addition all ω_k are known, the force constants of **F** may be calculated, as well as the elements of $\mathbf{\Sigma}$ and $\mathbf{J}_{\alpha\beta,a}$.

(b) If **F** is known, the **L** matrix and frequencies ω_k may be calculated by solving the secular-equation problem of molecular vibrations.[1] Hence, according to (a), the Coriolis constants, ζ^α, mean-square amplitude matrix elements, Σ_{ij}, and inertia tensor derivatives $J^{(k)}_{\alpha\beta,a}$, along with the t tensor of centrifugal distortion, may again be calculated.

1.4.2. Physical Constants

The normal coordinate analysis should, if possible, be based on a complete set of assigned vibrational frequencies from infrared and/or Raman measurements. Data for isotopic molecules use to be very helpful. In current analyses it is attempted to develop harmonic force fields which should reproduce the observed fundamentals and uncorrected experimental frequencies in general. In such cases the assigned vibrational frequencies are identified with the normal frequencies, ω_k. Sometimes the observed fundamentals must be "corrected for anharmonicity" in order to establish the validity of various isotope rules. The calculation with an anharmonic potential function is something quite different, and has so far been performed only for a limited number of very small polyatomic molecules.

Coriolis constants, $\zeta_{ij}{}^\alpha$ ($\alpha = x$, y, or z) are dimensionless quantities of magnitudes $-1 \leq \zeta \leq 1$, which pertain to the Coriolis coupling of vibration-rotation. They are more or less obtainable from an analysis of the fine structure or band contour of spectral bands. The most important ones are the first-order Coriolis constants, which combine degenerate normal modes (with the same frequency). Many Coriolis constants of this type have been obtained experimentally for symmetrical and spherical top molecules.

Every interatomic distance for bonded or nonbonded atom pairs $(i-j)$ is associated with a mean amplitude of vibration, l_{ij}. Its square value, l_{ij}^2, may be evaluated in terms of Σ matrix elements. In practical computations it is convenient to use the Cartesian displacement coordinates in order to obtain the coefficients of the appropriate linear combinations, or to make explicit use of the mean binary products of Cartesian displacements. For further details Ref. 9 should be consulted. Experimental mean amplitudes are obtainable from the interpretation of gaseous electron diffraction data.

Microwave spectral measurements may give us the centrifugal distortion constants of D_J, D_{JK}, and possibly D_K. These quantities may be calculated as linear combinations of the $t_{\alpha\beta\gamma\delta}$ tensor components.

REFERENCES

1 E.B.Wilson,Jr., J.C.Decius, and P.C.Cross: Molecular Vibrations, McGraw-Hill, New York 1955.

2 M.V.Volkenstein, M.A.Eliashevich, and B.I.Stepanov: Kolebanyia molekul (Vibrations of molecules), Vol.I., GITTL, Moscow 1949.

3 E.B.Wilson,Jr., J.Chem.Phys. 7, 1047 (1939).

4 E.B.Wilson,Jr., J.Chem.Phys. 9, 76 (1941).

5 M.A.Eliashevich, Compte rendus, URSS. 28, 604 (1940).

6 J.H.Meal and S.R.Polo, J.Chem.Phys. 24, 1119 (1956).

7 J.H.Meal and S.R.Polo, J.Chem.Phys. 24, 1126 (1956).

8 H.C.Allen,Jr. and P.C.Cross: Molecular Vib-Rotors, Wiley, New York 1963.

9 S.J.Cyvin: Molecular Vibrations and Mean Square Amplitudes, Universitetsforlaget, Oslo, and Elsevier, Amsterdam, 1968.

10 S.J.Cyvin, Spectrochim.Acta 15, 828 (1959).

11 D.Kivelson and E.B.Wilson,Jr., J.Chem.Phys. 20, 1575 (1952).

12 D.Kivelson and E.B.Wilson,Jr., J.Chem.Phys. 21, 1229 (1953).

13 J.E.Wollrab: Rotational Spectra and Molecular Structure, Academic Press, New York 1967.

14 S.J.Cyvin and G.Hagen, Chem.Phys.Letters 1, 645 (1968).

15 P.Pulay and W.Sawodny, J.Mol.Spectry. 26, 150 (1968).

16 K.Klauss and G.Strey, Z.Naturforschg. 22a, 1308 (1968).

17 S.J.Cyvin, B.N.Cyvin, and G.Hagen, Z.Naturforschg. 23a, 1649 (1968).

18 B.N.Cyvin, I.Elvebredd, and S.J.Cyvin, Z.Naturforschg. 24a, 139 (1969).

19 H.H.Nielsen, Revs. Modern Phys. 23, 90 (1951).

20 G.O.Sørensen, G.Hagen, and S.J.Cyvin, J.Mol.Spectry. 35, 489 (1970).

21 B.L.Crawford,Jr. and W.H.Fletcher, J.Chem.Phys. 19, 141 (1951).

CHAPTER 2

Derivation of the Vibration-Rotation Hamiltonian by Matrix Methods

G. O. SØRENSEN

The classical Hamiltonian is derived for the motion of the nuclei in a molecule presupposing that large amplitude vibrations can be ruled out. By using matrix notation it is attempted to make the derivation easier to grasp as well as to obtain simple expressions for the spectroscopic parameters. Further the Eckart conditions are discussed particularly emphasizing the reason for introducing them.

Den klassiske Hamilton-funktion udledes for kærnebevægelserne i et molekyle, hvori vibrationer med store amplituder kan udelukkes. Det er forsøgt ved anvendelse af matrix notation at skabe et bedre overblik over såvel udledningen som de resulterende udtryk til beregning af spektroskopiske parametre. Endvidere diskuteres Eckart-betingelsernes oprindelse.

2.1. Introduction

The molecules to be dealt with in this chapter are assumed to be of the general quasi-rigid nonlinear type where the atoms vibrate with small amplitudes only in the potential field introduced by the Born-Oppenheimer approximation. For this class of molecules Wilson and Howard in 1936 presented a general method[1] for the separation of the nuclear motions in translational, rotational and vibrational types based upon the introduction of a suitable set of generalized coordinates. The method is reproduced in later textbooks.[2,3]

It is essential to the investigation of perturbational effects in vibrational and rotational spectroscopy that the individual steps in Wilson's approach, and thereby the importance of the individual terms arising in the Hamiltonian, are fully realized. The study of these problems may be considerably facilitated by the use of the compact and transparent matrix notation. It will be shown below how the separation in coordinates is rewritten and how the most important spectroscopic parameters are expressed when matrices are rigorously applied.

2.2. Reference System and Eckart Conditions

In the study of vibration rotation coupling it is convenient to describe the over-all rotation and translation of a molecule in terms of the motion of a Cartesian reference system fixed in some way to the molecule. The motion of the atoms relative to such a moving coordinate system may be described as vibrational provided the moving system is defined appropriately.

As a consequence of the introduction of a moving coordinate system we may first notice, that six of the $3N$ dynamical variables must describe the motion of this system relative to a space fixed coordinate system. Hence only $3N - 6$ degrees of freedom are left, and the $3N$ Cartesian coordinates of the atoms referring to the moving system - the elements of the column matrix $\mathbf{R}$,

$$\mathbf{R}' = [X_1\ Y_1\ Z_1\ X_2\ Y_2\ Z_2\ \ldots\ldots\ Z_N] \tag{2.1}$$

- are not independent dynamical variables. Six equations of constraint in the coordinates of $\mathbf{R}$ must follow from requirements to the moving system. These requirements may be chosen in a variety of ways, but for the molecules discussed in this chapter the Eckart conditions are usually quoted[1-3] when assuming six equations of constraint. However, six equivalent equations result as a consequence of applying a conventional normal coordinate analysis to the pure vibrational motion of a particle system.

Hence the Eckart conditions must be inherent in the assumptions of the normal coordinate analysis.

The concept of vibrational motion belongs of course to classical mechanics. In molecular dynamics, however, the choice of generalized coordinates is most advantageously discussed before a quantum mechanical Hamilton operator is formed, and we shall therefore proceed by classical mechanical reasoning.

A pure vibrational motion appears when a classical particle moves in a potential well with a single minimum and a smooth surface. For the molecular model treated here only one relative arrangement of the atoms corresponds to a minimum in potential energy, V_{min}. That this equilibrium configuration corresponds to a single specific arrangement of the atoms relative to the moving system, given by $\mathbf{R} = \mathbf{R}^e$, is therefore a first condition in order that a vibration of atoms relative to this system can take place. We usually say that the molecular coordinate system is attached to the equilibrium configuration.

As intermediate vibrational coordinates we may now define the Cartesian displacements, $\mathbf{x} = \mathbf{R} - \mathbf{R}^e$, and expand the potential energy around the potential minimum corresponding to $\mathbf{x} = \mathbf{0}$. It is not always realized that the coefficients of this expansion may depend on the rotational state of the moving reference system, but here we shall assume as a second condition that we may use Cartesian force constants. In this way we have defined potential wells in which the atoms may vibrate relative to the moving system. Only in the harmonic approximation can the vibrational problem be solved by exact quantum mechanical methods and higher order terms are therefore neglected in the expansion given by the matrix product

$$2V = \mathbf{x}'\mathbf{f}\mathbf{x} = \mathbf{q}'\mathbf{f}^q\mathbf{q} . \qquad (2.2)$$

The value $V_{min} = 0$ has been chosen arbitrarily, and $\mathbf{q} = \mathbf{m}^{\frac{1}{2}}\mathbf{x}$ designates a column of mass-weighted displacements while $\mathbf{f}$ and $\mathbf{f}^q$ are square matrices of the appropriate expansion coefficients.

The kinetic energy depends on coordinates as well as on velocities. With $S_1 - S_6$ designating the dynamical variables of the moving coordinate system we may write

$$T = T(\mathbf{x}, \dot{\mathbf{x}}, \mathbf{S}, \dot{\mathbf{S}}).$$

Now, according to previous assumptions the partial derivatives $\frac{\partial V}{\partial \dot{S}_i}$ and $\frac{\partial V}{\partial S_i}$, vanish for all $i = 1 - 6$. Therefore six of the Lagrangian equations of motion are of the form

$$\frac{d}{dt}\frac{\partial L}{\partial \dot{S}_i} + \frac{\partial L}{\partial S_i} = \frac{d}{dt}\frac{\partial T}{\partial \dot{S}_i} + \frac{\partial T}{\partial S_i} = 0\ ,$$

which may be fulfilled by $\dot{S}_i = 0$ for all $i = 1 - 6$.

The solutions of the remaining $3N - 6$ Lagrangian equations obtained by assuming $\dot{S}_i = 0$ defines pure vibrational motions. In this special case the molecular coordinate system is at rest and consequently the kinetic energy is given by

$$2T = \dot{\mathbf{R}}'\mathbf{m}\dot{\mathbf{R}} = \dot{\mathbf{x}}'\mathbf{m}\dot{\mathbf{x}} = \dot{\mathbf{q}}'\dot{\mathbf{q}}. \tag{2.3}$$

Further progress requires the fundamental transformation from mass-weighted displacements to generalized vibrational coordinates ($3N - 6$ vibrational normal coordinates). This is achieved by an orthonormal transformation followed by a truncation of 6 columns. The transformation is chosen orthonormal so that the simple, purely quadratic form of the kinetic energy expression is retained. Further it may be chosen so that $\mathbf{f}^q$ is diagonalized and a purely quadratic form is obtained for the potential energy expression as well. Hence the columns of the transformation matrix are "eigenvectors" of $\mathbf{f}^q$. These normalized eigenvectors may be looked upon as mass-weighted displacements, and conversely, the columns of the transformation matrix can be derived from displacements which are eigenvectors of $\mathbf{f}^q$. The rigid displacements, rotations and translations, are particular eigenvectors corresponding to the eigenvalue zero. They form an infinite set from which it is possible to select six orthonormal columns. The transformation matrix can therefore be separated into two parts, represented by $\boldsymbol{l}$ and $\boldsymbol{l}_{tr}$, built up by $3N - 6$ and 6 columns respectively, conforming to the scheme

$$\begin{bmatrix} \boldsymbol{l}' \\ \boldsymbol{l}'_{tr} \end{bmatrix} \mathbf{f}^q \begin{bmatrix} \boldsymbol{l} & \boldsymbol{l}_{tr} \end{bmatrix} = \begin{bmatrix} \lambda & 0 \\ 0 & 0 \end{bmatrix}. \tag{2.4}$$

Among the individual block products the following equations are adequate representatives:

$$\boldsymbol{l}'\mathbf{f}^q\boldsymbol{l} = \lambda \qquad \text{dim: } (3N-6)\times(3N-6) \qquad (2.5a)$$

$$\mathbf{f}^q\boldsymbol{l}_{tr} = 0 \qquad '' \quad 3N \times 6 \qquad (2.5b)$$

Notice that the orthonormality of $[\boldsymbol{l}\ \ \boldsymbol{l}_{tr}]$ in a similar way leads to the relations

$$\boldsymbol{l}'\boldsymbol{l} = \mathbf{E} \qquad \text{dim: } (3N-6)\times(3N-6) \qquad (2.6a)$$

$$\boldsymbol{l}_{tr}'\boldsymbol{l} = 0 \qquad '' \quad 6\times(3N-6) \qquad (2.6b)$$

$$\boldsymbol{l}_{tr}'\boldsymbol{l}_{tr} = \mathbf{E} \qquad '' \quad 6\times 6 \qquad (2.6c)$$

while

$$\boldsymbol{l}\boldsymbol{l}' + \boldsymbol{l}_{tr}\boldsymbol{l}_{tr}' = \mathbf{E} \qquad \text{dim: } 3N\times 3N \qquad (2.7)$$

These relations in combination with the conditions above on the form of the potential energy expression are now used for the introduction of $3N - 6$ vibrational normal coordinates, defined by*

$$\mathbf{Q} = \boldsymbol{l}'\mathbf{q}. \qquad (2.8)$$

First Eq. (5a) is transformed in the following way:

$$\boldsymbol{l}\lambda\boldsymbol{l}' = \boldsymbol{l}\boldsymbol{l}'\mathbf{f}^q\boldsymbol{l}\boldsymbol{l}'.$$

Application of Eq. (7) and then Eq. (5b) to the right side in this relation now gives

$$\boldsymbol{l}\lambda\boldsymbol{l}' = \mathbf{f}^q, \qquad (2.9)$$

and the potential energy expression, Eq. (2), can be transformed by substitution, applying first Eq. (9) and then Eq. (8):

---

* $\boldsymbol{l}'$ corresponds to $\mathbf{L}$ in the notation of Cyvin.[4]

$$2V = \mathbf{Q}'\boldsymbol{\lambda}\mathbf{Q} = \sum_{k=1}^{3N-6} \lambda_k Q_k^{\;2}. \tag{2.10}$$

Since all eigenvalues λ_k are positive quantities it is obvious that $V = V_{\text{min}} = 0$ is obtained when $\mathbf{Q} = \mathbf{0}$ only, and from the first condition it is concluded that $\mathbf{Q} = \mathbf{0}$ implies $\mathbf{q} = \mathbf{0}$. This means that the $3N$ displacements in $\mathbf{q}$ depend on the $3N - 6$ coordinates $\mathbf{Q}$ only, and that

$$\mathbf{q} = \boldsymbol{l}\mathbf{Q} \tag{2.11}$$

is the proper transformation inverse to Eq. (8). From these equations and Eq. (7) it therefore follows

$$\mathbf{q} - \boldsymbol{l}\mathbf{Q} = (\mathbf{E} - \boldsymbol{l}\boldsymbol{l}')\mathbf{q} = \boldsymbol{l}_{\text{tr}}\boldsymbol{l}'_{\text{tr}}\mathbf{q} = \mathbf{0}$$

equivalent to

$$\boldsymbol{l}'_{\text{tr}}\mathbf{q} = \boldsymbol{l}'_{\text{tr}}\mathbf{m}^{\frac{1}{2}}\mathbf{x} = \mathbf{0}. \tag{2.12}$$

This matrix equation contains six requirements which will be met by any vibrational displacement. That the equation is equivalent to the six Eckart conditions may be shown by explicit evaluation of the individual columns of $\boldsymbol{l}_{\text{tr}}$ as rigid massweighted displacements. Examples of this will be given in the subsequent section.

Eq. (11) can be substituted into Eq. (3) to give a transformed expression for the pure vibrational kinetic energy

$$2T_{\text{vib}} = \dot{\mathbf{Q}}'\dot{\mathbf{Q}}, \tag{2.13}$$

but, Eq. (11) is very important for other purposes as well and an easy method for the calculation of the matrix $\boldsymbol{l}$ for specific molecules is desirable. This may be based upon the matrices appearing in the general normal coordinate analysis by Wilson's **GF**-method[2] (cf. Chapter 1). If $\mathbf{K}$ denotes a left inverse of $\mathbf{L}$ ($\mathbf{KL} = \mathbf{E}$)[5] a convenient expression for $\boldsymbol{l}$ is

$$\boldsymbol{l} = \mathbf{m}^{-\frac{1}{2}}\mathbf{B}\mathbf{K}. \tag{2.14}$$

2.3. Total Kinetic Energy

Assuming the coordinate system defined in the preceding section to be fixed in space the total kinetic energy may be given by Eq. (3) or Eq. (13). With a moving reference system however, the vibrational kinetic energy remains a separate term, provided that the $3N - 6$ normal coordinates are included in the set of generalized coordinates.

It is convenient to start the derivation from the basic vector expression of the kinetic energy in a system of N particles with masses m_a and velocities $\vec{V}_a$ ($a = 1, 2, \ldots, N$):

$$2T = \sum_a m_a \vec{V}_a \cdot \vec{V}_a \, . \qquad (2.15)$$

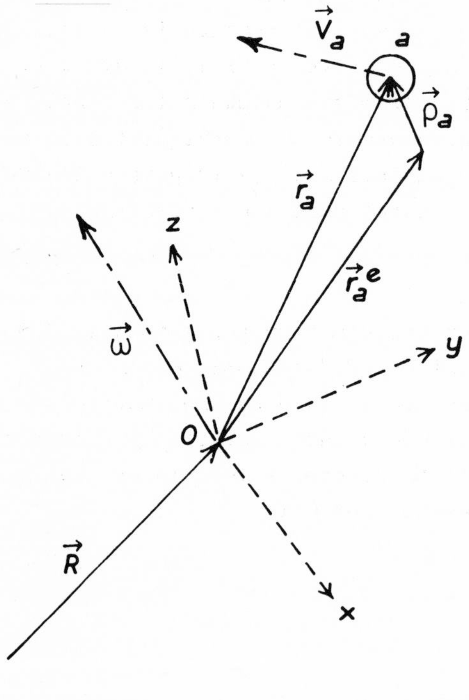

Fig. 2-1. Position and velocity vectors of the moving coordinate system and atom number a.

The vectors shown in Fig. 1 are introduced for the purpose of separating the velocities $\vec{V}_a$ into contributions from the motion of the coordinate system as a whole and from the motion of the individual atoms relative to the coordinate system.

$\vec{R}$ is the position vector of the origin O in the moving coordinate system relative to a point fixed in space. $\vec{R}$ should not be confused with the matrix $\mathbf{R}$ of Eq. (1).

$\vec{\omega}$ is the angular velocity vector of the moving coordinate system. A component of $\vec{\omega}$ in this system is ω_g, labelled with $g = x, y$ or z.

$\vec{r}_a$ is a position vector of an atom relative to the origin O, and the components X_a, Y_a and Z_a in the moving system for all atoms constitute the matrix $\mathbf{R}$; compare Eq. (1).

$\vec{r}_a^{\,e}$ is the position vector of the equilibrium position of atom a. The components are elements of $\mathbf{R}^e$ and as emphasized in Section 2 they are by definition constants of motion.

$\vec{\rho}_a$ is a displacement vector equal to $\vec{r}_a - \vec{r}_a^{\,e}$, and hence the components of the N displacement vectors are the elements of $\mathbf{X} = \mathbf{R} - \mathbf{R}^e$, the Cartesian displacements.

$\vec{v}_a$ is the velocity of atom a relative to the moving coordinate system. Therefore the components in this system are the elements $\dot{x}_a$, $\dot{y}_a$, $\dot{z}_a$ of $\dot{\mathbf{X}}$, and the pure vibrational kinetic energy given in Eq. (3) in matrix form can be expressed by the sum

$$2T_{\text{vib}} = \sum_a m_a \vec{v}_a \cdot \vec{v}_a . \tag{2.16}$$

The displacements $\mathbf{X}$ are subject to the conditions of Eq. (12). Since these are relations between the components of the displacement vectors, $\vec{\rho}_a$, relations between the vectors as such may also be derived. These are the Eckart conditions in vector form:

$$\sum_a m_a \vec{\rho}_a = \vec{0} \, , \quad \sum_a m_a \vec{v}_a = \vec{0} , \tag{2.17a}$$

$$\sum_a m_a \vec{r}_a^{\,e} \times \vec{\rho}_a = \vec{0} \, , \quad \sum_a m_a \vec{r}_a^{\,e} \times \vec{v}_a = \vec{0} . \tag{2.17b}$$

If the origin O is chosen to be the center of mass of the

equilibrium configuration we have in addition

$$\sum_a m_a \vec{r}_a^{\,e} = \vec{0}$$

and consequently

$$\sum_a m_a (\vec{r}_a^{\,e} + \vec{\rho}_a) = \sum_a m_a \vec{r}_a = \vec{0}. \tag{2.18}$$

The conditions (17a) therefore fix the origin 0 to the center of mass irrespective of the vibrational distortions in the molecule.

Since the position vector of an atom is $\vec{R} + \vec{r}_a$, the total velocity vector can be expressed as the sum

$$\vec{V}_a = \dot{\vec{R}} + \dot{\vec{r}}_a = \dot{\vec{R}} + \vec{\omega} \times \vec{r}_a + \vec{v}_a \tag{2.19}$$

of a translational, a rotational and a vibrational component, and by substitution into Eq. (15) the kinetic energy is also separated:

$$2T = \sum m_a \dot{\vec{R}} \cdot \dot{\vec{R}} + \sum m_a (\vec{\omega} \times \vec{r}_a) \cdot (\vec{\omega} \times \vec{r}_a)$$

$$+ \sum m_a \vec{v}_a \cdot \vec{v}_a + 2 \sum m_a \dot{\vec{R}} \cdot (\vec{\omega} \times \vec{r}_a) + 2 \sum m_a \dot{\vec{R}} \cdot \vec{v}_a$$

$$+ 2 \sum m_a (\vec{\omega} \times \vec{r}_a) \cdot \vec{v}_a \, .$$

The last three terms can be reduced by applying the Eckart conditions, Eqs. (17),(18) and the vector relation

$$\vec{a} \cdot (\vec{b} \times \vec{c}) = (\vec{a} \times \vec{b}) \cdot \vec{c}.$$

The fourth term vanishes, because

$$\sum m_a \dot{\vec{R}} \cdot (\vec{\omega} \times \vec{r}_a) = (\dot{\vec{R}} \times \vec{\omega}) \cdot \sum m_a \vec{r}_a = 0 \, ,$$

and so does the fifth term

$$\sum m_a \dot{\vec{R}} \cdot \vec{v}_a = \dot{\vec{R}} \cdot \sum m_a \vec{v}_a = 0.$$

The sixth term, however, vanishes only in the equilibrium configuration (all $\vec{\rho}_a = 0$) or if the coordinate system is non-rotating,

$$\sum m_a (\vec{\omega} \times \vec{r}_a) \cdot \vec{v}_a = \vec{\omega} \cdot \sum m_a \vec{r}_a \times \vec{v}_a$$

$$= \vec{\omega} \cdot \sum m_a (\vec{r}_a^{\,e} + \vec{\rho}_a) \times \vec{v}_a = \vec{\omega} \cdot \sum m_a \vec{\rho}_a \times \vec{v}_a \, .$$

By these reductions the total kinetic energy becomes

$$2T = \dot{\vec{R}}\cdot\dot{\vec{R}}\sum_a m_a + \sum_a m_a(\vec{\omega}\times\vec{r}_a)\cdot(\vec{\omega}\times\vec{r}_a)$$

$$+\sum_a m_a\vec{v}_a\cdot\vec{v}_a + 2\vec{\omega}\cdot\sum_a m_a\vec{\rho}_a\times\vec{v}_a\,, \tag{2.20}$$

a sum of a pure translational contribution and three terms which may be called the rotational, the vibrational [compare Eq. (16)] and the Coriolis coupling contribution. Since the position vector of the center of mass appears in the first term only, the problem of the translational motion can be treated separately by the theory of free particles. The translational term is therefore neglected throughout.

The vector products in the three remaining terms may now be replaced by equivalent expressions in their components. At this stage the advantages of the matrix notation will become apparent, since the complex product sums arising in the components can be written as rather simple matrix products in the matrices $\mathbf{R}^e$, $\mathbf{X}$ and $\dot{\mathbf{X}}$. Without further complications the normal coordinates can subsequently be introduced by the use of Eq. (11).

The most arduous task is the transcription of the cross products. This requires the introduction of the auxiliary matrices $\mathbf{i}^g$ and $\mathbf{I}^g$ (compare Section 1.2);

$$(\mathbf{i}^g)_{hf} = e_{ghf} \tag{2.21}$$

defined in terms of the Levi-Chivita symbol

$$e_{fgh} = 1\,,\quad e_{fhg} = -1\,,\quad e_{fgg} = 0 \tag{2.22}$$

with $f, g, h = (x, y, z)_{\text{cycl}}$, which means that f, g, h may represent any sequence obtained by cyclic permutation of x, y, z. Furthermore it is necessary to define the 3×1 column matrices $\boldsymbol{\omega}$ and $\mathbf{r}_a$ of the components ω_g and $r_{a,g}$ corresponding to the vectors $\vec{\omega}$ and $\vec{r}_a$. The component $r_{a,g}$ equals X_a, Y_a or Z_a of $\mathbf{R}$ according to the definition of $\vec{r}_a$ given at the beginning of this section. But, this intermediate change in notation is convenient in the following context.

The cross product $\vec{\omega}\times\vec{r}_a$ from the rotational term in Eq.(20) is considered first. The x-component is

$$(\vec{\omega}\times\vec{r}_a)_x = \omega_y r_{a,z} - \omega_z r_{a,y} = \sum_{gh} e_{xgh}\,\omega_g\, r_{a,h}\,,$$

and a general component may therefore be written

$$(\vec{\omega}\times\vec{r}_a)_f = \sum_{gh} e_{fgh}\,\omega_g\, r_{a,h}\,.$$

In the summations g,h run through x, y and z. The square sum of these components is equal to the dot product

$$(\vec{\omega}\times\vec{r}_a)\cdot(\vec{\omega}\times\vec{r}_a) = \sum_f (\vec{\omega}\times\vec{r}_a)_f^2$$

$$= \sum_f \Big(\sum_{gh} e_{fgh}\,\omega_g\, r_{a,h}\Big)\Big(\sum_{g'h'} e_{fg'h'}\,\omega_{g'}\, r_{a,h'}\Big)$$

$$= \sum_{gg'} \omega_g\,\omega_{g'} \sum_{hh'} r_{a,h}\, r_{a,h'} \sum_f e_{fgh}\, e_{fg'h'}\,.$$

As the last step a change in the order of summation has been made, so that it may easily be realized that the factors $\omega_g\,\omega_{g'}$ are common to all atomic contributions in the total sum forming the rotational term

$$2T_{\mathrm{rot}} = \sum_{gg'} \omega_g\,\omega_{g'} \sum_a m_a \sum_{hh'f} r_{a,h}\, r_{a,h'}\, e_{fgh}\, e_{fg'h'}\,.$$

The sum over the atoms (a) is recognized as a component $I_{gg'}$ of the inertia tensor $\mathbf{I}$, since the general form of the rotational energy is

$$2T_{\mathrm{rot}} = \sum_{gg'} \omega_g\,\omega_{g'}\, I_{gg'} = \boldsymbol{\omega}'\mathbf{I}\,\boldsymbol{\omega}\,, \tag{2.23}$$

and it is shown below how this sum can be reduced to a simple matrix product. First the definitions in Eqs. (21) and (22) are applied to the sum

$$\sum_{fhh'} r_{a,h}\, r_{a,h'}\, e_{fgh}\, e_{fg'h'} = \sum_{hh'} r_{a,h}\, r_{a,h'} \sum_f (\mathbf{i}^g)_{hf} (\mathbf{i}^{g'})'_{fh'}$$

$$= \sum_{h h'} r_{a,h} [\mathbf{i}^g(\mathbf{i}^{g'})']_{hh'} r_{a,h'} = \mathbf{r}_a' \mathbf{i}^g (\mathbf{i}^{g'})' \mathbf{r}_a ,$$

and thus we can write

$$I_{gg'} = \sum_a m_a \mathbf{r}_a' \mathbf{i}^g (\mathbf{i}^{g'})' \mathbf{r}_a .$$

If we finally introduce the $3N \times 3N$ square matrices $\mathbf{m}$ and $\mathbf{I}^g$, the latter composed of N diagonal blocks $\mathbf{i}^g$, we can express the general inertia tensor component as the product

$$I_{gg'} = \mathbf{R}'\mathbf{I}^g \mathbf{m} (\mathbf{I}^{g'})' \mathbf{R}. \tag{2.24}$$

This equation is equivalent to the wellknown sum formulae of the instantaneous moments of inertia (I_{xx}) and products of inertia (I_{xy}) with respect to the moving axes:

$$I_{xx} = \sum_a m_a (Y_a^2 + Z_a^2), \qquad I_{xy} = -\sum_a m_a X_a Y_a . \tag{2.25}$$

But, Eq. (24) is particularly suited to the evaluation of several important vibration-rotation parameters, as it is immediately seen from the ease with which the dependence on the vibrational coordinates is expressed. This only needs the substitution

$$\mathbf{R} = \mathbf{R}^e + \mathbf{x} = \mathbf{R}^e + \mathbf{m}^{-\frac{1}{2}} \boldsymbol{l} \mathbf{Q} ,$$

in which Eq. (11) has been used, and Eq. (24) becomes

$$I_{gg'} = I_{gg'}^e + \mathbf{Q}'\boldsymbol{l}'\mathbf{m}^{\frac{1}{2}} [\mathbf{I}^g(\mathbf{I}^{g'})' + \mathbf{I}^{g'}(\mathbf{I}^g)']\mathbf{R}^e + \mathbf{Q}'\boldsymbol{l}'\mathbf{I}^g(\mathbf{I}^{g'})'\boldsymbol{l}\mathbf{Q}.$$

Here $I_{gg'}^e$ denotes a component of the inertia tensor $\mathbf{I}^e$ corresponding to the equilibrium configuration, $\mathbf{R}^e$. It is presupposed throughout that the constant tensor $\mathbf{I}^e$ has been diagonalized by a proper choice of axes in the molecular reference system, and the symbol I_g^e is adopted for a principal moment of inertia on the diagonal;

$$\mathbf{I}^e = \mathrm{diag}(I_x^e, I_y^e, I_z^e). \tag{2.26}$$

An abbreviation in the expression of the instantaneous components of the inertia tensor is obtained by introducing the $(3N-6)\times 1$ column matrices $\mathbf{J}_{gg',\alpha}$

$$I_{gg'} = I_g^{\,e}\delta_{gg'} + \mathbf{Q}'\mathbf{J}_{gg',\alpha} + \mathbf{Q}'\boldsymbol{l}'\mathbf{I}^g(\mathbf{I}^{g'})'\boldsymbol{l}\mathbf{Q} \tag{2.27}$$

with

$$\mathbf{J}_{g'g,\alpha} = \mathbf{J}_{gg',\alpha} = 2\boldsymbol{l}'\mathbf{m}^{\frac{1}{2}}\mathbf{I}^g(\mathbf{I}^{g'})'\mathbf{R}^e. \tag{2.28}$$

If g' differs from g the former equality of Eq. (28) should be proved, while the latter is the definition. However, from the fundamental properties of the matrices $\mathbf{I}^g$, Eq. (21), it is easily seen that certain simple commutation relations are valid.

$$\mathbf{I}^f\mathbf{I}^g - \mathbf{I}^g\mathbf{I}^f = -\mathbf{I}^h, \tag{2.29}$$

and as $(\mathbf{I}^g)' = -\mathbf{I}^g$ this leads to

$$\mathbf{J}_{fg,\alpha} - \mathbf{J}_{gf,\alpha} = 2\boldsymbol{l}'\mathbf{m}^{\frac{1}{2}}\mathbf{I}^h\mathbf{R}^e. \tag{2.30}$$

From the product $\mathbf{I}^h\mathbf{R}^e$ we may form a rigid displacement

$$\mathbf{dx} = d\alpha(\mathbf{I}^h)'\mathbf{R}^e$$

corresponding to an infinitesimal rotation about the axis h through the angle $d\alpha$ $(d\vec{\rho}_a = d\alpha\, \vec{e}_h \times \vec{r}_a)$. The column of the corresponding mass-weighted displacement can therefore, according to Section 1, be normalized to give a column of $\boldsymbol{l}_{\mathrm{tr}}$. This will be denoted $\boldsymbol{l}_{\mathrm{r}}^{\,h}$,

$$\boldsymbol{l}_{\mathrm{r}}^{\,h} = (I_h^{\,e})^{-\frac{1}{2}}\mathbf{m}^{\frac{1}{2}}(\mathbf{I}^h)'\mathbf{R}^e. \tag{2.31}$$

When this equation and Eq. (6b) are used in Eq. (30) we finally get

$$\mathbf{J}_{fg,\alpha} - \mathbf{J}_{gf,\alpha} = -2\boldsymbol{l}'\boldsymbol{l}_{\mathrm{r}}^{\,h}(I_h^{\,e})^{\frac{1}{2}} = 0.$$

Notice that the three Eckart conditions fixing the direction of the axes [compare Eqs. (12) and (17b)] may now be expressed in the matrix form

$$(\mathbf{R}^e)'\mathbf{I}^g\mathbf{m}\mathbf{x} = 0. \tag{2.32}$$

The matrices $\mathbf{J}_{gg',\alpha}$ are very important to the theory of centrifugal distortion (cf. Sections 1.2 and 1.3). From Eq. (27) it follows that the elements of $\mathbf{J}_{gg',Q}$ are the first derivatives of an inertia tensor component with respect to the normal coordinates referred to the equilibrium configuration.

$$\left(\frac{\partial I_{gg'}}{\partial Q_k}\right)^e = (\mathbf{J}_{gg',Q})_k. \tag{2.33}$$

In a normal coordinate analysis these may be calculated from a relation obtained by substituting Eq. (14) into Eq. (28);

$$\mathbf{J}_{gg',Q} = 2\mathbf{K}\mathbf{B}\mathbf{I}^g(\mathbf{I}^{g'})'\mathbf{R}^e. \tag{2.34}$$

Eq. (31) may be applied to obtain another useful relation. Substitution of $(\mathbf{I}^{g'})'\mathbf{R}^e$ in Eq. (28) thus yields

$$\mathbf{J}_{gg',Q} = 2(I_g{}^e)^{\frac{1}{2}}\boldsymbol{l}'\mathbf{I}^{g'}\boldsymbol{l}_r{}^g. \tag{2.35}$$

Returning to the kinetic energy expression in Eq. (20), only the last term, the Coriolis contribution, remains to be rewritten. It is left to the reader to show that the methods presented above may be applied to this problem as well. Hereby the following expression for a component of the cross product sum is obtained.

$$\sum_a m_a(\vec{\rho}_a \times \vec{v}_a)_g = \mathbf{x}'\mathbf{m}\mathbf{I}^g\dot{\mathbf{x}} = \mathbf{Q}'\boldsymbol{l}'\mathbf{I}^g\boldsymbol{l}\dot{\mathbf{Q}}.$$

At this stage it is advantageous to introduce the ζ matrices of Meal and Polo[6] defined by

$$\boldsymbol{\zeta}^g = \boldsymbol{l}'\mathbf{I}^g\boldsymbol{l} = \mathbf{K}\mathbf{C}^g\mathbf{K}'. \tag{2.36}$$

(see also Sections 1.2 and 1.3), since this yields a reduction of the expression

$$\sum_a m_a(\vec{\rho}_a \times \vec{v}_a)_g = \mathbf{Q}'\boldsymbol{\zeta}^g\dot{\mathbf{Q}}. \tag{2.37}$$

Thus the dot product of the Coriolis term becomes

$$T_{Cor} = \sum_g \omega_g \mathbf{Q}'\boldsymbol{\zeta}^g\dot{\mathbf{Q}}. \tag{2.38}$$

For the formation of the classical Hamiltonian it is convenient to introduce an auxiliary coupling matrix $\mathbf{Z}$. It is dimensioned $3 \times (3N-6)$ and the 3 rows are formed by the products $\mathbf{Q}'\boldsymbol{\zeta}^g$

$$\mathbf{Z} = \begin{bmatrix} \mathbf{Q}'\boldsymbol{\zeta}^x \\ \mathbf{Q}'\boldsymbol{\zeta}^y \\ \mathbf{Q}'\boldsymbol{\zeta}^z \end{bmatrix} . \tag{2.39}$$

With this Eq. (38) becomes

$$T_{\mathrm{Cor}} = \boldsymbol{\omega}'\mathbf{Z}\dot{\mathbf{Q}} , \tag{2.40}$$

and collecting the results from Eqs. (13) and (23) the final matrix form of the rotation-vibration kinetic energy is found to be

$$2T = \boldsymbol{\omega}'\mathbf{I}\boldsymbol{\omega} + 2\boldsymbol{\omega}'\mathbf{Z}\dot{\mathbf{Q}} + \dot{\mathbf{Q}}'\dot{\mathbf{Q}} . \tag{2.41}$$

We may end this section by summarizing that the quantities of Eq. (41) are the columns of angular velocities $\boldsymbol{\omega}$ and velocities in the normal coordinates $\dot{\mathbf{Q}}$ in addition to the two matrices $\mathbf{I}$ and $\mathbf{Z}$, both depending on the instantaneous molecular configuration. The tensor of inertia $\mathbf{I}$ is given in terms of the components by Eq. (27) while the coupling matrix $\mathbf{Z}$ is given by Eq. (39).

2.4. Hamiltonian Form of the Kinetic Energy Expression

The quantization theorems apply to the classical Hamiltonian only and thus it is necessary in the kinetic energy expression to replace the velocities by proper angular or linear momenta. The matrix expression of the kinetic energy, Eq. (41), is very convenient for this substitution.

In a conservative physical system the momentum conjugated to a generalized coordinate Q_k is given by

$$P_k = \frac{\partial T}{\partial \dot{Q}_k} . \tag{2.42}$$

But, this definition does not include the angular velocities ω_g

because of the fact that not all of these correspond to generalized angular coordinates. However, such a set is formed by the Eulerian angles φ, θ, χ, and since these may be chosen so that $\omega_z = \dot{\chi}$ we may define

$$P_z = \frac{\partial T}{\partial \omega_z} . \tag{2.43}$$

The direction of the z-axis is not unique, and hence we can extend the validity of Eq. (43) to any direction including those of the x- and y-axes. Now differentiation in Eq. (41) leads to

$$\mathbf{P} = \mathbf{I}\boldsymbol{\omega} + \mathbf{Z}\dot{\mathbf{Q}} , \tag{2.44}$$

$$\mathbf{p} = \dot{\mathbf{Q}} + \mathbf{Z}\boldsymbol{\omega} , \tag{2.45}$$

where $\mathbf{P}$ is a column of the angular momenta P_x, P_y and P_z and $\mathbf{p}$ is the column formed by the $3N - 6$ vibrational momenta p_k .

The substitution is initiated by eliminating $\dot{\mathbf{Q}}$ as obtained from Eq. (45)

$$\dot{\mathbf{Q}} = \mathbf{p} - \mathbf{Z}'\boldsymbol{\omega} . \tag{2.46}$$

In Eq. (41) the two last terms become

$$2T_{\mathrm{Cor}} = 2\boldsymbol{\omega}'\mathbf{Z}\dot{\mathbf{Q}} = 2\boldsymbol{\omega}'\mathbf{Z}\mathbf{p} - 2\boldsymbol{\omega}'\mathbf{Z}\mathbf{Z}'\boldsymbol{\omega} ,$$

$$2T_{\mathrm{vib}} = \dot{\mathbf{Q}}'\dot{\mathbf{Q}} = \mathbf{p}'\mathbf{p} + \boldsymbol{\omega}'\mathbf{Z}\mathbf{Z}'\boldsymbol{\omega} - 2\boldsymbol{\omega}'\mathbf{Z}\mathbf{p} ,$$

and for the total kinetic energy we thereby obtain

$$2T = \boldsymbol{\omega}'(\mathbf{I} - \mathbf{Z}\mathbf{Z}')\boldsymbol{\omega} + \mathbf{p}'\mathbf{p} . \tag{2.47}$$

Notice that the tensor of inertia is modified by the term $\mathbf{Z}\mathbf{Z}'$. It is convenient to introduce the 3×3 square matrix $\boldsymbol{\mu}$ as the inverse

$$\boldsymbol{\mu} = (\mathbf{I} - \mathbf{Z}\mathbf{Z}')^{-1} . \tag{2.48}$$

The elements of $\boldsymbol{\mu}$ will be discussed in detail below. Elimination of $\dot{\mathbf{Q}}$ in Eq. (44) by Eq. (46) now yields

$$\mathbf{P} = (\mathbf{I} - \mathbf{Z}\mathbf{Z}')\boldsymbol{\omega}' + \mathbf{Z}\mathbf{p} ,$$

which is easily solved for $\boldsymbol{\omega}$:

$$\boldsymbol{\omega} = \boldsymbol{\mu}(\mathbf{P} - \mathbf{Z}\mathbf{p}). \tag{2.49}$$

This is finally used for substitution in Eq. (47) and the desired Hamiltonian form becomes

$$2T = (\mathbf{P}' - \mathbf{p}'\mathbf{Z}')\boldsymbol{\mu}(\mathbf{P} - \mathbf{Z}\mathbf{p}) + \mathbf{p}'\mathbf{p}. \tag{2.50}$$

The matrix expression given here is equivalent to and easily converted to the sum expression generally cited in the literature.[1-3] This requires that components of vibrational angular momenta are introduced by

$$p_g = (\mathbf{Z}\mathbf{p})_g = \dot{\mathbf{Q}}\boldsymbol{\zeta}^g\mathbf{p}, \tag{2.51}$$

and the matrix products of Eq. (50) may be expanded to give

$$2T = \sum_{gg'} \mu_{gg'}(P_g - p_g)(P_{g'} - p_{g'}) + \sum_k p_k^2. \tag{2.52}$$

The coefficients $\mu_{gg'}$ obtained by inversion of the modified inertial tensor, Eq. (48), are functions of the normal coordinates, and the corresponding derivatives are therefore important vibration-rotation interaction constants. These derivatives may be expressed in terms of the already established matrices in a very simple way as it will be outlined below.

It is convenient to separate the modified inertia tensor into terms that depend on $\mathbf{Q}$ to the zeroth, first and second degree:

$$\mathbf{I} - \mathbf{Z}\mathbf{Z}' = \mathbf{I}^e + \mathbf{I}^* + \mathbf{I}^{**}, \tag{2.53}$$

and in the same way we will express the inverse tensor by

$$\boldsymbol{\mu} = \boldsymbol{\mu}^e + \boldsymbol{\mu}^* + \boldsymbol{\mu}^{**} + \cdots; \qquad \boldsymbol{\mu}^e = (\mathbf{I}^e)^{-1}, \tag{2.54}$$

where terms of order higher than the second are neglected. From Eqs. (27) and (39) it follows that

$$I_{gg'}^* = \mathbf{Q}'\mathbf{J}_{gg',Q}, \tag{2.55}$$

$$I_{gg'}^{**} = \mathbf{Q}'\boldsymbol{l}'\mathbf{I}^g(\mathbf{I}^{g'})'\boldsymbol{l}\mathbf{Q} - \mathbf{Q}'\boldsymbol{\zeta}^g(\boldsymbol{\zeta}^{g'})'\mathbf{Q}. \tag{2.56}$$

Eq. (56) may be considerably reduced if the properties of $\boldsymbol{l}$ and $\boldsymbol{l}_{tr}$ discussed in Sections 1 and 2 are taken into account. For this purpose the ζ matrices are substituted using the definition, Eq. (36), and after a rearrangement Eq. (7) implies that

$$I_{gg'}^{**} = \mathbf{Q}'\boldsymbol{l}'\mathbf{I}^{g}\boldsymbol{l}_{tr}\boldsymbol{l}_{tr}'(\mathbf{I}^{g'})'\boldsymbol{l}\,\mathbf{Q}. \tag{2.57}$$

Since a column of mass-weighted displacements corresponding to a rigid translation is annihilated or transformed into another translation by multiplication with $\mathbf{I}^{g}$ only the rotational columns, $\boldsymbol{l}_{r}^{f}$, of $\boldsymbol{l}_{tr}$ can contribute in Eq. (57). The second order terms may therefore be expanded as

$$I_{gg'}^{**} = \sum_{f} \mathbf{Q}'\boldsymbol{l}'\mathbf{I}^{g}\boldsymbol{l}_{r}^{f}(\boldsymbol{l}_{r}^{f})'(\mathbf{I}^{g'})'\boldsymbol{l}\,\mathbf{Q},$$

and a subsequent substitution by means of Eq. (35) yields

$$I_{gg'}^{**} = \frac{1}{4}\sum_{f} \mathbf{Q}'\mathbf{J}_{gf,a}\frac{1}{I_{f}^{e}}\mathbf{J}_{fg',a}'\,\mathbf{Q}. \tag{2.58}$$

Notice, that this equation expresses the second order dependence in terms of the first order derivatives. The simplicity of the relation is emphasized if Eq. (55) is used for substitution in Eq. (58);

$$I_{gg'}^{**} = \frac{1}{4}\sum_{f} I_{gf}^{*}\frac{1}{I_{f}^{e}}I_{fg'}^{*} = \frac{1}{4}\sum_{f} I_{gf}^{*}\mu_{f}^{e}I_{fg'}^{*}.$$

In matrix form this result may be expressed in an exceptionally simple equation:

$$\mathbf{I}^{**} = \frac{1}{4}\mathbf{I}^{*}\boldsymbol{\mu}^{e}\mathbf{I}^{*}. \tag{2.59}$$

The elements of the inverse tensor of inertia are now determined by comparing terms of the same order in the product

$$\boldsymbol{\mu}(\mathbf{I}-\mathbf{Z}\mathbf{Z}') = \mathbf{E} = (\boldsymbol{\mu}^{e}+\boldsymbol{\mu}^{*}+\boldsymbol{\mu}^{**})(\mathbf{I}^{e}+\mathbf{I}^{*}+\mathbf{I}^{**}),$$

as obtained from Eqs. (53) and (54). To the right all terms depending on $\mathbf{Q}$ must vanish, so the following relations in the first and second order terms arise.

$$\mathbf{O} = \boldsymbol{\mu}^{e}\mathbf{I}^{*} + \boldsymbol{\mu}^{*}\mathbf{I}^{e},$$

$$0 = \mu^e I^{**} + \mu^* I^* + \mu^{**} I^e .$$

These are easily solved for μ^* and μ^{**}, and by use of Eq. (59) the results may be expressed in terms of the zeroth and first order matrices I^e, μ^e and I^* only:

$$\mu^* = -\mu^e I^* \mu^e , \tag{2.60}$$

$$\mu^{**} = \frac{3}{4} \mu^e I^* \mu^e I^* \mu^e = \frac{3}{4} \mu^* I^e \mu^* . \tag{2.61}$$

The individual elements of μ^* and μ^{**} become

$$\mu^*_{gg'} = - \frac{1}{I_g^e I_{g'}^e} \mathbf{Q}'\mathbf{J}_{gg',Q} , \tag{2.62}$$

$$\mu^{**}_{gg'} = \frac{3}{4 I_g^e I_{g'}^e} \sum_f \mathbf{Q}'\mathbf{J}_{gf,Q} \frac{1}{I_f^e} \mathbf{J}'_{fg',Q} \mathbf{Q} , \tag{2.63}$$

and the corresponding first and second derivatives with respect to the normal coordinates are given by

$$\frac{\partial \mu_{gg'}}{\partial Q_k} = - \frac{1}{I_g^e I_{g'}^e} (\mathbf{J}_{gg',Q})_k , \tag{2.64}$$

$$\frac{\partial^2 \mu_{gg'}}{\partial Q_k \partial Q_l} = \frac{3}{4 I_g^e I_{g'}^e} \sum_f \frac{1}{I_f^e} (\mathbf{J}_{gf,Q})_k (\mathbf{J}_{fg',Q})_l$$

$$= \frac{3}{4} \sum_f I_f^e \frac{\partial \mu_{gf}}{\partial Q_k} \frac{\partial \mu_{fg'}}{\partial Q_l} . \tag{2.65}$$

The derivation of the basic vibration-rotation interaction constants is hereby completed. Examples of application can be found in the extensive review by Nielsen,[7] discussing the interaction effects in infrared spectra, and in basic papers on the effects in pure rotational spectra presented by Herschbach and Laurie[8,9] and Oka and Morino.[10] A simple expression for the Hamilton operator has been presented by Watson.[11] It should be noted that some of the symbols used in this book deviate from those introduced by Nielsen. Since a large number of articles on vibration-rotation interaction are based on the formalism of Nielsen the most essential deviations are discussed below. In labelling normal coordinates Nielsen generally uses double indices, $s\sigma$, instead of a single k. σ is applied to distinguish members of degenerate sets. Also α, β, γ and δ are in common use as alternative to g, h and f, the running labels of the molecular axes. With these general deviations omitted the relations between the symbols are as follows.

The first derivatives of the inertia tensor are designated

$$a_k^{(gg')} = (\mathbf{J}_{gg',Q})_k = \frac{\partial I_{gg'}}{\partial Q_k} . \tag{2.66}$$

The second derivatives of the unmodified inertia tensor appear as explicit parameters in the Nielsen formalism:

$$A_{kl}^{(gg')} = [\boldsymbol{l}'\mathbf{I}^g(\mathbf{I}^{g'})'\boldsymbol{l}]_{kl} = \frac{\partial^2 I_{gg'}}{\partial Q_k \partial Q_l} . \tag{2.67}$$

They have been eliminated in the present treatment, since a simpler expression for the inverse tensor of inertia is thereby obtained. To the elimination applied here corresponds a relation between the interaction constants frequently cited in the literature[10-12] as belonging to the set of "sum rules". Comparing the evaluation of $I_{gg'}$ Eqs. (56)-(58) this relation is found to

$$\boldsymbol{l}'\mathbf{I}^g(\mathbf{I}^{g'})'\boldsymbol{l} = \boldsymbol{\zeta}^g(\boldsymbol{\zeta}^{g'})' + \frac{1}{4}\sum_f \frac{1}{I_f^e}\mathbf{J}_{gf,Q}\mathbf{J}'_{fg',Q} . \tag{2.68}$$

For convenience the general sum rules are summarized in matrix form:

$$\mathbf{J}'_{gg,Q}\mathbf{J}_{gg,Q} = 4\,I_g^e , \tag{2.69a}$$

$$\mathbf{J}'_{gg,Q}\mathbf{J}_{hh,Q} = 2\,(I_g^e + I_h^e - I_f^e) , \tag{2.69b}$$

$$\mathbf{J}'_{gh,Q}\mathbf{J}_{gh',Q} = \frac{(I_f^e)^2 - (I_g^e - I_h^e)^2}{I_f^e}\,\delta_{hh'} , \tag{2.69c}$$

$$\zeta^f J_{gh,Q} = \frac{1}{2}\left(J_{gg,Q} - J_{hh,Q} - \frac{I_g^e - I_h^e}{I_f^e} J_{ff,Q}\right) e_{fgh}, \qquad (2.70a)$$

$$\zeta^f J_{gg,Q} = -J_{gh,Q}\, e_{fgh}, \qquad (2.70b)$$

$$\zeta^g J_{gh,Q} = -\frac{1}{2}\left(1 + \frac{I_g^e - I_h^e}{I_f^e}\right) J_{gf,Q}\, e_{fgh}. \qquad (2.70c)$$

REFERENCES

1 E.B.Wilson,Jr. and J.B.Howard, J.Chem.Phys. 4, 260 (1936).

2 E.B.Wilson,Jr., J.C.Decius, and P.C.Cross: Molecular Vibrations, McGraw-Hill, New York 1955.

3 H.C.Allen,Jr. and P.C.Cross: Molecular Vib-Rotors, Wiley, New York 1963.

4 S.J.Cyvin: Molecular Vibrations and Mean Square Amplitudes, Universitetsforlaget, Oslo, and Elsevier, Amsterdam, 1968.

5 G.O.Sørensen, J.Mol.Spectry. 36, 359 (1970).

6 J.H.Meal and S.R.Polo, J.Chem.Phys. 24, 1119 (1956).

7 H.H.Nielsen, Handbuch der Physik, 37, 173; Springer Verlag, Berlin 1959.

8 D.R.Herschbach and V.W.Laurie, J.Chem.Phys. 37, 1668 (1962).

9 D.R.Herschbach and V.W.Laurie, J.Chem.Phys. 40, 3142 (1964).

10 T.Oka and Y.Morino, J.Mol.Spectry. 6, 472 (1961).

11 J.K.G.Watson, Mol.Phys. 15, 479 (1968).

12 G.Amat and L.Henry, Cah.Phys. 12, 273 (1958).

CHAPTER 3

Contributions to Vibrational Perturbation Theory - Part I

S. J. CYVIN and G. HAGEN

The Jacobian matrix in the vibrational perturbation theory is shortly reviewed. It contains the Jacobian elements for frequencies (in terms of λ parameters) with respect to force constants. New relations are developed for the Jacobians of mean-square amplitude quantities including their classical limits, and for compliants. They are useful in the study of changes in mean amplitudes of vibration with small variations of force constants.

3.1. Introduction

The vibrational perturbation theory has been advanced by many investigators. One of the early applications is due to Higgs.[1-3] A part of the modern developments of the theory consists of expressions for certain Jacobian elements, viz. partial derivatives with respect to force constants. Such Jacobian elements for vibrational frequencies and isotopic frequency shifts, Coriolis constants, and centrifugal distortion constants, have been reported by many authors.[4-18] One of the purposes of the present work is to report the Jacobian elements of mean-square amplitudes. To our knowledge the expressions have never been published before, although one related investigation has been communicated.[19] (See also Section 3.4.4.)

Perhaps the best treatment of Jacobian elements in the vibrational perturbation theory among the above cited works is the one due to Mills.[16]

3.2. Jacobian Matrix

Miyazawa and Overend[20] have advanced the vibrational perturbation theory further in the direction of systematization by introducing the Jacobian $\mathbf{J}$ matrix and its inverse, $\mathbf{J}^{-1}$. We are considering small changes in matrix elements according to

$$\boldsymbol{\lambda} = \boldsymbol{\lambda}^{\circ} + \Delta\boldsymbol{\lambda}, \quad \mathbf{L} = \mathbf{L}^{\circ} + \Delta\mathbf{L}, \tag{3.1}$$

and

$$\mathbf{G} = \mathbf{G}^{\circ} + \Delta\mathbf{G}, \quad \mathbf{F} = \mathbf{F}^{\circ} + \Delta\mathbf{F}. \tag{3.2}$$

Then the $\mathbf{J}$ matrix and its inverse are defined by[20]

$$\{\Delta\boldsymbol{\lambda}, \Delta\mathbf{L}\} = \mathbf{J}\{\Delta\mathbf{G}, \Delta\mathbf{F}\}, \quad \{\Delta\mathbf{G}, \Delta\mathbf{F}\} = \mathbf{J}^{-1}\{\Delta\boldsymbol{\lambda}, \Delta\mathbf{L}\}. \tag{3.3}$$

In the case of two-dimensional blocks (secular equations of second order) the column matrices $\{\Delta\boldsymbol{\lambda}, \Delta\mathbf{L}\}$ and $\{\Delta\mathbf{G}, \Delta\mathbf{F}\}$ contain six elements each, viz. $\{\Delta\lambda_1, \Delta\lambda_2, \Delta L_{11}, \Delta L_{22}, \Delta L_{12}, \Delta L_{21}\}$ on one hand and $\{\Delta G_{11}, \Delta G_{22}, \Delta G_{12}, \Delta F_{11}, \Delta F_{22}, \Delta F_{12}\}$ on the other. Within the first-order perturbation theory the elements of $\mathbf{J}$ (and $\mathbf{J}^{-1}$) are the appropriate Jacobians, viz. $\partial\lambda_1/\partial G_{11}$, $\partial\lambda_1/\partial G_{22}$, ... etc.

In the course of their developments Miyazawa and Overend[20] redefined the $\mathbf{J}$ matrix by substituting $\boldsymbol{P}$ in place of $\Delta\mathbf{L}$. The matrix was proposed by Nakagawa and Shimanouchi,[7] and is given by*

$$\mathbf{L} = \mathbf{L}^{\circ}(\mathbf{E} - \boldsymbol{P}), \quad \Delta\mathbf{L} = -\mathbf{L}^{\circ}\boldsymbol{P}. \tag{3.4}$$

It is convenient to introduce

$$\mathbf{K} = \mathbf{K}^{\circ} + \Delta\mathbf{K} \tag{3.5}$$

in addition to Eqs. (1) and (2). Here $\mathbf{K} = \mathbf{L}^{-1}$ and $\mathbf{K}^{\circ} = (\mathbf{L}^{\circ})^{-1}$. Then one has in terms of the $\boldsymbol{P}$ matrix:

$$\mathbf{K} = (\mathbf{E} + \boldsymbol{P})\mathbf{K}^{\circ}, \quad \Delta\mathbf{K} = \boldsymbol{P}\mathbf{K}^{\circ}. \tag{3.6}$$

The important paper of Miyazawa and Overend[20] contains

---

* When $\Delta\mathbf{G} = \mathbf{0}$, $\boldsymbol{P}$ is a skew-symmetric matrix. Then it is equal to $-\mathbf{B}$ in the notation of Mills,[16] and his $\mathbf{A}$ is equal to $\mathbf{E} - \boldsymbol{P}$.

explicit general expressions for all elements of **J** and $\mathbf{J}^{-1}$, and special attention has been offered to the case of two-dimensional blocks.

3.3. Jacobian Elements for Two-Dimensional Blocks

3.3.1. Jacobians for Vibrational Frequencies

Here we want to show an extremely elementary, but exact method for derivation of the $\partial\lambda_r/\partial F_{ij}$ quantities (i, j, ... are used to designate the internal coordinates; r, s the normal coordinates). They are the Jacobian elements for vibrational frequencies (represented by the λ_r parameters) in the case of two-dimensional blocks. These elements are part of the Jacobian **J** matrix of the preceding section.

From the familiar secular equation one obtains

$$\frac{\partial\lambda_1}{\partial F_{ii}} + \frac{\partial\lambda_2}{\partial F_{ii}} = G_{ii}\ , \quad \frac{\partial\lambda_1}{\partial F_{12}} + \frac{\partial\lambda_2}{\partial F_{12}} = 2G_{12}\ , \tag{3.7}$$

$$\lambda_2\frac{\partial\lambda_1}{\partial F_{ii}} + \lambda_1\frac{\partial\lambda_2}{\partial F_{ii}} = F_{jj}\,G\ , \quad \lambda_2\frac{\partial\lambda_1}{\partial F_{12}} + \lambda_1\frac{\partial\lambda_2}{\partial F_{12}} = -2F_{12}\,G\ ; \tag{3.8}$$

where $i, j = 1, 2$, and $i \neq j$. G is the determinant of the **G** matrix block, i.e. $G = G_{11}G_{22} - G_{12}^2$. The equations (7),(8) may be solved for the partial derivatives with the results

$$\frac{\partial\lambda_r}{\partial F_{ii}} = \frac{\lambda_r G_{ii} - F_{jj}G}{\lambda_r - \lambda_s}\ , \quad \frac{\partial\lambda_r}{\partial F_{12}} = \frac{2(\lambda_r G_{12} + F_{12}G)}{\lambda_r - \lambda_s}\ ; \tag{3.9}$$

where r and s should obey the same restrictions as i and j; $r, s = 1, 2$, and $r \neq s$. When **G** and **F** matrix elements are expressed in terms of $\boldsymbol{\lambda}$ and **L** matrix elements according to (1.25) and (1.26), the expressions (9) reduce to

$$\frac{\partial\lambda_r}{\partial F_{ii}} = L_{ir}^2\ , \quad \frac{\partial\lambda_r}{\partial F_{12}} = 2L_{1r}L_{2r}\ , \tag{3.10}$$

which are in accord with the known general expressions. The expressions equivalent to Eq. (10) for the two-dimensional case have also been given by Miyazawa and Overend.[20]

3.3.2. Jacobians for Mean-Square Amplitude Quantities

In this section a similar exact derivation of Jacobians for $\boldsymbol{\Sigma}$ matrix elements is outlined. Eqs. (7.18), (7.9), and (7.31) of Ref. 21 applied to two-dimensional blocks read:

$$\delta_1 + \delta_2 = M_{11}\Sigma_{11} + M_{22}\Sigma_{22} + 2M_{12}\Sigma_{12}\,, \tag{3.11}$$

$$\delta_1\delta_2 = (M_{11}M_{22} - M_{12}^2)(\Sigma_{11}\Sigma_{22} - \Sigma_{12}^2)\,, \tag{3.12}$$

$$\lambda_1\delta_1 + \lambda_2\delta_2 = F_{11}\Sigma_{11} + F_{22}\Sigma_{22} + 2F_{12}\Sigma_{12}\,. \tag{3.13}$$

By taking the partial derivatives one obtains:

$$M_{11}\frac{\partial\Sigma_{11}}{\partial F_{ij}} + M_{22}\frac{\partial\Sigma_{22}}{\partial F_{ij}} + 2M_{12}\frac{\partial\Sigma_{12}}{\partial F_{ij}} = \frac{\partial\delta_1}{\partial F_{ij}} + \frac{\partial\delta_2}{\partial F_{ij}}\,, \tag{3.14}$$

$$\Sigma_{22}\frac{\partial\Sigma_{11}}{\partial F_{ij}} + \Sigma_{11}\frac{\partial\Sigma_{22}}{\partial F_{ij}} - 2\Sigma_{12}\frac{\partial\Sigma_{12}}{\partial F_{ij}} = G\left(\delta_2\frac{\partial\delta_1}{\partial F_{ij}} + \delta_1\frac{\partial\delta_2}{\partial F_{ij}}\right)\,, \tag{3.15}$$

where $i\,,j$ = 1, 2, and i may be equal to j ; and:

$$F_{11}\frac{\partial\Sigma_{11}}{\partial F_{ii}} + F_{22}\frac{\partial\Sigma_{22}}{\partial F_{ii}} + 2F_{12}\frac{\partial\Sigma_{12}}{\partial F_{ii}}$$

$$= \delta_1\frac{\partial\lambda_1}{\partial F_{ii}} + \delta_2\frac{\partial\lambda_2}{\partial F_{ii}} + \lambda_1\frac{\partial\delta_1}{\partial F_{ii}} + \lambda_2\frac{\partial\delta_2}{\partial F_{ii}} - \Sigma_{ii}\,,$$

$$F_{11}\frac{\partial\Sigma_{11}}{\partial F_{12}} + F_{22}\frac{\partial\Sigma_{22}}{\partial F_{12}} + 2F_{12}\frac{\partial\Sigma_{12}}{\partial F_{12}}$$

$$= \delta_1\frac{\partial\lambda_1}{\partial F_{12}} + \delta_2\frac{\partial\lambda_2}{\partial F_{12}} + \lambda_1\frac{\partial\delta_1}{\partial F_{12}} + \lambda_2\frac{\partial\delta_2}{\partial F_{12}} - 2\Sigma_{12}\,. \tag{3.16}$$

Equations (14)-(16) may be solved for the desired partial derivatives. The resulting expressions become considerably simplified when the $\mathbf{G}$, $\mathbf{M}$, $\mathbf{F}$, and $\boldsymbol{\Sigma}$ matrix elements are expressed in terms of $\mathbf{L}$ matrix elements and frequency parameters according to Eqs. (1.25), (1.26), and (1.28). The final results read:

$$\frac{\partial\Sigma_{ij}}{\partial F_{kk}} = L_{i1}L_{j1}\frac{\partial\delta_1}{\partial F_{kk}} + L_{i2}L_{j2}\frac{\partial\delta_2}{\partial F_{kk}}$$

$$+ (L_{i1}L_{j2} + L_{i2}L_{j1}) L_{k1} L_{k2} \frac{\delta_1 - \delta_2}{\lambda_1 - \lambda_2} ,$$

$$\frac{\partial \Sigma_{ij}}{\partial F_{12}} = L_{i1} L_{j1} \frac{\partial \delta_1}{\partial F_{12}} + L_{i2} L_{j2} \frac{\partial \delta_2}{\partial F_{12}}$$

$$+ (L_{i1}L_{j2} + L_{i2}L_{j1})(L_{11}L_{22} + L_{12}L_{21}) \frac{\delta_1 - \delta_2}{\lambda_1 - \lambda_2} , \qquad (3.17)$$

where $i, j, k = 1, 2$ independently of each other.

3.4. Jacobian Elements for Mean Amplitudes

3.4.1. Jacobians for Mean Amplitudes

The Jacobian elements for a mean amplitude of vibration (l) are

$$\frac{\partial l}{\partial F_{ij}} = \frac{1}{2l} \frac{\partial l^2}{\partial F_{ij}} = \frac{1}{2l} \sum_{m \leq n} \sum c_{mn} \frac{\partial \Sigma_{mn}}{\partial F_{ij}} \qquad (3.18)$$

where c_{mn} are constants. Here it has been made use of the fact that the mean-square amplitudes (l^2) may be evaluated as linear combinations of $\mathbf{\Sigma}$ matrix elements; cf. Eq. (11.11) of Ref. 21. Hence we may in the following concentrate our attention upon the Jacobians for the mean-square amplitude quantities Σ_{ij}.

3.4.2. General Expressions for Jacobians of Mean-Square Amplitude Quantities

Write in addition to Eqs. (1), (2), and (5):

$$\boldsymbol{\delta} = \boldsymbol{\delta}^\circ + \Delta\boldsymbol{\delta} \qquad (3.19)$$

for the diagonal matrix of frequency parameters (1.24), and

$$\mathbf{\Sigma} = \mathbf{\Sigma}^\circ + \Delta\mathbf{\Sigma}. \qquad (3.20)$$

Then with the aid of Eq. (1.28) and Eq. (4) one obtains

$$\mathbf{\Sigma} = \mathbf{L}^\circ(\mathbf{E} - \boldsymbol{P})(\boldsymbol{\delta}^\circ + \Delta\boldsymbol{\delta})(\mathbf{E} - \boldsymbol{P}')(\mathbf{L}^\circ)'. \qquad (3.21)$$

Using $\mathbf{\Sigma}^\circ = \mathbf{L}^\circ \boldsymbol{\delta}^\circ (\mathbf{L}^\circ)'$ and neglecting all higher-order terms the following result was obtained.

$$\Delta\boldsymbol{\Sigma} = \mathbf{L}^\circ(\Delta\boldsymbol{\delta} - \boldsymbol{P}\boldsymbol{\delta}^\circ - \boldsymbol{\delta}^\circ\boldsymbol{P}')(\mathbf{L}^\circ)'. \tag{3.22}$$

When $\Delta\mathbf{G} = \mathbf{O}$, $\boldsymbol{P}$ is a skew-symmetric matrix with off-diagonal elements ($r \neq s$):[16]

$$P_{rs} = \sum_i \sum_j \frac{L_{ir}^\circ L_{js}^\circ}{\lambda_r - \lambda_s} \Delta F_{ij}$$

$$= \sum_i \frac{L_{ir}^\circ L_{is}^\circ}{\lambda_r - \lambda_s} \Delta F_{ii} + \sum_{i<j}\sum \frac{L_{ir}^\circ L_{js}^\circ + L_{is}^\circ L_{jr}^\circ}{\lambda_r - \lambda_s} \Delta F_{ij}. \tag{3.23}$$

Under this condition Eq. (22) may be written

$$\Delta\boldsymbol{\Sigma} = \mathbf{L}^\circ(\Delta\boldsymbol{\delta} + \boldsymbol{\delta}^\circ\boldsymbol{P} - \boldsymbol{P}\boldsymbol{\delta}^\circ)(\mathbf{L}^\circ)' \tag{3.24}$$

or

$$\Delta\boldsymbol{\Sigma} = \mathbf{L}^\circ\boldsymbol{Q}(\mathbf{L}^\circ)', \tag{3.25}$$

where $\boldsymbol{Q}$ appears to be the symmetrical matrix

$$\boldsymbol{Q} = \begin{bmatrix} \Delta\delta_1 & P_{12}(\delta_1^\circ - \delta_2^\circ) & \cdots \\ P_{12}(\delta_1^\circ - \delta_2^\circ) & \Delta\delta_2 & \cdots \\ P_{13}(\delta_1^\circ - \delta_3^\circ) & P_{23}(\delta_2^\circ - \delta_3^\circ) & \cdots \\ \cdots & \cdots & \cdots \end{bmatrix}. \tag{3.26}$$

Hence for the elements of $\Delta\boldsymbol{\Sigma}$ one obtains

$$\Delta\Sigma_{ij} = \sum_r L_{ir}^\circ L_{jr}^\circ \Delta\delta_r$$

$$+ \sum_{r<s}\sum (L_{ir}^\circ L_{js}^\circ + L_{is}^\circ L_{jr}^\circ) P_{rs}(\delta_r^\circ - \delta_s^\circ), \tag{3.27}$$

where i may be equal to j. Finally with the aid of Eq. (23) these elements may be expanded in terms of the changes in force constants, ΔF_{ij}, and the relation yields the desired Jacobian elements,

$$\frac{\partial \Sigma_{ij}}{\partial F_{kl}} = \sum_r L_{ir} L_{jr} \frac{\partial \delta_r}{\partial F_{kl}}$$

$$+ \mathfrak{H}_{kl} \sum_{r<s}\sum (L_{ir}L_{js} + L_{is}L_{jr})(L_{kr}L_{ls} + L_{ks}L_{lr}) \frac{\delta_r - \delta_s}{\lambda_r - \lambda_s} , \qquad (3.28)$$

where $\mathfrak{H}_{kl} = 1$ for $k \neq l$, and $\mathfrak{H}_{kk} = \frac{1}{2}$.* Eq. (28) is a generalization of Eq. (17).

3.4.3. Jacobians for δ_r Frequency Parameters

In the expression (28) for Jacobians of mean-square amplitude quantities the partial derivatives of δ_r appear. Hence these quantities need to be evaluated. Let Eqs. (1.23) and (1.24) be written

$$\lambda_r = k_1 \omega_r^2 , \qquad \sigma_r = \frac{1}{k_1 \omega_r^2} \qquad (3.29)$$

and

$$\delta_r = \frac{k_2}{\omega_r} \coth(k_3 \omega_r) , \qquad (3.30)$$

where k_1, k_2, and k_3 are constants; the latter one, however, being temperature-dependent.** From Eqs. (29) and (30) one obtains

$$\frac{d\delta_r}{d\lambda_r} = - \frac{k_2}{2\lambda_r} \left[\frac{1}{\omega_r \tanh(k_3\omega_r)} + \frac{k_3}{\sinh^2(k_3\omega_r)} \right] . \qquad (3.31)$$

The partial derivatives of δ_r which appear in Eq. (28) are

$$\frac{\partial \delta_r}{\partial F_{ij}} = \frac{d\delta_r}{d\lambda_r} \frac{\partial \lambda_r}{\partial F_{ij}} , \qquad (3.32)$$

where the partial derivatives of λ_r are known. They are[16,20]

---

* In terms of Kronecker's delta: $\mathfrak{H}_{kl} = 1 - \frac{\delta_{kl}}{2}$.

** When the usual units are applied (cf. Section 7.13 of Ref. 21) the numerical forms of these constants are: $k_1 = 0.589141 \times 10^{-6}$, $k_2 = 16.85748$, $k_3 = 0.719399/T$.

$$\frac{\partial \lambda_r}{\partial F_{ij}} = 2 \mathfrak{H}_{ij} L_{ir} L_{jr} \; ; \tag{3.33}$$

cf. also Eqs. (10).

3.4.4. Concluding Remark

The results of the preceding sections were derived independently of the work by Kukina,[22] who had developed the equivalent results. Her work is based on a representation of mean-square amplitude quantities in the formalism of Sverdlov and Kukina.[23]

3.5. Jacobian Elements for Classical Limits of Mean-Square Amplitudes and for Compliants

3.5.1. Jacobians for δ_r^{cl} and σ_r

For a survey of the theory of classical limits of mean-square amplitude quantities, see Section 7.10 of Ref. 21. To find the classical limit of $\partial\delta_r/\partial F_{ij}$ when $T \to \infty$ ($k_3 \to 0$), we utilize $\tanh x \to x$ and $\sinh x \to x$ for small x. Hence

$$\frac{d\delta_r}{d\lambda_r} \to -\frac{k_1 \; k_2}{k_3 \lambda_r^2} = -\frac{kT}{\lambda_r^2} . \tag{3.34}$$

The same result is obtained directly from $\delta_r^{cl} = kT/\lambda_r$; cf. Eq. (7.42) of Ref. 21. Hence

$$\frac{d\delta_r^{cl}}{d\lambda_r} = -\frac{kT}{\lambda_r^2} = -kT\sigma_r^2 . \tag{3.35}$$

With the aid of Eq. (33) it follows

$$\frac{\partial \delta_r^{cl}}{\partial F_{ij}} = -2\mathfrak{H}_{ij} kT L_{ir} L_{jr} \sigma_r^2 . \tag{3.36}$$

For the corresponding Jacobians of $\sigma_r = 1/\lambda_r = \delta_r^{cl}/kT$ it is found:

$$\frac{\partial \sigma_r}{\partial F_{ij}} = -2\mathfrak{H}_{ij} L_{ir} L_{jr} \sigma_r^2 . \tag{3.37}$$

3.5.2. Jacobians for Classical Limits of Mean-Square Amplitude Quantities

From Eq. (28) we find the corresponding expression for the classical limit when $\delta_r \to \delta_r^{cl} = kT\sigma_r$. With the aid of Eq. (36) it has been found:

$$\frac{\partial \Sigma_{ij}^{cl}}{\partial F_{kl}} = -\mathfrak{H}_{kl}\, kT \sum_r \sum_s L_{ir} L_{js} (L_{kr} L_{ls} + L_{ks} L_{lr}) \sigma_r \sigma_s \,. \qquad (3.38)$$

3.5.3. Jacobians for Compliance Constants

The classical limits of mean-square amplitudes are known to be proportional with the compliants.[21] From Eq. (38) one obtains for the corresponding Jacobians of compliance constants:

$$\frac{\partial N_{ij}}{\partial F_{kl}} = -\mathfrak{H}_{kl} \sum_r \sum_s L_{ir} L_{js} (L_{kr} L_{ls} + L_{ks} L_{lr}) \sigma_r \sigma_s \,. \qquad (3.39)$$

In the case of two-dimensional blocks (cf. also Section 3) it is easy to evaluate these quantities in an elementary way. From the relations expressing that the **F** and **N** blocks are inverse of each other[21] it is obtained:

$$\frac{\partial N_{ij}}{\partial F_{kk}} = -N_{ik} N_{jk} \,, \qquad \frac{\partial N_{ij}}{\partial F_{12}} = -(N_{i1} N_{j2} + N_{i2} N_{j1}) \,. \qquad (3.40)$$

When the **N** elements in Eqs. (40) are expressed in terms of $\boldsymbol{\sigma}$ and **L** matrix elements according to (1.26), which gives $N_{ij} = L_{i1} L_{j1} \sigma_1 + L_{i2} L_{j2} \sigma_2$, the resulting equations are found to be consistent with the general form of Eq. (39). For the sake of clarity the nine Jacobians in question as obtained according to Eqs. (40) and (39) are shown in Table I.

3.6. Alternative Approach to Calculation of Mean-Square Amplitudes

From Eqs. (1.28) and (1.24) it is obtained [see also Eq. (7.13) of Ref. 21]

Table 3-I. Jacobian elements for compliants in the case of two-dimensional blocks

	ΔF_{11}	ΔF_{22}	ΔF_{12}
ΔN_{11}	$-N_{11}^2$	$-N_{12}^2$	$-2N_{11}N_{12}$
ΔN_{22}	$-N_{12}^2$	$-N_{22}^2$	$-2N_{12}N_{22}$
ΔN_{12}	$-N_{11}N_{12}$	$-N_{12}N_{22}$	$-(N_{11}N_{22}+N_{12}^2)$
ΔN_{11}	$-(L_{11}^2\sigma_1+L_{12}^2\sigma_2)^2$	$-(L_{11}L_{21}\sigma_1+L_{12}L_{22}\sigma_2)^2$	$-2[L_{11}^3L_{21}\sigma_1^2+L_{12}^3L_{22}\sigma_2^2$ $+L_{11}L_{21}(L_{11}L_{22}+L_{12}L_{21})\sigma_1\sigma_2]$
ΔN_{22}	$-(L_{11}L_{21}\sigma_1+L_{12}L_{22}\sigma_2)^2$	$-(L_{21}^2\sigma_1+L_{22}^2\sigma_2)^2$	$-2[L_{11}L_{21}^3\sigma_1^2+L_{12}L_{22}^3\sigma_2^2$ $+(L_{11}L_{22}+L_{12}L_{21})L_{21}L_{22}\sigma_1\sigma_2]$
ΔN_{12}	$-[L_{11}^3L_{21}\sigma_1^2+L_{12}^3L_{22}\sigma_2^2$ $+L_{11}L_{12}(L_{11}L_{22}+L_{12}L_{21})\sigma_1\sigma_2]$	$-[L_{11}L_{21}^3\sigma_1^2+L_{12}L_{22}^3\sigma_2^2$ $+(L_{11}L_{22}+L_{12}L_{21})L_{21}L_{22}\sigma_1\sigma_2]$	$-[2L_{11}^2L_{21}^2\sigma_1^2+2L_{12}^2L_{22}^2\sigma_2^2$ $+(L_{11}L_{22}+L_{12}L_{21})^2\sigma_1\sigma_2]$

$$\langle S_i^2 \rangle = \sum_k L_{ik}^2 \frac{h}{8\pi^2 c\omega_k} \coth \frac{hc\omega_k}{2kT} . \qquad (3.41)$$

On introducing the Jacobian element from Eq. (10) one obtains

$$\langle S_i^2 \rangle = \sum_k \frac{h}{8\pi^2 c\omega_k} \frac{\partial \lambda_k}{\partial F_{ii}} \coth \frac{hc\omega_k}{2kT} , \qquad (3.42)$$

which by means of Eq. (1.23) simplifies to

$$\langle S_i^2 \rangle = \sum_k \frac{\partial \omega_k}{\partial F_{ii}} hc \coth \frac{hc\omega_k}{2kT} \qquad (3.43)$$

This relation was first given by Mayants;[24] see also Ref. 25. The quantity of Eq. (43) is equivalent to Σ_{ii} . A generalization to Σ_{ij} reads

$$\langle S_i S_j \rangle = \frac{1}{2\mathfrak{H}_{ij}} \sum_k \frac{\partial \omega_k}{\partial F_{ij}} hc \coth \frac{hc\omega_k}{2kT} . \qquad (3.44)$$

REFERENCES

1 P.W.Higgs, J.Chem.Phys. 21, 1131 (1953).
2 P.W.Higgs, J.Chem.Phys. 23, 1448 (1955).
3 P.W.Higgs, J.Chem.Phys. 23, 1450 (1955).
4 T.Miyazawa, Nippon Kagaku Zasshi 76, 1132 (1955).
5 D.E.Mann, T.Shimanouchi, J.H.Meal, and L.Fano, J.Chem.Phys. 27, 43 (1957).
6 W.T.King, I.M.Mills, and B.L.Crawford,Jr., J.Chem.Phys. 27, 455 (1957).
7 I.Nakagawa and T.Shimanouchi, Nippon Kagaku Zasshi 80, 128 (1959).
8 J.Overend and J.R.Scherer, J.Chem.Phys. 32, 1289 (1960).
9 I.M.Mills, Spectrochim.Acta 16, 35 (1960).
10 J.Aldous and I.M.Mills, Spectrochim.Acta 18, 1073 (1962).
11 R.C.Lord and I.Nakagawa, J.Chem.Phys. 39, 2951 (1963).
12 J.H.Schachtschneider and R.G.Snyder, Spectrochim.Acta 19, 117 (1963).
13 J.Aldous and I.M.Mills, Spectrochim.Acta 19, 1567 (1963).

14 T.Shimanouchi and I.Suzuki, J.Chem.Phys. 42, 296 (1965); erratum: ibid. 43, 1854 (1965).

15 I.W.Levin and S.Abramowitz, J.Chem.Phys. 43, 4213 (1965).

16 I.M.Mills, J.Mol.Spectry. 5, 334 (1960); erratum: ibid. 17, 164 (1965).

17 M.Tsuboi, J.Mol.Spectry. 19, 4 (1966).

18 A.A.Chalmers and D.C.McKean, Spectrochim.Acta 22, 251 (1966).

19 D.Papoušek and J.Plíva, Spectrochim.Acta 21, 1147 (1965).

20 T.Miyazawa and J.Overend, Bull.Chem.Soc.Japan 39, 1410 (1966).

21 S.J.Cyvin: Molecular Vibrations and Mean Square Amplitudes, Universitetsforlaget, Oslo, and Elsevier, Amsterdam, 1968.

22 V.S.Kukina, Opt.Spektroskopiya 26, 111 (1969).

23 L.M.Sverdlov and V.S.Kukina, Opt.Spektroskopiya 23, 172 (1967).

24 L.S.Mayants, Dokl.Akad.Nauk SSSR 151, 624 (1963).

25 L.S.Mayants, Zh.Fiz.Khim. 38, 623 (1964).

-- Part II: Jacobian Elements for Centrifugal Distortion Constants

G. O. SØRENSEN

Some of the matrices discussed in Chapter 2 are applied in evaluating partial derivatives of centrifugal distortion constants with respect to the force constants of a harmonic force field.

This part represents another contribution to the vibrational perturbation theory and deals with the centrifugal distortion constants. The approach is different from that of the preceding part.

Accurate centrifugal distortion constants are often produced by microwave investigations of rotational spectra, and their usefulness in force constant calculations is well established. The necessary Jacobian elements have been evaluated by Mills[1] for the case of an ordinary set of independent internal coordinates, while the implications of redundancies have been discussed by Strey.[2] However, the general expressions for Jacobian elements, valid when dependent internal coordinates are applied as well, are more easily derived and expressed by means of the matrices $\mathbf{V}$ and $\mathbf{K}_s$ discussed in a recent paper by Sørensen.[3]

The centrifugal distortion constants defined by Kivelson and Wilson[4] are given by[5,6]

$$\tau_{\alpha\beta\gamma\delta} = -\frac{1}{2}\sum_i \frac{\partial\mu_{\alpha\beta}}{\partial Q_i}\frac{\partial\mu_{\gamma\delta}}{\partial Q_i}\lambda_i^{-1} = -\frac{t_{\alpha\beta\gamma\delta}}{2I_\alpha^{e}I_\beta^{e}I_\gamma^{e}I_\delta^{e}}\ , \tag{3.45}$$

$$t_{\alpha\beta\gamma\delta} = \sum_i \frac{\partial I_{\alpha\beta}}{\partial Q_i}\frac{\partial I_{\gamma\delta}}{\partial Q_i}\lambda_i^{-1} = \mathbf{J}_{\alpha\beta,Q}'\,\boldsymbol{\sigma}\,\mathbf{J}_{\gamma\delta,Q}\ , \tag{3.46}$$

in which $I_{\alpha\beta}$ and $\mu_{\alpha\beta}$ are elements of the instantaneous tensor of inertia $\mathbf{I}$ and its inverse $\boldsymbol{\mu}$. I_α^e is an element of the diagonal tensor of inertia corresponding to the equilibrium configuration, and $\mathbf{J}_{\alpha\beta,Q}$ is the matrix of first derivatives dicussed in Chapter 2. $\boldsymbol{\sigma} = \boldsymbol{\lambda}^{-1}$; cf. also Eq.(1.23).

The matrices $\mathbf{V}$, $\mathbf{F}_s$, $\mathbf{L}_s$, $\mathbf{K}_s$ and $\mathbf{K}$ are defined by the following relations*:[3]

$$\mathbf{S} = \mathbf{V}\mathbf{s}, \quad \mathbf{s} = \mathbf{K}_s\mathbf{S} = \mathbf{L}_s\mathbf{Q}, \tag{3.47}$$

$$\mathbf{V}\mathbf{V}' = \mathbf{G}, \quad \mathbf{F}_s = \mathbf{V}'\mathbf{F}\mathbf{V}, \tag{3.48}$$

$$\mathbf{L}_s'\mathbf{F}_s\mathbf{L}_s = \boldsymbol{\lambda}, \quad \mathbf{L} = \mathbf{V}\mathbf{L}_s, \tag{3.49}$$

$$\mathbf{K}_s\mathbf{V} = \mathbf{E}, \quad \mathbf{K} = \mathbf{L}_s'\mathbf{K}_s. \tag{3.50}$$

$\mathbf{K}_s$ and $\mathbf{K}$ are 'left' inverses of $\mathbf{V}$ and $\mathbf{L}$ respectively, while $\mathbf{L}_s$ is orthonormal. If redundancies are not present the following relations not included in Eq. (50) are valid:

$$\mathbf{K}_s = \mathbf{V}^{-1}, \quad \mathbf{K} = \mathbf{L}^{-1}, \tag{3.51}$$

but, since the purpose is to derive general expressions, we will refrain from using them below.

The matrix $\mathbf{J}_{\alpha\beta,Q}$ may be expressed in several ways[5,6] (see Section 1.3), but for the derivation of Jacobians the following form is most convenient.

$$\mathbf{J}_{\alpha\beta,Q} = \mathbf{K}\mathbf{T}_{\alpha\beta,S} = 2\mathbf{K}\mathbf{B}\mathbf{I}^\alpha(\mathbf{I}^\beta)'\mathbf{R}^e. \tag{3.52}$$

We now introduce the matrices $\mathbf{I}_R^\alpha$ defined by

$$\mathbf{I}_R^\alpha = \mathbf{I}^\alpha 2^{-\frac{1}{4}}(I_\alpha^e)^{-1}, \tag{3.53}$$

and use these for the definition of

$$\bar{\mathbf{T}}_{\alpha\beta,S} = 2\mathbf{B}\mathbf{I}_R^\alpha(\mathbf{I}_R^\beta)'\mathbf{R}^e. \tag{3.54}$$

---

* The notation from Ref. 3 is adopted, except for $\mathbf{U}_s$, which here is denoted $\mathbf{L}_s$.

Hereby a simple expression for the centrifugal distortion constants can be obtained,

$$\tau_{\alpha\beta\gamma\delta} = -\bar{\mathbf{T}}'_{\alpha\beta,S}\mathbf{K}'\boldsymbol{\sigma}\mathbf{K}\bar{\mathbf{T}}_{\gamma\delta,S}\ , \tag{3.55}$$

or by using Eqs. (50) and (49)

$$\tau_{\alpha\beta\gamma\delta} = -\bar{\mathbf{T}}'_{\alpha\beta,S}\mathbf{K}_s'\mathbf{N}_s\mathbf{K}_s\bar{\mathbf{T}}_{\gamma\delta,S}\ , \tag{3.56}$$

where $\mathbf{N}_s$ is the inverse of $\mathbf{F}_s$. It must be emphasized that Eq. (56) cannot be further reduced, since the product $\mathbf{K}_s'\mathbf{N}_s\mathbf{K}_s$ only equals $\mathbf{F}^{-1}$ if there are no redundancies.

Now consider the effect of a small change $d\mathbf{F}$ of $\mathbf{F}$, namely the small changes $d\mathbf{F}_s$, $d\mathbf{N}_s$ and particularly the changes $d\tau_{\alpha\beta\gamma\delta}$. First the relation

$$d\mathbf{N}_s = -\mathbf{N}_s\, d\mathbf{F}_s\, \mathbf{N}_s \tag{3.57}$$

may be obtained from the equality $(\mathbf{N}_s + d\mathbf{N}_s)(\mathbf{F}_s + d\mathbf{F}_s) = \mathbf{E}$, and from Eq. (48) it follows that

$$d\mathbf{N}_s = -\mathbf{N}_s\mathbf{V}'d\mathbf{F}\mathbf{V}\mathbf{N}_s\ . \tag{3.58}$$

Following Mills[1] we introduce the column matrices $\boldsymbol{\pi}^{\alpha\beta}$ defined so that the changes in centrifugal distortion constants may be expressed

$$d\tau_{\alpha\beta\gamma\delta} = (\boldsymbol{\pi}^{\alpha\beta})'d\mathbf{F}\boldsymbol{\pi}^{\gamma\delta}. \tag{3.59}$$

From Eqs. (56) and (58) it is easily seen that

$$\boldsymbol{\pi}^{\alpha\beta} = \mathbf{V}\mathbf{N}_s\mathbf{K}_s\bar{\mathbf{T}}_{\alpha\beta,S}\ , \tag{3.60}$$

and we shall notice that this equation is general in the sense that redundancies may or may not appear in the set of internal coordinates. Hence Jacobian elements may always be calculated using Eq. (60) and the expression

$$\frac{\partial\tau_{\alpha\beta\gamma\delta}}{\partial F_{ij}} = (\pi_i^{\alpha\beta}\pi_j^{\gamma\delta} + \pi_j^{\alpha\beta}\pi_i^{\gamma\delta})(1 - \frac{\delta_{ij}}{2}). \tag{3.61}$$

If independent symmetry coordinates have been chosen we may apply Eq. (51) and obtain

$$\boldsymbol{\pi}^{\alpha\beta} = \mathbf{F}^{-1}\mathbf{G}^{-1}\overline{\mathbf{T}}_{\alpha\beta,S} , \qquad (3.62)$$

in agreement with the result obtained by Mills.[1] However, this transformation offers only small advantages as to ease of computation, and considering the loss in generality one must conclude that Eq. (60) is the more favourable.

The computational advantages or disadvantages of Eq. (60) should rather be discussed in relation to an expression obtained by substituting $\mathbf{N}_S$ with the aid of Eqs. (49) and (50):

$$\boldsymbol{\pi}^{\alpha\beta} = \mathbf{L}\boldsymbol{\sigma}\mathbf{K}\overline{\mathbf{T}}_{\alpha\beta,S} . \qquad (3.63)$$

An inversion is usually a simpler procedure than a diagonalization, but, if the diagonalization for some reason must be performed the inversion is unnecessary since the inverse may be obtained by transforming the reciprocal diagonal matrix. Hence, if $\boldsymbol{\lambda}$ and $\mathbf{L}$ have been evaluated for other purposes Eq. (63) should be applied for the calculation of $\boldsymbol{\pi}^{\alpha\beta}$. In other cases Eq. (60) must be prefered.

REFERENCES

1 I.M.Mills, J.Mol.Spectry. 5, 334 (1960); erratum: ibid. 17, 164 (1965).

2 G.Strey, J.Mol.Spectry. 17, 265 (1965).

3 G.O.Sørensen, J.Mol.Spectry. 36, 359 (1970).

4 D.Kivelson and E.B.Wilson, J.Chem.Phys. 21, 1229 (1953).

5 G.O.Sørensen, G.Hagen, and S.J.Cyvin, J.Mol.Spectry. 35, 489 (1970).

6 S.J.Cyvin, B.N.Cyvin, and G.Hagen, Z.Naturforschg. 23a, 1649 (1968).

CHAPTER 4

Mean Amplitudes of Vibration Compatible with Measured Normal Frequencies

F. TÖRÖK and G. KOVÁCS

The parameter method elaborated for computation of force constants has proved to be useful in the theory of mean amplitudes. It enables us to investigate mean amplitudes compatible with normal frequencies and to compare them with mean amplitudes originating from electron diffraction. Iteration processes are given, some of them for calculating maximum and minimum values of mean amplitudes compatible with normal frequencies.

Az erőállandók meghatározására kidolgozott paraméteres módszer a közepes rezgési amplitúdók elméletében is előnyösen alkalmazható. Lehetővé teszi hogy a normál frekvenciákkal összeegyeztethető közepes amplitúdókat vizsgáljuk és összehasonlítsuk őket az elektrondiffrakcióból eredő közepes rezgési amplitúdókkal. Iterációs eljárásokat ismertetünk, melyekkel kiszámíthatjuk a normál frekvenciákkal összhangban lévő közepes rezgési amplitúdók szélső értékeit.

4.1. Introduction

It is well known that the force constants of a molecule are not determined by normal frequencies. The infinite sets of force constant matrices reproducing the normal frequencies can be constructed by the parameter method.[1,2] This method has also proved to be useful in the theory of mean amplitudes of vibration.[3] The following problems can be solved.

(1) Calculation of all mean amplitudes of vibration compatible with the normal frequencies (compatible mean amplitudes).

(2) Calculation of minimum and maximum values (extreme values) of the compatible mean amplitudes using the partial derivatives of the mean amplitudes with respect to the parameters.

(3) Starting from a given force constant matrix which reproduces the normal frequencies, but not the mean amplitudes of vibration originating from electron diffraction, a force constant matrix can be calculated compatible with the normal frequencies and reproducing the given mean amplitudes as far as possible.

4.2. Determination of all Mean Amplitudes of Vibration Compatible with the Normal Frequencies

Let us consider

$$\mathbf{G} = \boldsymbol{g}\boldsymbol{g}, \tag{4.1}$$

where $\mathbf{G}$ is supposed to be based on a chosen set of independent symmetry coordinates, $S_1, S_2, \ldots, S_n$. In the decomposition (1) one has $\boldsymbol{g} = \mathbf{G}^{\frac{1}{2}}$. From the basic equations (1.25),(1.26) is obtained

$$\boldsymbol{g}\mathbf{F}\boldsymbol{g}\boldsymbol{U} = \boldsymbol{U}\boldsymbol{\lambda}, \tag{4.2}$$

where*

$$\boldsymbol{U} = \boldsymbol{g}^{-1}\mathbf{L}. \tag{4.3}$$

In Eq. (2) $\boldsymbol{g}\mathbf{F}\boldsymbol{g}$ is a real symmetric matrix; hence $\boldsymbol{U}$, which consists of its eigenvectors, can be made orthogonal by suitable normalization. If $\mathbf{L}$ is normalized by the usual requirement of Eq. (1.25) $\boldsymbol{U}$ becomes automatically orthogonal. From Eq. (2) we can express $\mathbf{F}$ explicitly as

---

* The $\boldsymbol{U}$ matrix should not be confused with $\mathbf{U}$ of $\mathbf{S} = \mathbf{U}\mathbf{R}$ [cf. Eq.(1.21)] applied later in this chapter.

$$\mathbf{F} = \boldsymbol{g}^{-1}\boldsymbol{U}\boldsymbol{\lambda}\boldsymbol{U}'\boldsymbol{g}^{-1} = \mathbf{K}'\boldsymbol{\lambda}\mathbf{K}; \qquad (4.4)$$

cf. Eq.(1.26). Any $\mathbf{F}$ satisfying the secular equation of the vibrational problem can be written in the form of Eq. (4), where $\boldsymbol{U}$ is any real orthogonal matrix with the determinant +1.[4-6] Eq. (4) was first derived by Taylor,[4] and independently of the work in our laboratory[1] by Toman and Plíva.[6]

It is known that any real n-dimensional orthogonal matrix with the determinant +1 (real proper orthogonal matrix) can be expressed as a function of $\binom{n}{2} = n(n-1)/2$ parameters. There are several parameter representations of real proper orthogonal matrices. According to our experience the antisymmetric parameters are very advantageous for solving numerical problems. The elements of an antisymmetric matrix $\mathbf{A}$ are the α_{ij} ($i < j = 2, 3, \ldots, n$) parameters; $A_{11} = A_{22} = \ldots = A_{nn} = 0$, $A_{ij} = \alpha_{ij}$, $A_{ji} = -\alpha_{ij}$. The antisymmetric parameters can take any value between $+\infty$ and $-\infty$. The real proper orthogonal matrices are formed from the $\mathbf{A}$ matrices by

$$\boldsymbol{U} = (\mathbf{E} - \mathbf{A})(\mathbf{E} + \mathbf{A})^{-1} = 2(\mathbf{E} + \mathbf{A})^{-1} - \mathbf{E}, \qquad (4.5)$$

where $\mathbf{E}$ is the unit matrix. From a given matrix $\boldsymbol{U}$ the corresponding antisymmetric matrix can easily be calculated by the formula

$$\mathbf{A} = 2(\boldsymbol{U} + \mathbf{E})^{-1} - \mathbf{E}. \qquad (4.6)$$

In order to find a connection between the force constant matrix given by Eq. (4) and the mean amplitudes of vibration let us consider a molecule with N atoms and n degrees of vibrational freedom ($n = 3N-5$ or $3N-6$). We adhere to the notation of Chapter 1. In addition let $\mathbf{d}$ denote the column matrix of interatomic distance deviations from equilibrium for various atom pairs. In general $N(N-1)/2$ pairs of atoms are present. Introduce the transformations*

---

* In the notation of Cyvin[7] this equation would read

$$\mathbf{d} = \mathbf{L}^{d}\mathbf{Q} = \mathbf{T}^{x}\mathbf{x};$$

cf. Eqs.(11.4) and (13.19) therein. Török et al.[3] would write

$$\mathbf{r} = \mathbf{K}\mathbf{Q} = \mathbf{B}_m\mathbf{X}.$$

$$\mathbf{d} = \mathbf{L}_d \mathbf{Q} = \mathbf{B}_d \mathbf{x}. \tag{4.7}$$

Let the matrix $\mathbf{P}$ be defined by the mean values of matrix elements according to

$$\mathbf{P} = \langle \mathbf{d}\mathbf{d}' \rangle. \tag{4.8}$$

Consequently

$$\mathbf{P} = \langle \mathbf{d}\mathbf{d}' \rangle = \mathbf{L}_d \langle \mathbf{Q}\,\mathbf{Q}' \rangle \mathbf{L}_d' = \mathbf{L}_d \boldsymbol{\delta} \mathbf{L}_d', \tag{4.9}$$

where $\boldsymbol{\delta}$ is the diagonal matrix with elements given in Eq. (1.24). Using the matrices defined previously we obtain

$$\mathbf{d} = \mathbf{L}_d \mathbf{Q} = \mathbf{L}_d \mathbf{L}^{-1} \mathbf{S} = \mathbf{L}_d \mathbf{L}^{-1} \mathbf{U} \mathbf{B}_R \mathbf{x} = \mathbf{B}_d \mathbf{x}, \tag{4.10}$$

from which

$$\mathbf{L}_d = \mathbf{B}_d \mathbf{m}^{-1} \mathbf{B}_R' \mathbf{G}_R^{-1} \mathbf{U}' \mathbf{L}. \tag{4.11}$$

Here $\mathbf{B}_R$ is defined by $\mathbf{R} = \mathbf{B}_R \mathbf{x}$ similarly to Eq. (1.5), and $\mathbf{G}_R$ is constructed on the basis of the valence coordinates ($\mathbf{R}$). The Crawford-Fletcher formula [cf. Eq.(1.11)] applied to the valence coordinates was utilized in the above derivation.

According to Eq. (3) the matrix $\mathbf{L}$ can be expressed as

$$\mathbf{L} = \mathbf{G}^{\frac{1}{2}} \boldsymbol{U}, \tag{4.12}$$

where $\mathbf{G}$ is based on the symmetry coordinates. $\boldsymbol{U}$, when based on the symmetry coordinates too, is partitioned into diagonal blocks of orthogonal matrices corresponding to the appropriate irreducible representations; the block structure is the same as for $\mathbf{G}$, $\mathbf{F}$, and $\mathbf{L}$. Each element in a given orthogonal block of dimension n_i can be constructed by $\binom{n_i}{2}$ parameters. On inserting into (9) it is seen that also the $\mathbf{L}_d$ matrix can be given as a function of parameters:

$$\mathbf{L}_d = \mathbf{B}_d \mathbf{m}^{-1} \mathbf{B}_R' \mathbf{G}_R^{-1} \mathbf{U} \mathbf{G}^{\frac{1}{2}} \boldsymbol{U} = \mathbf{B}_d \mathbf{m}^{-1} \mathbf{B}' \mathbf{G}^{-\frac{1}{2}} \boldsymbol{U}. \tag{4.13}$$

It follows that the $\mathbf{L}_d$ matrix (and consequently also $\mathbf{P}$) is additive in the sense that

$$L_d = \sum_i B_d m^{-1} B_i' G_i^{-\frac{1}{2}} U_i , \tag{4.14}$$

where i designates matrices belonging to the i-th irreducible representation.

Eqs. (13) and (8) give the possibility for 'mapping' the compatible mean amplitudes, i.e. calculation of the elements

$$l_t = \sqrt{P_{tt}} \quad ; \quad t = 1, 2, \ldots, \binom{N}{2} \tag{4.15}$$

by a computer with some systematically varied parameter values. Examples of this method are found in the last section.

4.3. Iteration Processes

4.3.1. Partial Derivatives of Mean Amplitudes with Respect to the Parameters

In most practical cases the dimension of the problem exceeds 3, and the mapping becomes cumbersome. Then the possible minimum and maximum values of compatible mean amplitudes give useful information. Iteration processes have been constructed in order to find such extreme values.

After the p-th step of the iteration the matrix $\mathbf{P}$ can be written in the form

$$\mathbf{P}^{(p)} = L_d^{(p)} \delta L_d^{(p)\prime} . \tag{4.16}$$

Since the multiplication of two orthogonal matrices results in an orthogonal matrix, the matrix $L_d^{(p+1)}$ in

$$\mathbf{P}^{(p+1)} = L_d^{(p+1)} \delta L_d^{(p+1)\prime} \tag{4.17}$$

can be expressed as

$$L_d^{(p+1)} = L_d^{(p)} U^{(p+1)} , \tag{4.18}$$

where $U^{(p+1)}$ is a proper orthogonal matrix for which the parameters have to be chosen according to the purpose of the iteration process. Eqs. (17) and (18) can be used to calculate the partial derivatives

$\frac{\partial P_{tt}}{\partial \alpha_{ij}}$, which are needed in order to find the direction of the next step of the iteration.

$$\frac{\partial \mathbf{P}^{(p)}}{\partial \alpha_{ij}} = \mathbf{L}_d^{(p)} \frac{\partial \boldsymbol{U}^{(p+1)}}{\partial \alpha_{ij}}\bigg|_0 \boldsymbol{\delta} \mathbf{L}_d^{(p)\prime} + \mathbf{L}_d^{(p)} \frac{\partial \boldsymbol{U}^{(p+1)\prime}}{\partial \alpha_{ij}}\bigg|_0 \boldsymbol{\delta} \mathbf{L}_d^{(p)\prime}$$

$$= \mathbf{L}_d^{(p)} \frac{\partial \boldsymbol{U}^{(p+1)}}{\partial \alpha_{ij}}\bigg|_0 \boldsymbol{\delta} \mathbf{L}_d^{(p)\prime} + \left[\mathbf{L}_d^{(p)} \frac{\partial \boldsymbol{U}^{(p+1)}}{\partial \alpha_{ij}}\bigg|_0 \boldsymbol{\delta} \mathbf{L}_d^{(p)\prime}\right]' . \qquad (4.19)$$

With the antisymmetric parameter representation of $\boldsymbol{U}$ [cf. Eq.(5)] a simple formula can be obtained for

$$\frac{\partial \boldsymbol{U}^{(p+1)}}{\partial \alpha_{ij}}\bigg|_0 = 2 \frac{\partial [\mathbf{E} + \mathbf{L}_d^{(p+1)}]^{-1}}{\partial \alpha_{ij}}\bigg|_0 = 2(\mathbf{E}_{ij} - \mathbf{E}_{ji}), \qquad (4.20)$$

where $\mathbf{E}_{ij}$ is a matrix with all elements equal to zero except the ij-th, which is +1. Eq. (20) was obtained by means of the relation

$$\frac{\partial \mathbf{A}^{-1}}{\partial \alpha_{ij}} = -\mathbf{A}^{-1} \frac{\partial \mathbf{A}}{\partial \alpha_{ij}} \mathbf{A}^{-1}, \qquad (4.21)$$

which follows from

$$\frac{\partial}{\partial \alpha_{ij}}(\mathbf{A}\mathbf{A}^{-1}) = \frac{\partial \mathbf{A}}{\partial \alpha_{ij}} \mathbf{A}^{-1} + \mathbf{A} \frac{\partial \mathbf{A}^{-1}}{\partial \alpha_{ij}} . \qquad (4.22)$$

From Eqs. (19) and (20) one obtains

$$\frac{\partial l_t^{(p)}}{\partial \alpha_{ij}} = \frac{\partial}{\partial \alpha_{ij}}[P_{tt}^{(p)}]^{\frac{1}{2}} = \frac{1}{2}[P_{tt}^{(p)}]^{-\frac{1}{2}} \frac{\partial \mathbf{P}^{(p)}}{\partial \alpha_{ij}}$$

$$= \frac{2}{l_t^{(p)}}(\delta_i - \delta_j)[\mathbf{L}_d^{(p)}]_{ti}[\mathbf{L}_d^{(p)}]_{tj} . \qquad (4.23)$$

The only difference between Eq. (4) for the $\mathbf{F}$ matrix and Eq. (9) for the matrix $\mathbf{P}$ is that in Eq. (4) we find the matrices $\mathbf{K}'$ and $\boldsymbol{\lambda}$ instead of $\mathbf{L}_d$ and $\boldsymbol{\delta}$, respectively. Therefore the partial derivatives of the force constants can be obtained almost in the same way. The result is:

$$\frac{\partial F_{kl}^{(p)}}{\partial \alpha_{ij}} = 2(\lambda_i - \lambda_j)([\mathbf{K}^{(p)\prime}]_{ki}[\mathbf{K}^{(p)\prime}]_{lj} + [\mathbf{K}^{(p)\prime}]_{kj}[\mathbf{K}^{(p)\prime}]_{li}). \quad (4.24)$$

Let the parameters be changed with the same small values

$$\Delta\alpha_{ij} = \Delta\alpha \; ; \quad i<j = 2, 3, \ldots, n \, . \quad (4.25)$$

Then

$$\Delta F_{kl} = \Delta\alpha \sum_{i<j=2}^{n} \frac{\partial F_{kl}}{\partial \alpha_{ij}} \quad (4.26)$$

and

$$\Delta l_t = \Delta\alpha \sum_{i<j=2}^{n} \frac{\partial l_t}{\partial \alpha_{ij}} \, . \quad (4.27)$$

From the above formulas we obtain

$$\frac{\Delta l_t}{\Delta F_{kl}} = \frac{\sum_{i<j=2}^{n} (\delta_i - \delta_j)\mathcal{L}_{ti}\mathcal{L}_{tj}}{\sum_{i<j=2}^{n} (\lambda_i - \lambda_j)(\mathcal{K}_{ki}'\mathcal{K}_{lj}' + \mathcal{K}_{kj}'\mathcal{K}_{li}')} \, , \quad (4.28)$$

where $\mathcal{L}_{ij}$ designates the ij-th element of the $\mathbf{L}_d$ matrix, and $\mathcal{K}_{ij}'$ of the $\mathbf{K}'$ matrix. The expression (28) gives information concerning the sensibility of mean amplitudes to the changes of force constants, when both the mean amplitudes and force constants are compatible with the normal frequencies.

4.3.2. Iteration to Approach Given Mean Amplitudes of Vibration

Let $\boldsymbol{l}$ be a column matrix of which the elements are the given mean amplitudes (e.g. mean amplitudes consistent with electron diffraction measurements), and let the column matrix $\boldsymbol{l}^{(0)}$ be constructed from the mean amplitudes of vibration consistent with a given $\mathbf{F}^{(0)}$ matrix reproducing the normal frequencies. Suppose that $\boldsymbol{l} \neq \boldsymbol{l}^{(0)}$, and we wish to find compatible mean amplitudes which approximate $\boldsymbol{l}$ as much as possible.

Let $\boldsymbol{\sigma}$ and $\boldsymbol{\sigma}^{(0)}$ be column matrices of which the elements are $\sigma_t = P_{tt} = l_t^2$ and $\sigma_t^{(0)} = P_{tt}^{(0)} = [l_t^{(0)}]^2$, respectively; $t = 1, 2,$

..., $\binom{N}{2}$. The consideration of $\boldsymbol{\sigma}$ column matrices rather than $\boldsymbol{l}$ is motivated by the fact that only the elements of $\boldsymbol{\sigma}$ (and not $\boldsymbol{l}$) are additive; i.e.

$$\boldsymbol{\sigma} = \sum_i \boldsymbol{\sigma}_i \ , \tag{4.29}$$

where i designates the i-th irreducible representation. The first problem is how to divide the column matrix $\boldsymbol{\sigma}$ among the irreducible representations. The distribution is sometimes determined unambiguously by the symmetry. In the opposite case the distribution is the better the smaller is the difference (a) between $\boldsymbol{l}$ and the calculated mean amplitudes and (b) between the starting $\mathbf{F}^{(o)}$ and the calculated $\mathbf{F}$.

Suppose that $\boldsymbol{\sigma}_i$ is fixed and $\boldsymbol{\sigma}_i^{(o)}$ is calculated, and $\boldsymbol{\sigma}_i \neq \boldsymbol{\sigma}_i^{(o)}$. Then an iteration process can be applied in order to approximate the $\binom{N}{2}$ components of the column matrix $\boldsymbol{\sigma}_i$. Let us suppose that $N > n_i$, i.e. the number of mean amplitudes is greater than the dimension of the vibrational equation belonging to the i-th irreducible representation. Then according to the well-known method of least squares let us define the vector

$$\boldsymbol{r}^{(p)} = \boldsymbol{\sigma} - \boldsymbol{\sigma}^{(p)}, \tag{4.30}$$

where the species label i has been omitted, and p indicates the p-th step of the iteration. Let $\mathbf{J}^{(p)}$ be the Jacobian matrix of derivatives obtained in the p-th step:

$$J_{t,ij}^{(p)} = \frac{\partial \sigma_t^{(p)}}{\partial \alpha_{ij}} \ . \tag{4.31}$$

Let the $\binom{n_i}{2}$ - dimensional vector $\boldsymbol{\alpha}^{(p)}$ represent the parameters necessary for constructing $\boldsymbol{U}^{(p+1)}$. In the first approximation the unknown values are obtained by solving the normal equation

$$[\mathbf{J}^{(p)\prime}\mathbf{W}\,\mathbf{J}^{(p)}]\boldsymbol{\alpha}^{(p+1)} = \mathbf{J}^{(p)\prime}\mathbf{W}\boldsymbol{r}^{(p+1)}, \tag{4.32}$$

where $\mathbf{W}$ is the diagonal matrix of weights. After normalizing the vector $\boldsymbol{\alpha}^{(p+1)}$ properly $\boldsymbol{U}^{(p+1)}$ can be formed, and the use of Eqs. (18) and (17) gives mean amplitudes which lie closer to the prescribed ones. The process is to be continued until the prescribed approximation is reached, or it must be finished because further improvement is impossible.

4.3.3. Maximum and Minimum Values of Mean Amplitudes in a Given Parameter Domain

Sometimes, especially in connection with electron-diffraction measurements, the rational limits of the possible mean amplitudes give useful information. In order to solve this problem two methods have been elaborated, which can be applied in each case separately.

Let us find the maximum value of l_t under the conditions

$$\sum_{i=1}^{\binom{n}{2}} a_{ij}\,\alpha_i \leq b_j\,; \quad j = 1,\ 2,\ \ldots,\ s\,, \tag{4.33}$$

where a_{ij} and b_j are given constants; for the sake of simplicity each parameter α has a simple subscript here. The conditions (33) confine the search for a part of the whole parameter space, e.g. for a given assignment of frequencies. The problem of finding the minimum value of l_t is equivalent to finding the maximum value of $-l_t$. Accordingly the condition

$$\sum_{i=1}^{\binom{n}{2}} a_{ij}\,\alpha_i \geq b_j \tag{4.34}$$

can easily be transformed into the form of inequalities similar to (33);

$$-\sum_{i=1}^{\binom{n}{2}} a_{ij}\alpha_i \leq b_j\,. \tag{4.35}$$

The problem can be solved by an iteration process worked out by Rosen.[8,9] The starting parameters have to fulfil the inequality system (33). In the p-th step the parameters will be changed by

$$\Delta \boldsymbol{\alpha}^{(p)} = \lambda \boldsymbol{d}^{(p)}, \tag{4.36}$$

where

$$\Delta \boldsymbol{\alpha}^{(p)} = \{ \Delta \alpha_1^{(p)}, \Delta \alpha_2^{(p)}, \ldots, \Delta \alpha_{\binom{n}{2}}^{(p)} \}, \tag{4.37}$$

λ is a normalization constant, and $\boldsymbol{d}^{(p)}$ denotes the gradient vector

$$\boldsymbol{d}^{(p)} = \left\{ \frac{\partial l_1^{(p)}}{\partial \alpha_1}, \frac{\partial l_2^{(p)}}{\partial \alpha_2}, \ldots, \frac{\partial l_t^{(p)}}{\partial \alpha_{\binom{n}{2}}} \right\}. \tag{4.38}$$

The change in the parameters after the p-th step is given by

$$\Delta \boldsymbol{\alpha} = \sum_{k=1}^{p} \Delta \boldsymbol{\alpha}^{(k)}. \tag{4.39}$$

If the maximum of l_t belongs to a point within the parameter domain given by the inequalities (33), the iteration will terminate when $\boldsymbol{d}$ approximates $\mathbf{0}$ to a prescribed limit.

In cases which are not so simple, some of the conditions are only fulfilled as equations,

$$\sum_{i=1}^{\binom{n}{2}} a_{ij} \alpha_i = b_j = \mathbf{a}_j \boldsymbol{\alpha} \; ; \; j = 1, 2, \ldots, m, \tag{4.40}$$

where

$$\mathbf{a}_j = \{ a_{1j}, a_{2j}, \ldots, a_{\binom{n}{2} j} \}. \tag{4.41}$$

This means that we are on the border of the allowed domain. Then, according to Rosen's method, one has to construct a matrix $\mathbf{M}^{(p)}$, of which the rows consist of the row vectors $\mathbf{a}_1', \mathbf{a}_2', \ldots, \mathbf{a}_m'$. The matrix $\mathbf{M}^{(p)}$ enables us to find the direction of the $(p+1)$-th step, which cannot be made in the direction $\boldsymbol{d}^{(p+1)}$ because the system of inequalities (33) would be violated. $\boldsymbol{d}^{(p+1)}$ can always be written in the form

$$\boldsymbol{d}^{(p+1)} = \boldsymbol{t}^{(p+1)} + \boldsymbol{s}^{(p+1)}, \tag{4.42}$$

where $\mathbf{s}^{(p+1)}$ is a vector from the subspace defined by the rows of the matrix $\mathbf{M}^{(p)}$, and the allowed optimal direction $\mathbf{t}^{(p+1)}$ is orthogonal to this subspace, i.e.

$$\mathbf{s}^{(p+1)} = \mathbf{M}^{(p)\prime}\mathbf{c}^{(p+1)}, \qquad \mathbf{M}^{(p)}\mathbf{t}^{(p+1)} = 0, \tag{4.43}$$

and

$$\Delta\boldsymbol{\alpha}^{(p+1)} = \lambda\,\mathbf{t}^{(p+1)}. \tag{4.44}$$

In order to determine the elements of the vector $\mathbf{c}^{(p+1)}$ one substitutes the new form of $\mathbf{s}^{(p+1)}$ into Eq. (42).

$$\mathbf{d}^{(p+1)} = \mathbf{t}^{(p+1)} + \mathbf{M}^{(p)\prime}\mathbf{c}^{(p+1)}. \tag{4.45}$$

From Eqs. (43) and (45) one obtains

$$\mathbf{M}^{(p)}\mathbf{d}^{(p+1)} = \mathbf{M}^{(p)}\mathbf{M}^{(p)\prime}\mathbf{c}^{(p+1)}. \tag{4.46}$$

From the last formula the unknown constants can be expressed as

$$\mathbf{c}^{(p+1)} = [\mathbf{M}^{(p)}\mathbf{M}^{(p)\prime}]^{-1}\mathbf{M}^{(p)}\mathbf{d}^{(p+1)}. \tag{4.47}$$

From Eqs. (44), (45), and (47)

$$\Delta\boldsymbol{\alpha}^{(p+1)} = \lambda\{\mathbf{E} - \mathbf{M}^{(p)\prime}[\mathbf{M}^{(p)}\mathbf{M}^{(p)\prime}]^{-1}\mathbf{M}^{(p)}\}\mathbf{d}^{(p+1)}. \tag{4.48}$$

Rosen[8,9] has proved that in the case when m conditions are fulfilled only as equations in the p-th step, the matrix $\mathbf{M}^{(p)}$ need not have m rows. If the vector $\mathbf{c}^{(p+1)}$ has negative components, one row corresponding to the index of one of the negative components (e.g. the one with the greatest absolute value) can be omitted from the matrix $\mathbf{M}^{(p)}$. The system of inequalities (33) remains valid if

$$\Delta\boldsymbol{\alpha}^{(p+1)} = \lambda\{\mathbf{E} - \bar{\mathbf{M}}^{(p)\prime}[\bar{\mathbf{M}}^{(p)}\bar{\mathbf{M}}^{(p)\prime}]^{-1}\}\mathbf{d}^{(p+1)}, \tag{4.49}$$

where the matrix $\bar{\mathbf{M}}^{(p)}$ is formed from $\mathbf{M}^{(p)}$ as described above.

As has been pointed out, this method can be used to determine the extreme values of mean amplitudes of vibration in a given parameter domain. Let us start from a probable force constant matrix $\mathbf{F}$ and from the mean amplitudes consistent with $\mathbf{F}$. This starting point can be chosen as the zero point in the parameter space, from where we start to change the parameters in order to obtain the extreme values, e.g. in a parameter domain defined by the inequality system

$$m_i \leq \alpha_i \leq M_i \; ; \qquad i = 1, 2, \ldots, \binom{n}{2} \; . \tag{4.50}$$

This is a simple version of the conditions (33). If we wish to remain close to the initial matrix $\mathbf{F}$, the absolute values of m_i and M_i should not be greater than 0.1 .

4.3.4. Maximum and Minimum Values of Mean Amplitudes in a Given Force Constant Domain

In most cases probable limits of some force constants can be given as

$$m_{kl} \leq F_{kl} \leq M_{kl} \; ; \quad k < l = 2, 3, \ldots, n \; . \tag{4.51}$$

For the force constants for which such limits are not at our disposal, we can take for example $m_{kl} = -100$, $M_{kl} = 100$ (in usual units).

In order to determine the extreme values of mean amplitudes under the condition that the system of inequalities (51) should be fulfilled our previous method is not suitable since the inequalities (51) are not linear functions of the parameters. However, for small changes of parameters the linear equations

$$\Delta F_{kl} = \sum_{i=1}^{\binom{n}{2}} \frac{\partial F_{kl}}{\partial \alpha_i} \Delta \alpha_i \tag{4.52}$$

is valid to a good approximation. Therefore when we arrive at the border defined by (51), the following restrictions must be satisfied for some force constants in the next step.

$$\Delta F_{kl} = \sum_{j=1}^{\binom{n}{2}} \frac{\partial F_{kl}}{\partial \alpha_j} \Delta \alpha_j = \mathbf{f}_{kl}' \Delta \boldsymbol{\alpha} = 0 \; . \tag{4.53}$$

This means that the matrix $\mathbf{M}$ defined above has to be constructed. Its rows are the row vectors $\mathbf{f}_{kl}'$ defined in Eq. (53). The second process is somewhat more complicated because one has to calculate for many steps the partial derivatives of force constants with respect to the parameters with the aid of Eq. (24).

4.4. Examples

In Table I the mean amplitudes of the NO_2 molecule are shown

Table 4-I. The mean amplitudes (in Å) at T = 380 K and force constants of the NO_2 molecule. F_{11} in mdyne/Å, F_{12} in mdyne, and F_{22} in mdyne Å. F_{33} = 8.399 mdyne/Å.

Parameter	l (N-O)	l (O...O)	F_{11}	F_{12}	F_{22}
0	0.03888	0.04842	12.172	0.529	1.100
0.01	0.03885	0.04810	12.342	0.607	1.102
0.02	0.03883	0.04778	12.507	0.688	1.095
0.03	0.03882	0.04746	12.668	0.770	1.091
0.04	0.03881	0.04714	12.824	0.854	1.088
0.05	0.03881	0.04741	12.975	0.940	1.088
0.08	0.03884	0.04588	13.394	1.206	1.096
0.10	0.03888	0.04527	13.644	1.388	1.111
0.15	0.03907	0.04382	14.152	1.852	1.177
0.20	0.03938	0.04256	14.485	2.312	1.282
0.30	0.04024	0.04082	14.611	3.145	1.582
0.40	0.04127	0.04027	14.096	3.778	1.951
-0.01	0.03891	0.04874	11.999	0.452	1.119
-0.02	0.03894	0.04905	11.821	0.379	1.131
-0.03	0.03898	0.04936	11.641	0.307	1.144
-0.04	0.03902	0.04967	11.457	0.238	1.159
-0.05	0.03907	0.04998	11.271	0.172	1.176
-0.08	0.03925	0.05088	10.699	-0.008	1.236
-0.10	0.03938	0.05145	10.311	-0.113	1.284
-0.15	0.03980	0.05277	9.337	-0.318	1.428
-0.20	0.03980	0.05391	8.390	-0.435	1.599
-0.30	0.04138	0.05552	6.722	-0.409	1.993
-0.40	0.04248	0.05622	5.522	-0.085	2.397
Electr. diffr.*	0.0398	0.0486			

*From Ref. 7, p. 208.

Table 4-II. Responsiveness of mean amplitudes of SO_2 to changes in force constants. For the applied units, see Table I. The data are from Ref. 3.*

$\frac{\Delta l(XY)}{\Delta F(A_1)}$	F_{11}	F_{12}	F_{22}
S-O	-1.34×10^{-4}	-1.00×10^{-4}	1.99×10^{-3}
O...O	-4.93×10^{-3}	-3.71×10^{-3}	7.37×10^{-2}

* F_{11} = 10.036, F_{12} = 0.357, F_{22} = 1.621, F_{33} = 9.996.

at some parameter values. All mean amplitudes here are compatible with the measured normal frequencies. The corresponding **F** matrices are also given.

The sensibilities of mean amplitudes to the changes of force constants are indicated in Table II, using SO_2 as an example.[3] The values were calculated by Eq. (28) in the point of the parameter space characterized by the listed force constants.

The maximum and minimum values of mean amplitudes have been calculated for the ethylene molecule. The internal and symmetry coordinates, and the general features of the initial force constants were taken from the work of Pulay and Meyer.[10] In-plane force constants and corresponding frequencies are given below. (Applied units are mdyne/Å, mdyne, and mdyne Å for the stretch-stretch, stretch-bend, and bend-bend force constants, respectively. Frequencies in cm^{-1}. For detailed specifications, see the cited reference.[10])

Species A_g:	5.17	0.16	-0.11	3026
		8.54	0.50	1633
			1.50	1342
Species B_{1g}:		5.15	0.20	3102
			0.634	1236
Species B_{2u}:		5.11	0.08	3106
			0.44	810
Species B_{3u}:		5.19	-0.19	3020
			1.31	1444

Table 4-III. Maximum and minimum values of mean amplitudes (in Å) for ethylene at T = 298 K.

Distance	Method (1)		Method (2)	
	max	min	max	min
C=C	0.0426	0.0423	0.0439	0.0413
C–H	0.0790	0.0769	0.0776	0.0770
C•••H	0.1018	0.0960	0.1003	0.0951
cis(H•••H)	0.1716	0.1633	0.1716	0.1623
trans(H•••H)	0.1217	0.1162	0.1237	0.1135
gem(H•••H)	0.1223	0.1178	0.1268	0.1153

The extremal values of mean amplitudes were calculated in two ways: (1) Within a domain defined by the prescription that each parameter should remain between 0.02 and -0.02. (At the value 0 of the parameters we obtain the initial force constants.) (2) Within a domain which was defined by the following force constant limits. (A_g) $5.07 \leq F_{11} \leq 6.07$, $8.54 \leq F_{22} \leq 9.74$, $1.40 \leq F_{33} \leq 1.60$; ($B_{1g}$) $0.1 \leq F_{12} \leq 0.3$; (B_{2u}) $0.0 \leq F_{12} \leq 0.20$; $-0.3 \leq F_{12} \leq -0.1$. Table III shows the obtained results. The electron-diffraction values are (in Å):[7,11] l(C=C) = 0.0443 ± 0.002, l(C–H) = 0.0798 ± 0.002, l(C···H) = 0.0993 ± 0.004.

Further applications of the method outlined here are in progress.

Acknowledgment: The authors wish to express their gratitude to Professor S. Cyvin for valuable discussions and for supplying the basic data of the NO_2 molecule.

REFERENCES

1 P.Pulay and F.Török, Acta Chim.Hung. 44, 287 (1965).
2 F.Török and P.Pulay, J.Mol.Structure 3, 1 (1969).
3 F.Török and Gy.B.Hun, Acta Chim.Hung. 59, 303 (1969).
4 W.J.Taylor, J.Chem.Phys. 18, 1301 (1950).
5 F.R.Gantmacher: Matrizenrechnung, Deutscher Verlag der Wissenschaften, Berlin 1958.
6 S.Toman and J.Plíva, J.Mol.Spectry. 21, 362 (1966).

7 S.J.Cyvin: Molecular Vibrations and Mean Square Amplitudes, Universitetsforlaget, Oslo, and Elsevier, Amsterdam, 1968.

8 J.B.Rosen, J.Soc.Industr.Appl.Mathem. 8, 181 (1960).

9 J.B.Rosen, J.Soc.Industr.Appl.Mathem. 9, 514 (1961).

10 P.Pulay and W.Meyer, J.Mol.Spectry. 40, 59 (1971).

11 L.S.Bartell, E.A.Roth, C.D.Hollowell, K.Kuchitsu, and J.E. Young,Jr., J.Chem.Phys. 42, 2683 (1965).

CHAPTER 5

The Monte Carlo Method for the Complete Solution of the Operator Equation

VLADIMÍR ŠPIRKO

The completeness of the solution of the general operator equation, as obtained by the method of statistical simulations (Monte Carlo method), is discussed. The method is applied to the calculations of simple Urey-Bradley force fields of COF_2, $COCl_2$, and $COBr_2$.

Je diskutována míra úplnosti řešení obecné operátorové rovnice metodou statistických simulací (Monte Karlo metoda). Dále je zde tato metoda použita k výpočtu jednoduchých Urey-Bradley silových polí molekul COF_2, $COCl_2$ a $COBr_2$.

5.1. Introduction

In order to extract physically meaningful information from an experimental molecular spectrum a system of nonlinear equations must in general be solved. To find all the possible mathematically acceptable solutions of the system is a very complex and difficult problem, which has not yet been solved in the general case.[1] The deterministic methods which have been described so far are limited in their applicability to either simple[2,3] or special cases.[4,5]

On the other hand the only generally applicable approach, the Monte Carlo method,[6] is time indetermined, i.e. the method yields all the solutions of the given problem only when an infinite number of Monte Carlo simulations have been carried out. However, the experience gained in application of this method [cf. Ref. 6 and unpublished results by V. Špirko] indicates that a certain class of solutions belonging to the set of all possible solutions of a real problem can be obtained with a high probability at the cost of an amount of numerical work, which already is technically feasible. It is a priori obvious that the information obtained by this means is of considerable physical importance. Therefore we shall consider this method in some detail here, and we shall demonstrate its potential applicability to the calculations of simple Urey-Bradley force fields of some carbonyl halides.

5.2. Theoretical Considerations

Consider a set of n nonlinear equations in m variables ($n \geq m$) described by

$$f_i(x_1, x_2, \ldots, x_m) = 0; \quad i = 1, 2, \ldots, n, \tag{5.1}$$

or in vector form

$$\mathbf{f}(\mathbf{x}) = \mathbf{0}, \tag{5.2}$$

where $\mathbf{x}$ is an m-dimensional vector, $\mathbf{f}$ is the mapping from the set of all m-dimensional vectors (R^m) into itself, and $\mathbf{0}$ is the m-dimensional null vector. Assume that Eq. (2) has r solutions ($\mathbf{x}_1$, $\mathbf{x}_2$, ..., $\mathbf{x}_r$) lying in the m-dimensional rectangular parallelepipe I defined by

$$I = \langle a_1, b_1 \rangle \times \langle a_2, b_2 \rangle \times \cdots \times \langle a_m, b_m \rangle. \tag{5.3}$$

Assume also that a standard iteration procedure[1] (i.e. a procedure which yields one solution only) for the solution of Eq. (2) has been chosen. This procedure may be described by a symbolic operator $\mathbf{T}_\varepsilon$ ($\varepsilon > 0$), which has the following property. If $\| \mathbf{x}_\rho - \mathbf{y}_\rho \| < \varepsilon$ holds for $\mathbf{y}_\rho \in R^m$, then the sequence $\{\mathbf{y}^{(k)}\}_{k=0}^{\infty}$ defined by

$$\mathbf{y}^{(0)} = \mathbf{y}_\rho, \qquad \mathbf{y}^{(k+1)} = \mathbf{T}_\varepsilon(\mathbf{y}^{(k)}) \tag{5.4}$$

converges to the point $\mathbf{x}_\rho$, i.e.

$$\lim_{k\to\infty} \mathbf{y}^{(k)} = \mathbf{x}_\rho . \tag{5.5}$$

According to the Monte Carlo method we generate a sequence of random vectors $\mathbf{y}_1, \mathbf{y}_2, \ldots, \mathbf{y}_N$ uniformly distributed in I. For any of the vectors $\mathbf{y}_j$ we define in our standard iteration procedure the sequence $\{\mathbf{x}^{(k)}\}_{k=0}^{\infty}$ by putting $\mathbf{x}_j^{(0)} = \mathbf{y}_j$ and $\mathbf{x}_j^{(k+1)} = \mathbf{T}_\varepsilon(\mathbf{x}_j^{(k)})$ in accordance with Eq. (4). In all cases where the sequence converges to some solution $\mathbf{x}_\rho$ of the equation (2) let us determine the limit

$$\lim_{k\to\infty} \mathbf{x}_j^{(k)} = \mathbf{x}(\mathbf{y}_j). \tag{5.6}$$

It is possible to prove the following theorem.[6] If $N \to \infty$, the set of points $\{\mathbf{x}(\mathbf{y}_j)\}$ of I obtained in the described way is identical with the set $\{\mathbf{x}_1, \mathbf{x}_2, \ldots, \mathbf{x}_r\}$ of all solutions of Eq. (2) in I with the probability equal to one.

Owing to the principal technical limitations we cannot choose $N \to \infty$ in real calculations, i.e. we cannot state "all the solutions of a given problem have been found" with certainty. We can, however, assert such a statement with a certain probability, which depends on the number of Monte Carlo simulations, i.e. on the number N and on the relative magnitudes of the Lebesque measures $\mu(K_i)$ of individual disjunctive* Borel subsets of the set I [with the measure $\mu(I)$], from which the iterative process converges to the individual solutions of the problem studied. It was found[6]

$$P_N(i) = \left[1 - \frac{\mu(K_i)}{\mu(I)}\right]^N , \tag{5.7}$$

where $P_N(i)$ is the probability for the event that the i-th solution of the problem has not been obtained after having carried out N times the above specified simulations.

If the chosen problem has altogether r solutions the set I can obviously be decomposed into $r + 1$ disjunctive subsets, of

---

* The iterative process converges (if it does at all) always to one solution within I, depending upon the standard iteration procedure employed, and it is always the same solution if the same initial point is used.

which r subsets K_i correspond to the events that the applied iterative procedure converges to some of the solutions. The remaining subset K_{r+1}, given by the relation

$$K_{r+1} = I - \bigcup_{i=1}^{r} K_i \tag{5.8}$$

corresponds to the opposite event.

Since the events corresponding to the probabilities $P_N(i)$ are pairwise mutually exclusive (in consequence of disjunctivity of the subsets K_i) it is found

$$P_N(\text{tot}) = \sum_{i=1}^{r+1} P_N(i) = \sum_{i=1}^{r+1} \left[1 - \frac{\mu(K_i)}{\mu(I)} \right]^N , \tag{5.9}$$

where $P_N(\text{tot})$ is the probability for the event that all solutions have not been obtained after having carried out N simulations. The number of necessary Monte Carlo simulations corresponding to the chosen value of the probability $P_N(\text{tot})$ might be determined according to the obtained relationship (9) if the individual Lebesque measures $\mu(K_i)$ and the total number of solutions were known. Since this information is not available in general the relation cannot be applied with exactness a priori. But we can perform an approximate analysis of the desired solution in every real case.

(a) In the cases where the studied nonlinear system consists of polynomial equations only, the Bezout theorem[7] can be applied, and the upper bound R for the number of respective real solutions r is thus found; in general $R \geq r$. Furthermore it is a priori possible to confine the search for solutions to such a class of them for which

$$\frac{\mu(K_i)}{\mu(I)} \geq \eta . \tag{5.10}$$

These limitations obviously permit a pessimistic estimate of a number N^* of Monte Carlo simulations, which would yield a postulated probability P_{N*} for the event that not all the solutions satisfying the relation (10) have been found. The following relations are valid.

$$P_{N^*} \leq \sum_{i=1}^{R+1} (1-\eta)^{N^*} = (R+1)(1-\eta)^{N^*}, \tag{5.11}$$

and consequently

$$N^* = [\ln P_{N^*} - \ln(R+1)]/\ln(1-\eta). \tag{5.12}$$

(b) In the case of a more general system where no reliable estimate of the number of solutions is available the considered analysis of the system in question can be performed a posteriori. This is the case of transcendental or polynomial-transcendental equations. Assume that a certain number N^+ of simulations of the first approximations of solutions have yielded $S + 1$ solutions ($\mathbf{x}_1, \mathbf{x}_2, \ldots, \mathbf{x}_{S+1}$) occurring with frequencies $f_{11}, f_{12}, \ldots, f_{1\,S+1}$. This means f_{11} solutions $\mathbf{x}_1$, f_{12} solutions $\mathbf{x}_2$, etc. The $(S+1)$-st solution corresponds to the case where the applied iterative process does not converge. Assume also that the next $k-1$ sequencies of N^+ simulations always yield the same solutions $\mathbf{x}_j$ with the frequencies $f_{2j}, f_{3j}, \ldots, f_{kj}$. It is obvious that the number N^+ can be chosen in such a way that the sequence of the frequencies f_{pq} represents almost constant vectors. With this assumption we can put[8]

$$\mu(K_i) \approx \frac{1}{kN^+} \sum_{l=1}^{k} f_{li}\,; \quad i = 1, 2, \ldots, S+1 \tag{5.13}$$

for $\mu(I) = 1$. Thus we obtain the estimates of relative magnitudes of the measures $\mu(K_i)$ of individual Borel subsets K_i of the set I. The estimates permit us to assert that we have found, with the probability close to one, all those solutions of the problem studied for which the relative Lebesque measure is not smaller than $\min\{\mu(K_i) \mid i = 1, 2, \ldots, S+1\}$.

5.3. Application of the Random Search Method

In the above discussion it was assumed that an unspecified iterative procedure was available. The numerical properties of this procedure determine the numerical requirements and hence the practical applicability of the method considered. The criterion for choosing a procedure from the extensive number of described iterative procedures (see, e.g., Refs.1,9) is evident: we are looking for a procedure which maximizes the Lebesque measure of Borel subsets, from the points of which we get the corresponding solutions. At the same time the procedure should yield the solutions within minimal computing time.

Practical experience gained in testing of various described procedures [V.Špirko, thesis, Institute of Chemical Technology, Prague 1968; see also Ref. 9] indicates that the method of damped least squares[10,11] is probably the most convenient one. The method can be characterized by

$$\mathbf{y}_{k+1} = \mathbf{y}_k + (\mathbf{J'WJ} + \rho\mathbf{E})^{-1}\mathbf{J'WR}, \tag{5.14}$$

where $\mathbf{J}$ is the Jacobian matrix of the system studied, $\mathbf{W}$ is a diagonal matrix of statistical weights of individual equations, $\mathbf{R}$ is a vector of residuals, $\mathbf{E}$ is the unit matrix, and ρ is a positive constant, the so-called damping factor. If we define ρ by the relation[12]

$$\rho = \text{const} \sum_i W_i R_i^2, \tag{5.15}$$

where 'const' is an empirical constant, the method can be interpreted as a hybrid of the gradient method and the Newton method. In the cases where (a) $(\mathbf{J'WJ})_{ii} \ll \rho$ for all i's or (b) $\rho \to 0$, Eq. (14) represents the gradient and Newton method, respectively. Because of this feature the method yields the desired solutions even when rough initial estimates of the solutions are used. Then for some time the method proceeds as the gradient method. When the estimates get closer to the exact solutions ($\rho \to 0$) the speed of convergence is greatly increased due to the general properties of the Newton method. Nevertheless the method of damped least squares has a certain weak point: the convergence is relatively slow in the case of extremely coarse initial estimates of solutions. In view of the frequent occurrences of such cases of slow convergence in our Monte Carlo scheme this is not a negligible problem.

The entire process of finding all the solutions by the discussed Monte Carlo method can be speeded up by two means: (a) interrupting each slowly converging process after a few steps and considering it as a nonconvergent one, or (b) choosing another iterative procedure which in such a case is more efficient than the simple gradient method defined by relation (14). The latter possibility is more convenient since each elimination of the converging iterative process results in a decrease of a certain Lebesque measure $\mu(K_i)$. The so-called creeping random search method appears to be convenient in this connection. Since this method has been critically analysed by Rastrigin,[13] we shall describe how it can be incorporated into the scheme of the damped least squares method, while the method itself will be outlined only briefly.

The search for one solution of the system (2) can be interpreted as a search for the corresponding minimum of the function defined by the relation [see Eq. (15)]

$$S(\mathbf{x}) = \sum_i W_i R_i^2. \tag{5.16}$$

If we again use the above described Monte Carlo vectors $\mathbf{y}_k$ as the first approximation of the vectors $\mathbf{x}_l$, for which the function reaches a minimum, then one of the simplest forms of the discussed method can be described by the formulae $\mathbf{x}_l^{(0)} = \mathbf{y}_l$ and $\mathbf{x}_l^{(k+1)} = \mathbf{x}_l^{(k)} + \Delta\mathbf{x}_l^{(k)}$ for

$$\Delta\mathbf{x}_l^{(k)} = \begin{cases} a\boldsymbol{\xi} & \text{if } S(\mathbf{x}_l^{(k+1)}) < S(\mathbf{x}_l^{(k)}) \\ \mathbf{O} & \text{otherwise,} \end{cases} \tag{5.17}$$

where $\boldsymbol{\xi}$ is a random vector (of the same dimension as the vectors $\mathbf{x}_l$ and the null vector $\mathbf{O}$) of uniformly distributed components in the interval (-1, 1). a is a multiplicative constant which determines the optimal length of a random correction $\Delta\mathbf{x}_l$. This form of multiparameter optimalization by sequential random perturbations has several indisputable advantages. (a) It operates with a minimum of control logic, (b) the convergence of the iterative process is not affected by ugly parameter-space terrain features, such as ridges and canyons, and (c) it has minimal computer memory requirements. Furthermore Rastrigin[13] has shown that the method even in this simple form is numerically more efficient than the classical gradient method. This is already true in the cases of systems with more than three unknown parameters; the relative efficiency increases rapidly with the number of parameters. In spite of being generally more efficient the described process still bears all the convergence features of the classical gradient method, i.e. the convergence slows down after relatively fast converging iterative steps have been executed. At this stage the method of damped least squares can usually be applied with sufficient efficiency. Hence it is clear that the two iterative procedures can be combined in an optimal way if proper attention is paid to their properties as discussed above.

5.4. Simple Urey-Bradley Force Fields of Some Carbonyl Halides

In order to exemplify the discussed stochastic method and to demonstrate the necessity of considering the nonlinearity of the problem studied we shall consider the solution of a classical spectroscopical problem: the calculation of the force fields for COF_2, $COCl_2$, and $COBr_2$ from the fundamental frequencies of the planar vibrations.

If the real force fields of the molecules in question are approximated by the Urey-Bradley force fields which include[14] the stretching constants K_{CO} and K_{CX}, the bending constants H_{OX} and H_{XX}, and the repulsive constant F_{OX} (where X = F, Cl, Br), then each of these problems represents a system of five equations in five unknowns: two linear, two quadratic, and one cubic equation. In other words we are solving the inverse secular equation problem for two matrices of dimensions two and three. It follows from the Bezout theorem that the upper bound for the total number of solutions is $R = 1 \times 1 \times 2 \times 2 \times 3 = 12$. If we only search for the solutions lying in the subsets of which the Lebesque measures are at least equal to one-tenth of the measure of the whole region, in which we try to find the solution, the probability of not finding all such solutions is given by [see Eq. (11)]

$$P_{N^*} = 13 \times 0.9^{N^*}. \qquad (5.18)$$

This means that if we put $P_{N^*} = 0.1$ or 0.01, the numbers of necessary Monte Carlo simulations are 46 and 68, respectively.*

Table 5-I. Urey-Bradley force constants of some carbonyl halides (in mdyne/Å).

Set		I	II	III	IV
COF_2	K_{CO}	12.69	11.23	5.179	-0.264
	K_{CF}	4.355	1.580	1.515	4.698
	H_{FF}	0.101	1.782	4.418	5.299
	H_{OF}	0.143	0.843	0.841	-0.298
	F_{OF}	1.554	1.636	1.690	2.506
$COCl_2$	K_{CO}	12.33	12.18	0.805	-0.598
	K_{CCl}	1.886	0.971	0.839	2.537
	H_{ClCl}	0.047	0.369	6.233	6.284
	H_{OCl}	0.166	0.289	0.292	-0.082
	F_{OCl}	0.956	1.208	1.306	1.443
$COBr_2$	K_{CO}	12.50	12.48	0.325	-0.917
	K_{CBr}	1.402	1.170	0.729	2.600
	H_{BrBr}	0.059	0.125	7.320	7.251
	H_{OBr}	0.143	0.168	0.186	-0.213
	F_{OBr}	0.728	0.784	1.030	1.523

The given numbers of simulations, which are easy to realize, were carried out for the problems mentioned. The solutions were sought in the parallelepiped $I = \langle -3, 15\rangle \times \langle -3, 15\rangle \times \langle -3, 15\rangle \times \langle -3, 15\rangle \times \langle -3, 15\rangle$. For each of the two numbers N^* we obtained four different exact solutions (see Table I). The result permits us to claim, at least with 99 percent of certainty, that all the solutions characterized above have been obtained.

* If we want to find all the solutions of which the measure is not smaller than one-hundredth of the measure of the whole region, then $P_{N^*} = 0.1$ or 0.01 requires $N^* = 484$ and 713, respectively.

REFERENCES

1 T.L.Saaty and J.Bram: Nonlinear Mathematics, McGraw-Hill, New York 1964.

2 J.M.Ruth and R.J.Phillipe, J.Chem.Phys. 41, 1492 (1964).

3 S.Castellano and J.S.Waugh, J.Chem.Phys. 34, 295 (1961).

4 P.Pulay and F.Török, Acta Chim.Hung. 44, 287 (1965).

5 S.Toman and J.Plíva, J.Mol.Spectry. 21, 362 (1966).

6 V.Špirko and J.Morávek, J.Mol.Spectry. 33, 368 (1970).

7 J.L.Coolidge: A Treatise on Algebraic Plane Curves, Oxford University Press, New York 1931.

8 J.M.Hammersley and D.C.Handscomb: Monte Carlo Methods, Methuen, London 1967.

9 J.Pitha and R.N.Jones, Can.J.Chem. 44, 3031 (1966).

10 K.Levenberg, Quart.Appl.Math. 2, 164 (1944).

11 D.Papoušek, S.Toman, and J.Plíva, J.Mol.Spectry. 15, 502 (1965).

12 J.Plíva, V.Špirko, and S.Toman, J.Mol.Spectry. 21, 106 (1966).

13 A.A.Rastrigin: Statisticheskie metody poiska (Random search methods), Izd. Nauka, Moscow 1968.

14 J.Štokr and B.Schneider, Coll.Czechoslov.Chem.Commun. 26, 1489 (1961).

CHAPTER 6

Definition and Applications of a Transformation Matrix T

S. J. CYVIN

The $\boldsymbol{T}$ matrix connects two sets of internal coordinates, of which at least one is a complete set of independent coordinates for molecular vibrations. The $\boldsymbol{T}$ matrix is used to formulate a familiar approach to the determination of force constants. It may also be used to develop Urey-Bradley force constants within certain limitations. Finally the $\boldsymbol{T}$ matrix gives the clue to a method referred to as 'specific imposition of a potential parameter'.

6.1. The $\boldsymbol{T}$ Matrix

6.1.1. Definition of $\boldsymbol{T}$

Assume two sets of internal coordinates, S and $\hat{\mathsf{S}}$, connected by

$$\hat{\mathsf{S}} = \boldsymbol{T}\mathsf{S}, \qquad (6.1)$$

where S is supposed to represent a complete set of independent coordinates for molecular vibrations. Hence the transformation matrix $\boldsymbol{T}$ may be determined uniquely even if $\hat{\mathsf{S}}$ contains an arbitrary number of internal coordinates. If $\hat{\mathsf{S}}$ is not a complete set of independent coordinates the $\boldsymbol{T}$ matrix cannot be inverted; it may even not be a square matrix.

6.1.2. Derivation of $\boldsymbol{T}$

The $\boldsymbol{T}$ matrix may be derived in the following way, which is convenient for computer solution. Let the S and $\hat{\mathsf{S}}$ coordinates be given in terms of the $\mathbf{B}$ and $\hat{\mathbf{B}}$ matrices; Eq. (1.5) may be used to define $\mathbf{B}$, and

$$\hat{\mathrm{S}} = \hat{\mathbf{B}}\mathbf{x} \tag{6.2}$$

On inserting $\mathbf{X}$ from Eq. (1.10), in which the expression for $\mathbf{A}$ from Eq. (1.11) has been introduced, one obtains

$$\hat{\mathrm{S}} = \hat{\mathrm{B}}\,\mathrm{m}^{-1}\mathrm{B}'\mathrm{G}^{-1}\mathrm{S}. \tag{6.3}$$

Hence

$$\boldsymbol{T} = \hat{\mathrm{B}}\,\mathrm{m}^{-1}\mathrm{B}'\mathrm{G}^{-1}. \tag{6.4}$$

6.2. Force-Constant Calculations

6.2.1. Theory

A harmonic force field is defined by F based on the S coordinates. If a force field is given as $\hat{\mathrm{F}}$ in terms of the arbitrary set of internal coordinates $\hat{\mathrm{S}}$, the proper F matrix is found from

$$\mathrm{F} = \boldsymbol{T}'\hat{\mathrm{F}}\boldsymbol{T}. \tag{6.5}$$

Notice that this relation does not imply any inversion of $\boldsymbol{T}$. The opposite transformation, viz. from F to $\hat{\mathrm{F}}$, is not always possible. If $\hat{\mathrm{S}}$ represents a complete set containing redundant coordinates the transformation from F to $\hat{\mathrm{F}}$ is not unique; different $\hat{\mathrm{F}}$ matrices may be physically equivalent and correspond to the same F .

6.2.2. General Applications

It is frequently useful to set up an initial force field for a molecule on the basis of a coordinate set containing redundancies. The procedure which utilizes the $\boldsymbol{T}$ matrix, applied to planar five-membered ring molecules,[1] was described in some details. The same method was applied to naphthalene,[2,3] oxides of phosphorus,[4] hexamethylene tetramine (Chapter 17, Part I), adamantane (Chapter 19, Part III), and other molecules. Again some details of the application are reported in one of the subsequent chapters on four-atomic alkali halide molecules (Chapter 15, Part II).

6.2.3. Simulation of Urey-Bradley Force Field

The formalism of the $\boldsymbol{T}$ matrix may be used to develop some (but not all!) of the essential features of the Urey-Bradley force field; see, e.g., Nakamoto[5] and references cited therein.

Take the planar symmetrical XY_3 (D_{3h}) model as the first example. The standard symmetry coordinates[6,7] may be supplemented by including the non-bonded distance displacements: $\hat{S}_1(A_1') = S_1(A_1')$, and

$$\hat{S}_2(A_1') = 3^{-\frac{1}{2}}(d_1+d_2+d_3) = 3^{\frac{1}{2}}S_1(A_1'). \tag{6.6}$$

For the E' species $\hat{S}_1(E') = S_1(E')$, $\hat{S}_2(E') = S_2(E')$ for both of the a and b members, and

$$\hat{S}_{3a}(E') = 6^{-\frac{1}{2}}(2d_1-d_2-d_3) = -\frac{1}{2}3^{\frac{1}{2}}S_{1a}(E') + \frac{1}{2}S_{2a}(E'),$$

$$\hat{S}_{3b}(E') = 2^{-\frac{1}{2}}(d_2-d_3) = -\frac{1}{2}3^{\frac{1}{2}}S_{1b}(E') + \frac{1}{2}S_{2b}(E'). \tag{6.7}$$

The appropriate $\boldsymbol{T}$ matrix blocks are

$$\boldsymbol{T}(A_1') = \begin{bmatrix} 1 \\ 3^{\frac{1}{2}} \end{bmatrix}, \qquad \boldsymbol{T}(E') = \begin{bmatrix} 1 & 0 \\ 0 & 1 \\ -\frac{1}{2}3^{\frac{1}{2}} & \frac{1}{2} \end{bmatrix}. \tag{6.8}$$

Assume now an in-plane force field represented by a diagonal $\hat{\mathbf{F}}$ matrix with elements K, H, and F in accord with the conventional notation for Urey-Bradley force constants:

$$\hat{\mathbf{F}}(A_1') = \text{diag}(K, F), \qquad \hat{\mathbf{F}}(E') = \text{diag}(K, H, F). \tag{6.9}$$

Then the transformation of Eq. (5) leads to the following expressions for symmetry force constants.

$$F(A_1') = K + 3F;$$

$$F_1(E') = K + \frac{3}{4}F, \quad F_2(E') = H + \frac{1}{4}F,$$

$$F_{12}(E') = -\frac{1}{4}3^{\frac{1}{2}}F. \tag{6.10}$$

These expressions are identical to those from a standard Urey-Bradley force field analysis,[8] except for the tension constant, F', which does not come in through the present simple considerations.

The tetrahedral XY_4 (T_d) model[6,9] was chosen as the next example. The same procedure gave the result:

$$F(A_1) = K + 4F\,; \qquad F(E) = H + \tfrac{1}{3}F\,;$$

$$F_1(F_2) = K + \tfrac{4}{3}F, \qquad F_2(F_2) = H + \tfrac{1}{3}F,$$

$$F_{12}(F_2) = -\tfrac{2}{3}F. \tag{6.11}$$

These expressions again are identical with those from a standard treatment[10] if the F' terms are neglected.*

Finally let us consider the bent XY_2 (C_{2v}) model,[6,7] and include k as the interaction constant between the two X-Y stretchings. It might be included in a modified Urey-Bradley force field. Assume the force field to be given by

$$\hat{\mathbf{F}} = \begin{bmatrix} K & k & 0 & 0 \\ & K & 0 & 0 \\ & & H & 0 \\ & & & F \end{bmatrix} \tag{6.12}$$

in terms of the coordinates $\hat{\mathbf{S}} = \{r_1, r_2, R\alpha, d\}$. The appropriate $\boldsymbol{T}$ matrix is

$$\boldsymbol{T} = \begin{bmatrix} 2^{-\frac{1}{2}} & 0 & 2^{-\frac{1}{2}} \\ 2^{-\frac{1}{2}} & 0 & -2^{-\frac{1}{2}} \\ 0 & 1 & 0 \\ 2^{\frac{1}{2}}\sin A & \cos A & 0 \end{bmatrix}\,; \tag{6.13}$$

---

*The apparent discrepancy in sign for $F_{12}(F_2)$ arises from different definitions of symmetry coordinates. In Ref. 10 $G_{12}(F_2) < 0$, but according to the present conventions[6,9] $G_{12}(F_2) > 0$.

cf. Ref. 6, especially Section 8.6.2. The resulting expressions for symmetry force constants read:

$$F_1(A_1) = K + k + 2F\sin^2 A\ , \quad F_2(A_1) = H + F\cos^2 A\ ,$$

$$F_{12}(A_1) = 2^{\frac{1}{2}} F \sin A \cos A\ ;$$

$$F(B_2) = K - k\ . \tag{6.14}$$

6.2.4. Specific Imposition of a Potential Parameter

In Chapter 3 the variations of mean amplitudes with force constants are studied in terms of Jacobian elements. The derived expressions may be useful in refinements of force fields with the aid of observed mean amplitudes. A different method, referred to as 'specific imposition of a potential parameter on non-bonded distances', has proved to be convenient for numerical solutions of this kind of problems.

Let $\mathbf{F}$ (based on $\mathbf{S}$) be the initial force constant matrix. Introduce a redundant stretching coordinate, say p, which corresponds to a non-bonded interatomic distance. A new force field is constructed by adding one row and one column to $\mathbf{F}$, and in the simplest case introducing one non-vanishing constant, say h (the potential parameter), which corresponds to p. The coordinate p may be expressed in terms of the $\mathbf{S}$ coordinates. Hence the new force field may be represented as $\hat{\mathbf{F}}$ in the formalism of the $\boldsymbol{T}$ matrix.

In symmetrical molecules allowance must be made for a multiplicity > 1 of the interatomic (non-bonded) distance. Then there is a set of symmetrically equivalent coordinates p_i ($i = 1,2,\ldots$). When $\mathbf{F}$ is a symmetry force constant matrix one must construct the appropriate combinations of p_i belonging to the various symmetry species and add the h constant to the corresponding blocks of the $\mathbf{F}$ matrix. When the same value (h) is used for the potential parameter in every block it means that all interaction terms between the different p_i's are neglected.

A rough estimate of the variation in a non-bonded amplitude (l) with the variation of the corresponding potential parameter (h) may be given as

$$\Delta l = - \frac{l^3}{2kT} \Delta h . \qquad (6.15)$$

The imposition of potential parameters also affects the vibrational frequencies. The effect on the mean amplitudes is expected to be considerably damped if the new force field is refined so as to restore the agreement with given (observed) frequencies. This feature may place serious limitations on the applicability of the method.

In one of the subsequent chapters the specific imposition of a potential parameter is applied to the hydrogen bond of formic acid monomer (see Chapter 14, Part II).

REFERENCES

1 B.N.Cyvin and S.J.Cyvin, Acta Chem.Scand. 23, 3139 (1969).
2 S.J.Cyvin, B.N.Cyvin, and G.Hagen, Chem.Phys.Letters 2, 341 (1968).
3 B.N.Cyvin and S.J.Cyvin, J.Phys.Chem. 73, 1430 (1969).
4 S.J.Cyvin and B.N.Cyvin, Z.Naturforschg. 26a, 901 (1971).
5 K.Nakamoto: Infrared Spectra of Inorganic and Coordination Compounds, 2. Ed., Wiley-Interscience, New York 1970.
6 S.J.Cyvin: Molecular Vibrations and Mean Square Amplitudes, Universitetsforlaget, Oslo, and Elsevier, Amsterdam, 1968.
7 S.J.Cyvin, J.Brunvoll, B.N.Cyvin, I.Elvebredd, and G.Hagen, Mol.Phys. 14, 43 (1968).
8 E.Meisingseth, Acta Chem.Scand. 16, 1601 (1962).
9 S.J.Cyvin, B.N.Cyvin, I.Elvebredd, G.Hagen, and J.Brunvoll, Kgl. Norske Videnskab. Selskabs Skrifter, No. 22 (1972).
10 T.Shimanouchi, J.Chem.Phys. 17, 245 (1949).

CHAPTER 7

Vibrational Symmetry from Computers

M. GUSSONI, G. DELLEPIANE, and G. ZERBI

A method is outlined for the automatic evaluation of internal vibrational symmetry coordinates divided into symmetry species. The few complications which may arise are discussed and methods for overcoming them are presented. It is moreover shown how to derive automatically the number and type of independent elements of the potential energy matrix. The method presented is particularly useful in molecular dynamics of large molecules with high symmetry or when dealing with a large set of molecules.

Abbiamo riportato un metodo per il calcolo automatico delle coordinate di simmetria interne raggruppate in specie di simmetria, ed abbiamo dimostrato come è possibile trattare i casi in cui possono sorgere alcune complicazioni. E' inoltre possibile ricavare automaticamente il numero e il tipo di elementi indipendenti della matrice dell'energia potenziale. Questo metodo è particolarmente utile nel caso di molecole grandi ad alta simmetria o per calcoli di coordinate normali su un grande numero di molecole.

7.1. Introduction

When dealing with the problem of molecular vibrations it is very useful and sometimes necessary to employ symmetry coordinates. These are coordinates that form the basis for a completely reduced representation of the molecular symmetry group. Each symmetry coordinate belongs to one of the irreducible representations of this group, i.e. to one of the symmetry species into which the normal modes of vibrations may be divided. There must be as many independent symmetry coordinates as there are normal vibrations belonging to it.

The usefulness of symmetry coordinates lies chiefly in the factorization of the secular equation into as many blocks as the number of symmetry species covered by the normal modes of the molecule. This factorization makes the computation easier by reducing the dimensions of the matrices to be diagonalized, and makes the assignment of experimental frequencies possible by providing the eigenvalues collected into symmetry species.

Symmetry coordinates are usually constructed with the aid of the projection operator of Wigner,[1] but many other authors[2-5] (cf. also Chapter 20) have developed methods for the construction of various kinds of symmetry coordinates to be used in the study of normal vibrations of molecules with symmetry. All these methods ar usually applied by deriving manually the transformation matrices with the use of character tables. In order to decrease the probabi lity of human errors some methods[6,7] have been proposed which can be automatized for computers.

We have independently developed[8,9] a method which requires only the knowledge of $\mathbf{G}$, the inverse of the kinetic energy matrix as expressed in the space of internal coordinates[10] for a given molecule. The algorithm has been translated into a computer progra which provides: (i) a set of symmetry coordinates conveniently divided into symmetry species, (ii) the number and type of independent elements of the potential energy matrix $\mathbf{F}$ in the space of internal coordinates, and (iii) occasionally the cartesian symmetry coordinates.[10]

In what follows we discuss the principles of our method, its applications, and the associated peculiarities.

7.2. Theory

Let the kinetic and potential energy (K.E. and P.E.) in internal coordinates[10] $\mathbf{R}$ be written as

$$2T = \dot{\mathbf{R}}'(\mathbf{G}^R)^{-1}\dot{\mathbf{R}}\ , \quad 2V = \mathbf{R}'\mathbf{F}^R\mathbf{R}\ . \tag{7.1}$$

The superscript R will be omitted in the remainder of this chapter. The matrix $\mathbf{G}$ can be computed either directly from available tables[12-14] or from the equation

$$\mathbf{G} = \mathbf{B}_R \mathbf{m}^{-1} \mathbf{B}_R' \tag{7.2}$$

which is equivalent to Eq. (1.6) in terms of the $\mathbf{R}$ coordinates, and

$$\mathbf{R} = \mathbf{B}_R \mathbf{x}. \tag{7.3}$$

Let the number of internal coordinates $\mathbf{R}$ defined for a molecule be $M \geq 3N - 6$ and the $(\mathbf{G})_{M \times M}$ matrix be known. Let the point group of the molecule be $\mathcal{G}$ and its h elements $\mathcal{R}_1, \mathcal{R}_2, \ldots, \mathcal{R}_h$. From the invariance of the K.E. and P.E. under a symmetry operation it follows that

$$\mathbf{P}(\mathcal{R}_i)\mathbf{G}\mathbf{F} = \mathbf{G}\mathbf{F}\mathbf{P}(\mathcal{R}_i) \tag{7.4}$$

for each symmetry operator $\mathcal{R}_i$ which can be applied to the molecule, namely for each $\mathcal{R}_i \in \mathcal{G}$. In Eq. (4) each matrix $\mathbf{P}(\mathcal{R}_i)$ is the permutation or pseudopermutation[9] matrix which represents $\mathcal{R}_i$ on the basis $\mathbf{R}$.

It is possible to construct by computer all the possible permutation or pseudopermutation matrices and to find those which satisfy Eq. (4) even when the group $\mathcal{G}$ and its order are unknown. The number of these matrices equals h and they provide the reducible representation of $\mathcal{G}$ on the basis of $\mathbf{R}$. The use of Eq. (4) is complicated by the fact that the matrix $\mathbf{F}$ is not always known a priori. Since the same invariance must contemporarily and separately hold for K.E. and P.E., instead of using Eq. (4) we use

$$\mathbf{P}(\mathcal{R}_i)\mathbf{G} = \mathbf{G}\mathbf{P}(\mathcal{R}_i). \tag{7.5}$$

When M is large it may become too cumbersome to fit into Eq. (5) all the possible M-dimensional permutation and pseudopermutation matrices in order to choose the good ones. A partitioning technique then becomes necessary. The $\mathbf{R}$ coordinates could be divided into symmetrically equivalent sets by the following necessary condition. "Equivalent coordinates R_i and R_j have equal diagonal elements $g_{ii} = g_{jj}$ in the $\mathbf{G}$ matrix." Of course the above condition may not be sufficient (example: the C-H stretchings in CH_3-CH_2Cl), thus in some cases the sets considered will contain coordinates which are not equivalent by symmetry. This fact does not affect what follows because the only aim of the partitioning is to solve Eq. (5) on smaller matrices. It is unimportant whether the blocks can be further partitioned.

Let us assume the M coordinates can be partitioned into two sets containing m and n coordinates respectively ($m+n=M$). Eq. (5) reads now

$$\begin{bmatrix} \mathbf{P}_m(\mathcal{R}_i) & 0 \\ 0 & \mathbf{P}_n(\mathcal{R}_i) \end{bmatrix} \begin{bmatrix} \mathbf{G}_m & \mathbf{G}_{mn} \\ \mathbf{G}'_{mn} & \mathbf{G}_n \end{bmatrix} = \begin{bmatrix} \mathbf{G}_m & \mathbf{G}_{mn} \\ \mathbf{G}'_{mn} & \mathbf{G}_n \end{bmatrix} \begin{bmatrix} \mathbf{P}_m(\mathcal{R}_i) & 0 \\ 0 & \mathbf{P}_n(\mathcal{R}_i) \end{bmatrix} . \tag{7.6}$$

Among all the possible m-dimensional permutation and pseudopermutation matrices $\mathbf{P}_m(\mathcal{R}_i)$ only those which satisfy

$$\mathbf{P}_m(\mathcal{R}_i)\mathbf{G}_m = \mathbf{G}_m\mathbf{P}_m(\mathcal{R}_i) \tag{7.7}$$

are to be chosen. The corresponding $\mathbf{P}_n(\mathcal{R}_i)$ may be derived from

$$\mathbf{P}_m(\mathcal{R}_i)\mathbf{G}_{mn} = \mathbf{G}_{mn}\mathbf{P}_n(\mathcal{R}_i). \tag{7.8}$$

The solution of this equation requires the inversion of an off-diagonal submatrix $\mathbf{G}_{mn}$, which may be singular or even rectangular. The inversion can be performed following Lanczos' method.[15]

Let us write Eq. (8) as

$$\overline{\mathbf{P}}_n(\mathcal{R}_i) \equiv \mathbf{G}_{mn}^{-1}\mathbf{G}_{mn}\mathbf{P}_n(\mathcal{R}_i) = \mathbf{G}_{mn}^{-1}\mathbf{P}_m(\mathcal{R}_i)\mathbf{G}_{mn}, \tag{7.9}$$

where $\mathbf{G}_{mn}^{-1}$ is the generalized[15] inverse of $\mathbf{G}_{mn}$ and has the same rank as $\mathbf{G}_{mn}$. $\overline{\mathbf{P}}_n(\mathcal{R}_i)$ can be evaluated from the last equality in Eq. (9). If the rank of $\mathbf{G}_{mn}$ (and $\mathbf{G}_{mn}^{-1}$) and/or the rank of $\mathbf{P}_m(\mathcal{R}_i)$ (which is necessarily equal to m) are lower than n, it follows that $|\overline{\mathbf{P}}_n(\mathcal{R}_i)| = 0$. In this case $\overline{\mathbf{P}}_n(\mathcal{R}_i)$ cannot be a permutation or pseudopermutation matrix. We have shown[9,11] that the true permutation or pseudopermutation matrix $\mathbf{P}_n(\mathcal{R}_i)$ can be derived from $\overline{\mathbf{P}}_n(\mathcal{R}_i)$ using the following equation.

$$\mathbf{P}_n(\mathcal{R}_i) = \bar{\mathbf{P}}_n(\mathcal{R}_i) + \mathbf{U}_0^{(i)}(\mathcal{R}_i)[\mathbf{V}_0^{(i)}(\mathcal{R}_i)]' , \qquad (7.10)$$

where $\mathbf{U}_0^{(i)}(\mathcal{R}_i)$ and $\mathbf{V}_0^{(i)}(\mathcal{R}_i)$ are the zero combinations on the rows and columns of $\bar{\mathbf{P}}_n(\mathcal{R}_i)$ respectively:

$$[\bar{\mathbf{P}}_n(\mathcal{R}_i)]'\mathbf{U}_0^{(i)}(\mathcal{R}_i) = 0, \quad \bar{\mathbf{P}}_n(\mathcal{R}_i)\mathbf{V}_0^{(i)}(\mathcal{R}_i) = 0. \qquad (7.11)$$

The matrices $\mathbf{U}_0^{(i)}(\mathcal{R}_i)$ and $\mathbf{V}_0^{(i)}(\mathcal{R}_i)$ with dimensions $n \times p$, where $n-p$ is the rank of $\bar{\mathbf{P}}_n(\mathcal{R}_i)$, are not necessarily the same because in general $[\bar{\mathbf{P}}_n(\mathcal{R}_i)]' \neq \bar{\mathbf{P}}_n(\mathcal{R}_i)$. They can be easily computed as eigenvectors belonging to the zero eigenvalues of the symmetrical matrices[15] $\bar{\mathbf{P}}_n(\mathcal{R}_i)[\bar{\mathbf{P}}_n(\mathcal{R}_i)]'$ and $[\bar{\mathbf{P}}_n(\mathcal{R}_i)]'\bar{\mathbf{P}}_n(\mathcal{R}_i)$ respectively:

$$\begin{aligned} \bar{\mathbf{P}}_n(\mathcal{R}_i)[\bar{\mathbf{P}}_n(\mathcal{R}_i)]'\mathbf{U}_0^{(i)}(\mathcal{R}_i) &= 0, \\ [\bar{\mathbf{P}}_n(\mathcal{R}_i)]'\bar{\mathbf{P}}_n(\mathcal{R}_i)\mathbf{V}_0^{(i)}(\mathcal{R}_i) &= 0. \end{aligned} \qquad (7.12)$$

Thus, when all the possible $\mathbf{P}_m(\mathcal{R}_i)$ satisfying Eq. (8) have been computed and the corresponding $\mathbf{P}_n(\mathcal{R}_i)$ have been derived through Eqs. (9) and (10), all the possible $\mathbf{P}(\mathcal{R}_i)$ satisfying Eq. (5) are known. This procedure can be repeated when the number of sets into which the coordinates have been divided is larger than two.

The number h of different $\mathbf{P}_n(\mathcal{R}_i)$ so computed gives the order of the unknown group $\mathcal{G}$, and the set of $\mathbf{P}(\mathcal{R}_i)$ gives the reducible representation of $\mathcal{G}$ in terms of the $\mathbf{R}$ coordinates.

The completely reduced representation $\mathbf{P}^{irr}(\mathcal{R}_i)$ of the elements of $\mathcal{G}$ is now easily obtained by diagonalizing $\mathbf{G}$. Indeed the same matrix $\mathbf{D}$ which reduces $\mathbf{G}$ to diagonal form $\mathbf{\Gamma}$;

$$\mathbf{D}'\mathbf{G}\mathbf{D} = \mathbf{\Gamma}, \qquad (7.13)$$

reduces the representation:[1]

$$\mathbf{P}^{irr}(\mathcal{R}_i) = \mathbf{D}'\mathbf{P}(\mathcal{R}_i)\mathbf{D} . \qquad (7.14)$$

The h matrices $\mathbf{P}^{irr}(\mathcal{R}_i)$ provide the completely reduced representation of $\mathcal{G}$ in terms of the symmetry coordinates*:[8]

$$\mathbf{S} = \mathbf{D}'\mathbf{R}. \qquad (7.15)$$

Γ is the inverse of the kinetic energy matrix as expressed in the space $\mathbf{G}$;

$$2T = \dot{\mathbf{G}}' \Gamma^{-1} \dot{\mathbf{G}} , \tag{7.16}$$

and its structure gives the number of irreducible representations in $\mathbf{P}^{irr}$ and the dimensions of each one of them. For example if Γ has the pattern

$$\Gamma = \text{diag}(\, \gamma_1, \gamma_2, \gamma_2, \gamma_3, \gamma_3, \gamma_3, \gamma_4, 0, 0)$$

it means that there are two one-dimensional (γ_1 , γ_4), one two-dimensional (γ_2), one three-dimensional (γ_3) irreducible representations and two redundancies. Any $\mathbf{P}^{irr}(\mathcal{R}_i)$ has the corresponding pattern

$$\mathbf{P}^{irr}(\mathcal{R}_i) = \text{diag}\left(a_{11}^{\,i}, \begin{bmatrix} b_{11}^{\,i} & b_{12}^{\,i} \\ b_{21}^{\,i} & b_{22}^{\,i} \end{bmatrix}, \begin{bmatrix} c_{11}^{\,i} & c_{12}^{\,i} & c_{13}^{\,i} \\ c_{21}^{\,i} & c_{22}^{\,i} & c_{23}^{\,i} \\ c_{31}^{\,i} & c_{32}^{\,i} & c_{33}^{\,i} \end{bmatrix}, d_{11}^{\,i}, \begin{bmatrix} e_{11}^{\,i} & e_{12}^{\,i} \\ e_{21}^{\,i} & e_{22}^{\,i} \end{bmatrix} \right) .$$

The characters of each $\mathcal{R}_i$ in each irreducible representation can be obtained by adding the diagonal elements of the corresponding submatrix of $\mathbf{P}^{irr}(\mathcal{R}_i)$. Once the characters of the irreducible representations corresponding to all the $\mathbf{P}^{irr}(\mathcal{R}_i)$ have beer derived, the $\mathbf{P}^{irr}(\mathcal{R}_i)$ are divided into classes and the $\mathbf{G}$ into species.

The amount of information so far derivable, via computer, from the only knowledge of the $\mathbf{G}$ matrix, is: (i) the order h of the group, (ii) the number of its classes, (iii) the number of the

---

* In the original publications[8,9] $\mathbf{G}$ is denoted Σ. In the present context the notation has been changed in order to avoid confusion with Σ as the mean-square amplitude matrix.

species which contain the normal modes of the molecule studied, (iv) the symmetry coordinates $\mathbf{G}$ divided into species, and (v) the symmetrized inverse kinetic energy matrix Γ.

At this stage we know everything we need about the symmetry of the vibrational problem. It must be realized that in many cases these pieces of information are not sufficient to identify and to label the species of the group $\mathcal{G}$. Indeed we may not have arrived at the complete character table, but only at that portion which describes our molecule. This is not essential to the solution of the vibrational problem.

In the next section we discuss methods of overcoming possible complications which may arise by using the present technique.

7.3. Possible Complications

7.3.1. Accidental Higher Symmetry of the G Matrix

Strictly speaking, the use of Eq. (5) leads us to the detection of all and only all the operators $\mathbf{P}(\mathcal{R}_i)$ which leave $\mathbf{G}$ invariant. It can easily be shown that these operators contain among themselves the representations in the $\mathbf{R}$ space of all the symmetry operations which can be performed on the molecule. If $\mathcal{R}_i$ is an operation of the symmetry point group to which the molecule belongs, each set of equivalent coordinates $\mathbf{R}_k$ will transform into itself by application of $\mathbf{P}(\mathcal{R}_i)$ on them. Namely, the transformation

$$\hat{\mathbf{R}} = \mathbf{P}(\mathcal{R}_i)\mathbf{R} \tag{7.17}$$

does not change the K.E.;

$$2T = \dot{\hat{\mathbf{R}}}'\hat{\mathbf{G}}^{-1}\dot{\hat{\mathbf{R}}} = \dot{\mathbf{R}}'\mathbf{G}^{-1}\dot{\mathbf{R}}. \tag{7.18}$$

Moreover the matrices are the same[16]

$$\hat{\mathbf{G}}^{-1} = \mathbf{G}^{-1}. \tag{7.19}$$

By introducing Eq. (17) into Eq. (18), imposing Eq. (19) and making use of the orthogonality of permutation and pseudopermutation matrices

$$\mathbf{P}^{-1} = \mathbf{P}' , \tag{7.20}$$

Eq. (5) follows immediately.

However, it is not always true that the operators $\mathbf{P}$ obeying Eq. (5) represent a symmetry operation which could be performed on the molecule. That is, it may happen that the $\mathbf{G}$ matrix, for numerical reasons, may show a symmetry higher than that of the molecule. It is expedient to verify whether or not all the $\mathbf{P}(\mathcal{R}_i)$ computed through Eq. (5) correspond to symmetry operations feasible on the molecule and to find how to choose the good ones. Let us consider the example of Fig. 1. The $\mathbf{G}$ matrix takes the form

$$\mathbf{G} = \begin{bmatrix} a & 0 & 0 & d & d \\ 0 & b & 0 & e & e \\ 0 & 0 & b & e & e \\ d & e & e & c & f \\ d & e & e & f & c \end{bmatrix} \begin{matrix} \Delta R_1 \\ \Delta R_2 \\ \Delta R_3 \\ \Delta \alpha_1 \\ \Delta \alpha_2 \end{matrix} \tag{7.21}$$

where

$$a = \frac{2}{M} , \quad b = \frac{m+M}{mM} , \quad c = \frac{2}{MR^2} + \frac{m+M}{mMr^2} ,$$

$$d = -\frac{1}{Mr} , \quad e = -\frac{1}{MR} , \quad f = \frac{2}{MR^2} . \tag{7.22}$$

The point group of the molecule is $\mathcal{C}_2\{E, C_2\}$ and its representation in the $\mathbf{R}$ space is given by the matrices

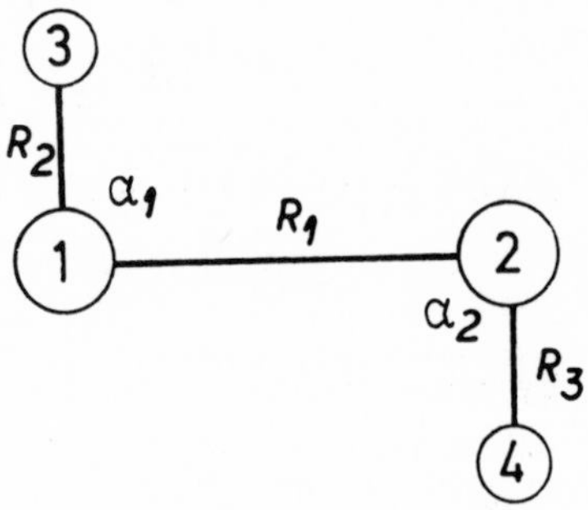

Fig. 7-1. Hypothetical molecule model. The masses of the atoms are $m_1 = m_2 = M$ and $m_3 = m_4 = m$. The equilibrium parameters are $R_1^{eq} = R$ and $R_2^{eq} = R_3^{eq} = r$; $\alpha_1^{eq} = \alpha_2^{eq} = 90^\circ$.

$$\mathbf{P}_1 = \begin{bmatrix} 1 & 0 & 0 & 0 & 0 \\ 0 & 1 & 0 & 0 & 0 \\ 0 & 0 & 1 & 0 & 0 \\ 0 & 0 & 0 & 1 & 0 \\ 0 & 0 & 0 & 0 & 1 \end{bmatrix}, \quad \mathbf{P}_2 = \begin{bmatrix} 1 & 0 & 0 & 0 & 0 \\ 0 & 0 & 1 & 0 & 0 \\ 0 & 1 & 0 & 0 & 0 \\ 0 & 0 & 0 & 0 & 1 \\ 0 & 0 & 0 & 1 & 0 \end{bmatrix}, \tag{7.23}$$

which satisfy Eq. (5). However, the matrices

$$\mathbf{P}_3 = \begin{bmatrix} 1 & 0 & 0 & 0 & 0 \\ 0 & 0 & 1 & 0 & 0 \\ 0 & 1 & 0 & 0 & 0 \\ 0 & 0 & 0 & 1 & 0 \\ 0 & 0 & 0 & 0 & 1 \end{bmatrix}, \quad \mathbf{P}_4 = \begin{bmatrix} 1 & 0 & 0 & 0 & 0 \\ 0 & 1 & 0 & 0 & 0 \\ 0 & 0 & 1 & 0 & 0 \\ 0 & 0 & 0 & 0 & 1 \\ 0 & 0 & 0 & 1 & 0 \end{bmatrix} \tag{7.24}$$

also satisfy the same equation even if they do not correspond to any symmetry operation feasible on the molecule. This is due to the fact that the interactions $\Delta R_2 \Delta\alpha_1$ and $\Delta R_2 \Delta\alpha_2$ are numerically equal because of the peculiar value of α^{eq}, thus increasing the number of symmetry operators on $\mathbf{G}$.

We overcame this difficulty in the following way. Let $\mathbf{P}_i^{\alpha}$ ($\alpha = 1,\ldots,n$) be the blocks of $\mathbf{P}_i$ corresponding to different kinds of internal coordinates. Compute from each $\mathbf{P}_i^{\alpha}$ the corresponding permutation matrix $\mathbf{p}_i^{\alpha}$ among the nuclei. Refuse those $\mathbf{P}_i$ which lead to different $\mathbf{p}_i$ for $\alpha = 1,\ldots,n$. In Fig. 2 the permutations on the nuclei computed as described above are given. It is easily seen that only $\mathbf{P}_1$ and $\mathbf{P}_2$ correspond to feasible operations, and only they must be retained.

7.3.2. Accidental Degeneracy in the G Matrix

The detection of the number of irreducible representations in the completely reduced representation $\mathbf{P}^{irr}$ follows from the number of different eigenvalues of the $\mathbf{G}$ matrix. However, it may happen that the $\mathbf{G}$ matrix shows an accidental degeneracy, namely that some of its eigenvalues are degenerate even if the corresponding eigenvectors (and therefore the corresponding $\mathcal{G}$'s) do not belong to the same irreducible representation.

For instance a hypothetical molecule like that of Fig. 3 has a Γ matrix with the following pattern.

$$\Gamma = \mathrm{diag}(\gamma_1, \gamma_1, \gamma_2, \gamma_2, \gamma_3, \gamma_4, \gamma_5, \gamma_5, \gamma_6,$$

$$\gamma_7, \gamma_7, \gamma_8, \gamma_8, \gamma_8, \gamma_8, \gamma_8, \gamma_8, \gamma_9, 0, 0, 0, 0), \tag{7.25}$$

where accidental degeneracy occurs at γ_8. Thus the completely reduced representation $\mathbf{P}^{irr}$, neglecting the four redundancies, should consist of 4 one-dimensional ($\gamma_3, \gamma_4, \gamma_6, \gamma_9$), 4 two-

	P_i^{str}	P_i^{bend}
P_1	$\begin{bmatrix} 1 & 0 & 0 & 0 \\ 0 & 1 & 0 & 0 \\ 0 & 0 & 1 & 0 \\ 0 & 0 & 0 & 1 \end{bmatrix}$	$\begin{bmatrix} 1 & 0 & 0 & 0 \\ 0 & 1 & 0 & 0 \\ 0 & 0 & 1 & 0 \\ 0 & 0 & 0 & 1 \end{bmatrix}$
P_2	$\begin{bmatrix} 0 & 1 & 0 & 0 \\ 1 & 0 & 0 & 0 \\ 0 & 0 & 0 & 1 \\ 0 & 0 & 1 & 0 \end{bmatrix}$	$\begin{bmatrix} 0 & 1 & 0 & 0 \\ 1 & 0 & 0 & 0 \\ 0 & 0 & 0 & 1 \\ 0 & 0 & 1 & 0 \end{bmatrix}$
P_3	$\begin{bmatrix} 0 & 1 & 0 & 0 \\ 1 & 0 & 0 & 0 \\ 0 & 0 & 0 & 1 \\ 0 & 0 & 1 & 0 \end{bmatrix}$	$\begin{bmatrix} 1 & 0 & 0 & 0 \\ 0 & 1 & 0 & 0 \\ 0 & 0 & 1 & 0 \\ 0 & 0 & 0 & 1 \end{bmatrix}$
P_4	$\begin{bmatrix} 1 & 0 & 0 & 0 \\ 0 & 1 & 0 & 0 \\ 0 & 0 & 1 & 0 \\ 0 & 0 & 0 & 1 \end{bmatrix}$	$\begin{bmatrix} 0 & 1 & 0 & 0 \\ 1 & 0 & 0 & 0 \\ 0 & 0 & 0 & 1 \\ 0 & 0 & 1 & 0 \end{bmatrix}$

Fig. 7-2. Permutations on the nuclei (see the text).

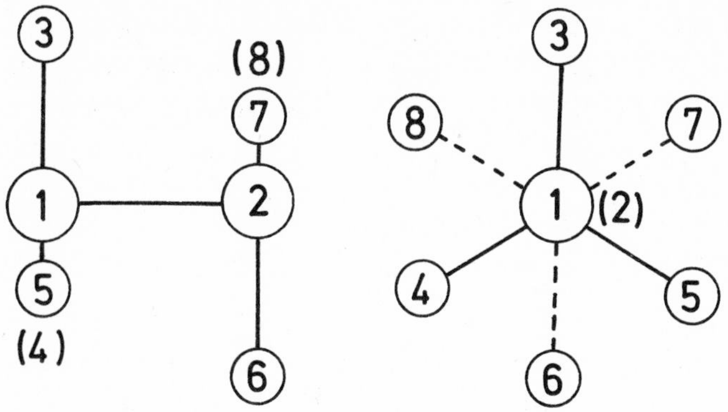

Fig. 7-3. Hypothetical molecule model. Masses: $m_1 = m_2$, $m_3 = m_4 = m_5 = m_6 = m_7 = m_8$.

dimensional (γ_1, γ_2, γ_5, γ_7), and one six-dimensional (γ_8) irreducible representations. To ascertain whether this is true or if accidental degeneracies have occurred, one must remember that, if two or more $\mathfrak{G}_i$'s are accidentally degenerate, the corresponding blocks in the $\mathbf{P}^{irr}$'s are further reducible. This can easily be verified because in this case the following relation holds:

$$\sum_{\mathcal{R} \in \mathcal{G}} \chi^*(\mathcal{R}) \chi(\mathcal{R}) \neq h$$

[$\chi(\mathcal{R})$ is the character of the reducible block of $\mathbf{P}^{irr}(\mathcal{R})$].

We have not yet found a general method to overcome this difficulty. However, the only accidental degeneracies we have found in real molecules are among redundancies. Fortunately we need not separate the redundant symmetry coordinates since they can be neglected. A case like that reported in Eq. (25) does not correspond to any real molecule.

7.3.3. Proper Order of the G Matrix

Once the $\mathbf{G}$ matrix has been partitioned into submatrices corresponding to the sets $\mathbf{R}_1$, $\mathbf{R}_2$, ..., $\mathbf{R}_k$, ..., $\mathbf{R}_p$ and once all the $\mathbf{P}_k(\mathcal{R}_i)$ have been computed according to Eq. (7) for a set k, the corresponding $\mathbf{P}_{k+1}(\mathcal{R}_i)$ can be computed from Eqs. (9) and (10) provided that the submatrix $\mathbf{G}_{k,k+1} \neq \mathbf{O}$. If $\mathbf{G}_{k,k+1} = \mathbf{O}$ the sets must be rearranged (and also the rows and columns of $\mathbf{G}$) such that the submatrix $\mathbf{G}_{k,k+1} \neq \mathbf{O}$. For instance, in the case of CH_3-CD_3 the 22 coordinates can be divided into the sets $\{R_1^{CH}, R_2^{CH}, R_3^{CH}\}$, $\{R_1^{CD}, R_2^{CD}, R_3^{CD}\}$, $\{R^{CC}\}$, $\{\beta_1^{HCC}, \beta_2^{HCC}, \beta_3^{HCC}\}$, $\{\tau_1^{HD}, \tau_2^{HD}, \tau_3^{HD}\}$, $\{\beta_1^{DCC}, \beta_2^{DCC}, \beta_3^{DCC}\}$, $\{\alpha_1^{HCH}, \alpha_2^{HCH}, \alpha_3^{HCH}\}$, $\{\alpha_1^{DCD}, \alpha_2^{DCD}, \alpha_3^{DCD}\}$. However, in this case $\mathbf{G}(R^{CH}, R^{CD}) = \mathbf{O}$ and we cannot derive $\mathbf{P}_{CD}(\mathcal{R}_i)$ from $\mathbf{P}_{CH}(\mathcal{R}_i)$; but we can reshuffle the sets, for instance, in the order $\mathbf{R}^{CH}$, $\boldsymbol{\beta}^{HCC}$, $\mathbf{R}^{CD}$, and consequently obtain $\mathbf{G}(\beta^{HCC}, R^{CD}) \neq \mathbf{O}$ and $\mathbf{G}(R^{CH}, \beta^{HCC}) \neq \mathbf{O}$.

Sometimes it may happen that one or more sets have no interactions with the other sets. For instance it happens for in-plane and out-of-plane coordinates of a planar molecule. No reshuffling can help in this case. One must compute separately the permutation matrices for in-plane and out-of-plane coordinates, and then combine them into $\mathbf{P}(\mathcal{R}_i)$. For other details of these computational difficulties we refer the reader to Ref. 11.

7.3.4. Cyclic Groups

The cyclic groups are Abelian, and therefore any completely reduced representation of them contains only one-dimensional irreducible representations[17] with complex characters. Let us take as an example the point group $\mathcal{C}_3$. The character table is[10]

$\mathcal{C}_3$		E	C_3	C_3^2
A		1	1	1
E	{	1	ε	ε^*
		1	ε^*	ε

$$\varepsilon = e^{2\pi i/3}$$

The reason why the last two irreducible representations are lumped together as a two-dimensional representation E lies in their mixing under a time reversal operation. In the absence of external fields the Hamiltonian is invariant under time reversal

$$\mathcal{H}\psi = i\hbar \frac{\partial \psi}{\partial t}, \quad \mathcal{H}\psi^* = i\hbar \frac{\partial \psi^*}{\partial(-t)}. \tag{7.26}$$

ψ and ψ^* are eigenfunctions having the same energy, and the time reversal operator exchanges them as if they were degenerate. However, no spatial operator in the group will transform ψ to ψ^* and vice versa. The two operations are therefore one-dimensional from the viewpoint of spatial symmetry, but they may be considered as degenerate owing to the existence of the reversal operation.

However, diagonalization of the **G** matrix of a molecule belonging to the $\mathcal{C}_3$ group (e.g. CH_3-CD_3 partially rotated as in Fig. 4) gives only real linear combinations of the internal coordinates such as, for instance,

$$\mathfrak{S}_1 = 6^{-\frac{1}{2}}(2\Delta R_1 - \Delta R_2 - \Delta R_3), \quad \mathfrak{S}_2 = 2^{-\frac{1}{2}}(\Delta R_2 - \Delta R_3). \tag{7.27}$$

These coordinates are really degenerate from the viewpoint of spatial symmetry, since an operator of the group transforms each of them into a linear combination of the two. This is due to the fact that the representation Γ_{real} provided by the $\mathfrak{S}$'s of Eq. (27) is not in a completely reduced form[18] with respect to the set of all the symmetry (time and space) operators. A linear complex transformation could be operated on them yielding a complex, but completely reduced representation $\Gamma_{complex}$ of $\mathcal{C}_3$. The matrices representing the elements of $\mathcal{C}_3$ in Γ_{real} and $\Gamma_{complex}$ are given below.

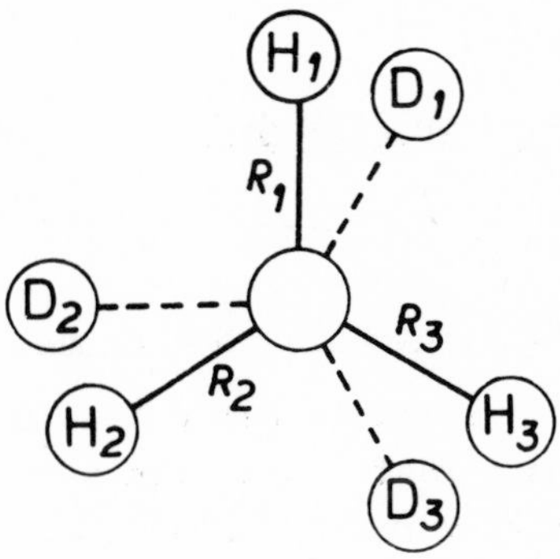

Fig. 7-4. Partially rotated (hypothetical) CH_3-CD_3.

$$\Gamma_{\text{real}} \quad \overset{E}{\begin{bmatrix} 1 & 0 \\ 0 & 1 \end{bmatrix}} \quad \overset{C_3}{\begin{bmatrix} \cos\frac{2\pi}{3} & \sin\frac{2\pi}{3} \\ -\sin\frac{2\pi}{3} & \cos\frac{2\pi}{3} \end{bmatrix}} \quad \overset{C_3^{-1}}{\begin{bmatrix} \cos\frac{2\pi}{3} & -\sin\frac{2\pi}{3} \\ \sin\frac{2\pi}{3} & \cos\frac{2\pi}{3} \end{bmatrix}} \tag{7.28}$$

$$\Gamma_{\text{complex}} \quad \overset{E}{\begin{bmatrix} 1 & 0 \\ 0 & 1 \end{bmatrix}} \quad \overset{C_3}{\begin{bmatrix} \varepsilon^* & 0 \\ 0 & \varepsilon \end{bmatrix}} \quad \overset{C_3^{-1}}{\begin{bmatrix} \varepsilon & 0 \\ 0 & \varepsilon^* \end{bmatrix}} \tag{7.29}$$

We do not use the representation (29), but are content to keep the representation in its semireduced form (28), because the reduction from Eq. (28) to (29) gives us no further information on the problem. Indeed we would reduce it from a two-dimensional real representation to the sum of two complex one-dimensional representations if we were referring to two complex conjugate quantities ψ and ψ^*, which in the physical problem would have to be considered together. In the present case the representation (28) is used as it comes out from the computer, and the character table computed is

$\mathcal{C}_3$	E	C_3, C_3^2
A	1	1
E	2	-1

7.4. Simplifying Features in the Solution of the Vibrational Problem

7.4.1. Potential Energy Matrices in the Spaces R and G

In the construction of the matrix $\mathbf{F}$, viz. $\mathbf{F}^R$ of Eq. (1), considerable attention must be given in order to preserve the symmetry of the molecule. Many of the elements of the $\mathbf{F}$ matrix must be equal (or opposite) for symmetry reasons. The usual procedure[19] in computations is to set up a three-index matrix $\mathbf{Z}$ such that

$$\mathbf{F} = \mathbf{Z}\boldsymbol{\Phi}, \tag{7.30}$$

where $\boldsymbol{\Phi}$ is the vector of the independent valence force constants, and the matrix $\mathbf{Z}$ contains all the relations among them dictated by the symmetry. The $\mathbf{Z}$ matrix can easily be constructed by computer when the representation $\mathbf{P}$ is known.

Indeed, an equation similar to Eq. (5) must hold also for the $\mathbf{F}$ matrix, namely

$$\mathbf{P}(\mathcal{R}_i)\mathbf{F} = \mathbf{F}\,\mathbf{P}(\mathcal{R}_i), \tag{7.31}$$

where the matrices $\mathbf{P}(\mathcal{R}_i)$ represent the elements of $\mathcal{G}$ in the $\mathbf{R}$ space. Once all the $\mathbf{P}(\mathcal{R}_i)$ obeying Eq. (5) have been computed and only those corresponding to feasible symmetry operations have been retained, their introduction into Eq. (31) allows the detection of the number of independent force constants to be entered in the vector $\boldsymbol{\Phi}$ and the construction of the matrix $\mathbf{Z}$.

It has to be pointed out that, in the Born-Oppenheimer approximation, isotopically substituted molecules have the same $\mathbf{F}$ matrix, but have different $\mathbf{G}$ matrices, possibly corresponding to different point groups. It is clear that the matrices $\mathbf{P}(\mathcal{R}_i)$ to be introduced in Eq. (31) must be those obtained from the isotopic compound of higher symmetry while the matrices $\mathbf{P}(\mathcal{R}_i)$ computed and used to derive the point group for each molecule must be those obtained from its $\mathbf{G}$. For instance if we want to know the point group, symmetry coordinates, and independent $\mathbf{F}$ matrix elements for the molecules of Fig. 5 we must first solve Eq. (5) for $\mathbf{G}^{(1)}$. From $\mathbf{P}^{(1)}$ we obtain the symmetry coordinates ($\mathcal{D}_{3d}$) for molecule (1) and the equivalent elements in the $\mathbf{F}$ matrix. From $\mathbf{P}^{(2)}$ and $\mathbf{P}^{(3)}$ we obtain the symmetry coordinates for molecules (2) and (3) - $\mathcal{C}_{2h}$ and $\mathcal{C}_{3v}$ respectively - but the $\mathbf{F}$ matrix will always be the same as that computed for molecule (1).

The P.E. matrix in the space $\mathfrak{G}$ is now

$$\mathbf{F}_{\mathfrak{G}} = \mathbf{D}'\mathbf{F}\mathbf{D}, \tag{7.32}$$

where $\mathbf{D}$ is the matrix defined in Eq. (13), and where the proper

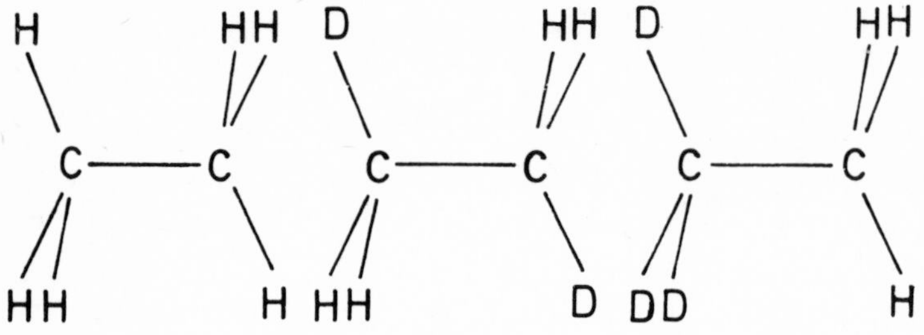

Fig. 7-5. Isotopic ethane molecules.

D matrix referring to the proper isotopic compound must be used.

7.4.2. Factorization of the Secular Equation

The secular equation in the **R** space follows from Eq. (1).

$$\mathbf{G}\mathbf{F}\mathbf{L}_R = \mathbf{L}_R \boldsymbol{\lambda}\,, \tag{7.33}$$

where $\mathbf{L}_R$ is the transformation to normal coordinates $\mathbf{Q}$:

$$\mathbf{R} = \mathbf{L}_R \mathbf{Q}\,. \tag{7.34}$$

The corresponding equation in the $\mathfrak{G}$ space follows from Eqs.(13) and (32).

$$\boldsymbol{\Gamma}\mathbf{F}_{\mathfrak{G}}\mathbf{L}_{\mathfrak{G}} = \mathbf{L}_{\mathfrak{G}}\boldsymbol{\lambda}\,, \tag{7.35}$$

where

$$\mathfrak{G} = \mathbf{L}_{\mathfrak{G}}\mathbf{Q}\,, \tag{7.36}$$

and the eigenvalues are the same as in Eq. (33) because the transformation operated on **GF** is orthogonal ($\mathbf{D}^{-1} = \mathbf{D}'$).

The secular equation (35) cannot be solved by usual computing techniques because $\boldsymbol{\Gamma}\mathbf{F}_{\mathfrak{G}}$ is not symmetrical. Let us rewrite it as

$$\boldsymbol{\Gamma}\mathbf{F}_{\mathfrak{G}}\mathbf{L}_{\mathfrak{G}} = \boldsymbol{\Gamma}^{\frac{1}{2}}\boldsymbol{\Gamma}^{\frac{1}{2}}\mathbf{F}_{\mathfrak{G}}\boldsymbol{\Gamma}^{\frac{1}{2}}\boldsymbol{\Gamma}^{-\frac{1}{2}}\mathbf{L}_{\mathfrak{G}} = \mathbf{L}_{\mathfrak{G}}\boldsymbol{\lambda}\,.$$

If we introduce

$$\mathbf{C}_{G} = \mathbf{\Gamma}^{-\frac{1}{2}} \mathbf{L}_{G} \tag{7.37}$$

we obtain

$$(\mathbf{\Gamma}^{\frac{1}{2}} \mathbf{F}_{G} \mathbf{\Gamma}^{\frac{1}{2}}) \mathbf{C}_{G} = \mathbf{C}_{G} \lambda . \tag{7.38}$$

The matrix $(\mathbf{\Gamma}^{\frac{1}{2}} \mathbf{F}_{G} \mathbf{\Gamma}^{\frac{1}{2}})$ is now symmetrical and has the same eigenvalues as the matrix $\mathbf{\Gamma} \mathbf{F}_{G}$ because a similarity transformation connects them; its eigenvectors $\mathbf{C}_{G}$ provide the $\mathbf{L}_{G}$ matrix through Eq. (37).

When the $\mathbf{R}$ coordinates involve redundancies, only that portion of the $\mathbf{D}$ matrix referring to nonvanishing G's is used; thus $\mathbf{\Gamma}$ and $\mathbf{F}_{G}$ come out as nonsingular $(3N-6) \times (3N-6)$ matrices. Both $\mathbf{F}_{G}$ and its transformed $\mathbf{\Gamma}^{\frac{1}{2}} \mathbf{F}_{G} \mathbf{\Gamma}^{\frac{1}{2}}$ are block diagonal, each block referring to a different symmetry species. Thus the vibrational problem may be solved for each species separately.

7.5. Conclusions

From the above discussion it is apparent that it is presently possible to obtain the maximum information about symmetry with as little human contribution as possible. Computers have been widely used for the automatic handling of most of the matrices and equations necessary in molecular dynamics. The portion of work dealing with the symmetry properties of the molecular system and the consequent factorization of the necessary matrices has so far required the judgement and the skilled experience of the worker. The method proposed in this chapter fills the gap of an almost complete automatization of normal coordinate calculations. This procedure is of particular use when dealing with large molecules of high symmetry (and the consequent large number of degenerate modes) or when calculations must be performed on a large number of molecules involved in a simultaneous least-squares refinement of the force field.

A similar method can also be applied to obtain cartesian symmetry coordinates as shown in Ref. 20.

The same concepts can be applied to problems other than

those of molecular vibrations. In order to obtain symmetry adapted functions one must recognize in the problem a quadratic form (with constant coefficients) which is invariant under all the symmetry operators.

The extension of this technique to simple crystals, complex molecular crystals, as well as to polymers is being carried out in the laboratory of the authors.

REFERENCES

1 E.P.Wigner: Group Theory, Academic Press, New York 1959.
2 J.R.Nielsen and L.H.Berryman, J.Chem.Phys. 17, 659 (1949).
3 B.Crawford, J.Chem.Phys. 21, 1108 (1953).
4 J.E.Kilpatrick, J.Chem.Phys. 16, 749 (1948).
5 S.J.Cyvin, J.Brunvoll, B.N.Cyvin, I.Elvebredd, and G.Hagen, Mol.Phys. 14, 43 (1968).
6 N.K.Morosov and V.P.Morosova, Sov.Phys.Dokl. 10, 326 (1964).
7 R.Moccia, Theor.Chim.Acta 7, 85 (1967).
8 M.Gussoni and G.Zerbi, J.Mol.Spectry. 26, 485 (1968).
9 G.Dellepiane, M.Gussoni, and G.Zerbi, J.Chem.Phys. 53, 3450 (1970).
10 E.B.Wilson,Jr., J.C.Decius, and P.C.Cross: Molecular Vibrations, McGraw-Hill, New York 1955.
11 L. Degli Antoni Ferri, G.Dellepiane, M.Gussoni, and G.Zerbi, Istituto di Chimica delle Macromolecole Milano, Techn. Report No. 3 (1970).
12 J.C.Decius, J.Chem.Phys. 16, 1025 (1948).
13 J.B.Lohman, Office Naval Research, Techn. Report 17 (1951).
14 S.M.Ferigle and A.G.Meister, J.Chem.Phys. 19, 982 (1951).
15 C.Lanczos: Linear Differential Operators, van Nostrand, London 1964.
16 R.McWeeny: The International Encyclopaedia of Physical Chemistry and Chemical Physics, Vol. 3, MacMillan (Pergamon), New York 1963.
17 M.Tinkham: Group Theory and Quantum Mechanics, McGraw-Hill, New York 1964.
18 D.S.Schonland: Molecular Symmetry, van Nostrand, London 1965.

19 J.H.Schachtschneider, Shell Development Company, Techn. Report 57 (1965).
20 M.Gussoni and G.Dellepiane, Chem.Phys.Letters, 10, 559 (1971).

CHAPTER 8

Molecular Compliants as Double-Time Thermal Green's Functions of Amplitudes

ØYSTEIN RA

In this chapter attention is drawn to the possibility of regarding the molecular compliance matrix as a collection of double-time Green's functions of amplitudes. The identification can be effected by solving Heisenberg's equation for the time development of operators as well as by resorting to perturbation theory.

8.1. Introduction

In recent years transcriptions of the Green's function and propagator techniques of quantum field theory have become next to ubiquitous in crystal dynamics. By contrast, many research chemists dealing with the related subject of molecular vibrations seem to have remained aloof. It is perhaps not surprising that this should be so since molecular mechanics still is, in large measure, a theory of harmonic oscillations in systems comprising comparatively few atoms. In the past there has been little cause for deviations from the molecular one-body procedure wherein the transition to quantum mechanics usually awaits the completion of a classical normal mode analysis. However, there is a growing awareness of the necessity of letting additional effects (anharmonicity, interaction between quanta of molecular vibrations and lattice phonons, etc.) come more actively into play. Hence it would seem profitable to relate the traditional theory to a general many-body formalism within the framework of which elaborate approximation procedures already exist.

Also, on the principal level there are useful dividends which accrue from applying many-body concepts when scrutinizing even small and medium size harmonic model molecules. After these prefatory remarks we state the purpose of this chapter, which is to point out that many-body methodology offers a simple explanation of some remarkable and seemingly puzzling properties of $\mathbf{N}$ (the compliance matrix), and its intimate connection with $\mathbf{\Sigma}$ (the mean-square amplitude matrix) in particular.

8.2. Definition of Double-Time Molecular Compliants

Consider a free molecule whose state vector components in oscillatory space evolve independently of rotatory space events. That is, we overlook centrifugal distortion in this chapter. Furthermore, to simplify the bookkeeping we endow the $3N - 6$ dimensional manifold with a metric tensor equal to $\mathbf{E}$ and choose some suitable set of orthonormal basis vectors $\{\chi\}$, the χ's having real-valued components. This frame is merely a convenience and involves no real restrictions. Matrix equations relating the usual $\{S\}$-basis[1] versions of important quantities to their $\{\chi\}$-basis counterparts are listed in Chapter 1. The vibrational Hamiltonian may now be written as

$$\mathcal{H}_\chi = \tfrac{1}{2}\mathbf{P}_\chi'\mathbf{P}_\chi + \tfrac{1}{2}\boldsymbol{\chi}'\mathbf{F}_\chi\boldsymbol{\chi}\,. \tag{8.1}$$

With a slight deviation from Zubarev's notation[2] we write for the retarded and advanced Green's functions of amplitudes

$$N_\chi(\chi_i,\chi_j;tt')_{r,a} = -2\pi\langle\langle\chi_i(t);\chi_j(t')\rangle\rangle_{r,a}\,, \tag{8.2}$$

where

$$\langle\langle\chi_i(t);\chi_j(t')\rangle\rangle_r = i\theta(t-t')\langle[\chi_i(t),\chi_j(t')]\rangle\,, \tag{8.3}$$

$$\langle\langle\chi_i(t);\chi_j(t')\rangle\rangle_a = -i\theta(t'-t)\langle[\chi_i(t),\chi_j(t')]\rangle\,, \tag{8.4}$$

$$[\chi_i,\chi_j] = \chi_i\chi_j - \eta\chi_j\chi_i\,, \tag{8.5}$$

and where θ is the unit step function. A bracket $\langle\ldots\rangle$ signifies the expectation value of an operator averaged over a suitable ensemble. For our purpose a canonical ensemble is the most propitious choice. Furthermore, being interested in the Bose case we set $\eta = +1$. For typographic simplicity we use a system of units in which $\hbar = 1$. The symbols $\chi_i(t)$ and $\chi_j(t')$ denote displacement operators in the Heisenberg picture. Henceforth we shall feel free to refer to the above constructs as generalized compliants. The reason for this apparent abuse of terminology will become clear shortly. Introducing also correlation functions

$$\Sigma_\chi^<(\chi_i, \chi_j; tt') = \langle \chi_j(t')\chi_i(t)\rangle , \tag{8.6}$$

$$\Sigma_\chi^>(\chi_i, \chi_j; tt') = \langle \chi_i(t)\chi_j(t')\rangle , \tag{8.7}$$

we have

$$\langle[\chi_i(t), \chi_j(t')]\rangle = \Sigma_\chi^>(\chi_i, \chi_j; tt') - \Sigma_\chi^<(\chi_i, \chi_j; tt') . \tag{8.8}$$

By drawing on the crystal work of Elliott and Taylor,[3] the thermalized Heisenberg equation for $N_\chi(\chi_i, \chi_j; tt')_{r,a}$ may, in the harmonic case, be solved exactly with the result that

$$\mathbf{N}_\chi(\omega) = -(\mathbf{E}\,\omega^2 - \mathbf{F}_\chi)^{-1} , \tag{8.9}$$

where $\mathbf{N}_\chi(\omega, \mathrm{Im}\,\omega > 0)$ is the time Fourier transform of $\mathbf{N}_\chi(tt')_r$ continued analytically into the upper half-plane, while $\mathbf{N}_\chi(\omega, \mathrm{Im}\,\omega < 0)$ is the time Fourier transform of $\mathbf{N}_\chi(tt')_a$ continued analytically into the lower half-plane. Note that $\mathbf{N}_\chi(\omega = 0)$ coincides with Decius' compliance matrix[4] written in the $\{\chi\}$-basis. We adopt the convention that $\mathbf{N}_\chi$ stands for $\mathbf{N}_\chi(\omega = 0)$, thus regarding the compliance matrix of Decius as a particular case of a double-time thermal Green's function matrix. As for the phrase 'thermal' we hurry to state that it is peculiar to a harmonic system that the temperature dependence disappears. Although we shall not make explicit use of it here, the possibility of considering (Fourier transformed) advanced and retarded Green's functions to be defined on the complex frequency plane is of great convenience. For instance, advanced and retarded Green's functions are frequently recovered from propagators defined at an infinite set of points along (a line parallel to) the imaginary axis.[5,6]

8.3. Connection with Mean-Square Amplitudes

To establish contact with the $\mathbf{\Sigma}$ matrix, as normally defined in molecular spectroscopy,[1] we introduce the spectral intensity matrix by writing

$$\mathbf{\Sigma}_\chi^{<}(tt') = \int_{-\infty}^{\infty} \mathbf{J}_\chi(\omega)\ \exp[-i\omega(t-t')]\,d\omega . \tag{8.10}$$

Then, with recourse to Zubarev's paper,[2] one has

$$\mathbf{N}_\chi(\omega+io) - \mathbf{N}_\chi(\omega-io) = 2\pi i[\exp(\omega/kT)-1]\ \mathbf{J}_\chi(\omega), \tag{8.11}$$

whence follows for equal-time correlation functions that

$$\mathbf{\Sigma}_\chi = \frac{1}{2}\sum_{p=1}^{3N-6} \mathbf{W}_p \mathbf{W}_p' \lambda_p^{-\frac{1}{2}} \coth(\lambda_p^{\frac{1}{2}}/2kT) , \tag{8.12}$$

where $\mathbf{W}_p$ is the p-th column of $\mathbf{W}$ which, in turn, is specified by $\mathbf{W}'\mathbf{F}_\chi\mathbf{W} = \boldsymbol{\lambda}$. In arriving at Eq. (12) one takes into consideration the distribution relation $(x \pm io)^{-1} = \mp i\pi\delta(x) + \mathcal{P}x^{-1}$, where $\mathcal{P}$ denotes a Cauchy principal value. Eq. (12) is readily recognized as the standard $\mathbf{\Sigma}$-equation transformed to $\{\chi\}$. In Chapter 9 we shall put the Green's function concept to use in derivation of alternative expressions involving pure imaginary frequency values.

8.4. Compliants as Response Functions

To reconcile Eqs.(2)-(5) with Decius' original definition[4] in an alternative way that further brings out the significance of $\mathbf{N}_\chi(tt')$ we invoke standard time-dependent perturbation theory.[7] Provisionally we take the system to be in a stationary state $|\psi(t_o)\rangle$ prior to time t_o. Switching on at t_o a weak external disturbance

$$\mathcal{H}' = -\sum_p \chi_p \mathfrak{f}_p(t) , \tag{8.13}$$

the system behaviour at subsequent times will be governed by

$$|\psi(t)\rangle = \hat{T}\exp[-i\int_{t_o}^{t}\mathcal{H}'(t')dt']|\psi(t_o)\rangle\ , \tag{8.14}$$

where $\hat{T}$ is the (positive) Dyson time-ordering operator, $|\psi(t)\rangle$ is the state vector in the interaction representation, and where

$$\mathcal{H}'(t') = -\sum_{p}\chi_p(t')\mathfrak{f}_p(t') \tag{8.15}$$

with

$$\chi_p(t') = \exp[i\mathcal{H}_\chi t']\chi_p\exp[-i\mathcal{H}_\chi t']\ . \tag{8.16}$$

Truncating the series expansion of the time evolution operator afte the first order term we get

$$|\psi(t)\rangle_{\mathrm{I}} = |\psi(t_o)\rangle - i\int_{t_o}^{t}\mathcal{H}'(t')|\psi(t_o)\rangle dt'\ . \tag{8.17}$$

Then, retaining only contributions which are at most linear in $\mathcal{H}'$, and bearing in mind that for a free harmonic molecule $\langle\psi(t_o)|\chi_q|\psi(t_o)\rangle$ vanishes identically, one has for the response of χ_q

$$\langle\psi(t)|\chi_q(t)|\psi(t)\rangle_{\mathrm{I}} = -i\int_{t_o}^{t}\sum_{p}\langle\psi(t_o)|[\chi_q(t),\chi_p(t')\mathfrak{f}_p(t')]|\psi(t_o)\rangle dt'\ . \tag{8.18}$$

Assuming henceforth that the perturbation was turned on adiabatical in the infinite past, and taking a thermal average to allow for the fact that the physical system started at a finite temperature rathe than in a fixed stationary state, we have

$$\langle\chi_q(t)\rangle_{\mathrm{I}} = (2\pi)^{-1}\sum_{p}\int_{-\infty}^{\infty}N_\chi(\chi_q,\chi_p;t-t')_r\,\mathfrak{f}_p(t')dt'\ , \tag{8.19}$$

which expression is typical of the response functions frequently encountered in irreversible thermodynamics.[8] We see from Eq. (19) that the generalized compliants have a rather direct physical significance in view of their role as after-effect functions relati a disturbance to its (linear order) effect on the molecule. After all, when measuring a molecular property we invariably expose the molecule to an external probe and focus on the response. Eq. (19), it should be added, is equally valid in the anharmonic case provided one includes on the right hand side a term $\langle\psi(t_o)|\chi_q|\psi(t_o)\rangle$.

We now specialize to the case of $\mathfrak{f}_p(t')$ exhibiting a simple periodic time dependence, viz.

$$\mathfrak{f}_p(t') = \mathfrak{f}_p \cos(\omega t') e^{\varepsilon t'} \quad ; \quad \varepsilon \to 0+ \tag{8.20}$$

($e^{\varepsilon t'}$ is just an adiabatic turning-on factor[7]), whereupon

$$\langle\chi_q(t)\rangle_{\mathrm{I}} =$$

$$(2\pi)^{-1} \sum_p \mathrm{Re} \int_{-\infty}^{\infty} N_\chi(\chi_q,\chi_p;t-t')_r \exp[(i\omega-\varepsilon)(t-t')]\mathfrak{f}_p \exp[-(i\omega-\varepsilon)t]dt' \tag{8.21}$$

or

$$\langle\boldsymbol{\chi}(t)\rangle_{\mathrm{I}} = -\mathrm{Re}\,\mathbf{N}_\chi(\omega+io)\,\mathfrak{f}\exp[-i(\omega+io)t]. \tag{8.22}$$

If, for the moment, we let the molecule be illuminated by a monochromatic light beam, then we may take $\mathfrak{f}_p \exp[-i\omega t]$ to be a force induced by the electric field. We see that $\mathbf{N}_\chi(\omega)$ is, essentially, the dielectric susceptibility. Leaving the restriction to electric fields and setting $\omega = 0$ to get at the static case we find

$$\langle\boldsymbol{\chi}\rangle_{\mathrm{I}} = \mathbf{N}_\chi \mathfrak{f}, \tag{8.23}$$

which is the 'strain equals compliant times stress' analogue we set out to recover. In the harmonic approximation Eq. (23) concerns, up to a linear transformation, an assembly of independent oscillators; each oscillator being subjected to a constant external force. The resultant effect on an oscillator does not encompass any alteration of its characteristic frequency, but merely entails a force and frequency dependent space displacement of each state, as evidenced by Eq. (23). This translation in configuration space shows up in function space as a rotation caused by an operator the natural logarithm of which is a scalar multiplum of the momentum operator. The static external forces are of course counteracted by distortion induced restoring forces set up in the molecule, and at equilibrium one finds $\mathbf{N}_\chi\mathfrak{f} - \mathbf{F}_\chi^{-1}\mathfrak{f} = 0 \Leftrightarrow \mathbf{N}_\chi = \mathbf{F}_\chi^{-1}$.

In the standard theory[1,4] molecular compliants are defined much the same way compliants are introduced in the continuum theory of elasticity. Imposing the harmonic approximation by demanding that a nonequilibrium gain in potential energy be represented by a quadratic form in generalized forces one finds, for a set of independent generalized coordinates, $\mathbf{N} = \mathbf{F}^{-1}$. Thus, on the face of it there is nothing to distinguish $\mathbf{N}$ from $\mathbf{F}$, save a trivial set of linear equations. Nevertheless, one subsequently proceeds by matrix manipulations to deduce equations (invariance relations and expressions relating $\mathbf{N}$ to $\mathbf{\Sigma}$) which appear to attribute to $\mathbf{N}$ more direc importance than can be ascribed to $\mathbf{F}$. This gain in physical signif cance, apparently obtained by the mere inversion of a matrix, has occasionally precipitated some amazement. We now see, however, that there is no real cause for surprise. When identifying (generalized) compliants with double-time thermal amplitude Green's functions, a close relationship between $\mathbf{N}$ and $\mathbf{\Sigma}$ is (almost) a matter of defini tion. On the other hand, the original relations defining $\mathbf{N}$ turn int a corollary to a well known principle stating that the linear response of a system disturbed from equilibrium is characterized by th expectation value in the equilibrium ensemble of a product of two time dependent operators. This principle, formalized and extended b Kubo,[8] can be traced back, via Callen and Welton,[9] to the Nyquist theorem[10] relating the random noise in an electric circuit to the response elicited by an applied voltage.

8.5. Concluding Remarks

A final comment on the relative merits of $\mathbf{F}$ and $\mathbf{N}$ is in order. The author agrees with Decius' statement "Note that $\mathbf{C}$ [read $\mathbf{N}$] has a perfectly respectable physical interpretation - it is not merely a mathematical artifice".[4] Furthermore, $\mathbf{N}(tt')$, and thereby $\mathbf{N}(\omega)$ and $\mathbf{N}$, retains a perfectly respectable physical interpretation even when the harmonic approximation becomes relaxed. Although $\mathbf{N}(\omega)$ no longer is temperature independent in the anharmonic case, it keeps on governing amplitude-amplitude correlation functions according to Eqs. (10) and (11). Also, the role of $\mathbf{N}(tt')$ as a response function in no way hinges on the potential energy being a quadratic form in displacements. A measureable $\mathbf{F}$, by contrast, can only be approximately defined when anharmonic effects receive recognition. Apart from the indeterminacy encountered when reconstructing a matrix from its eigenvalues, the physical contents of a fit to $\mathbf{F}$, obtained on the basis of measured frequencies, depend on the validity of renormalizing quanta of vibrations. For instance, if the normal modes represent physical occurences which are sufficiently independent one may attribute to each mode its one-body density matrix.[11] A renormalized vibration mode then experiences a time-smeared force field depending on the density matrices of the remaining modes. Assuming self-consistency one enters a quasi-harmonic temperature dependent Hartree approximation implicitly at the heart of most numerical work on vibrating molecules thus far reported.

As for the prospects of applying many-body approximation devices[6,11-15] in numerical calculations on molecular dynamics, we leave this for further contemplation.

Acknowledgment: Valuable discussions with Dr. S.J.Cyvin are gratefully acknowledged.

REFERENCES

1 S.J.Cyvin: Molecular Vibrations and Mean Square Amplitudes, Universitetsforlaget, Oslo, and Elsevier, Amsterdam, 1968.
2 D.N.Zubarev, Sov.Phys.Uspekhi 3, 320 (1960).
3 R.J.Elliott and D.W.Taylor, Proc.Phys.Soc. 83, 189 (1964).
4 J.C.Decius, J.Chem.Phys. 38, 241 (1963).
5 G.Baym and N.D.Mermin, J.Mathm.Phys. 2, 232 (1961).
6 W.Parry and R.Turner, Repts.Prog.Phys. 27, 23 (1964).
7 A.Messiah: Quantum Mechanics, North-Holland, Amsterdam 1962.
8 R.Kubo, J.Phys.Soc.Japan 12, 570 (1957).
9 H.B.Callen and T.A.Welton, Phys.Rev. 83, 34 (1951).
10 H.Nyquist, Phys.Rev. 32, 110 (1928).
11 P.Choquard: The Anharmonic Crystal, Benjamin, New York 1967.
12 A.I.Alekseev, Sov.Phys.Uspekhi 4, 23 (1961).
13 A.A.Maradudin and A.E.Fein, Phys.Rev.128, 2589 (1962).
14 R.A.Cowley, Advan.Phys. 12, 421 (1963).

15 R.D.Mattuck: A Guide to Feynmann Diagrams in the Many-Body Problem, McGraw-Hill, London 1967.

CHAPTER 9

Contribution to the Theory of the Σ Matrix

ØYSTEIN RA

Owing to the advent of fast digital computers the interest in calculation methods for **Σ** avoiding the solution of secular equations waned abruptly a decade ago. However, since at present large molecules and molecular crystals are coming into play, the author deems it timely to resume the search for an alternative recipe. Here, using Cauchy's residue theorem to expand **Σ** in generalized compliants on the imaginary frequency axis and subsequently invoking conformal mapping techniques, we arrive at a scheme which expresses **Σ** as the classical limit augmented with Tchebycheff polynomials in **GF**. Formulae for upper error bounds are included.

9.1. Introduction

From the standpoint of current statistical thermodynamics[1-3] the procedure normally employed to obtain **Σ** is that of integrating the spectral density of displacement correlation functions. Let $\mathbf{N}(z)$; $\mathrm{Im}\, z > 0$, denote the matrix composed of Fourier transforms of retarded Green's functions of displacement continued analytically into the upper half-plane. For $\mathrm{Im}\, z < 0$ we use the advanced thermal Green's function instead. Then the spectral density, and thereby **Σ**, can be calculated from the discontinuity in $\mathbf{N}(z)$ on the real axis. This results in the customary expression for **Σ**. In keeping with a statement made in Chapter 8 we now attempt to uncover further information on **Σ** by relating it to the behaviour of $\mathbf{N}(z)$ for z-values not constricted to the immediate neighbourhood of the real frequency axis.

9.2. Expansion of $\boldsymbol{\Sigma}$ in Residue Matrices

Consider the expression

$$\boldsymbol{\Sigma}_\chi = -\frac{\hbar}{2\pi i}\oint_{C_I}[2\mathcal{N}(z)+1]\,\mathbf{N}_\chi(z)\,dz\ , \qquad (9.1)$$

where C_I signifies a counterclockwise contour encircling all the $3N - 6$ mode frequencies (for a nonlinear molecule) without enclosin any of the poles $\{\omega_n = iy_n = i\,2\pi nkT/\hbar\}$ of the function

$$\mathcal{N}(z) = [\exp(\hbar z/kT) - 1]^{-1}\ . \qquad (9.2)$$

The subscript χ appearing in Eq. (1) refers to a $3N - 6$ dimensiona real coordinate space with $\mathbf{E}$ for its metric tensor. In this basis, which serves the purpose of simplifying formal arguments, the perti nent matrix quantities are obtainable from their well known $\mathbf{S}$-basi counterparts according to

$$\chi = \mathbf{V}^{-1}\mathbf{S}\ , \quad \mathbf{M} = (\mathbf{V}^{-1})^\dagger\mathbf{V}^{-1}\ , \quad \mathbf{G} = \mathbf{V}\mathbf{V}^\dagger\ , \qquad (9.3)$$

$$\mathbf{W} = \mathbf{V}^{-1}\mathbf{L}\ , \quad \mathbf{W}'\mathbf{F}_\chi\mathbf{W} = \mathbf{KGFL} = \lambda\ , \qquad (9.4)$$

$$\boldsymbol{\Sigma}_\chi = \mathbf{V}^{-1}\boldsymbol{\Sigma}(\mathbf{V}^{-1})^\dagger, \quad \mathbf{N}_\chi = \mathbf{V}^{-1}\mathbf{N}(\mathbf{V}^{-1})^\dagger, \quad \mathbf{F}_\chi = \mathbf{V}^\dagger\mathbf{F}\mathbf{V}. \qquad (9.5)$$

As is easy to verify from Cauchy's residue theorem Eq. (1) readily follows from the fact that the harmonic $\mathbf{N}_\chi(z)$ turns out to be the resolvent of $\mathbf{F}_\chi$, viz.

$$\mathbf{N}_\chi(z) = -(z^2\mathbf{E} - \mathbf{F}_\chi)^{-1}\ , \qquad (9.6)$$

and, similarly in terms of symmetry coordinates

$$\mathbf{N}(z) = -(z^2\mathbf{M} - \mathbf{F})^{-1}\ . \qquad (9.7)$$

It is evident that to recover the usual compliance matrix $\mathbf{N}$ we nee only take the value of $\mathbf{N}(z)$ at the origin. Accordingly, we feel justified in occasionally referring to double-time thermal Green's functions of displacement as generalized compliants. To establish contact between Eq. (1) and the usual expression for $\boldsymbol{\Sigma}$, which we write as

$$\boldsymbol{\Sigma}_\chi = \frac{\hbar}{2} \sum_{p=1}^{3N-6} \mathbf{W}_p \mathbf{W}_p' \lambda_p^{-\frac{1}{2}} \coth(\hbar\lambda_p^{\frac{1}{2}}/2kT) \tag{9.8}$$

or

$$\boldsymbol{\Sigma} = \frac{\hbar}{2} \sum_{p=1}^{3N-6} \mathbf{L}_p \mathbf{L}_p^\dagger \lambda_p^{-\frac{1}{2}} \coth(\hbar\lambda_p^{\frac{1}{2}}/2kT) , \tag{9.9}$$

where $\mathbf{W}_p$ stands for the vector formed by the p-th column of $\mathbf{W}$, it suffices to note that contributions from individual modes can be selected on the basis of Eq. (1) by simply breaking C_I into several contours, each one enclosing a $\lambda_p^{\frac{1}{2}}$ on the real axis while including none of the poles of $\mathcal{N}(z)$. However, our primary goal is not the recovery of a standard result of long standing in molecular dynamics. To arrive at additional formulae in supplement to Eqs.(8) and (9), it proves useful to deform C_I differently. Let C_{II}^m designate a circle centered on the origin, and let $R^m = (2\pi kT/\hbar)(m + \frac{1}{2})$ be the radius of C_{II}^m. Then, by choosing m sufficiently large for R^m to exceed the largest $\lambda_p^{\frac{1}{2}}$, we have (by virtue of the Cauchy theorem)

$$\boldsymbol{\Sigma}_\chi - kT\sum_{n=-m}^{m} \mathbf{N}_\chi(\omega_n) = -\frac{\hbar}{2\pi i} \oint_{C_{II}^m} \mathcal{N}(z)\, \mathbf{N}_\chi(z)\, dz , \tag{9.10}$$

where we have taken into consideration that the residue of $\mathcal{N}(z)$ at ω_n is $kT/\hbar$, and also that

$$2\mathcal{N}(z) + 1 = \mathcal{N}(z) - \mathcal{N}(-z) , \tag{9.11}$$

as is apparent from Eq. (2). Finally, bearing in mind that $\mathbf{N}_\chi(z)$ has no branch points or essential singularities anywhere, that $\mathcal{N}(z)$ is bounded at infinity except on the imaginary axis, and that $|z N_\chi(z)_{ij}| \to 0;\ |z| \to \infty$ for all ij, one finds

$$\oint_{C_{II}^m} \mathcal{N}(z)\, \mathbf{N}_\chi(z)\, dz \to 0; \quad m \to \infty , \tag{9.12}$$

and consequently

$$\boldsymbol{\Sigma}_\chi = kT\mathbf{N}_\chi + 2kT\sum_{n=1}^{\infty} \mathbf{N}_\chi(\omega_n) \ , \tag{9.13}$$

which expression is true even if we transform back to $\mathbf{S}$-space and exhibit this by suppressing the subscript χ. Notice that the classical limit emerges here as a residue at the origin. Also, in view of the definition of ω_n, Eq. (13) may well be regarded as a relatic ship between $\boldsymbol{\Sigma}$ and the set of finite-temperature-propagator matric (see Chapter 8). Eq. (13) furnishes the basis for the material of t succeeding sections.

9.3. Upper Bound on $\|\boldsymbol{\Delta}\|$

Judged by empirical evidence* the classical limit $kT\mathbf{N}$ usuall is a poor approximation even for comparatively low frequencies in the temperature range normally of interest.[4] Now we add to the empirical assessment studies by deriving an upper bound on $\boldsymbol{\Delta} = \boldsymbol{\Sigma}_\chi - kT\mathbf{N}_\chi$, or rather $\|\boldsymbol{\Delta}\|$, where for an arbitrary square $n \times n$ $\mathbf{A}$

$$\|\mathbf{A}\| = n \max_{ij} |A_{ij}| . \tag{9.14}$$

Henceforth, when displaying a matrix norm we are considering the cubic vector norm, matrices being viewed as vectors in $n\times n$-dimensional arithmetic space. As is easy to see from Eqs. (3)-(6) one has

$$\mathbf{N}_\chi(\omega_n) = \sum_{p=1}^{3N-6} (y_n^{\,2} + \lambda_p)^{-1}\mathbf{W}_p\mathbf{W}_p' \tag{9.15}$$

and

$$\boldsymbol{\Delta} = 2kT\sum_{n=1}^{\infty}\sum_{p=1}^{3N-6} (y_n^{\,2} + \lambda_p)^{-1}\mathbf{W}_p\mathbf{W}_p' . \tag{9.16}$$

Using the triangle inequality twice in succession, invoking the Cauchy-Bunyakovsky-Schwarz inequality, and bearing in mind that $\mathbf{W}$ is orthogonal, there is no difficulty in establishing that for arbitrary index pair (ij)

---

*In the present paper the phrase 'empirical' generally refers to numerical tes

$$|\Delta_{ij}| \leq \frac{\hbar^2}{2\pi^2 kT} \sum_{n=1}^{\infty} (n^2 + \tau^<)^{-1}, \tag{9.17}$$

where $\tau^<$ is a lower bound on $\{(\hbar/2\pi kT)^2\lambda_p ; p = 1, 3N-6\}$. A first estimate is obtained by equating $\tau^<$ with zero, whereupon

$$|\Delta_{ij}| < (\hbar^2/12k)T^{-1}, \tag{9.18}$$

$$\|\mathbf{\Delta}\| < (3N - 6)(\hbar^2/12k)T^{-1}, \tag{9.19}$$

since $\zeta(2) = \sum_{n=1}^{\infty} n^{-2} = \pi^2/6$.[5] Hence there is an upper bound, which depends only on T, for the $\mathbf{\Sigma}$-deviation from the classical limit. With information on a nonvanishing $\tau^<$ one can improve the estimate by using the partial fraction decomposition

$$\coth(\pi x) = (\pi x)^{-1} + 2(x/\pi) \sum_{n=1}^{\infty} (n^2 + x^2)^{-1} \tag{9.20}$$

to find

$$|\Delta_{ij}| \leq \frac{\hbar^2}{4\pi^2 kT\tau^<}\left[\pi\sqrt{\tau^<}\coth(\pi\sqrt{\tau^<}) - 1\right]. \tag{9.21}$$

By way of illustration we shall dwell briefly on the case of a diatomic molecule. If we take $(\hbar/2\pi kT)^2\lambda$ for $\tau^<$, we can use the equality sign in Eq. (21); the right hand side coincides with $(\hbar/2\lambda^{\frac{1}{2}})\coth(\hbar\lambda^{\frac{1}{2}}/2kT) - (kT/\lambda)$ in this case. This finding may serve as a check on the derivation of Eq. (21). Moreover, identifying the diatomic test molecule with $^{32}S_2$ ($\lambda^{\frac{1}{2}}/2\pi \approx 726\ \text{cm}^{-1}$),[4] application of Eq. (18) yields

$$|\Delta_{11}|/\Sigma_{11} < \hbar\lambda^{\frac{1}{2}}/[6kT\coth(\hbar\lambda^{\frac{1}{2}}/2kT)]$$

$$= 0.86,\ 0.38,\ 0.20,\ 0.12 \tag{9.22}$$

for T = 200, 400, 600, and 800 K, respectively. According to Cyvin's tables[4] the true relative errors at the above temperatures are

$$|\Delta_{11}|/\Sigma_{11} = 0.62,\ 0.34,\ 0.19,\ 0.12. \tag{9.23}$$

From this we gather that the simple formula (18) does not necessarily lead to overly pessimistic estimates of errors incurred by use of the classical limit. However, it may become rather crude even in the diatomic case if $(\hbar/2\pi kT)^2\lambda$ be large. It is obvious from Eq.(1 that both estimates, Eqs. (18) and (21), are better for diagonal $\boldsymbol{\Sigma}$-elements than for off-diagonal ones. When comparing two molecul the upper error bounds are more suitable for the molecule with the lesser spread in frequency values. Indeed, in the extreme case of $3N - 6$ equal frequencies we see directly from Eq. (16) that $\Delta_{ij} = 0$ for $i \neq j$, so that we need the estimates only for diagonal elements. For $i = j$ Eq. (21) approaches an equality relation when $\tau^<$ approaches the unique mode frequency from below.

9.4. Numerical Treatment of Residue Matrices

9.4.1. Introductory Remarks

We now explore the prospect of using the residue formula to circumvent the process of solving secular equations in $\boldsymbol{\Sigma}$-computations pertaining to large molecules of low symmetry. The spectacular advance of computer capability notwithstanding, the implementa tion of a fast procedure aimed at the delivery of a complete set o eigenvalues and eigenvectors is a non-trivial matter for the above mentioned class of molecules. Therefore it seems justified to look for an alternative recipe for $\boldsymbol{\Sigma}$-values. To the author's knowledge most previous work in this direction relates to the series

$$\boldsymbol{\Sigma} = kT\left\{\mathbf{N} + \sum_{m=0}^{\infty} (\hbar/kT)^{2m+2}[(2m+2)!]^{-1}B_{2m+2}(\mathbf{GF})^m\mathbf{G}\right\}, \qquad (9.24$$

where B_{2m+2} are Bernoulli numbers. In the χ-basis Eq. (24) reads

$$\boldsymbol{\Sigma}_\chi = kT\left\{\mathbf{N}_\chi + \sum_{m=0}^{\infty} (\hbar/kT)^{2m+2}[(2m+2)!]^{-1}B_{2m+2}\mathbf{F}_\chi^{\,m}\right\}. \qquad (9.25$$

In the molecular context this expansion has been discussed by Morino et al.,[6] Cyvin,[7] Decius,[8] and by Bartell.[9] However, as point out by Born,[10] it is remarkable that this formula, written as a power series in matrices is already contained in Waller's principa paper.[11] In regard to the practical usefulness of this series it should be noted that for a given molecule convergence is secured only above a certain critical temperature, T_c. In fact, for the series to be summable it is necessary (and sufficient) that all eigenvalues of $(\hbar/2\pi kT)^2\mathbf{GF}$ be less than unity in modulus. This connects with the existence of a finite convergence radius ($|z| < \pi$) for the Maclaurin series for $z\coth z$. The foregoing word of caution, although trivial from a mathematical viewpoint, is not entirely unwarranted here, since a cursory search in the literature brought to light several haphazard applications of Eq. (24). Clearly, divergence trouble detracts from the usefulness of the hyperbolic cotangent Laurent series, the room temperature critical spectral radius of $\mathbf{GF}$ coming out at ~ 1314 cm^{-1} on the wavenumber scale. Moreover, even when T exceeds T_c, convergence usually is not impressive, as brought to attention by Cyvin.[4]

9.4.2. Connection Between the two Expansions of $\boldsymbol{\Sigma}$

When reconsidering the problem with a view to finding a more efficient approach it is instructive to study first the way Eqs. (24), (25) and Eq. (13) are interrelated. This will help us avoid carrying along the pitfalls inherent in the former equations. Assuming for the moment that $T_c < T$, we rewrite Eq. (25) as

$$\boldsymbol{\Sigma}_\chi = kT\left[\mathbf{N}_\chi + 2\sum_{m=0}^{\infty}(\hbar/2\pi kT)^{2m+2}\zeta(2m+2)(-1)^m\mathbf{F}_\chi{}^m\right], \tag{9.26}$$

where $\zeta(x)$ is the Riemann zeta function; whence follows

$$\boldsymbol{\Sigma}_\chi = kT\left[\mathbf{N}_\chi + 2\sum_{m=0}^{\infty}\left(\sum_{n=1}^{\infty}y_n{}^{-2m-2}\right)(-1)^m\mathbf{F}_\chi{}^m\right]. \tag{9.27}$$

On interchanging the order of the two summations we get

$$\boldsymbol{\Sigma}_\chi = kT\left[\mathbf{N}_\chi + 2\sum_{n=1}^{\infty}y_n{}^{-2}\mathbf{C}_n\right] \tag{9.28}$$

with

$$\mathbf{C}_n = \sum_{m=0}^{\infty}(-y_n{}^{-2}\mathbf{F})^m. \tag{9.29}$$

$\mathbf{C}_n$ is readily recognized as a Neumann expansion of the solution to

$$C_n(\mathbf{E} + y_n^{-2}\mathbf{F}_\chi) = \mathbf{E}\,. \tag{9.30}$$

The connection between Eqs. (29) and (30) becomes evident upon iteration of the identity

$$(\mathbf{E} + y_n^{-2}\mathbf{F}_\chi)^{-1} = \mathbf{E} - y_n^{-2}\mathbf{F}_\chi(\mathbf{E} + y_n^{-2}\mathbf{F}_\chi)^{-1}\,. \tag{9.31}$$

The assumption $T_c < T$ guarantees the convergence of the right hand side of Eq. (29). On combining Eq. (28) with Eq. (30) one recovers Eq. (13). The latter relation, therefore, may be regarded as the analytic continuation of Waller's series, the Riemann zeta functio[n] serving as pivot. Conversely, if we regard Eq. (25) as derived fro[m] Eq. (13) it will be recognized that each $\mathbf{N}_\chi(iy_n)$ becomes partitioned and distributed among all the terms of Eq. (25), the dissection being effected by Eq. (29). Hence from this viewpoint the difficulties normally encountered when applying Eq. (25) originate in divergence or slow convergence of geometric progressions associate[d] with low n values. In point of fact, when using truncated versions of Waller's formula at temperatures not very much higher than T_c o[ne] is implicitly treating the leading residue matrices inadequately while handling less important ones with care. Obviously Eq. (29) converges rapidly for sufficiently large n; this means that also when reckoned from the classical limit off-diagonal $\mathbf{\Sigma}_\chi$-elements a[re] determined primarily by the behaviour of $\mathbf{N}_\chi(z)$ for not too large $|z|$. In the diagonal cases large-n residue matrices are relatively more important than for off-diagonal elements; we are nevertheless not at liberty to maim say $\mathbf{N}_\chi(iy_1)$ without properly putting the pieces back together.

9.4.3. Basis for the Numerical Treatment

Turning to the task of casting Eq. (13) into a form which is convenient for numerical computations, we shelve direct inversion procedures at the outset. If applied to a few $\mathbf{N}_\chi(iy_n)$ such approaches would require the same computational volume as called for by a direct diagonalization of $\mathbf{F}_\chi$ or $\mathbf{GF}$. Rather, we would like to preserve the simple structure exhibited by Eqs. (24) and (25). With modest alertness mere multiplication is a tractable process even if large matrices be involved. To sidestep the shortcomings of Eqs. (24) and (25) we take Eq. (30) as a <u>defining</u> relation for $\mathbf{C}_n$ and proceed by rewriting $\mathbf{N}_\chi(iy_n)$ as

$$\mathbf{N}_\chi(iy_n) = y_n^{-2}\mathbf{C}_n = y_n^{-2}\hat{P}_n^{-1}(\mathbf{E}-\mathbf{Y}_n)^{-1}, \tag{9.32}$$

where

$$\mathbf{Y}_n = \hat{P}_n^{-1}[-y_n^{-2}\mathbf{F}_\chi + \mathbf{E}\{\hat{P}_n - 1\}] \tag{9.33}$$

and

$$\hat{P}_n = (2n^2+\tau^>+\tau^<)/(2n^2);\quad \tau_n^> = n^{-2}\tau^>,\quad \tau_n^< = n^{-2}\tau^<. \tag{9.34}$$

Here $\tau^>$ is an upper bound on $\{(\hbar/2\pi kT)^2\lambda_p;\ p = 1, 3N-6\}$. Since, in keeping with our previous definition, $\tau^<$ is a lower bound on the same set, clearly $\tau_n^>$ and $\tau_n^<$ are upper and lower bounds on the spectrum of $y_n^{-2}\mathbf{F}_\chi$. For the above defined matrix $\mathbf{Y}_n$ we have upper and lower bounds ε_n^{-1} and $-\varepsilon_n^{-1}$ on the spectrum $\{\mu_p^n;\ p = 1, 3N-6\}$, where $\varepsilon_n^{-1} = (\tau^>-\tau^<)/(2n^2+\tau^>+\tau^<) < 1$, and $\mu_p^n = 1 - \hat{P}_n^{-1}(1+y_n^{-2}\lambda_p)$. Consequently unity always exceeds the spectral radius of $\mathbf{Y}_n$ and, in contradistinction to Eq. (29), we are always justified in writing

$$\mathbf{C}_n = \hat{P}_n^{-1}\{\mathbf{E} + \mathbf{Y}_n + \mathbf{Y}_n^2 + \ldots + \mathbf{Y}_n^k + \ldots\}. \tag{9.35}$$

Thus, by replacing $-y_n^{-2}\mathbf{F}_\chi$ with $\mathbf{Y}_n$ we simultaneously translate and compress the spectrum to an extent sufficing for a disposal of divergence trouble.

At this point a simple case study is in order. Let us return to the diatomic case, choosing this time $^{16}O_2$ ($\lambda^{\frac{1}{2}}/2\pi \approx 1580\ \mathrm{cm}^{-1}$) at 200 K. Normally, only sheer obstinacy could make us refrain from solving the diatomic problem directly. However, at present O_2 serves to throw into relief the difference between Eq. (29) and Eq. (35). From Cyvin's tabulation[4] one sees that for O_2 at 200 K Eq. (24) produces results which markedly deteriorate upon increasing the number of included terms. This comes as no surprise, though, since for O_2 $T_c \approx 362$ K; the $n = 1$ version of Eq. (29) misbehaving as follows:

$$\mathbf{N}_\chi(iy_1) = N_\chi(iy_1) =$$

$$y_1^{-2}\{1 - 3.27 + 10.71 - 35.06 + 114.76 - \ldots\}. \tag{9.36}$$

The series (35), on the other hand, comes out like

$$\mathbf{C}_1 = C_1 = \hat{P}_1^{-1}\{1 + 0 + 0 + \ldots + 0 + \ldots\} \qquad (9.37)$$

and truncates itself as soon as the exact result $C_1 = \hat{P}_1^{-1}$ has been secured. In principle, this feature does not hinge on the molecule being diatomic, since, by construction, Eq. (35) always turns self-truncating when the spectrum condenses to a point. However, a one-point spectrum ($3N - 6$ equal frequencies) is hardly ever met with for polyatomic molecules, so that one would generally have to include more than one summand in Eq. (35). Nevertheless, although the ideal self-truncation exhibited by Eq. (37) is atypical, this examp does illustrate the convenience of a scheme which compresses the 'effective' spectrum in order to improve a width-depending convergence rate. By contrast, Eqs. (24) and (25) judge the situation by the maximum frequency value, and therefore go astray even in the simple O_2 case.

Despite the general validity of $\varepsilon_n^{-1} < 1$ Eq. (35) is not entirely satisfactory. After all, for an ε_n^{-1} close to 1 [Eqs.(24), (25) would be strongly divergent in such a case] use of this relation is analogous to a Maclaurin series calculation of $(1 - z)^{-1}$ in a case where z is barely inside the circle of convergence. Evidentl this is not an expedient way of computing $(1 - z)^{-1}$ for the z-value in mind. Since, for reasons previously detailed, it is inconvenient to resort to direct inversion in our matrix cases, we seek to remed the situation by altering the way of adding together the powers of $\mathbf{Y}_n$. Therefore, to find a polynomial substitute for Eq. (35), let

$$z_n(w) = \alpha_1^n w + \alpha_2^n w^2 + \ldots \qquad (9.38)$$

be a function which is meromorphic in $\{w : |w| < 1\}$, which assumes the value 1 for some $\eta_n \in \{w : |w| < 1\}$, and which has the property that if $\mathcal{D}^-$ be defined by $z_n(w) : \{w : |w| < 1\} \to \mathcal{D}^-$ then $\mathcal{D}^- \cap \{(\mu_p^n)^{-1}\} = \emptyset$. The elements of the matrix valued function $[\mathbf{E} - z_n(w)\mathbf{Y}_n]^{-1}$ will be regular functions of w in $\{w : |w| < 1\}$, and we may write

$$(\mathbf{E} - \mathbf{Y}_n)^{-1} = \mathbf{E} + \eta_n e_1^n(\mathbf{Y}_n) + \eta_n^2 e_2^n(\mathbf{Y}_n) + \ldots , \qquad (9.39)$$

where $e_i^n(\mathbf{Y}_n)$ is an i-th degree polynomial in $\mathbf{Y}_n$:

$$e_i^n(\mathbf{Y}_n) = \sum_{j=0}^{i} \gamma(n,i,j)\mathbf{Y}_n^j . \tag{9.40}$$

The convergence rate of Eq. (39) increases with decreasing η_n. Accordingly we look for a function $z_n(w)$ which takes on the value 1 for some small η_n. Having tacitly formulated the problem in a way which makes it very nearly coincide with the one being studied in the theory of $\mathcal{S}$- universal algorithms for systems of linear equations,[12,13] we may now lean on Kublanovskaya's result[12] to the effect that given a bounded covering $\mathcal{S}^n$ of $\{\mu_p^n;\ p = 1, 3N-6\}$ so that $\tilde{\mathcal{S}}^n$ <u>is simply connected and</u> $1 \in \tilde{\mathcal{S}}^n$; then, if $z^{-1}: \tilde{\mathcal{S}}^n \rightarrow \mathcal{D}^n$, one has that $\mathcal{D}^n$ <u>is simply connected</u>, $0 \in \mathcal{D}^n$, $1 \in \mathcal{D}^n$, $\{(\mu_p^n)^{-1}\} \cap \mathcal{D}^n = \emptyset$. Also, if $z_n(w): \{w : |w| < 1\} \rightarrow \mathcal{D}^- \subset \mathcal{D}^n$ be conformal, then the associated $|\eta_n|$ is the least common $|\eta_n|$ for all $\mathbf{Y}_n$ the spectra of which are in $\mathcal{S}^n$. The choice of $\mathcal{S}^n$ and a conformal mapping $z_n(w)$ remains. Firstly, with recourse to Kublanovskaya's work[12] it can be seen that if we take for $\mathcal{S}^n$ a circular disk of which the radius is ε_n^{-1}, and for $z_n(w)$ the function $z_n(w) = \varepsilon_n w$, then $\eta_n = \varepsilon_n^{-1}$, and we obtain the geometric series (35). Thus the latter expansion corresponds to treating $\mathbf{Y}_n$ as if it were a scalar multiplum of a unitary matrix. But $\mathbf{Y}_n$ is Hermitian (it is symmetric, in fact), and all its eigenvalues belong to the interval $(-\varepsilon_n^{-1}, \varepsilon_n^{-1})$. Consequently the natural thing to do is to pick a covering which is, in some sense, smaller* than the circular disk. We simply choose $(-\varepsilon_n^{-1}, \varepsilon_n^{-1})$. Attempts to further diminish the size of $\mathcal{S}^n$ [for instance, one might think of a collection of disjoint intervals inside $(-\varepsilon_n^{-1}, \varepsilon_n^{-1})$] would entail multiply connected sets $\tilde{\mathcal{S}}^n$, $\mathcal{D}^n$. This, in turn, would necessitate a mapping of the unit circle into a Riemann sheet covering of $\mathcal{D}^n$. Apart from such a mapping being likely to submerge us in the intricacies of elliptic functions, the gain in further reducing the measure of $\mathcal{S}^n$ would probably be largely offset by the fact that the conformal mapping method, as

---

* To state precisely the meaning of 'smaller' in the present context would carry us too far afield. Suffice it merely to say that we regard any line or line segment as smaller than every non-meagre subset of the complex plane.

such, is less efficient for multiply connected $\mathscr{D}^n$'s.[12,13] We therefore settle for $(-\varepsilon_n^{-1}, \varepsilon_n^{-1}) = \mathscr{S}^n$. Parenthetically, this gives us a scheme which comes close to the so-called universal trinomial algorithm with constant multiplier.[13] The foregoing $\mathscr{S}^n$ is compatible with

$$z_n(w) = 2\varepsilon_n w/(1+w^2), \quad \eta_n = \varepsilon_n - (\varepsilon_n^2-1)^{\frac{1}{2}}, \tag{9.41}$$

$\mathscr{D}^n$ being now the entire plane cut on the real axis from $-\varepsilon_n$ and $+\varepsilon_n$ to $-\infty$ and $+\infty$, respectively. Although Kublanovskaya lists the mapping (41) we could not find in her work explicit information on the coefficients denoted $\gamma(n,i,j)$ by us. However, taking refuge in the general rule for constructing recurrence relations[13] one easily finds

$$e_1^n(Y_n) = 2\varepsilon_n Y_n, \quad e_2^n(Y_n) = (2\varepsilon_n)^2 Y_n^2,$$

$$e_i^n(Y_n) = 2\varepsilon_n Y_n e_{i-1}^n(Y_n) - e_{i-2}^n(Y_n); \quad i > 2, \tag{9.42}$$

which, at once, points to an $e_i^n(Y_n)$ akin to some Jacobi-type polynomial. Indeed, from Eq. (42) one sees that all the $e_i^n(Y_n)$ are divisible by $2\varepsilon_n Y_n$; whence follows that if we define $U_i^n(\varepsilon_n Y_n)$ by writing

$$e_{i+1}^n(Y_n) = 2\varepsilon_n Y_n U_i^n(\varepsilon_n Y_n), \tag{9.43}$$

then

$$U_0^n(\varepsilon_n Y_n) = \mathbf{E}, \quad U_1^n(\varepsilon_n Y_n) = 2\varepsilon_n Y_n,$$

$$U_{i+1}^n(\varepsilon_n Y_n) = 2\varepsilon_n Y_n U_i^n(\varepsilon_n Y_n) - U_{i-1}^n(\varepsilon_n Y_n); \quad i \geq 1, \tag{9.44}$$

which shows that the $U_i^n(\varepsilon_n Y_n)$ are simply Tchebycheff polynomials of the second kind with the matrix $\varepsilon_n Y_n$ substituted for a scalar argument. This identification is readily vindicated by noting that Eq. (42) implies:

$$\text{odd } i \Rightarrow \gamma(n,i,j)/(2\varepsilon_n)^j = \begin{cases} 0 \text{ for even } j \\ 1 \text{ for } i=j=1 \\ (-1)^{[i+j-2]/2}\binom{[i+j-2]/2}{j-1} \text{ otherwise} \end{cases} \tag{9.45}$$

$$\text{even } i \Rightarrow \gamma(n,i,j)/(2\varepsilon_n)^j = \begin{cases} 0 \text{ for odd } j \\ 0 \text{ for } j = 0 \\ (-1)^{[i+j]/2} \binom{[i+j-2]/2}{j-1} \text{ otherwise} \end{cases} \tag{9.46}$$

On the practical level it is profitable to have some estimate of $\|\mathbf{L}_n{}^k\|$, where

$$\mathbf{L}_n{}^k = \sum_{i=k}^{\infty} \eta_n{}^i e_i{}^n(\mathbf{Y}_n) \ , \tag{9.47}$$

since this will permit us to determine beforehand the number of terms to be included in Eq. (39). Expanding $\mathbf{L}_n{}^k$ on the set of projectors of $\mathbf{Y}_n$ we find

$$\mathbf{L}_n{}^k = \sum_{i=k}^{\infty} \eta_n{}^i \sum_{p=1}^{3N-6} e_i{}^n(\mu_p{}^n)\mathbf{W}_p\mathbf{W}_p' \tag{9.48}$$

or

$$\mathbf{L}_n{}^k = \sum_{i=k}^{\infty} \eta_n{}^i \sum_{p=1}^{3N-6} e_i{}^n(|\mu_p{}^n|)\,\mathrm{sgn}(\mu_p{}^n)\,\mathbf{W}_p\mathbf{W}_p' . \tag{9.49}$$

To simplify the inner summation it is helpful to consider for a moment the one-dimensional oscillator version of Eq. (39),

$$\frac{\varepsilon_n}{\varepsilon_n - 1} = 1 + \sum_{i=1}^{\infty} \eta_n{}^i \sum_{j=0}^{i} \gamma(n,i,j)\,\varepsilon_n{}^{-j} . \tag{9.50}$$

It is fairly evident that

$$\frac{2\eta_n}{(1-\eta_n)^2} + 1 = 1 + \sum_{i=1}^{\infty} 2i\eta_n{}^i . \tag{9.51}$$

Also, it is a simple corollary to Eq. (41) that

$$\frac{\varepsilon_n}{\varepsilon_n - 1} = \frac{2\eta_n}{(1-\eta_n)^2} + 1 \ . \tag{9.52}$$

Bearing in mind the uniqueness of power series representations we conclude that

$$\sum_{j=0}^{i} \gamma(n,i,j)\varepsilon_n^{-j} = 2i. \tag{9.53}$$

We note in passing that Eq. (53) further corroborates the connectio between the e_i^n and Tchebycheff polynomials of the second kind sinc $U_i(1) = i+1$. Moreover since[5] $|U_i(x)| \le U_i(1)$ $(-1 \le x \le 1)$ and, by construction, $|\mu_p^n| \le \varepsilon_n^{-1}$, Eq. (43) along with (53) implies that $|e_i^n(|\mu_p^n|)|$ is bounded by $2i$ for all p. Consequently, reverting to Eq. (49) and calling to mind the Cauchy-Bunyakovsky-Schwarz inequality we find

$$\|\mathbf{L}_n^k\| \le (3N-6)\sum_{i=k}^{\infty} \eta_n^i(2i) \tag{9.54}$$

and

$$\|(\mathbf{E}-\mathbf{Y}_n)^{-1} - \mathbf{E} - \eta_n e_1^n(\mathbf{Y}_n) - \ldots - \eta_n^{k-1}e_{k-1}^n(\mathbf{Y}_n)\| \le$$

$$(3N-6)2k\eta_n^k[1-\eta_n(k+1)/k]^{-1} \tag{9.55}$$

for k sufficiently large that $\eta_n(k+1)/k < 1$. In other words, after a certain number of terms the number of correct decimal figures obtained in using Eq. (39) increases, essentially, in geometric progression; the series being characterized by a constant factor η_n. As is easily seen from Eq. (41) $\eta_n < \varepsilon_n^{-1}$ for $\varepsilon_n > 1$, which indicates that Eq. (39) is superior to Eq. (35). Furthermore we not that $(3N-6)\varepsilon_n/(\varepsilon_n - 1)$ constitues an upper bound on $\|(\mathbf{E}-\mathbf{Y}_n)^{-1}\|$. It should be emphasized that $|U_i(x)|$ rises steeply whenever $|x|$ become larger than 1, so that care should be taken to avoid using an overl optimistic estimate of ε_n. Also it is important that $\gamma(n,i,j)$ be calculated correctly. For checking purposes we supplement Eqs. (45),(46), and (53) with the formula

$$\sum_{j=0}^{i-1} \gamma(n, i+j, i-j)(2\varepsilon_n)^{-i+j} = 0, \tag{9.56}$$

which simply expresses the fact that a summation over the base of a Pascal triangle, the base having been modified so as to contain numbers which alternate in sign, yields zero.

We now use Eq. (39) in Eq. (32). Subsequently we insert the result into Eq. (13) to find

$$\mathbf{\Sigma}_\chi = kT\{\mathbf{N}_\chi + 2(\hbar/2\pi kT)^2\sum_{n=1}^{\infty}\hat{P}_n^{\,-1}n^{-2}$$

$$+ 2\sum_{i=1}^{\infty}[\sum_{r=0}^{i}\psi(\tau^>,\tau^<,i,r)(\hbar/2\pi kT)^{2r+2}(-1)^r\mathbf{F}_\chi^{\,r}]\}, \qquad (9.57)$$

where

$$\psi(\tau^>,\tau^<,i,r) =$$

$$2\{(\tau^>+\tau^<)/2\}^{-r}\{\sum_{j=r}^{i}\gamma(n,i,j)\varepsilon_n^{-j}\left[\frac{\tau^>+\tau^<}{\tau^>-\tau^<}\right]^j\binom{j}{j-r}\}\,\phi(\tau^>,\tau^<,i) \qquad (9.58)$$

and

$$\phi(\tau^>,\tau^<,i) = \sum_{n=1}^{\infty}\eta_n^{\,i}/(2n^2+\tau^>+\tau^<). \qquad (9.59)$$

Finally, transforming back to the $\mathbf{S}$-basis we have

$$\mathbf{\Sigma} = kT\{\mathbf{N} + 2(\hbar/2\pi kT)^2[\sum_{n=1}^{\infty}\hat{P}_n^{\,-1}n^{-2}]\mathbf{G}$$

$$+ 2\sum_{i=1}^{\infty}[\sum_{r=0}^{i}\psi(\tau^>,\tau^<,i,r)(\hbar/2\pi kT)^{2r+2}(-1)^r(\mathbf{GF})^r\mathbf{G}]\}, \qquad (9.60)$$

or, preferably in a symmetrized form, e.g.

$$\mathbf{\Sigma} = kT\{\mathbf{N} + 2(\hbar/2\pi kT)^2[\sum_{n=1}^{\infty}\hat{P}_n^{\,-1}n^{-2}]\mathbf{G}$$

$$+ 2\mathbf{G}^{\frac{1}{2}}\sum_{i=1}^{\infty}[\sum_{r=0}^{i}\psi(\tau^>,\tau^<,i,r)(\hbar/2\pi kT)^{2r+2}(-1)^r(\mathbf{G}^{\frac{1}{2}}\mathbf{F}\,\mathbf{G}^{\frac{1}{2}})^r]\mathbf{G}^{\frac{1}{2}}\} \qquad (9.61)$$

or

$$\mathbf{\Sigma} = kT\{\mathbf{N} + 2(\hbar/2\pi kT)^2[\sum_{n=1}^{\infty}\hat{P}_n^{\,-1}n^{-2}]\mathbf{G}$$

$$+ 2\mathbf{G}\mathbf{F}^{\frac{1}{2}}\sum_{i=1}^{\infty}[\sum_{r=0}^{i}\psi(\tau^>,\tau^<,i,r)(\hbar/2\pi kT)^{2r+2}(-1)^r(\mathbf{F}^{\frac{1}{2}}\mathbf{G}\,\mathbf{F}^{\frac{1}{2}})^r]\mathbf{F}^{-\frac{1}{2}}\}.$$

$$(9.62)$$

Eqs. (60)-(62), it is believed, are preferable to Eq. (24). To evaluate bounds on errors caused by truncating the summation over i one may apply a trivial modification of Eq. (21) in conjunction with Eqs. (54) and (55).

9.5. Concluding Remarks and Comparison with the Morino-Cyvin Approach

As an expedient way of summarizing the features of the above procedure (Section 9.4.3) we compare it to a method first set fort by Morino et al.[6] and later elaborated by Cyvin.[7] In brief, the gist of the Morino-Cyvin approach (MC) is to fit coth(x) on some interval containing all the (temperature weighted) frequencies wit a polynomial in x, and subsequently to insert the result into Eq. (9). Both positive and negative powers of x are admitted. As imple- mented by Cyvin[7] the unknown coefficients are determined numerical by the method of least-squares. Apart from 'a least absolute maxim deviation fit to x^{-1}coth(x)' being more helpful than 'a minimized squared mean deviation approximation to coth(x)' when assessment o uncertainties in amplitudes is called for [see Eqs. (8),(9)]; MC i to some extent incomplete in that it presupposes some ingenuity an quite a lot of preparatory work on the part of the user. This lack becomes aggravated by the fact that in cases where use of non-spectral methods (i.e. methods which avoid the solution of secular equations) is warranted wide intervals ($\tau^{<}$, $\tau^{>}$) are apt to occur. Large molecules, it will be recalled, usually have a large spread in mode frequencies. This, in turn, makes it imperative to optimiz curve-fits in order that inordinate large numbers of terms be avoi ded. MC would gain in practical value, no doubt, if one could remo some latitude by supplying: (i) A criterion which, from the requir accuracy in Σ and the available information of ($\tau^{<}$, $\tau^{>}$), would fi in advance the number of positive and the number of negative powers to be included. (ii) A determinate rather than a numerically defin procedure for generation of coefficients. When interpreted from a curve-fitting viewpoint the approach of the present work comes clo to furnishing just this reinforcement of MC. To bring about conta between the preceding paragraphs and the point of view adopted by

Morino et al. and by Cyvin[7] we note that Eq. (13) can be rederived by using Eq. (20) in Eq. (8) since x^{-2} and $(n^2+x^2)^{-1}$, by virtue of their being analytic in simply connected regions containing all x^2-values corresponding to mode frequencies, are admissable functions for $\mathbf{F}_\chi$ and $\mathbf{GF}$. Thus, the present method represents $x^{-1}\coth(x)$ by treating the x^{-2} term exactly and by simulating bounded and well-behaved partial fractions by linear combinations of Tchebycheff polynomials, the arguments of which are reckoned from adjusted origins. On the technical level the advantage of the present approach over MC stems from the fact that the constituent parts of $x^{-1}\coth(x)$ (partial fractions) are amenable to direct and detailed scrutiny, which the composite transcendental itself is not.

It might be argued that the present method lacks in flexibility since we have thus far left out other negative matrix powers than the classical limit. This is easily remedied. A simple stratagem is provided by the identity

$$(\mathbf{E} + y_n^{-2}\mathbf{F}_\chi)^{-1} = \mathbf{E} - (\mathbf{E} + y_n^2\mathbf{N}_\chi)^{-1}. \qquad (9.63)$$

By going back to Eq. (32) and subjecting $(\mathbf{E}+y_n^2\mathbf{N}_\chi)^{-1}$, for a certain set of n-values, to a treatment identical to the one given to $(\mathbf{E}+y_n^{-2}\mathbf{F}_\chi)^{-1}$, the resulting counterparts of Eqs. (60)-(62) will contain any negative power desired. Since the entire derivation carries over with obvious modifications it is superfluous to elaborate this point here, save to remark that the necessary condition for an introduction of Eq. (63) to make sense is that

$$\tau_n^{<} \in ([\tau_n^{>}]^{-1},\ \tau_n^{>}); \qquad (9.64)$$

only then would convergence become accelerated. Presumably, this could happen only for $n = 1$. If so, the $\mathbf{N}_\chi(iy_1)$ residue alone would call for special treatment which might be rendered in the form of direct numerical application of the recurrence relations (42). From the properties of Tchebycheff polynomials of the first and second kind[5] we do not expect any pathological accretion of round-off errors from forward recursion in this case.

To substantiate our previous contention to the effect that the present method is capable of predicting beforehand coefficients which are equivalent to those obtained empirically within the framework of MC we examine briefly the original formula of Morino et al.

$$\coth(x) \approx x^{-1} + 0.25x \, , \tag{9.65}$$

which, at room temperature, is intended to cover molecules whose frequencies range from 0 to 1200 cm^{-1}. Cyvin's empirically adjusted version[7] very nearly coincides with

$$\coth(x) \approx x^{-1} + 0.267x \, , \tag{9.66}$$

only negligibly small terms having been left out. If in Eq. (62) we discard all but the two first contributions and calculate $\left[\sum_{n=1}^{\infty} \hat{P}_n^{-1} n^{-2}\right]$ for the $(\tau^<, \tau^>)$ region in question [while making use of the result $\zeta(2) = \pi^2/6$], we do in fact obtain $\coth(x) \approx x^{-1} + 0.267x$.

Although the present work has been designed for molecules, large mass defected molecules in particular, it may possibly acquire some relevance in crystal dynamics as well. For instance, when computing lattice Debye-Waller factors one normally diagonalizes the (space) Fourier transformed force constant matrix for quite a few otherwise noninteresting wave vector values. Evidently the present method can be applied throughout dull regions in the Brillouin zone without further ado; the subspace of crystal displacement space spanned by discrete plane wave analogues of fixed $\mathbf{k}$ taking the place of the manifold of molecular amplitude. Conceivably it could be advantageous to generate $\mathbf{N}_k$ and some $\mathbf{N}_k(iy_n)$ by Hotelling iteration[13] from the corresponding matrices at an adjacent $\mathbf{k}$, while handling the sum of remaining residue matrices by the present method.

Acknowledgment: The author wishes to thank Dr. S.J.Cyvin for valuable discussions.

REFERENCES

1 D.N.Zubarev, Sov.Phys.Uspekhi 3, 320 (1960).

2 L.P.Kadanoff and G.Baym: Quantum Statistical Mechanics, Benjamin New York 1962.

3 W.Parry and R.Turner, Repts.Prog.Phys. 27, 23 (1964).
4 S.J.Cyvin: Molecular Vibrations and Mean Square Amplitudes, Universitetsforlaget, Oslo, and Elsevier, Amsterdam, 1968.
5 M.Abramowitz and I.A.Segun: Handbook of Mathematical Tables, Dover, New York 1965.
6 Y.Morino, K.Kuchitsu, A.Takahashi, and K.Maeda, J.Chem.Phys. 21, 1927 (1953).
7 S.J.Cyvin, Spectrochim.Acta 15, 56 (1959).
8 J.C.Decius, J.Chem.Phys. 38, 241 (1963).
9 L.S.Bartell, J.Chem.Phys. 42, 1681 (1965).
10 M.Born, Repts.Prog.Phys. 9, 249 (1942-43).
11 I.Waller, Dissertation, Uppsala 1925.
12 V.N.Kublanovskaya, Tr.Matem. in-ta AN SSSR 53, 145 (1959).
13 D.K.Faddeev and V.N.Faddeeva: Computational Methods of Linear Algebra, Freeman, San Francisco 1963.

CHAPTER 10

Experimental Errors in Gas Electron Diffraction

KOZO KUCHITSU

Principal sources and possible magnitudes of systematic and random experimental errors involved in the structural parameters determined by gas electron diffraction are discussed.

気体電子線回折の実験誤差

朽 津 耕 三

東京大学理学部化学教室

気体電子線回折の実験によって決定される分子構造に関するパラメーターに含まれる系統誤差と偶発誤差について，その主要な原因と大きさの程度を考察した。

10.1. Introduction

A structure study by gas electron diffraction is an extremel difficult experiment, being liable to more serious errors than is generally thought by non-specialists,[1] since it is based on a delicate measurement of relative intensities of scattered electron Careful consideration is necessary in order to eliminate or evalua errors from various experimental and theoretical sources. The pres chapter gives a general list of the principal sources of systemati errors, and then examines their disturbance on the measurements of each structural parameter. The description has to be less quantitative in order to be more general. A number of reference papers have been published on this subject with particular attention to individual apparatus, methods and samples.[1-22]

10.2. Principal Sources of Systematic Errors

10.2.1. Introductory Remarks

Systematic errors in our measurements arise mainly from the following origins. (Item 3 does not apply to a nonphotographic measurement[23] of electron intensity.*) (1) Sample impurity. (2) Scattering process. (3) Photographic process. (4) Measurement of scattered intensity. (5) Measurement of scale factor. (6) Theoretical interpretations. - Generally speaking, the errors which disturb the periods of molecular oscillations in the scattered intensity have much influence on the determination of internuclear distances and phase parameters, while mean amplitudes and indices of resolution are sensitive to the errors which deteriorate relative intensities.

10.2.2. Sample Impurity

The presence of impurity of unknown origin and quantity is a serious source of error. In a very fortunate case, an impurity can be detected and eliminated in the stage of analysis of diffraction intensities. This happens when a contaminating molecule has internuclear distances which are very different from any of the internuclear distances of the molecule under investigation. For example, an appreciable amount of a nitrogen impurity in a bromine gas can be discovered by a radial distribution analysis, and hence,

*A critical comparison of the data obtained by photographic and nonphotographic methods was made by Bonham and Fink.[23] They matched the total intensities for N_2 and O_2 (40 kV, $0.5 < s < 8$) with the corresponding absolute intensities (37 kV) measured by a counting method using a scintillator and a photomultiplier. The photographic data reproduced the shape of the absolute intensities with maximum deviations of less than 10% and average deviations of around 5%, which were comparable with the uncertainties in the counting data.

it may not cause a serious problem. In most other cases, however, all the impurities must be known quantitatively, say to one percen or less. (As is well known, the limit of resolving power of distan ces lies in the thermal motion of atoms, which cannot be controled experimentally, and this limits the application of gas electron diffraction to chemical analysis.) If impurities are identified, a correction can be made on the measured total intensity by subtracting the contributions from the intensities of the impurities, whi can be estimated either experimentally or theoretically.[24] The precision required for such a chemical analysis depends on the sca tering power of the impurity relative to that of the sample. When the sample has several heavy atoms, contamination of relatively weak scatterers (water, air, etc.) may not cause much trouble. On the other hand, extreme care has to be taken when the sample being examined is a weak scatterer. A possibility of chemical change of the sample and an air leak into the sample holder must also be tak into account in a similar fashion. For measurements of structural parameters in the presence of appreciable impurities, see, for example, Refs. 25 to 28, where the structures of S_2Br_2, B_2Cl_4, B_2Cl_4, and tri-t-butylmethane were determined when Br_2, BCl_3, SiCl and air, respectively, were present in the samples.

10.2.3. Scattering Process

10.2.3.1. Beam Characteristics

A primary electron beam with a diameter of a few hundredths of one millimeter on a detector can be regarded as tolerable; no correction for its size effect seems to be necessary for ordinary purposes.[6,7] An experiment with a much less focused beam would nee a correction by means of a proper 'slit function'. The beam must also be free from unnecessary electric or magnetic disturbance.[7] An electrostatic charge, which may be built up on an insulator sur face hit by electrons, a magnetostatic field from a permanent magnetization placed near the electron beam (e.g., from a ball-bearing race of a rotating sector),[29,30] and an alternating magnet field produced by an electric current around the diffraction unit are the most likely sources of such disturbance.

10.2.3.2. Extraneous Radiation

In an ideal experiment, nothing except those electrons which are scattered from a point source of sample gas molecules should fall onto the detector. If the detector is a photographic plate, it should be protected from visible or invisible light, X-rays, and extraneous electrons. Such extraneous radiation can be eliminated or decreased if a counting technique[17,23] is used with a good angular resolution [cf. S.Konaka, Jap.J.Appl.Phys. (in press)]. In addition to ordinary light sources (leaks of room light through a viewing window, through a camera box, etc.), a fluorescent substance placed in the unit, or a sample furnace heated to several hundred degrees may become sources of stray light. X-rays may emit from a metal plate bombarded by a strong electron beam.

Possible origins of stray electrons may be grouped into two categories: The first group consists of those extraneous electrons which are detected even without specimen injection. The primary electrons scattered by metal surfaces (slit edges and other pieces of instruments in the diffraction unit, including its walls) are the most probable sources. When the pressure of the chamber is higher than say 1×10^{-3} Torr, electrons scattered by residual air may also contribute. It is relatively easy, however, to keep the intensity of stray electrons of these kinds negligibly small by making a proper adjustment of the electron-optical and vacuum systems. Even if the effect is not negligible, it is possible to make a correction by making blank measurements.

10.2.3.3. Extraneous Scattering with Sample Injection (I). Scattering by Delocalized Samples

The extraneous scattering of the second group occurs in association with specimen injection. There is a finite probability that undiffracted or diffracted electrons are scattered by the sample molecules diffused into the diffraction chamber. The extent of this effect depends on the sample flow rate and the efficiency of the pumping system, and in particular, that of the cold trap. Most part of this effect, which usually decreases roughly in proportion to s^{-3}, can be corrected for at least approximately if the distribution function of the gas can be estimated experimentally.

This estimation can be made, for instance, by one of the following measurements: a measurement of the pressure in the diffraction chamber, the sample flow rate and the pumping speed,[12] a shadow microscope of the emitted gas,[6,7] extinction of the beam intensity on passing through the delocalized gas, a diffuse shadow cast by a sharp-cut diaphragm placed between the diffraction point and the photographic plate,[11,31] and the diffraction intensity of the delocalized gas.[6,7,12,32]

As one of the features of the effect in question, suppose tha the undeflected electron beam is scattered at a place different fr the defined diffraction point and somewhere between the nozzle and the detector by those sample molecules which are spread uniformly i the chamber. The influence of this 'delocalized-gas scattering' is relatively important when the sample is noncondensable at liquid-nitrogen temperature.[12] As a result, measured internuclear distance appear apparently longer than the true distances.[33] On the other hand, when the design of the nozzle is such that the emitted gas mo cules swarm around the nozzle forming an exponential or a Gaussian density distribution, the primary electrons scattered by such a distribution should cause the observed mean amplitudes to be larger than the true values.[11,29]

A secondary scattering of diffracted electrons by another molecule (intermolecular multiple scattering) is also expected to take place. Such a process can be dealt with theoretically by a formula similar to the Debye-Menke equation of the X-ray scattering from liquids in terms of the intermolecular potential function.[34] Except in a case of extreme density[35] however, such an intermolecular multiple scattering contributes mainly to a smooth background intensity[6] and hence, it causes the index of resolution to decrease in a nearly uniform manner. The correction for this effect can be estimated either analytically[6] or numerically[36] provided the distribution function is known.

10.2.3.4. Extraneous Scattering with Sample Injection (II). Secondary Scattering from Metals.

A more troublesome extraneous scattering related to specimen injection originates from those electrons which are first scattered by the sample molecules and subsequently by metals (nozzle, slits, sector surface, sector edge, ball-bearing race, beam stopper, etc.) before they strike the detector (Fig. 1). Since the intensities of such electrons fall approximately in proportion to s^{-1}, the effect is usually more significant toward larger scattering angles. A shadow cast by the nozzle and a reflection from the wall of the sector race and the beam stopper often cause significant troubles.[7,16] On the other hand, secondary scattering from the sector edge may sometimes be so intense as to affect the intensity measurement in smaller scattering angles. Since those extraneous scatterings raise the background intensity irregularly and confuse molecular oscillations, extremely careful compensation is necessary in order to make correct measurements of intensities at very small and very

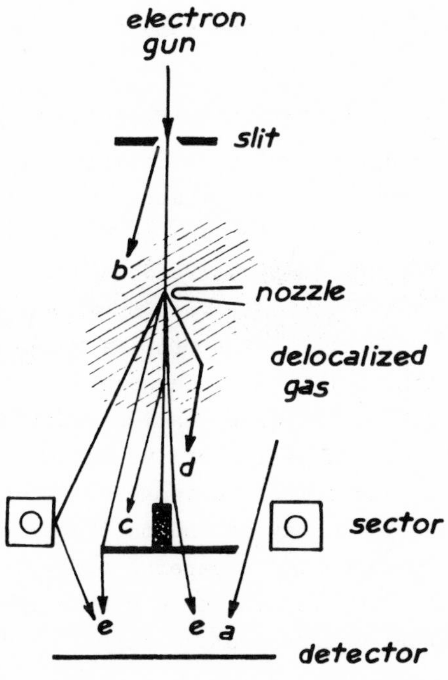

Fig. 10-1. Possible sources of extraneous scattering (schematic). (a) Extraneous radiation; (b) stray electrons; (c) delocalized-gas scattering; (d) intermolecular double scattering; (e) secondary scattering from metals.

large scattering angles. The extent of the disturbance can be estimated by a comparison of observed and calculated relative background intensities I_B or molecular intensities I_M, and by a decrease in the observed index of resolution.

10.2.3.5. Sector Function

A deviation of the sector opening from a defined function can be compensated for by a direct measurement of the sector shape under a microscope or by that of the diffraction intensity from gas molecules of known structure, say argon. A diffraction photograph from gas or thin metal film taken with a nonrotating sector may also be useful for examining the sector function and extraneous scattering from the sector edge.[7] The region closer to the center of the sector should be measured with more accuracy. The incident electron beam should be directed to hit the exact center of the sector, which should, at the same time, be exactly at the center of rotation.

10.2.3.6. Measurement of Vibrational Temperature

If a molecule under study is relatively rigid, one can practically assume a thermal equilibrium among vibrational states at the temperature of the nozzle tip, and an error of a few degrees in the measurement of this temperature does not influence seriously the interpretation of the thermal-average values of structural parameters.[27,37-39] However, if the problem under study is concerned with a vibration with a very large amplitude,[40-42] conformational equilibrium,[27,43,44] dissociation equilibrium,[45,46] etc., the measurement and the definition of 'vibrational temperature' poses a hard problem. The effective temperature of the system may not be defined uniquely and may depend on the sample, the nozzle shape,[27,47] and the pressure.[35] A careful study of a system with known equilibrium constants seems to be worthwhile for a calibration of the temperature measurement.

10.2.4. Photographic Process

The technical cautions to be exerted in ordinary photographic densitometry must be followed in obtaining good diffraction photographs. The exposure condition should be adjusted in such a way to make the densities fall within a proper range, say from 0.2 to 0.8 in the D scale. Only negligible shrinkage of the diffraction pattern recorded on photographic emulsion is assumed to occur during the process of development.[48] For eliminating local irregularities in the density, uniform and constant stirring of the developer is essential.[49] The effect of grains and minor irregularities in the emulsion may be averaged out by spinning the plate at an adequate speed about the center of the diffraction halos while being microphotometered[50-52] [cf. H.R.Foster, D.A.Kohl, R.A.Bonham, and M.L.Williams, unpublished data].

The sensitivity of the emulsion may vary with position on the plate; the edges are believed to be more sensitive than the center[53,54] [cf. also D.A.Kohl, thesis, Indiana University, 1966]. Although this effect is not serious as long as the variation is small, smooth, and uniform, a calibration may be necessary for intensity measurements with very high precision.

When a photographic plate is hit by a strong electron beam, electrostatic charges may be built up on the plate and distort the diffraction pattern. This effect sometimes needs a correction, particularly when a nonsector photograph is used for measuring the wavelength of electrons.[16,48,55]

10.2.5. Measurement of Scattered Intensity

The densitometry of a photographic plate by means of a microphotometer[50-52] involves the following sources of systematic errors.

(a) Optical problems. The light source should not fluctuate irregularly, although a small, reproducible drift in the light can be compensated for. A double-beam instrument may be used, for example. The optical system must be free from stray lights, and the effect of the slit width must be negligible.

(b) Electronic problems. The photoelectric detector and the amplifier system need careful adjustment in order to secure reliable operation. While accuracy in the absolute density scale is not required for usual measurements, a lack of linearity in the electronic system may cause a systematic error. Analog-to-digital conversion followed by electronic integration[51,52,56,57] has improved the accuracy of the measurement.

(c) Mechanical problems. Problems of this category are, for example, irregularity in the plate travelling device, failure in locating the exact center of the diffraction pattern when a photographic plate is mounted on a plate-spinner, and mechanical vibrations of the spinner transmitted to the optical system. A part of the errors can be eliminated by scanning the plate across the diameter and taking the average of both sides.

(d) Recording. By means of the digital recording of photocurrents, most problems associated with the pen-recording, such as the time lag in the record, pen friction, errors in the process of recording on and reading from the chart paper, have been eliminated.

(e) Conversion of photographic density to relative electron intensity. It is claimed that photographic density D is nearly proportional to relative electron intensity integrated over the exposure time, E, in a certain range of small D, which depends on the emulsion and the method of development (say $D < 0.5 \sim 1.0$).[37,58,59] [Cf. also D.A.Kohl, thesis, Indiana University, 1966]. Beyond this range, the $D-E$ relation shows a slight, uniform deviation from linearity.[53,58-61] A number of methods have been proposed for determining the $D-E$ relation, $E = f(D)$, experimentally on the basis of the reciprocity relation.[62] Any method of conversion may be used if it is simple and does not introduce systematic errors; in practice, one may use a simple analytical equation for $f(D)$ without significant loss of accuracy. Uncertainty in this calibration influences the accuracy of mean amplitudes and the index of resolution critically, but it does not interfere with the measurement of internuclear distances unless they are close to one another.

10.2.6. Measurement of Scale Factor

Accuracy of the scale factor (a constant proportional to the electron wavelength λ and the camera length L) has direct influence on the accuracy of experimental internuclear distances. If absolute measurements of λ and L are made separately, both should have sufficient accuracy. Suppose one wants to get an accuracy of 0.1% in λ and in L. Since λ is nearly proportional to the square root of the accelerating voltage V, one has to regulate and measure V to within 100V for an accelerating voltage of 50kV. The camera length has to be measured to within 0.1mm for an L of 10cm. With special designs for high-precision measurements, one can attain an accuracy higher than this standard by an order of magnitude. For instance, Witt[48] has reported the total systematic error of $\Delta\lambda/\lambda = \pm 1.3 \times 10^{-5}$ for V = 50kV and $\Delta L/L = \pm 1.4 \times 10^{-5}$ for L = 332mm in the measurement of the lattice constant of thallium(I)chloride. As a result, the lattice constant (CsCl type) at 20 °C has been measured to be 3.84145 Å with a systematic error of ± 0.00012 Å and a random error of ± 0.00005 Å.

When a solid reference substance with a known lattice constant is used for a relative measurement of the scale factor $L\lambda$, the diameters of the diffraction rings and the difference in the camera lengths of the sample and the reference must be measured precisely, together with a rough estimate of λ or L to be used for a correction for a plane detector.[63] Systematic errors may result from inadequate preparation and handling of the reference specimen,[37] instability of λ and L, and errors in the measurement of ring diameters.[10] It seems to be better to use a gas molecule with precisely known structural parameters for calibration of the scale factor.[16]

10.2.7. Theoretical Interpretations

Among the problems related to the statistical analysis of scattered electron intensity, ambiguity associated with the experimental background function I_B,[51,64] and the effect of interdependence among structural parameters are relatively more important sources of error. An inadequate fit of the background to an analytical function introduces systematic errors in the structural parameters.[36,51,65]

In a least-squares analysis of experimental data, a proper statistical weight should be assigned to each measurement[20,64,66-71] in order to derive 'most probable' parameter values and their 'standard errors'. The weight matrix may contain off-diagonal elements in consideration of correlation among the measurements;[72,73] cf. also Chapter 11.

Since it is a usual practice to assume some of the parameters when a complex molecule is under investigation, uncertainties originating from the assumptions must be examined carefully[74,75] (see Appendix).

In addition, uncertainties may originate from incorrect theoretical interpretations:

(a) Theoretical formulas. In the equations on which our ordinary analysis is based,[18-22] the following items are ignored.[1] (1) Effect of chemical binding[76,77] in elastic and inelastic scattering factors (for molecules containing only light atoms; important in small scattering angles). (2) Polarization[78] (for molecules containing heavy atoms; important in small scattering angles). (3) Intramolecular multiple scattering[79-83] (for molecules containing at least one heavy atom). (4) Higher-order effect of anharmonic vibrations (for molecules with very low frequencies and/or very large anharmonicity).[40-42,84-88]

At least approximate corrections can be made, if necessary, by means of analytical or numerical calculations.

(b) Atomic scattering factors. Use of incorrect scattering factors introduces serious error particularly in the mean amplitudes.[89] When the molecule under study has heavy and light atoms, correct phase differences must be taken into account.[1,13,90-93] In some cases, the total intensity for monoatomic gases measured by the same apparatus and the same procedure can be used for estimating the atomic background for molecular scattering (e.g., Xe and Ne atoms for XeF_6 [84]).

(c) Physical significance of structural parameters. In order that the experimental structural parameters have physical significance, they must be defined clearly[94] (cf. Chapter 12). With the aid of the theory of normal vibrations, the internuclear distances, mean amplitudes and phase parameters, which are derived directly from the analysis of experimental electron intensity (or from simple manipulations thereof) may be defined with little ambiguity as certain vibrational-average values.[95-97] For numerical calculations, the intramolecular force field needs to be known. The use of inaccurate force constants for such calculations may result in errors of a few thousandths of an Ångström in the distances and mean amplitudes.

10.2.8. Systematic Errors in Structural Parameters: Summary

As a summary of the above discussion, systematic errors which may have much influence on each structural parameter under normal experimental conditions are listed in the following in the order of probable importance.

(a) Internuclear distances
- Uncertainty in the scale factor
- Correlation with neighboring distances, if any
- Correlation with phase parameters
- Gas spread
- Ambiguity in the physical significance of distance parameters

(b) Mean amplitudes
- Extraneous scattering
- Correlation with index of resolution
- Uncertainty in the density-intensity calibration
- Uncertainty in the scattering factors

(c) Phase parameters
- Extraneous scattering
- Correlation with internuclear distances and mean amplitudes

If experiments are made under more difficult conditions, e.g., at high temperatures, with very low sample pressures, with unstable or impure samples, then more serious systematic errors may be introduced from other sources.

10.3. Random Experimental Errors

10.3.1. Orders of Magnitude of Random Errors

Random errors in the measurement of scattered electron intensity are likely to be within ± 0.1% in relative magnitude when digital densitometry is applied to a good photographic plate.[23,51,52] This corresponds to a random error in the reduced molecular intensity $M(s)$ (i.e., molecular scattering undulations divided by atomic background) of about $\pm$ 0.001 in absolute magnitude, in contrast to the peak value of $M(s)$ ranging from about $\pm$ 0.01 to 0.05 in a medium s region for molecules commonly studied by gas electron diffraction. This random error contributes to the weighted sum of residuals [observed minus calculated $M(s)$] and is propagated by way of the 'error matrix',[66-70] to the random errors in the structural parameters. The random errors estimated in this way should nearly agree with those obtained from the analysis by examination of the 'internal consistency' and the 'reproducibility' of the results.[64]

For an order-of-magnitude discussion of random errors in structural parameters, the following approximate evaluation may be useful. For a diatomic molecule, the molecular intensity is given by[95]

$$sM(s) = C \exp(-l^2 s^2/2) \sin s(r_a - \varkappa s^2). \qquad (10.1)$$

According to the law of propagation of errors[70] the standard errors in the parameters, r_a, l, and $\varkappa$, should approximately be

$$\sigma_r \sim \sigma_M / N^{\frac{1}{2}} \langle sM \cot sr \rangle \sim \sigma_M / N^{\frac{1}{2}} \langle sM \rangle, \qquad (10.2)$$

$$\sigma_l \sim \sigma_M / N^{\frac{1}{2}} l \langle sM \rangle, \qquad (10.3)$$

$$\sigma_\varkappa \sim \sigma_M / N^{\frac{1}{2}} \langle s^3 M \cot sr \rangle \sim \sigma_M / N^{\frac{1}{2}} \langle s^3 M \rangle, \qquad (10.4)$$

where σ_M and N represent the standard error in $M(s)$ and the effective number of observations, respectively, and the carets denote effective mean values. [Eqs. (2)-(4) result from taking appropriate derivatives of Eq. (1) with respect to the parameters.] Since the orders of σ_M, N, l, and s are 0.1M, 100, 0.05 Å, and 10 Å^{-1}, respectively, for ordinary experimental conditions, typical orders of magnitude of the parameters should be

$$\sigma_r \sim 0.001 \text{ Å}, \quad \sigma_l \sim 0.002 \text{ Å}, \quad \text{and} \quad \sigma_\varkappa \sim 1 \times 10^{-5} \text{ Å}^3.$$

These values are consistent with what we experience in favorable experimental conditions [e.g. Ref. 12]. The random error in the internuclear distance is often similar in magnitude to that in the mean amplitude of the same pair.[67] The above estimation of σ_κ implies that even in the best situation the phase parameter κ for a non-hydrogen atom pair is likely to be obscured by random experimental error, if not obscured by systematic error, since typical orders of magnitude of κ are 10^{-5} Å^3 and 10^{-6} Å^3 for atom pairs with and without a hydrogen atom, respectively.[12,95,96,98] It therefore seems quite difficult to vary non-hydrogen κ parameters in a least-squares analysis.

The above argument can be extended to polyatomic molecules if M is replaced by M_{ij}, a contribution to M from individual atom pairs. For atom pairs with relatively weak scattering powers, M_{ij} may be so small as to make the σ_r and σ_l an order of magnitude larger than the above estimates. In addition, the effect of correlations among the parameters may be more serious.

In a least-squares analysis, random errors in the 'measurements' are estimated on the spot from the 'residuals' (observed minus 'best fit' theoretical). The errors estimated in this way, which represent internal consistency,[64] are a useful but only limited indication of the quality of the measurements, since years of experience in one's laboratory about random and systematic errors (e.g., reproducibility[64] and general reliability of the measurements) does not appear explicitly in the estimation except in the process of assigning a set of a priori statistical weights. In principle, such experience can supply more important information on the plausible limits of error to be assigned to the parameters derived. For instance, a statistical treatment leads to a reasonable assessment of errors only if the residuals are of the same order of magnitude as that estimated from experience ('estimated noise level') and they have no definitely nonrandom features.

10.3.2. Estimation of Experimental Random Errors Prior to Experiment

Prior to an electron-diffraction study of a particular molecule, one often wishes to know with what precision the structural parameters can be determined as a final result of the study. This prediction is helpful to decide whether or not a laborious, time-consuming experiment is really worth while. Provided one has good knowledge of the random experimental error in the measurement of total electron intensity for the apparatus and method being used and an approximate set of structural parameters for the molecule being studied, this prediction can be made with a reasonable accuracy and ease by means of a computer simulation.

The random standard error of a structural parameter x_j is estimated to be[66,70]

$$\sigma_j^2 = (\mathbf{B}^{-1})_{jj}[\mathbf{V'PV}/(n-m)] = (\mathbf{B}^{-1})_{jj}\sigma_M^2. \qquad (10.5)$$

In the absence of systematic errors, the standard error in the molecular intensity is characteristic of the measurement and is independent of the molecule to be studied: σ_M can be predicted from experience for a suitably assigned weight function P, the number of independent measurements n, and the number of independent parameters m. For example, a typical order of magnitude of the residual (observed minus calculated molecular intensities), $V(s)$, for a medium s region is 1×10^{-3} in the $M(s)$ scale.[7,12,52] On the other hand, the elements of the $\mathbf{B}$ matrix are characteristic of the parameters to be determined in the analysis and can be estimated by means of an assumed set of the structural parameters (taking the derivatives of the molecular intensity with respect to the parameters). The rank and the eigenvectors of the $\mathbf{B}$ matrix provide instruction as to how many variable parameters can be determined and how they should be selected in the analysis in order to avoid singularity. If the measurements are under suspicion of mutual correlation, one may use a nondiagonal weight matrix for $\mathbf{P}$[72,73](see also Chapter 11). When the molecule is expected to have closely-spaced distances, and/or when the structural parameters are very uncertain, the estimates can be too optimistic since some of the $\mathbf{B}^{-1}$ elements can be larger. In order to be safe, one may multiply the standard errors estimated in this procedure by about three to get final estimates of the uncertainties.

10.4. Standard Gas Molecules

The diffraction instrument and the method of analysis have to be inspected occasionally on suspicion of systematic errors. It seems suitable to use several reference gas molecules for this purpose and for a calibration of the scale factor. The following are the necessary conditions for such a standard sample. (a) Precise geometrical structure is known by spectroscopy. (b) Precise quadratic and anharmonic potential constants are known by spectroscopy. (c) The reduced molecular intensity is strong enough so that the structural parameters derived from electron diffraction have good accuracy (of the order of 0.001 Å). (d) The molecule is chemically stable, and a pure sample can be obtained easily.

Structural parameters for several candidates, N_2, Cl_2, Br_2, CO_2, CS_2, SO_2, and CH_4, are listed in Table I.[95,96]

Table 10-I. Structural parameters of reference molecules.*

Molecule	Bond	r_a	l_m	κ
N_2	N-N	1.100_7 Å	0.031_8 Å	$0.4_0 \times 10^{-6}$ Å^3
Cl_2	Cl-Cl	1.993_1	0.044_5	1.5_5
Br_2	Br-Br	2.286_3	0.045_2	1.9_8
CO_2	C-O	1.164_6	0.034_8	0.5_1
	O•••O	2.324_4	0.039_5	0.4_9
CS_2	C-S	1.557_0	0.039_0	0.6_4
	S•••S	3.108_5	0.041_2	0.4_9
SO_2	S-O	1.435_2	0.034_8	0.5_8
	O•••O	2.475_9	0.054_4	0.7_3
CH_4	C-H	1.101_7	0.075	$(9._9)$
	H•••H	1.803	0.12_2	$(4._1)$

* At room temperature. The uncertainties in r_a and l_m are about 0.001 Å or somewhat less. Those in κ are about 0.1×10^{-6} Å^3.

10.5. Concluding Remarks

It must be kept in mind in dealing with experimental data of gas electron diffraction not to place blind reliance upon a statistical analysis, since systematic errors can be so much more important than random errors as to invalidate any statistical approach at all. For example, if alternative solutions are obtained from a least-squares analysis, one should not carelessly reject any of them on weak statistical grounds. It may well be the case that the experimental data at hand have no sufficient information to derive a unique, precise solution of the problem. Other independent experimental information must be consulted whenever possible (with adequate caution not to introduce an additional systematic error by such a procedure) to favor one of the solutions on physical reasons. For example, if the rotational constants of the molecule are known from rotational spectroscopy, they may be taken into the analysis as valuable additional observables.[99-106] For analyses of relatively complicated molecules, mean amplitudes calculated by the use of intramolecular force fields provide indispensable aid to a more unique solution of the problem.[74,106] *

Acknowledgments: The author is deeply indebted to all the researchers of gas electron diffraction with whom he has had stimulating and informative discussions on the present subject. Particularly, he wishes to thank Professor Emeritus Yonezo Morino and the members of his group and Professors L.S.Bartell and R.A.Bonham.

Appendix: Systematic Errors due to Fixed Parameters

In order that a least-squares analysis has full statistical significance, the following requirements must be satisfied. (a) The measurements, to which a known analytical function is fitted to determine their parameters, can be made many times and involve

* The author is aware of his omission of references to many significant Russian works of relevance to the topic of the present chapter. (The reader may refer to, e.g., Chapter 13 and Part I of Chapter 18, and references cited therein.) This omission was made only on the ground of practical limitations.

only random errors. (b) All the independent parameters in the analytical function are treated as variables.

In the problem of gas electron diffraction, however, these conditions are never met in a rigorous sense. Item (a) has been discussed in the text in relation to the possibility of systematic errors. As for Item (b), it is often compulsory to limit the number of variable parameters in order to avoid non-convergence of the solutions. In many cases, all the phase parameters $\boldsymbol{\kappa}$, all or a part of the mean amplitudes, and some of the bond distances or angles are 'frozen' as constants as a result of estimation or assumption. Even in that case, one may still apply the criterion of 'least-square-sum of the weighted residuals' to an automatic search for a 'plausible' set of parameter values, provided the assumed constants are 'reasonable'. It must be kept in mind, however, that the fixed constants should not be assumed as 'perfectly correct'. In other words, systematic errors due to this procedure should properly be taken into account in the estimation of errors. The following method for this estimation has been discussed in a recent paper.[74]

Suppose a set of parameters $x_1, \ldots, x_k$ (denoted, as a set, as $\mathbf{r}$) among the independent parameters $x_1, \ldots, x_m$ are taken as variables in a least-squares analysis, while the other parameters, $x_{k+1}, \ldots, x_m$ (denoted as $\mathbf{l}$) are kept constant ($\mathbf{l}_0 = x_{k+1}^0, \ldots, x_m^0$). Let the probability of finding a set of correct estimates of $\mathbf{l}$ at the assumed constants $\mathbf{l}_0$ be represented as $w(\mathbf{l})d\mathbf{l}$. This function is not necessarily a multi-dimensional Gaussian nor may be symmetric about $\mathbf{l}_0$. When a least-squares method is applied to this problem, the most probable set of the $\mathbf{r}$ parameters and their standard errors σ_r depend on $\mathbf{l}$ parametrically and constitute an error function $W(\mathbf{r}, \sigma_r; \mathbf{l})d\mathbf{r}$, which is presumed to be a multi-dimensional Gaussian. The probability function of errors, $P(\mathbf{r})$, which one wishes to estimate, is a folding of w and W functions:

$$P(\mathbf{r})d\mathbf{r} = \int W(\mathbf{r}, \sigma_r; \mathbf{l})\, w(\mathbf{l})d\mathbf{l}\, d\mathbf{r}. \qquad (10.6)$$

This integration can be made either analytically or numerically on the basis of the estimated W and w functions. In some cases (Fig. 2a), the $\mathbf{r}$ parameters and their standard errors are almost

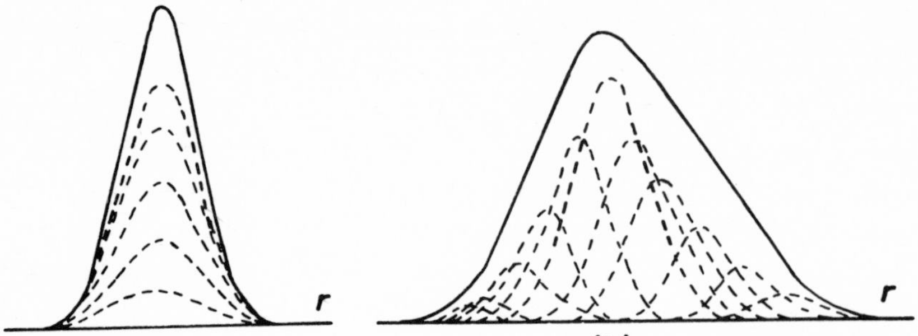

Fig. 10-2. Probability distribution functions of errors. Typical $W(r;l)$ functions for several fixed l constants with relative heights proportional to $w(l)$ are shown as broken curves. Solid curves represent the P functions, given in Eq. (6), in an arbitrary scale. The W functions are practically independent of l in case (a), whereas in case (b) the P function is broadened by the dependence of $W(r)$ on l.

independent of the l parameters, and hence, the uncertainties in r are essentially characterized by σ_r; but in other cases (Fig. 2b), the dependence of r on l is much more important than σ_r, and the uncertainties in r originate essentially from the systematic errors due to the fixed l constants. Examples of these extreme cases are given in Ref. 74 for norbornane and norbornadiene. A simple method of estimating $P(r)$ by means of successive variation in l has been used by Tsuchiya and Kimura.[75]

REFERENCES

1 R.L.Hilderbrandt and R.A.Bonham, Ann.Rev.Phys.Chem. 22, 279 (197
2 J.M.Hastings and S.H.Bauer, J.Chem.Phys. 18, 13 (1950).
3 I.L.Karle and J.Karle, J.Chem.Phys. 17, 1052 (1949).
4 I.L.Karle and J.Karle, J.Chem.Phys. 18, 565 (1950).
5 J.Karle and I.L.Karle, J.Chem.Phys. 18, 957 (1950).
6 I.L.Karle and J.Karle, J.Chem.Phys. 18, 963 (1950).

7 L.O.Brockway and L.S.Bartell, Rev.Sci.Instr. 25, 569 (1954).

8 O.Bastiansen, O.Hassel, and E.Risberg, Acta Chem.Scand. 9, 232 (1955).

9 P.A.Akishin, M.I.Vinogradov, K.D.Danilov, N.P.Levkin, E.N. Martinson, N.G.Rambidi, and V.P.Spiridonov, Pribory Tekhn. Eksperim. 3, 70 (1958).

10 H.Morimoto, J.Phys.Soc.Japan 13, 1015 (1958).

11 K.Kuchitsu, Bull.Chem.Soc.Japan 32, 748 (1959).

12 L.S.Bartell, K.Kuchitsu, and R.J. de Neui, J.Chem.Phys. 35, 1211 (1961).

13 H.M.Seip, Acta Chem.Scand. 19, 1955 (1965).

14 A.Almenningen, O.Bastiansen, A.Haaland, and H.M.Seip, Angew. Chem. Internat.Ed. 4, 819 (1965).

15 W.Zeil, J.Haase, and L.Wegmann, Z.Instrumentenk. 74, 84 (1966).

16 Y.Murata, K.Kuchitsu, and M.Kimura, Jap.J.Appl.Phys. 9, 591 (1970).

17 M.Fink and R.A.Bonham, Rev.Sci.Instr. 41, 389 (1970).

18 J.Karle, Determination of Organic Structures by Physical Methods, Vol. 5 (Ed. E.A.Braude and F.C.Nachod), 1972.

19 O.Bastiansen and P.N.Skancke, Advances Chem.Phys. 3, 323 (1960).

20 L.S.Bartell, Physical Methods in Chemistry (Ed. A.Weissberger and B.W.Rossiter), 4. Ed., Interscience, New York 1971.

21 S.H.Bauer, Physical Chemistry, Vol. 4, p.741, Academic Press, 1970.

22 M.I.Davis: Electron Diffraction in Gases, Marcel Dekker, New York 1971.

23 R.A.Bonham and M.Fink, J.Chem.Phys. 47, 3676 (1967).

24 K.W.Hansen and L.S.Bartell, Inorg.Chem. 4, 1775 (1965).

25 E.Hirota, Bull.Chem.Soc.Japan 31, 130 (1958).

26 K.Hedberg and R.Ryan, J.Chem.Phys. 41, 2214 (1964).

27 R.R.Ryan and K.Hedberg, J.Chem.Phys. 50, 4986 (1969).

28 H.B.Burgi and L.S.Bartell, J.Am.Chem.Soc. 94, (1972).

29 L.S.Bartell and H.K.Higginbotham, J.Chem.Phys. 42, 851 (1965).

30 L.S.Bartell and B.L.Carroll, J.Chem.Phys. 42, 1135 (1965).

31 Y.Morino and Y.Murata, Bull.Chem.Soc.Japan 38, 114 (1965).

32 T.Iijima, R.A.Bonham, C.Tavard, M.Roux, and M.Cornille, Bull. Chem.Soc.Japan 38, 1757 (1965).

33 L.S.Bartell, J.Appl.Phys. 31, 252 (1960).

34 Y.Morino, K.Kuchitsu, and R.A.Bonham, Bull.Chem.Soc.Japan 38, 1796 (1965).

35 P.Audit, J.Phys.(Paris) 30, 192 (1969).

36 D.A.Kohl, Trans.Am.Cryst.Assoc. 2, 124 (1966).

37 Y.Morino and T.Iijima, Bull.Chem.Soc.Japan 35, 1661 (1962).

38 T.Ukaji and K.Kuchitsu, Bull.Chem.Soc.Japan 39, 2153 (1966).

39 E.W.Ng, L.S.Su, and R.A.Bonham, J.Chem.Phys. 50, 2038 (1969).

40 A.Almenningen, S.P.Arnesen, O.Bastiansen, H.M.Seip, and R.Seip, Chem.Phys.Letters 1, 569 (1968).

41 A.Clark and H.M.Seip, Chem.Phys.Letters 6, 452 (1970).

42 M.Tanimoto, K.Kuchitsu, and Y.Morino, Bull.Chem.Soc.Japan 43, 2776 (1970).

43 O.Bastiansen, H.M.Seip, and J.E.Boggs, Perspectives in Structural Chemistry, Vol. 4 (Ed. J.D.Dunitz and J.A.Ibers), p. 60, Wiley, New York 1971.

44 A.Almenningen, O.Bastiansen, L.Fernholt, and K.Hedberg, Acta Chem.Scand. 25, 1946 (1971).

45 J.Janzen and L.S.Bartell, J.Chem.Phys. 50, 3611 (1969).

46 A.Almenningen, O.Bastiansen, and T.Motzfeldt, Acta Chem.Scand. 23, 2848 (1969).

47 R.B.Harvey, F.A.Keidel, and S.H.Bauer, J.Appl.Phys. 21, 860 (1950).

48 W.Witt, Z.Naturforschg. 19a, 1363 (1964).

49 H.R.Foster, J.Appl.Phys. 41, 5344 (1970).

50 I.L.Karle, D.Hoober, and J.Karle, J.Chem.Phys. 15, 765 (1947).

51 L.S.Bartell, D.A.Kohl, B.L.Carroll, and R.M.Gavin,Jr., J.Chem.Phys. 42, 3079 (1965).

52 Y.Morino, K.Kuchitsu, and T.Fukuyama, Bull.Chem.Soc.Japan 40, 423 (1967).

53 L.S.Bartell and L.O.Brockway, J.Appl.Phys. 24, 656 (1953).

54 R.C.Valentine, Advances in Optical and Electron Microscopy, Vol. 1 (Ed. R.Baerer and V.E.Coslett), p. 180, Academic Press, 1966.

55 T.B.Rymer and K.H.R.Wright, Proc.Roy.Soc.(London) A215, 550 (1952).

56 R.L.Hilderbrandt and S.H.Bauer, J.Mol.Structure 3, 325 (1969).

57 G.Gundersen and K.Hedberg, J.Chem.Phys. 51, 2500 (1969).
58 M.Takagi, N.Kitamura, and S.Morimoto, J.Phys.Soc.Japan 16, 792 (1961).
59 H.Morimoto, Oyo Buturi 31, 137 (1962).
60 J.Karle and I.L.Karle, J.Appl.Phys. 24, 1522 (1953).
61 L.S.Bartell and L.O.Brockway, J.Appl.Phys. 24, 1523 (1953).
62 A.Becker and E.Kipphan, Ann.Physik [5] 10, 15 (1931).
63 R.A.Bonham and L.S.Bartell, J.Chem.Phys. 31, 702 (1959).
64 Y.Morino, K.Kuchitsu, and Y.Murata, Acta Cryst. 18, 549 (1965).
65 G.Pauli, Trans.Am.Cryst.Assoc. 2, 126 (1966).
66 O.Bastiansen, L.Hedberg, and K.Hedberg, J.Chem.Phys. 27, 1311 (1957); erratum: ibid. 28, 512 (1958).
67 K.Hedberg and M.Iwasaki, Acta Cryst. 17, 529 (1964).
68 M.Iwasaki, F.N.Fritsch, and K.Hedberg, Acta Cryst. 17, 533 (1964).
69 O.Bastiansen, F.N.Fritsch, and K.Hedberg, Acta Cryst. 17, 538 (1964).
70 W.C.Hamilton: Statistics in Physical Science, Ronald Press, New York 1964.
71 B.Beagley, D.W.J.Cruickshank, P.M.Pinder, A.G.Robiette, and G.M.Sheldrick, Acta Cryst. B25, 737 (1969).
72 Y.Murata and Y.Morino, Acta Cryst. 20, 605 (1966).
73 H.M.Seip, T.G.Strand, and R.Stølevik, Chem.Phys.Letters 3, 617 (1969).
74 A.Yokozeki and K.Kuchitsu, Bull.Chem.Soc.Japan 44, 2356 (1971).
75 S.Tsuchiya and M.Kimura, Bull.Chem.Soc.Japan 45, 736 (1972).
76 R.A.Bonham, Record Chem.Progress 30, 185 (1969).
77 A.Jaeglé, A. Duguet and M.Rouault, J.Chim.Phys. 67, 687 (1970).
78 R.A.Bonham, J.Chem.Phys. 43, 1933 (1965).
79 P.J.Bunyan, Proc.Phys.Soc.(London) 82, 1051 (1963).
80 J.Gjønnes, Acta Cryst. 17, 1075 (1964).
81 R.A.Bonham, J.Chem.Phys. 43, 1103 (1965).
82 L.S.Bartell and T.C.Wong, J.Chem.Phys. 56, 2364 (1972).
83 R.A.Bonham and E.M.A.Peixoto, J.Chem.Phys. 56, 2377 (1972).
84 R.M.Gavin,Jr. and L.S.Bartell, J.Chem.Phys. 48, 2460 (1968).
85 E.J.Jacob and L.S.Bartell, J.Chem.Phys. 53, 2235 (1970).
86 W.J.Adams, H.B.Thompson, and L.S.Bartell, J.Chem.Phys. 53, 4040 (1970).

87 W.J.Adams, H.J.Geise, and L.S.Bartell, J.Am.Chem.Soc. 92, 5013 (1970).
88 A.Yokozeki, K.Kuchitsu, and Y.Morino, Bull.Chem.Soc.Japan 43, 2017 (1970).
89 S.Konaka, T.Ito, and Y.Morino, Bull.Chem.Soc.Japan 39, 1146 (1966).
90 J.A.Hoerni and J.A.Ibers, Phys.Rev. 91, 1182 (1953).
91 H.L.Cox and R.A.Bonham, J.Chem.Phys. 47, 2599 (1967).
92 H.M.Seip, Selected Topics in Structural Chemistry (Ed. P.Andersen, O.Bastiansen, and S.Furberg), p. 25, Universitetsforlaget, Oslo 1967.
93 B.Andersen, H.M.Seip, T.G.Strand, and R.Stølevik, Acta Chem. Scand. 23, 3224 (1969).
94 Y.Morino, K.Kuchitsu, and T.Oka, J.Chem.Phys. 36, 1108 (1952).
95 K.Kuchitsu, Bull.Chem.Soc.Japan 40, 498 (1967).
96 K.Kuchitsu, Bull.Chem.Soc.Japan 40, 505 (1967).
97 K.Kuchitsu, Bull.Chem.Soc.Japan 44, 96 (1971).
98 K.Kuchitsu, T.Shibata, A.Yokozeki, and C.Matsumura, Inorg.Chem. 10, 2584 (1971).
99 G.Dallinga and L.H.Toneman, J.Mol.Structure 1, 11 (1967-68).
100 K.Kuchitsu, T.Fukuyama, and Y.Morino, J.Mol.Structure 1, 463 (1967-68).
101 K.Kuchitsu, T.Fukuyama, and Y.Morino, J.Mol.Structure 4, 41 (1969).
102 T.Fukuyama, K.Kuchitsu, and Y.Morino, Bull.Chem.Soc.Japan 42, 379 (1969).
103 T.Fukuyama and K.Kuchitsu, J.Mol.Structure 5, 131 (1970).
104 E.J.Jacob, H.B.Thompson, and L.S.Bartell, J.Mol.Structure 8, 38 (1971).
105 R.L.Hilderbrandt and J.D.Wieser, J.Chem.Phys. 55, 4648 (1971).
106 K.Kuchitsu, MTP International Review of Science, Physical Chemistry Series One, Vol. 2, Molecular Structure and Propertie (Ed. G.Allen), Chapter 6, Medical and Technical Publ.Co., Oxford 1972.

CHAPTER 11

Error Analysis in Structure Determinations of Gaseous Molecules by Electron Diffraction

H. M. SEIP and R. STØLEVIK

The problem of obtaining realistic estimates of the errors in structure determinations by electron diffraction is considered. Particular attention is paid to an analytical expression which may be used to correct the error estimates obtained in least-squares refinements with a diagonal weight matrix for the effect of correlation between the intensity data.

11.1. Introduction

In structure determination by electron diffraction in gases intensity values may be obtained as closely spaced as wanted. The theoretical expression for the modified molecular intensity may be written[1*]

$$I^{\mathrm{T}}(s) = \sum_{l} M_l g_l(s)\exp(-\tfrac{1}{2}u_l^2 s^2)\sin(R_l s)/R_l \ . \qquad (11.1)$$

R_l and u_l are interatomic distances and corresponding root-mean-square amplitudes of vibration, and the function $g_l(s)$ depends on the scattering amplitudes for the atoms.[1]

---

*This expression corresponds to Eq. (11) of Ref. 1 where the denominator is given erroneously as $(R{\cdot}s)$. We have simplified the notation somewhat in this paper, and used M instead of A to avoid confusion in later expressions. u corresponds to the mean amplitude of vibration (denoted l elsewhere in this book); cf. also Part II of Chapter 18.

For more than a decade it has been common to refine the structural parameters by the least-squares procedure.[2-5] However, it is well known that the observations are highly correlated, and that this implies a weight matrix with off-diagonal elements different from zero.[5-9] The normal equations are then*

$$\mathbf{B}\hat{\mathbf{X}} = \mathbf{Y}, \tag{11.2}$$

where the elements of **B** and **Y** are

$$B_{ij} = \sum_k \sum_l P_{kl} A_{ki} A_{lj}, \quad Y_i = \sum_k \sum_l P_{kl} A_{li} V_k. \tag{11.3}$$

A is the design matrix [see Eq. (8) below] and **P** the weight matrix. $\hat{\mathbf{X}}$ is in the non-linear case the shifts in the parameter vector, and **V** represents the residuals.

11.2. Properties of the Weight Matrix

The elements of the weight matrix **P** may be written [Paper I, Eq. (13)]

$$P_{kl} = \sigma_0^2 \sqrt{w_l}\, (\varrho^{-1})_{kl} \sqrt{w_k}\,. \tag{11.4}$$

The elements of the correlation matrix for the observations (ϱ) depend to a first approximation only on the difference $|s_k - s_l|$; i.e. $\varrho_{kl} = \varrho(|s_k - s_l|)$.[8] A satisfactory approximation for ϱ^{-1} was found to be

$$\varrho^{-1} \sim \begin{bmatrix} 1 & p_2 & p_3 & 0 & 0 & . & . & \cdots & . & . & 0 \\ p_2 & 1 & p_2 & p_3 & 0 & 0 & . & \cdots & . & . & 0 \\ p_3 & p_2 & 1 & p_2 & p_3 & 0 & 0 & \cdots & . & . & 0 \\ 0 & p_3 & p_2 & 1 & p_2 & p_3 & 0 & \cdots & . & . & 0 \\ . & . & . & . & . & . & . & \cdots & . & . & . \\ 0 & 0 & . & . & . & . & . & \cdots & p_3 & p_2 & 1 \end{bmatrix}. \tag{11.5}$$

---

*The notation in this paper is nearly the same as in Ref. 8 (later referred to as Paper I) to which it is closely related.

We have found $-0.65 < p_2 < -0.5$ and p_3 small and positive for all the data we have investigated. ρ^{-1} is then positive definite if $p_2 + p_3 \geq -0.5$. This may be seen as follows.

Let $\mathbf{C}$ and $\mathbf{D}$ denote two matrices with diagonal elements equal to 1, $c_{i,i-1} = c_{i,i+1} = c$, $d_{i,i-1} = d_{i,i+1} = d$, and all other elements zero. These matrices are positive definite if c and d are in the range [-0.5, 0.5]. A matrix of the form (5) is now obtained by multiplication

$$\mathbf{G} \approx \frac{1}{1 + 2cd} \mathbf{C}\,\mathbf{D}.$$

Comparison to (5) shows that

$$p_2 = (c + d)/(1 + 2cd), \text{ and } p_3 = cd/(1 + 2cd).$$

With $c = -0.5$ it is seen that $p_2 + p_3 = -0.5$. Thus $\mathbf{G}$ may be positive definite and satisfy $p_2 + p_3 = -0.5$, if also $\mathbf{D}$ is positive definite, i.e. $-0.5 \leq d \leq 0.5$. With $d = -0.5$ we have

$$p_2 = -2/3 \quad \text{and} \quad p_3 = 1/6.$$

Larger values of d give larger p_2.

If $p_2 \leq -2/3$, the sum $p_2 + p_3$ must be greater than -0.5 to obtain a positive definite $\mathbf{G}$ matrix.

Fig. 1 shows the variation in the elements in the correlation matrix (ρ) corresponding to a fixed value of p_2 (-0.64) and various values of p_3. The effect of a small change in p_3 is especially large when $p_2 + p_3 \approx -0.5$, and decreases as this sum increases. Curve 5, corresponding to the largest p_3 value, has a small negative region.

11.3. Analytical Expression to Include Correlation

The results in Paper I[8] were based on least-squares refinements using weight matrices according to Eqs. (4) and (5). However, it is sometimes difficult to obtain convergence with this weight matrix even if the calculation with a diagonal weight matrix converges easily. The results obtained by the latter method may then be applied and the standard deviations for the parameters computed from Eq. (5) of Paper I,[8]

$$\mathbf{M}_x = \mathbf{B}^{-1}\mathbf{A}'\mathbf{P}\mathbf{M}\mathbf{P}\mathbf{A}\mathbf{B}^{-1}. \tag{11.6}$$

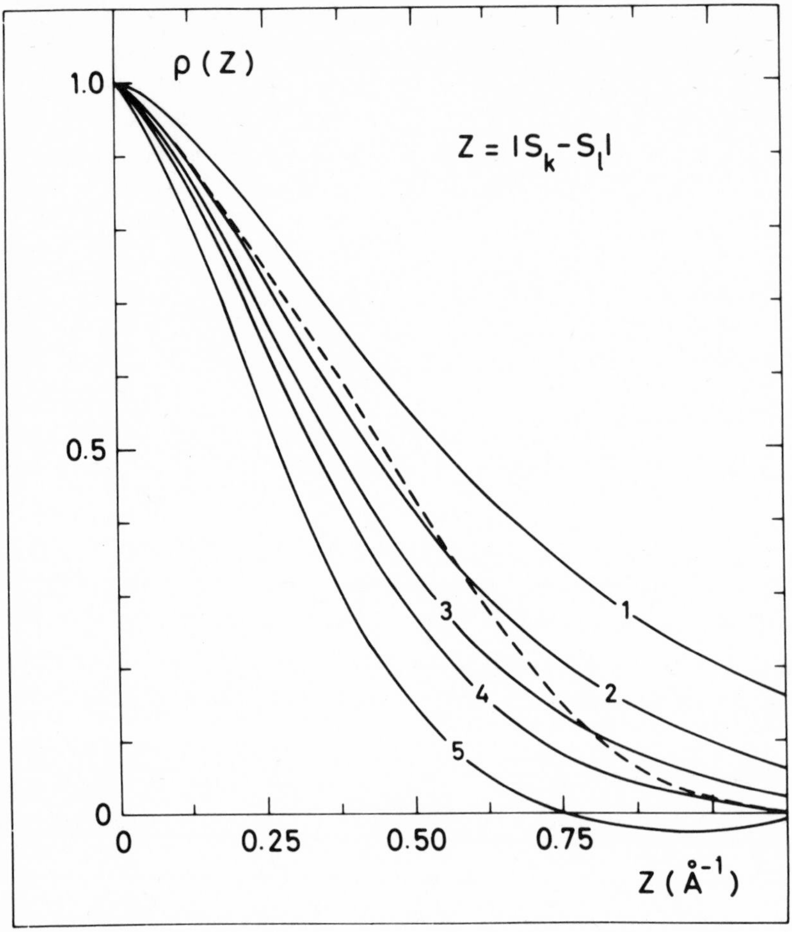

Fig. 11-1. The elements of ρ calculated by inversion of ρ^{-1} given by Eq. (5). The average of the elements in each array parallel to the diagonal is given. In all cases p_2 = -0.64, and p_3 = 0.142 (curve 1, a = 0.68), 0.144 (curve 2, a = 0.50), 0.146 (curve 3, a = 0.42), 0.148 (curve 4, a = 0.38), and 0.154 (curve 5, a = 0.29). The experimental curve is also shown.

M is here the moment matrix for the observations and $\mathbf{M}_x$ the moment matrix for the parameters. If **P** is equal to the unity matrix the expressions for the elements in **B** , **Y** [see Eq. (3)] and **A** may be written

$$B_{ij} = \sum_s A_i(s) A_j(s), \quad Y_i = \sum_s A_i(s) V_s, \tag{11.7}$$

where

$$A_k(s) = A_{sk} = \frac{\partial I^T(s)}{\partial(\text{parameter } k)}. \tag{11.8}$$

V_s is the difference between the observed and calculated intensity at the point s.

The expression (6) is not very convenient, and simple analytical formulas for estimating the standard deviations when all the distances are independent parameters have been derived.[7,8] The formula given in Paper I is in the following derived for a distance parameter. The procedure is very similar for a mean amplitude of vibration.

From the equations (1) and (8) we obtain for a distance parameter

$$A_k(s) \approx H_k(s) \cos(R_k s), \tag{11.9}$$

where

$$H_k(s) = (M_k/R_k) s g_k(s) \exp(-\tfrac{1}{2} u_k^2 s^2). \tag{11.10}$$

Let the standard deviation for the observation s be σ_s and $\rho_{s,s'}$ the correlation coefficient between two observations. An element in **M** is then

$$M_{s,s'} = \sigma_s \sigma_{s'} \rho_{s,s'} . \tag{11.11}$$

Experimental results have shown that the standard deviations may to a first approximation be assumed independent of s ($\sigma_s = \bar{\sigma}$). Further we assume (cf. Paper I)

$$\rho_{s,s'} = \exp(-\beta z^2), \tag{11.12}$$

where $z = s' - s$.

The solutions of the normal equations (2) may be written

$$X_i = \sum_{k=1}^{m} b_{ik} Y_k = \sum_{s} C_i(s) V_s , \tag{11.13}$$

where

$$b_{ik} = (\mathbf{B}^{-1})_{ik} , \quad C_i(s) = \sum_{k=1}^{m} b_{ik} A_k(s). \tag{11.14}$$

m is the number of parameters refined.

From the law of propagation of errors, i.e. when we express the diagonal elements of $\mathbf{M_x}$ in Eq. (6) (with $\mathbf{P} = \mathbf{E}$), it is obtai

$$\sigma_i^2 = \sum_{s} \sum_{s'} C_i(s) C_i(s') M_{s,s'} =$$

$$\bar{\sigma}^2 \sum_{k} \sum_{l} b_{ki} b_{il} \sum_{s} \sum_{s'} A_k(s) A_l(s') \rho_{s,s'} . \tag{11.15}$$

From Eq. (9) it follows that

$$A_l(s') = H_l(s') \cos[R_l(s+z)] =$$

$$H_l(s') \cos(R_l s) \cos(R_l z) - H_l(s') \sin(R_l s) \sin(R_l z).$$

In Eq. (15) $A_l(s')$ is multiplied by $\rho_{s,s'}$. Since $\rho_{s,s'}$ is smal for large z-values [cf. Eq. (12)], we have effectively

$$H_l(s') \approx H_l(s). \tag{11.16}$$

Thus

$$A_l(s') \approx A_l \cos(R_l z) - H_l(s) \sin(R_l s) \sin(R_l z). \tag{11.17}$$

Combining Eqs. (12), (15), and (17) we find that σ_i^2 consists of t terms

$$\sigma_i^2 = \bar{\sigma}^2 (K_1 + K_2), \tag{11.18}$$

where

$$K_1 = \sum_{k} \sum_{l} b_{ik} b_{il} \sum_{s} A_k(s) A_l(s) \sum_{z} \cos(R_l z) \exp(-\beta z^2), \tag{11.19}$$

and K_2 contains $\sin(R_l z)$.

From Eq. (7) we see that $\sum_s A_k(s)A_l(s) = B_{kl}$, and since $b_{ik} = (\mathbf{B}^{-1})_{ik}$ one has

$$\sum_k b_{ik} B_{kl} = \delta_{il} ,$$

which makes it possible to reduce Eq. (19) to

$$K_1 = b_{ii} \sum_z \cos(R_i z) \exp(-\beta z^2). \tag{11.20}$$

Except for very large values of β the summation in Eq. (20) may be replaced by the integral

$$\frac{1}{\Delta s} \int_{-\infty}^{+\infty} \cos(R_i z) \exp(-\beta z^2) dz,$$

which gives

$$K_1 = b_{ii} (\Delta s)^{-1} (\pi/\beta)^{\frac{1}{2}} \exp(-R_i^2/4\beta). \tag{11.21}$$

Since K_2 contains $\sin(R_i z) \exp(-\beta z^2)$ this term will vanish when a summation as in Eq. (20) is carried out.

Combining these results Eq. (18) may be written

$$\sigma_i^2 = \bar{\sigma}^2 b_{ii} F_i^2 , \tag{11.22}$$

where

$$F_i^2 = (\Delta s)^{-1} (\pi/\beta)^{\frac{1}{2}} \exp(-R_i^2/4\beta). \tag{11.23}$$

With no correlation $\beta \to \infty$, and the summation in Eq. (20) cannot be replaced by an integral. However, the summation reduces then to only one term $K_1 = b_{ii}$ which shows that

$$F_i \to 1 \ ; \qquad \sigma_i^2 = \bar{\sigma}^2 b_{ii} \cdot \tag{11.24}$$

Eq. (24) is the expected relation for uncorrelated observed data.

It is important to realize that b_{ii} in the derivation given above has been calculated with a unit weight matrix. If the correlation is included by using a weight matrix

$$\mathbf{P} = \sigma_0^2 \mathbf{M}^{-1}, \tag{11.25}$$

the elements in $\mathbf{B}$ are given by Eq. (3). Let the diagonal elements of $\mathbf{B}^{-1}$ in this case be b_{ii}'. Then

$$\sigma_i^2 = \sigma_0^2 b_{ii}' \approx \bar{\sigma}^2 b_{ii} F_i^2. \tag{11.26}$$

11.4. Properties of the Derived Correction Formula

The expression (22) was derived for the standard deviation of a distance parameter. The same relation with F_i given by Eq. (23) i valid for the mean amplitude (u_i) corresponding to the distance R_i (cf. Ref. 10). Usually all the distances are not considered as inde pendent parameters. However, if the independent geometrical parameters are bond distances and bond angles, the formula gives an upp limit to F_i, and may therefore still be a useful approximation.

Eq. (23) may be written in a slightly more convenient form by introducing the area (a) defined by the correlation curve and the coordinate axes (see Fig. 1). Since

$$a = \frac{1}{2}(\pi/\beta)^{\frac{1}{2}} \tag{11.27}$$

Eq. (23) becomes

$$F_i = \left(\frac{2a}{\Delta s}\right)^{\frac{1}{2}} \exp\left(-\frac{a^2}{2\pi} R_i^2\right). \tag{11.28}$$

Experimentally we have $a \approx 0.50$ Å^{-1} for our data corresponding to the final background as described on p. 618 of Paper I.

The factor F_i gives the ratio between the standard deviations when the correlation is included and the error estimates resulting from a least-squares refinement with a diagonal weight matrix. Fig. shows F_i as a function of a for three R values. The maximum in F ar the corresponding a value change considerably with R. For $R = 4.50$ Å F is less than 1 for the experimental a value of 0.50 Å^{-1}. When least-squares refinements are carried out, F is usually found to be close to 1 for distances of about this length, and smaller values have also been obtained. Eq. (28) (cf. Fig. 2) shows that $F \rightarrow 0$ whe

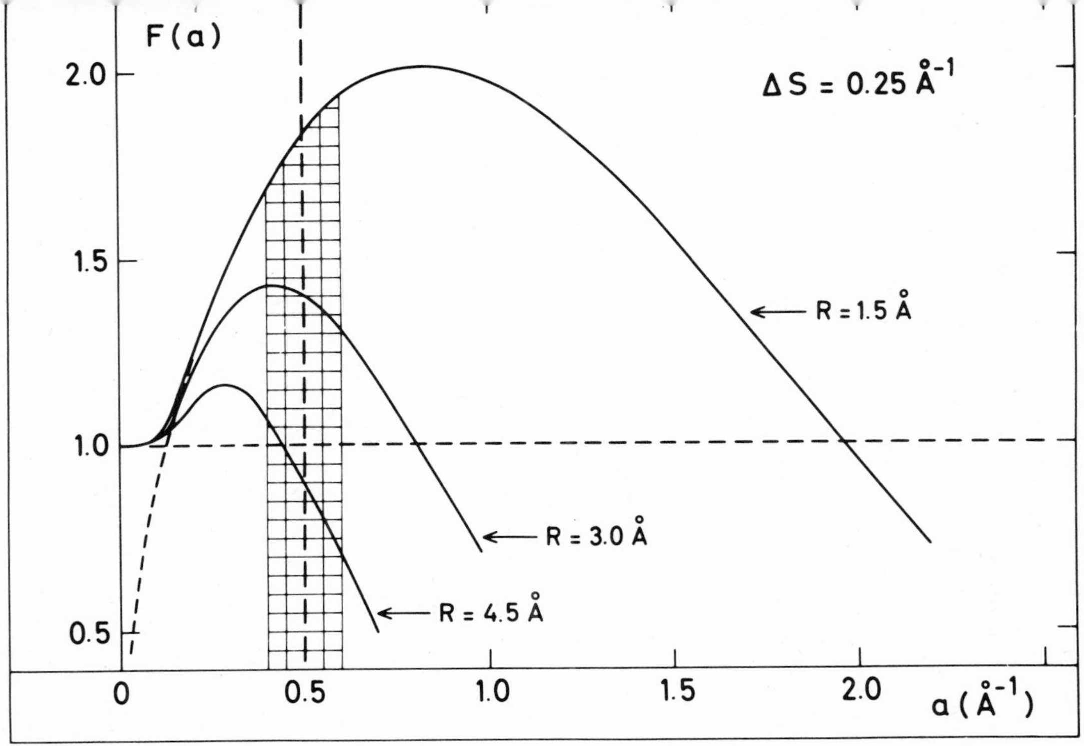

Fig. 11-2. The factor F in Eq. (26) as a function of a for three R values. The full lines have the correct behaviour for small a values, while the dotted lines show the result of using Eq. (28) when $a \rightarrow 0$. The shaded area around a = 0.50 $Å^{-1}$ indicates the range of a values obtained experimentally.

$a \to \infty$, which is at first rather surprising. However, Eq. (12) sho that all the ρ elements approach 1 when $a \to \infty$. All the difference between experimental and theoretical values must then be nearly equal; a case which will not occur in any actual refinement on ele tron diffraction data.

If the correlation between the observations is neglected, th error estimates obtained in the least-square refinement will go to zero when the interval between the points (Δs) goes to zero. σ_i in Eq. (26) is approximately independent of Δs if Δs is small enoug to allow the summation in the expression for B_{ij}, Eq. (7), to be replaced by an integral, i.e.

$$B_{ij} \approx \frac{1}{\Delta s} \int_{s_{min}}^{s_{max}} A_i(s) A_j(s)\, ds \ .$$

Then

$$b_{ii} = (\mathbf{B}^{-1})_{ii} = \frac{B^{ii}}{\det \mathbf{B}} \approx \mathrm{const} \cdot \Delta s \, ,$$

where B^{ii} is the cofactor of the element B_{ii} in $\det \mathbf{B}$.

Combination of these results with the equations (26) and (28 shows that σ_i is nearly independent of Δs as stated.

11.5. Discussion and Conclusions

The results obtained by least-squares refinements with a diagonal weight matrix and a weight matrix of the type given by Eqs. (4) and (5), have now been compared for a number of compounds viz. monobromodiacetylene,[10] hexafluorobutyne,[11] 1,3,4-thiadiazole oxalic acid,[13] disilylmethane,[14] tetrahydroselenophene,[15] tris(tri- fluoro)methylmethane,[16] π-allylcobalttricarbonyl,[17] and monomer an dimer trimethylaluminum.[18] In most cases some distances were treat as dependent parameters. The general experience is that the para- meters obtained by the two procedures are not uncomfortably differ though exceptions have been found, and that the change in the stan- dard deviations is qualitatively in agreement with Eq. (28) . The least-squares error estimates must of course be compared before an corrections, say for the uncertainty in the wave length, have been applied.

Does then the present procedure give satisfactory estimates of the standard deviations? For some molecules we have compared the standard deviations obtained using the results from 4 to 6 observed intensity curves, to the values obtained by least-squares refinements. The former values are in fact usually smaller than the standard deviations obtained with a weight matrix of the type (4) and often even smaller than the values obtained with a diagonal weight matrix.[11,19-22] The present procedure may therefore give results rather on the pessimistic side. This is probably an advantage, since more or less systematic errors, including errors in assumed mean amplitudes and shrinkage values, are usually not considered in the error estimates, though it may be done.

At present, the standard deviations reported in electron diffraction investigations are calculated in a variety of ways, which makes a comparison impossible. The correlation ought to be included, but even then it is a problem how to determine the necessary constant of proportionality [σ_0^2 or $\bar{\sigma}^2$ in Eq. (26)]. Usually σ_0 is estimated from

$$\sigma_0^2 = \mathbf{V}'\mathbf{P}\mathbf{V}/(n-m). \qquad (11.29)$$

However, $\mathbf{V}'\mathbf{P}\mathbf{V}$ depends on systematic errors and vary drastically with changes in the background, while experience shows that many parameters are rather insensitive to background changes. σ_0 may also be estimated from the experimental data, though this is seldom done. The experimental value of σ_0 (properly scaled) does probably not vary too much from investigation to investigation in one laboratory if the same procedure is applied. Thus as an alternative to substituting $\mathbf{V}'\mathbf{P}\mathbf{V}/(n-m)$ for σ_0^2, one may use a value for σ_0 found by experience from experimental data.

REFERENCES

1 B.Andersen, H.M.Seip, T.G.Strand, and R.Stølevik, Acta Chem. Scand. 23, 3224 (1969).

2 O.Bastiansen, L.Hedberg, and K.Hedberg, J.Chem.Phys. 27, 1311 (1957).

3 K.Hedberg and M.Iwasaki, Acta Cryst. 17, 529 (1964).

4 A.Almenningen, O.Bastiansen, R.Seip, and H.M.Seip, Acta Chem. Scand. 18, 2115 (1964).

5 W.C.Hamilton: Statistics in Physical Science, Ronald Press, New York 1964.

6 Y.Murata and Y.Morino, Acta Cryst. 20, 605 (1966).

7 L.S.Bartell, Physical Methods in Chemistry, 4. Ed. (Edit. A.Weissberger and B.W.Rossiter), Interscience, New York 1971.

8 H.M.Seip, T.G.Strand, and R.Stølevik, Chem.Phys.Letters 3, 617 (1969).

9 M.A. MacGregor and R.K.Bohn, Chem.Phys.Letters 11, 29 (1971).

10 A.Almenningen, I.Hargittai, E.Kloster-Jensen, and R.Stølevik, Acta Chem.Scand. 24, 3463 (1970).

11 K.Kveseth, H.M.Seip, and R.Stølevik, Acta Chem.Scand. 25, 2975 (1971).

12 P.Markov and R.Stølevik, Acta Chem.Scand. 24, 2525 (1970).

13 Z.Náhlovská, B.Náhlovský, and T.G.Strand, Acta Chem.Scand. 24, 2617 (1970).

14 A.Almenningen, H.M.Seip, and R.Seip, Acta Chem.Scand. 24, 1697 (1970).

15 Z.Náhlovská, B.Náhlovský, and H.M.Seip, Acta Chem.Scand. 24, 1903 (1970).

16 R.Stølevik and E.Thom, Acta Chem.Scand. 25, 3205 (1971).

17 R.Seip, Acta Chem.Scand. 26, (1972).

18 A.Almenningen, S.Halvorsen, and A.Haaland, Acta Chem.Scand. 25, 1937 (1971).

19 H.M.Seip and R.Stølevik, Acta Chem.Scand. 20, 1535 (1966).

20 H.M.Seip and R.Seip, Acta Chem.Scand. 20, 2698 (1966).

21 S.P.Arnesen and H.M.Seip, Acta Chem.Scand. 20, 2711 (1966).

22 H.M.Seip and R.Seip, Acta Chem.Scand. 24, 3431 (1970).

CHAPTER 12

Representation and Experimental Determination of the Geometry of Free Molecules

KOZO KUCHITSU and S. J. CYVIN

Various definitions of the geometry of free molecules (r_e, r_z, r_v, r_α, $r_\alpha{}^0$, r_g, r_0, r_s, and r_a) are contrasted with one another on the basis of their physical significance and in terms of the mean values of Cartesian displacement coordinates. Processes for determining these 'structures' by means of current experimental techniques (rotation or rotation-vibration spectroscopy and gas electron diffraction) and schemes for mutual conversions are surveyed. Relative merits and limitations of each experimental method are discussed with comments on the complementary use of these methods.

自由分子の幾何学的構造の表現法と実験的決定法

朽　津　耕　三

東京大学理学部化学教室

スベン　シュビーン

ノールウェー工科大学物理化学教室

自由分子の幾何学的構造を表わす種々の定義を示し，それらの物理的意味を基礎として，直交変位座標の平均値を用いて相互の対比を行なった。これらの「構造」を現在ひろく用いられている実験技術（回転または振動回転スペクトルと気体電子線回折）によって決定するプロセスと，それらの構造を相互に変換する方法を概説した。さらに上記の両方法の長所と欠点を比較し，それらを相補的に応用する方法について論じた。

12.1. Introduction

In the present chapter, various existing definitions of internuclear distances and angles, and their relationships with one another are discussed. These definitions have their origin in the two main experimental procedures, which are based on entirely different principles, for determining the 'geometry' of a free molecule: (i) rotation and rotation-vibration spectroscopy and (ii) gas electron diffraction. For a precise interpretation of molecular structure, it is often profitable to make a critical comparison and a joint use of the results obtained by these method A general scheme is outlined in Fig. 1: Besides several operationa

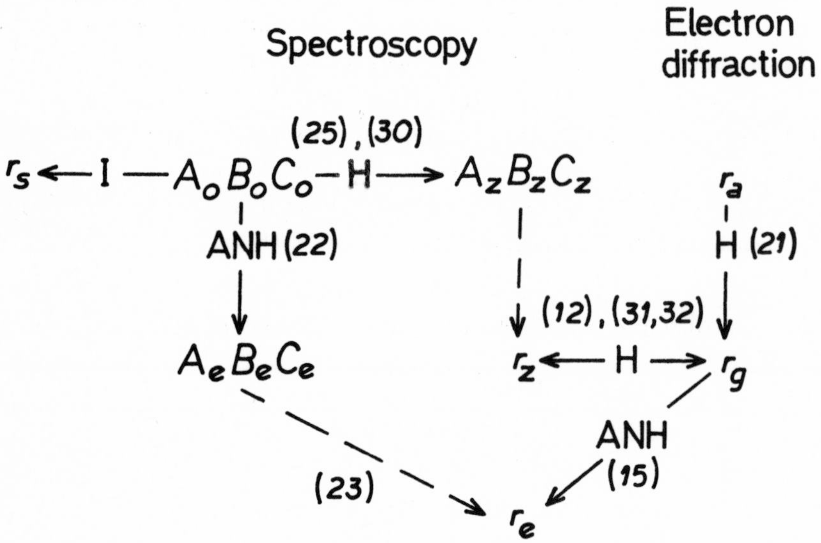

Fig. 12-1. Diagram showing the relationship among the rotational constants and distance parameters determined by spectroscopy and gas electron diffraction. Symbols H and ANH indicate harmonic and anharmonic corrections for vibrational effects, respectively; I stands for isotopic substitution. The numbers refer to the equations in the text.

Table 12-I. Section index to various distance parameters.

Symbol	Significance	Sections (Chapter 12)
r_e	Distance between equilibrium nuclear positions	2.1, 3.1.1
r_z, r_α^0	Distance between average nuclear positions (ground vibrational state)	2.2.1, 3.1.2, 3.1.3
r_v	Distance between average nuclear positions (excited vibrational state)	2.2.1, 3.1.2
r_α	Distance between average nuclear positions (thermal equilibrium)	2.2.1, 3.2
r_g	Thermal average value of internuclear distance	2.2.2, 3.2
r_0	Effective distance derived directly from ground state rotational constants	2.3.1
r_s	Effective distance derived from isotopic rotational constants by Kraitchman's method	2.3.2
r_a	Constant argument in molecular scattering intensity of electron diffraction	2.3.3
ϕ	Parameters representing bond angles	2.2.4

definitions (r_0, r_s, r_a, etc.), which can be derived directly from experiment, one can define structures which have more clear physical significance: i.e., the equilibrium structure (r_e) and several average structures (r_z, r_α, r_g, etc.). They can be reached from experimental measurements after appropriate harmonic (H) or anharmonic (ANH) treatments. An index of the contents is given in Table I.

Suggestive discussions on the present subject have been made in the review papers written by Lide[1] and Stoicheff;[2] see also Mulliken.[3]

12.2. Definitions of Structural Parameters

12.2.1. Equilibrium Structure

12.2.1.1. General Considerations

Among all the definitions of the geometrical structure of a molecule, the equilibrium structure seems to be the most fundamental. Since it is most easily accountable by molecular quantum theory, it is currently a subject of extensive ab initio or semi-empirical calculations.[4]

The r_e structure lays its basis on the theory of Born and Oppenheimer.[5] According to them, it is generally a good approximation to treat the motion of the electrons in a molecule separately from that of the nuclei.[6] The potential energy for nuclear vibrations can be described as a sum of electrostatic interactions among the nuclei plus electronic energy counted with respect to a fixed nuclear arrangement, while the electronic wavefunction involves nuclear coordinates as parameters.

12.2.1.2. Diatomic Molecule

For a bound state of a diatomic molecule, such a vibrational potential energy can be determined from spectroscopic experiments.[7] In fact, vibrational and rotational energy levels for a great many diatomic systems have been measured and used to evaluate the potential energy curves (Fig. 2) by means of the method of Rydberg, Klein and Rees (RKR)[8] or in terms of various empirical functions with adjustable parameters.[9]

There are three principal properties which represent the 'equilibrium structure': (i) the position of the potential minimum, (ii) the absolute value of the minimum energy with reference to an infinite nuclear separation, and (iii) the second (or any higher order) derivatives taken at the minimum position. They are

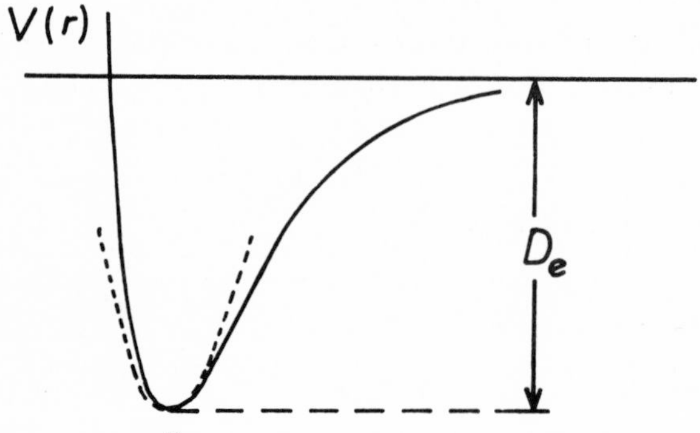

Fig. 12-2. Potential energy curve for a diatomic molecule (schematic).

called, respectively, the equilibrium internuclear distance (r_e), the dissociation energy (D_e), and the quadratic (harmonic) force constant (k_e) (or any of the anharmonic force constants). In this way, the structure is 'determined' provided the RKR potential curve or the above parameters are specified with sufficient accuracy.

12.2.1.3. Polyatomic Molecule

Likewise, the 'structure' of a polyatomic molecule can be determined if the potential energy hypersurface is specified as a function of the molecule-fixed nuclear coordinates, which satisfy the Eckart conditions,[10] of $3N-6$ degrees of freedom (or $3N-5$ for a linear system). The equilibrium position of the nuclei may be defined with reference to the principal axes of inertia. When the molecule in question has a low-frequency, large-amplitude vibration, e.g., when one of the vibrational quanta is much less than the thermal energy, the potential hypersurface in regard to that degree of freedom must be specified in detail. A variety of molecules having internal rotation, inversion, ring-puckering motion, etc., have been studied in this context.[11,12] If the molecule has no vibration of this sort, or (even if there is one) if one assumes that other degrees of freedom are practically unaffected by such a motion, then the nuclear motions can be treated as almost infinitesimal, so that one has only to determine nuclear positions corresponding to the minimum potential energy together with second and higher order 'force constants' in terms of the 'internal coordinates' (i.e., displacements in the bond distances, valence angles, and dihedral angles). This set is called the 'equilibrium structure',[13,14] and the distance between the equilibrium nuclear positions is called the r_e distance.

12.2.2. Average Structures

12.2.2.1. Average Nuclear Positions

In order to define the structure in the presence of vibrational motions, two alternative schemes are conceivable according to whether the average nuclear positions or the average internuclear distances are of primary concern.[15]

Molecular vibrations shift the average nuclear positions

from their equilibrium positions. A Cartesian displacement of a nucleus i from its equilibrium position at any moment can be represented in terms of the $\boldsymbol{l}$ matrix defined by Nielsen,[16,17]

$$\Delta\alpha_i = \alpha_i - \alpha_i^{(e)} = m_i^{-\frac{1}{2}} \sum_s l_{is}^{(\alpha)} Q_s , \qquad (12.1)$$

where m_i is the atomic mass, and Q_s for $s = 1$ to $3N-6$ (or $3N-5$) are the vibrational normal coordinates.* Three translational and three (or two) rotational coordinates may be included in the Q_s set so as to make the $\boldsymbol{l}$ matrix orthogonal;

$$\boldsymbol{l}\boldsymbol{l}' = \mathbf{E}, \qquad (12.2)$$

where $\mathbf{E}$ is the unit matrix of dimension $3N$. While the molecule-fixed Cartesian coordinates denoted as α may be taken in any direction, they are usually taken along the principal axes of inertia (a, b, c) with reference to the equilibrium structure. Explicit expressions and mutual relations of the $\boldsymbol{l}$ matrix elements are summarized in Refs. 18 and 19.

The Cartesian coordinates of a nucleus i in the principal axis system may be averaged over the vibrational wavefunction for a set of quantum numbers v: $\langle\alpha_i\rangle_v$. A set of the distances among these positions (r_v) and the geometrical angles defined in this system is called the r_v structure (vibrational average), and in a special case, when the average is taken for the ground vibrational state (for all v equal to zero), it is called the r_z structure** (zero-point average).[20-22] In spectroscopic experiments, one is usually concerned with the moments of inertia for the ground

---

* In the notation of Ref. 17 $\Delta\alpha_i = \alpha_i - \alpha_i^{(e)}$ corresponds to $x_i = X_i - X_i^e$, $y_i = Y_i - Y_i^e$ or $z_i = Z_i - Z_i^e$. The $\boldsymbol{l}$ matrix is identical with $\mathbf{l}$ of Ref. 18, but corresponds to $\mathbf{l}'$ in Refs. 17 and 19. See also Chapter 2, especially Eq. (2.8).

** The r_z structure was denoted as $\langle r\rangle$ by Herschbach and Laurie.[21,22] However, we shall here use r_z in order to avoid confusion with the average internuclear distance to be discussed in Section 12.2.2.2.

vibrational state, which is related to the r_z structure (see subsequent sections).

One can also define the thermal average nuclear positions at a certain vibrational temperature T,

$$\langle \alpha_i \rangle_T = \sum_v \langle \alpha_i \rangle_v \, e^{-E_v/kT} \Big/ \sum_v e^{-E_v/kT} , \qquad (12.3)$$

where the statistical weight is taken according to the Boltzmann distribution law. A set of the distances among the $\langle \alpha \rangle_T$ positions and their geometrical angles is called the r_α structure (thermal average) and is sometimes used in diffraction experiments.[23]

The r_v, r_z and r_α distances between the nuclei i and j can readily be compared with the $r_e(i-j)$ distance by an orthogonal transformation from the principal axis coordinates to local Cartesian coordinates.[24] Let the local z axis be taken in the direction from the equilibrium position of the nucleus i (arbitrarily taken as the origin) to that of j, and the x and y axes perpendicular to z. Then the average positions are shown in Table II, where the subscript ij is omitted from the differences for the sake of simplicity.* It may be shown that

$$r_v = [\langle \Delta x \rangle_v^2 + \langle \Delta y \rangle_v^2 + (r_e + \langle \Delta z \rangle_v)^2]^{\frac{1}{2}}$$

$$= r_e + \langle \Delta z \rangle_v + \frac{\langle \Delta x \rangle_v^2 + \langle \Delta y \rangle_v^2}{2 r_e} + \ldots . \qquad (12.4)$$

The order of magnitude of the linear averages,[24] $\langle \Delta x \rangle$, $\langle \Delta y \rangle$, and $\langle \Delta z \rangle$, is $\langle \Delta r^2 \rangle / r_e$. Therefore, the ratio of the square of a linear average and the quadratic average is of the order $\langle \Delta r \rangle^2 / \langle \Delta r^2 \rangle \approx \langle \Delta r^2 \rangle / r_e^2$. Since the linear averages are usually much less than

---

* The position coordinates (x_i, y_i, z_i) themselves were not introduced in Ref. 17. The displacement coordinate $z_{i(ij)}$ therein (see Chapter 13 of Ref. 17) corresponds to the present Δz_i. Δz or Δz_{ij} has the same meaning in both places. The same remarks apply to the corresponding x and y coordinates.

Table 12-II. Nuclear positions and their differences in the local Cartesian coordinate system.*

	Equilibrium	Momentary	Average
Nucleus i	0, 0, 0	Δx_i, Δy_i, Δz_i	$\langle\Delta x_i\rangle$, $\langle\Delta y_i\rangle$, $\langle\Delta z_i\rangle$
Nucleus j	0, 0, r_e	Δx_j, Δy_j, $r_e+\Delta z_j$	$\langle\Delta x_j\rangle$, $\langle\Delta y_j\rangle$, $r_e+\langle\Delta z_j\rangle$
Difference	0, 0, r_e	Δx, Δy, $r_e+\Delta z$	$\langle\Delta x\rangle$, $\langle\Delta y\rangle$, $r_e+\langle\Delta z\rangle$

* The origin of the coordinate system is temporarily taken at the equilibrium position of the nucleus i. By definition, z_j in equilibrium is denoted as r_e; see text.

0.01 Å, and r_e is of the order of 1 Å, the third term is less than 1% of the second term. Therefore, it is normally a good approximation to assume that

$$r_v = r_e + \langle\Delta z\rangle_v , \tag{12.5}$$

$$r_z = r_e + \langle\Delta z\rangle_0 , \tag{12.6}$$

$$r_\alpha = r_e + \langle\Delta z\rangle_T , \tag{12.7}$$

where 0 and T denote the zero-point and thermal averages, respectively. The average nuclear positions $\langle\alpha\rangle$ and the linear average displacements $\langle\Delta z\rangle$, etc., can be calculated if a set of linear average values of the normal coordinates $\langle Q_s\rangle$ are given. When the potential function V is expanded as

$$\frac{V}{hc} = \frac{1}{2}\sum_s \omega_s q_s^2 + \sum_{s\le s'\le s''}\sum\sum k_{ss's''}\, q_s q_{s'} q_{s''} + \ldots \tag{12.8}$$

in terms of the dimensionless normal coordinates,[16]

$$q_s = 2\pi(c\omega_s/h)^{\frac{1}{2}} Q_s , \tag{12.9}$$

it follows that $\langle Q_s\rangle_v$ and $\langle Q_s\rangle_T$ are expressed as linear functions of the cubic potential constants as[25]

$$\langle Q_s \rangle_v = -(h/4\pi^2 c\omega_s^3)^{\frac{1}{2}}[3k_{sss}(v_s + \tfrac{1}{2}) + \sum_{s' \neq s} g_{s'} k_{ss's'}(v_{s'} + \tfrac{1}{2} g_{s'})], \tag{12.10}$$

$$\langle Q_s \rangle_T = -(h/16\pi^2 c\omega_s^3)^{\frac{1}{2}}[3k_{sss} \coth(hc\omega_s/2kT)$$

$$+ \sum_{s' \neq s} g_{s'} k_{ss's'} \coth(hc\omega_{s'}/2kT)], \tag{12.11}$$

where $g_{s'}$ denotes the degeneracy of the s' normal mode, and a minor effect of the centrifugal force is left out.

12.2.2.2. Average Internuclear Distances

Another definition of the 'average structure' is to specify an average value of an internuclear distance. In a diffraction experiment, one can measure the thermal average internuclear distances (r_g) for bonded and nonbonded atom pairs.[26-28] For an $i-j$ pair shown in Table II, the r_g distance is given by

$$r_g = \langle [\Delta x^2 + \Delta y^2 + (r_e + \Delta z)^2]^{\frac{1}{2}} \rangle_T + \delta r$$

$$= r_e + \langle \Delta z \rangle_T + K + \ldots + \delta r = r_\alpha + K + \delta r + \ldots, \tag{12.12}$$

where

$$K = \frac{\langle \Delta x^2 \rangle_T + \langle \Delta y^2 \rangle_T}{2r_e}, \tag{12.13}$$

and δr represents a small effect of centrifugal force due to molecular rotation (proportional to the absolute temperature T).[25,29] For a system of low vibrational frequencies, temperature dependence of the r_g distances can be observed.[30-32] The r_g representation for a bonded atom pair has the following merits.[15]

(i) It is a direct measure of the bond length in thermal equilibrium. On the other hand, the r_v, r_z and r_α distances defined in Section 12.2.2.1 do not precisely correspond to any real 'bond lengths', since they essentially represent [according to Eqs. (6)-(8)] average projections of the bond lengths onto the line joining the equilibrium positions of the nuclei in question.

(ii) The r_e bond length can be estimated from r_g provided one has approximate knowledge about the anharmonicity of the bond-stretching vibration, since the average displacement of the bond length,

$$\langle \Delta r \rangle_T = r_g - r_e , \tag{12.14}$$

can be estimated by treating the bond as if it were a diatomic molecule.[26,33-35] For example, if the potential function has a term representing cubic anharmonicity, $a_3 \Delta r^3$, for a bond stretching displacement, then $\langle \Delta r \rangle_T$ is estimated to be[36]

$$\langle \Delta r \rangle_T \approx \frac{3}{2} a_3 \langle \Delta r^2 \rangle_T \approx \frac{3}{2} a_3 \langle \Delta z^2 \rangle_T . \tag{12.15}$$

Another equivalent definition of r_g in terms of the normalized probability distribution function $P(r)$ of an internuclear distance is given by Bartell:[26] With the $r_g(n)$ defined as 'the center of gravity of the $r^{-n}P(r)$ function',

$$r_g(n) = \int_0^\infty r^{-n+1} P(r)dr \Big/ \int_0^\infty r^{-n} P(r)dr . \tag{12.16}$$

r_g [representing the center of gravity of $P(r)$] may be redefined as $r_g(0)$.

12.2.2.3. Shrinkage Effects

A demerit of the r_g representation is that an r_g distance for a nonbonded atom pair, coupled with those for the corresponding bonded atom pairs, does not rigorously represent any physically significant molecular geometry. In order to define the bond angle, one must take into account corrections for linear or nonlinear shrinkage effects.[37,38]

Suppose three nuclei X, Y and Z are on a straight line in equilibrium. By symmetry, the r_v , r_z and r_α positions are also linear. However, r_g(X-Z) is in general less than the sum of r_g(X-Y) and r_g(Y-Z), and this failure of additivity, first observed by Bastiansen and his coworkers,[39] is called <u>linear shrinkage effect</u>. The following is a simple, qualitative explanation of this effect.

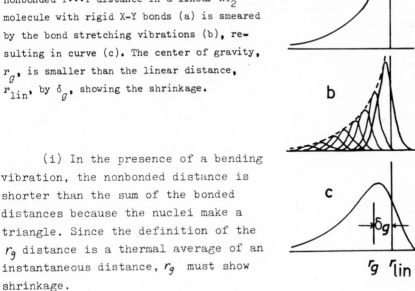

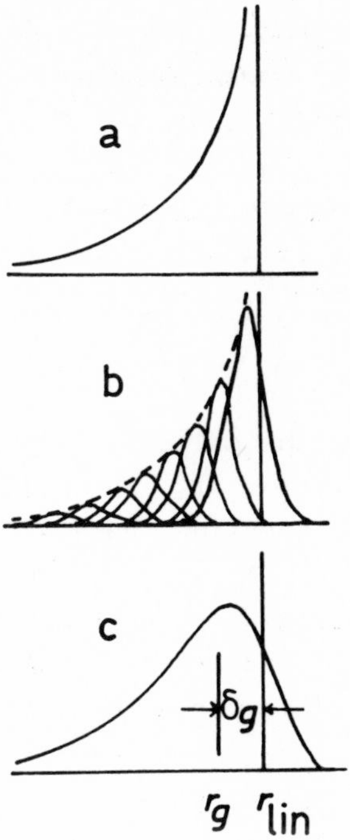

Fig. 12-3. Classical probability distribution curves (schematic) illustrating linear shrinkage effect. A distribution for the nonbonded Y•••Y distance in a linear XY_2 molecule with rigid X-Y bonds (a) is smeared by the bond stretching vibrations (b), resulting in curve (c). The center of gravity, r_g, is smaller than the linear distance, r_{lin}, by δ_g, showing the shrinkage.

(i) In the presence of a bending vibration, the nonbonded distance is shorter than the sum of the bonded distances because the nuclei make a triangle. Since the definition of the r_g distance is a thermal average of an instantaneous distance, r_g must show shrinkage.

(ii) Suppose a linear XY_2 molecule has rigid X-Y bonds. The probability distribution of the X-Y distance must vanish beyond r_{lin} (the distance corresponding to a linear arrangement); see Fig. 3a. If the bonds are flexible, the distribution curve is a superposition (c) of the curve (a) and a Gaussian-like distribution (b) representing a stretching vibration; see Fig. 3. The center of gravity (r_g) of the curve (c) is slightly smaller than r_{lin}, and this is the shrinkage effect.

An analytical expression of the shrinkage, δ_g, can be derived from Eq. (12).

$$\delta_g = r_g(\text{X-Y}) + r_g(\text{Y-Z}) - r_g(\text{X-Z}) \approx K(\text{X-Y}) + K(\text{Y-Z}) - K(\text{X-Z}). \tag{12.17}$$

The linear averages $\langle \Delta z \rangle$ cancel each other. Higher order terms in

Eq. (12) do not contribute to δ_g significantly unless the bending frequency is very low, as in the case of carbon suboxide.[32,40,41] Hence, only harmonic bending vibrations contribute to δ_g essentially.

The preceding argument may be readily extended to a nonlinear shrinkage effect.[38] It takes the simplest form for molecules with high symmetry. For example, the r_g(X-Y) and r_g(Y-Y) distances in a planar XY_3 molecule[17,42] are related to each other as

$$\delta_g = \sqrt{3}\, r_g(\text{X-Y}) - r_g(\text{Y-Y}) \approx \sqrt{3}\, K(\text{X-Y}) - K(\text{Y-Y}). \tag{12.18}$$

When the molecule has no totally-symmetric bending vibration, as in planar XY_3, in which case the r_e and r_v (r_z, r_α) structures make identical bond angles (say 120°) by symmetry, the $\langle \Delta z \rangle$ terms in Eq. (12) do not contribute to δ_g, and a calculation of the shrinkage can be reduced to that of the K terms.

Nonlinear shrinkage effect is especially important when the molecule has a torsion or puckering motion, which usually has a very low frequency and large vibrational amplitude. For instance, the r_g(trans C-C) distance in butane (3.91 Å) is found to be about 0.015 Å shorter than the predicted rigid trans distance because of the torsional motion around the C-C axis.[43,44] This effect is similar to that of the linear shrinkage illustrated in Fig. 3.[45] A conventional method of correction for the shrinkage effect of this sort has been discussed by Bartell and Kohl[46] in regard to the rotational isomerization of normal hydrocarbons. See Ref. 17 for further discussions of the shrinkage effect and experimental or theoretical values for specific molecules.

12.2.2.4. Average Values of Bond Angles

In parallel with internuclear distances, two alternative representations of 'average bond angles' are conceivable.[47]

(i) Since the r_v, r_z and r_α structures are defined as average nuclear positions, the angles made of the nuclear positions, ϕ_v, ϕ_z and ϕ_α, respectively, are defined unambiguously. If one starts from r_g distances, the ϕ_α angle can be obtained by subtracting theoretical estimates of K [Eqs. (12),(13)] from r_g, and this is equivalent to the correction for the shrinkage effect [Eqs.

(17) and (18)].

(ii) A thermal average value of an instantaneous bond angle, made of two bonded (r_1 and r_2) distances and one nonbonded (r_3) distance, may be defined by means of the cosine rule as

$$\phi_g = \langle \cos^{-1}[(r_1^2 + r_2^2 - r_3^2)/2r_1r_2] \rangle_T \, . \qquad (12.19)$$

Analytical relations among the ϕ_e, ϕ_z and ϕ_g angles for a nonlinear arrangement have been given.[47] According to numerical calculations, the differences among them for bonds involving, and not involving, hydrogen (or deuterium) atoms are, respectively, only a few tenths, and a few hundredths, of one degree. Hence, they are comparable with, or less than, current experimental uncertainties. Nevertheless, it may not necessarily be permissible to ignore these differences entirely. Particularly, the effect of anharmonicity in angle bending vibrations on the angle parameters have not been fully investigated.[26,34,35,48] For example, for some inorganic molecules with relatively low frequencies of bending vibration, ignorance of this effect may be one of the largest sources of systematic experimental error.[49,50]

12.2.3. Operational Definitions

12.2.3.1. r_0 Structure

The best-known operational definitions of molecular geometry may be r_0 and r_s. The r_0 structure is defined from the ground state rotational constants $B_0^{(\alpha)}$ (assumed to be inversely proportional to the effective moments of inertia) directly or with assumptions about a part of the structural parameters (e.g., Refs. 7, 51, 52). For a diatomic molecule, r_0 has a clear physical significance,[7]

$$r_0 = (h/8\pi^2 c\mu B_0)^{\frac{1}{2}} = \langle r^{-2} \rangle^{-\frac{1}{2}} = [r_g(1)\, r_g(2)]^{\frac{1}{2}}. \qquad (12.20)$$

It should be emphasized, however, that this equation should never be extended to an r_0 parameter in a polyatomic system, because Coriolis interactions are involved in the $B_0^{(\alpha)}$ constants.[1,16] A typical example of this problem is demonstrated in the inertia defect observed for a planar molecule.[18,53] Another example is given for methane, where the r_0(C-H) determined uniquely from B_0 is not equal to $\langle r^{-2}\rangle^{-\frac{1}{2}}$.[54] Appreciable uncertainties and inconsistencies have been observed for some of the r_0 parameters derived from isotopic substitutions, because of the effect of zero-point vibration, particularly for those related to hydrogen (deuterium) atoms.[55,56]

12.2.3.2. r_s Structure

When the rotational constants for various isotopic species are observed, the r_s structure (representing isotope substitution), i.e., a set of the nuclear coordinates being substituted by isotopes, can be determined by means of Kraitchman's method.[57-59] Many of the recent microwave studies report structures in this representation.[11,12,14] When Kraitchman's method is applied to the equilibrium rotational constants, it should give a correct r_e structure which is isotope invariant. In its application to the ground state rotational constants, one usually assumes that the (hypothetical) r_s structure is also isotope invariant. Since a major part of the vibrational effects are cancelled when the differences in the isotopic moments of inertia are taken, the r_s structure is presumed to be very close to the r_e structure.[1] Costain has shown that r_s structures for simple systems such as cyanides are remarkably consistent.[59] When one of the atoms is close to a principal axis, the precision of the atomic position can be improved by the method of double isotopic substitution.[60]

In this way, the r_s structure seems to give, in many circumstances, the best information on molecular geometry attainable by means of current spectroscopic methods. Even for a relatively large molecule, it seems possible to estimate the uncertainties in the r_s parameters[61] by means of the first-moment condition, for instance. However, one needs to be cautious in comparing the r_s parameters for one molecule with those for analogous molecules (e.g., comparing distance parameters to within $\pm$ 0.01 Å or better), particularly if one of the relevant atoms is close to a principal axis. Bond lengths for a number of aliphatic molecules in the r_s representation are compared in Table III with those in r_g and r_z representations, which seem to have clearer physical significance. The differences between r_g and r_s parameters are estimated

Table 12-III. Comparison of bond distances in r_s, r_g and r_z parameters.*

Molecule	Bond	r_s	Ref.	r_g	r_z	Ref.
Acetone	C=O	1.222(3)	62	1.212(4)	1.209(4)	63[a]
	C-C	1.507(3)		1.518(3)	1.515(3)	
Acetyl chloride	C=O	1.192(10)	64	1.187(3)		65
	C-C	1.499(10)		$1.507_5(5)$		
	C-Cl	1.789(5)		1.798(3)		
Acrolein	C=O	1.219(5)	66	1.217(3)	$1.212_5(3)$	67
	C=C	1.345(3)		1.345(3)	1.341(3)	
	C-C	1.470(3)		1.484(4)	1.482(4)	
Acrylonitrile	C≡N	1.163_7	68	1.167(4)	1.161(4)	69
	C=C	1.338_9		1.343(4)	1.341(4)	
	C-C	1.425_6		1.438(3)	1.435(3)	
t-Butyl chloride	C-C	1.530(2)	70	1.528(2)	1.525(3)	71,72
	C-Cl	1.803(2)		1.828(5)	1.827(5)	
Isobutene	C=C	1.330(4)	73	1.342(3)	1.339(3)	74
	C-C	1.507(3)		1.507(2)	1.504(2)	
Propane	C-C	1.526(2)	75	1.532(3)	1.533_7	76
Propene	C=C	1.336(4)	77	1.341(3)	$1.338_5(3)$	78
	C-C	1.501(4)		1.504(3)	1.502(3)	
Propynal	C≡C	$1.208_9(1)$	79	1.211(4)	1.205(6)	80
	C=O	$1.215_0(1)$		1.214(5)	1.212(5)	
	C-C	$1.444_6(1)$		1.453(3)	1.449(2)	

* In Å units. Uncertainties quoted in the original references, to be attached to the last significant digits, are given in parentheses; e.g., 1.507(3) represents 1.507 ± 0.003 Å. Most r_z parameters have been derived from the joint analysis of electron diffraction and microwave data and are denoted as r_{av} in the references.

[a] Also private communication (1972).

to be of the order of 0.01 ± 0.01 Å, but they sometimes deviate from this range with no apparent systematic trends. The origin of the largest discrepancy observed in the r_g and r_s parameters of the C-Cl bond in t-butyl chloride (0.025 Å) has been discussed by Hilderbrandt and Wieser[72] in terms of the proximity of the tertiary carbon atom to the b axis (0.46 Å). It seems worthwhile to make further experimental and theoretical studies of the structural parameters in different definitions for simple molecules.

The method of isotope substitution can also be used to evaluate the r_z structure. However, the r_z structure thus determined may have appreciable systematic error due to strong dependence on small isotopic differences in the r_z structure (of the order of 1×10^{-4} Å).[15,61,67,69,71,72]

12.2.3.3. r_a Structure

The molecular intensity measured by gas electron diffraction consists of contributions proportional to $\sin s[r_a - \kappa s^2 + o(s^4)]$ from all the atom pairs.[26-28,81] The function in brackets of the argument depends only very slightly on the scattering variable, $s = 4\pi\sin(\theta/2)/\lambda$, under normal experimental conditions. On a suitable assumption about the asymmetry parameter κ, one can determine the distance parameter r_a (representing 'argument') by means of, for instance, a least-squares analysis.[82-84] It has been shown[27,81] that r_a is rigorously equal to the $r_g(1)$ defined in Eq. (16), the center of gravity of $P(r)/r$. For a Gaussian (or near-Gaussian) probability distribution, it is further shown that r_a is related to r_g in terms of a mean-square amplitude, l^2,

$$r_a = r_g - \frac{l^2}{r} . \qquad (12.21)$$

Rigorously speaking, the parameters in the second term should be specified in more detail in terms of the parameters characterizing the $P(r)$ function. The mean-square amplitudes, $l^2 = \langle \Delta r^2 \rangle$, can be defined in several different ways: $l_e^2 = \langle (r - r_e)^2 \rangle$, $l_g^2 = \langle (r - r_g)^2 \rangle$, etc.; their analytical expressions have been given.[27,28,81] However, since the differences depending on alternative definitions of r and l are usually much smaller than the experimental uncertainties in r_a, it is a current practice of gas electron diffraction to convert r_a into r_g by the use of either experimental or theoretical (harmonic) mean-square amplitudes.

12.2.4. Summary

The r_e representation of molecular geometry is unambigous and is suitable for comparison with that calculated by molecular quantum theory. If the r_e structure is not available, the r_g parameter seems to be one of the best possible choices for representing bond distances, while the ϕ_z , ϕ_α and ϕ_g parameters may be used for defining angles. These 'average structures' are subject to vibrational effects, and for a system with a large-amplitude vibration an experimental determination of these parameters may result in appreciable uncertainties. This problem is left for future studies (see Section 12.3.3).

12.3. Determination of Structural Parameters from Experiments

12.3.1. Spectroscopic Methods

12.3.1.1. Rotational Constants

Spectroscopic information about molecular geometry ordinarily comes from the rotational constants, $B_v^{(\alpha)}$ ($\alpha = a, b, c$), which are defined as the coefficients of the quadratic operators representing the components of the total angular momentum (P_α^2) in the Hamiltonian.[16] The rotational constants depend on a set of vibrational quantum numbers v_s as[51]

$$B_v^{(\alpha)} = B_e^{(\alpha)} - \sum_s \alpha_s^{(\alpha)}(v_s + \tfrac{1}{2}g_s) + \sum_s \sum_{s'} \beta_{ss'}^{(\alpha)}(v_s + \tfrac{1}{2}g_s)(v_{s'} + \tfrac{1}{2}g_{s'}) + \ldots , \qquad (12.22)$$

where s and g_s represent the number and the degeneracy of the normal modes.

12.3.1.2. r_e Structure

When all the independent $\alpha_s^{(\alpha)}$ constants of the molecule are observed, $B_0^{(\alpha)}$ can be extrapolated into the equilibrium rotational constants $B_e^{(\alpha)}$. For this purpose, however, one must make complete measurements of the rotational constants for (at least) all the first excited vibrational states as well as for the ground state. The second-order effect, β, seems to be of minor importance[85] unless the molecule has a large-amplitude vibration.[86] The equilibrium rotational constants $B_e^{(\alpha)}$ (in cm^{-1}) thus determined correspond, to a high degree of accuracy, to the moments of inertia (in amu Å^2) for the r_e structure defined in Section 12.2.1, $I_e^{(\alpha)}$, as

$$B_e^{(\alpha)} = \frac{Nh}{8\pi^2 c I_e^{(\alpha)}} \ . \qquad (12.2$$

Equilibrium structures have been determined mainly by microwave or infrared spectroscopy for a number of simple molecules, e.g. H_2O, NH_3, CH_4, H_2S, H_2Se, HNO, NF_3, SO_2, OF_2, O_3, SeO_2, SiF_2, GeF_2, CO_2, CS_2, HCN, ClCN, HCO, C_2H_2, and their deuterides. A list of their equilibrium bond lengths and bond angles are given in Ref. 14.

12.3.1.3. r_z Structure

According to the theory of rotation-vibration interaction, the constants $\alpha_s^{(\alpha)}$ which appear in Eq. (22) are composed of harmonic and anharmonic terms,

$$\alpha_s^{(\alpha)} = \alpha_s^{(\alpha)}(\text{harmonic}) + \alpha_s^{(\alpha)}(\text{anharmonic}), \qquad (12.2$$

where a minor correction for centrifugal distortion is left out.[16]

The harmonic terms are shown to depend on the vibrational frequencies ω_s and the elements of the $\boldsymbol{l}$ matrix, which also appear in Eq. (1),

$$\alpha_{s\sigma}^{(\alpha)}(\text{harmonic}) = -\frac{2[B_e^{(\alpha)}]^2}{g_s\,\omega_s}\Big\{3A_{s\sigma s\sigma}^{(\alpha\alpha)} + 4\sum_{\substack{s'\sigma'\\ s'\neq s}}[\zeta_{s\sigma s'\sigma'}^{(\alpha)}]^2\omega_s^2(\omega_s^2-\omega_{s'}^2)^{-1} - 3\sum_{\sigma'}[\zeta_{s\sigma s\sigma'}^{(\alpha)}]^2\Big\}, \qquad (12.2$$

where σ and σ' represent one of the g_s degenerate modes of s, and

$$A_{s\sigma s\sigma}^{(\alpha\alpha)} = 1 - \sum_i [l_{is\sigma}^{(\alpha)}]^2 \ , \qquad (12.2$$

$$\zeta_{s\sigma s'\sigma'}^{(\alpha)} = \sum_i [l_{is\sigma}^{(\beta)} l_{is'\sigma'}^{(\gamma)} - l_{is'\sigma'}^{(\beta)} l_{is\sigma}^{(\gamma)}];$$

$$(\alpha \neq \beta \neq \gamma,\ \alpha \neq \gamma). \qquad (12.27)$$

On the other hand, the anharmonic terms depend linearly on the cubic constants of the potential function, k, given in Eq. (8),[16]

$$\alpha_{s\sigma}^{(\alpha)}(\text{anharmonic}) = -4\pi\left(\frac{c}{h}\right)^{\frac{1}{2}} \frac{[B_e^{(\alpha)}]^2}{g_s}\left[3k_{sss}\, a_{s\sigma}^{(\alpha\alpha)}\, \omega_s^{-\frac{3}{2}} + \sum_{\substack{s'\sigma' \\ s' \neq s}} k_{sss'}\, a_{s'\sigma'}^{(\alpha\alpha)}\, \omega_{s'}^{-\frac{3}{2}}\right], \qquad (12.28)$$

where

$$a_{s\sigma}^{(\alpha\alpha)} = 2\sum_i m_i^{\frac{1}{2}}[\beta_i^{(e)} l_{is\sigma}^{(\beta)} + \gamma_i^{(e)} l_{is\sigma}^{(\gamma)}]. \qquad (12.29)$$

Except for very simple molecules, it is difficult to make either experimental determination or theoretical evaluation[34,48,87] of $\alpha_s^{(\alpha)}$(anharmonic).

When the quadratic force field of the molecule is known (even approximately) one can calculate $\alpha_s^{(\alpha)}$(harmonic), by which the rotational constants $B_0^{(\alpha)}$ are corrected for the effect of rotation-vibration interaction,

$$B_z^{(\alpha)} = B_0^{(\alpha)} + \frac{1}{2}\sum_s g_s\, \alpha_s^{(\alpha)}(\text{harmonic}). \qquad (12.30)$$

In parallel with Eq. (23), it is shown in general that $B_z^{(\alpha)}$ defined above is related, to a close approximation, to the moments of inertia $I_\alpha^{(z)}$ for the r_z structure defined in Section 12.2.2.1. In other words, the $\alpha_s^{(\alpha)}$(anharmonic) correction fully accounts for the differences in the nuclear positions in the r_e and r_z structures.[25] If desired, a similar correction can be applied to the rotational constants for excited vibrational states to determine r_v structures.

12.3.1.4. Limitations

The number of independent geometrical parameters (N) which is equal to that of the totally symmetric normal modes, is compared in Table IV with the number of independent rotational constants (n) for a number of simple molecular types. In case (a), the numbers are equal to each other, and a unique structure can be determined directly. One more observable, say one of the isotopic rotational constants, is needed in case (b). The difference $N - n$ increases as the molecule becomes more complicated. If a sufficient number of isotopic rotational constants are available, they provide a practical solution of this problem. According to whether B_e, B_z, or B_0 are used in the process, the r_e, r_z, or r_s (r_0) structures are derived. Isotopic differences in the r_z structure due to the zero-point vibration, which are very small but cannot be estimated with sufficient accuracy, limit the use of isotope substitution for determining the r_z structure.[15,21,22,67,71,72,88] The uncertainties from this origin can be much larger than the experimental errors in the moments of inertia $I_\alpha^{(z)}$, as shown in Fig. 4. Other principal limitations of the spectroscopic method are the following:

(i) It is hard to determine accurate rotational constants for

Table 12-IV. Examples of molecular types.*

		n	(a)	(b)
Linear		1	CO_2	<u>OCS</u>, C_2N_2
Planar	sym.top	1	BF_3	C_6H_6
	asym.top	2	<u>SO_2</u>	<u>ONF</u>, <u>cis-N_2F_2</u>, C_2H_4, <u>H_2CO</u>, trans-N_2F_2
Nonplanar	sph.top	1	P_4, CH_4, SF_6	-
	sym.top	2	<u>NH_3</u>	C_2H_6, <u>CH_3F</u>
	asym.top	3	-	<u>H_2O_2</u>, <u>CH_2F_2</u>, <u>OSF_2</u>

*Simple molecules are here classified according to the number of independent rotational constants (n) and the number of parameters sufficient to describe equilibrium molecular geometry (N); typical molecules with $n = N$ are grouped in (a), and those with $n = N - 1$ are grouped in (b). Polar molecules, which can be studied by the ordinary techniques of pure rotational spectroscopy, are underlined.

relatively complicated nonpolar molecules, because microwave spectroscopy is not generally applicable and the resolution of infrared or Raman spectrometers is not sufficient to observe rotational fine structure.

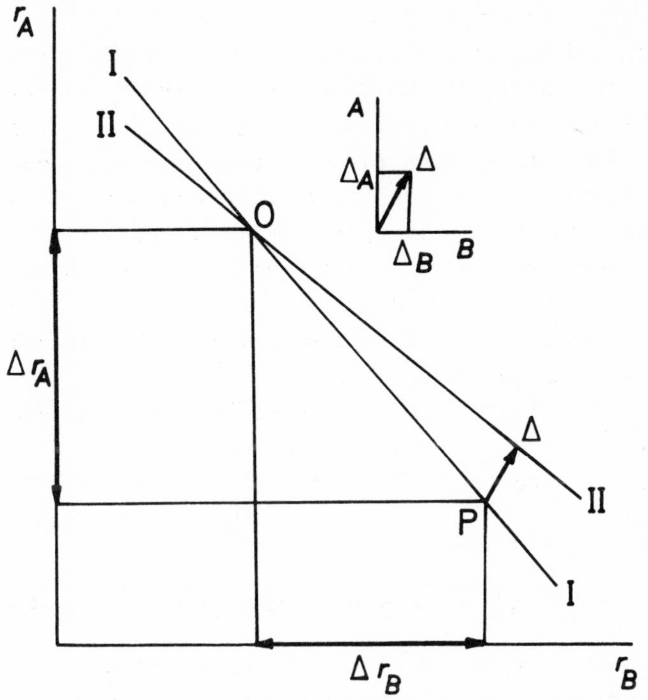

Fig. 12-4. A schematic two-dimensional diagram illustrating the influence of uncertainties in the isotopic differences on the uncertainties in the distance parameters. Suppose the rotational constants for isotopic species I and II measured by spectroscopy represent nearly identical curves for the distance parameters r_A and r_B. If the r_A and r_B distances for II and I differ by Δ_A and Δ_B, then the parameter set satisfying the condition is given by P, where the difference makes the vector Δ. If this difference is ignored, the parameter set shifts to O. The differences in the distances, Δr_A and Δr_B, are several fold larger than Δ_A and Δ_B; thus small uncertainties in Δ cause large uncertainties in the distance parameters.

(ii) It is not always possible to observe a sufficient set of independent rotational constants. For example, some of the atoms have only one stable nuclide (say fluorine), and for many of the symmetric-top molecules it has often been difficult to determine very accurate axial rotational constants A or C.

12.3.2. Electron Diffraction

With the aid of the theoretical corrections for the parallel and perpendicular mean-square amplitudes, the r_g and r_α distances can be derived from a diffraction experiment [Eqs. (12) and (21)]. In order to relate r_α with the r_z determined from spectroscopy, r_α must be extrapolated to zero Kelvin.[23] In principle, this step can be followed experimentally by measuring the r_α parameters at different temperatures. However, no successful measurements of this sort have yet been reported because of the limit of experimental accuracy, whereas temperature dependence of the distance parameters have sometimes been observed.[30-32]

It is therefore necessary to estimate the temperature effect by theory.[15,28,89] For bonded distances, this may be done by assuming that the bond is a diatomic system, i.e. vibrating independently from the rest of the molecule. It follows from Eq. (15 that

$$r_g - r_g{}^0 = \langle \Delta r \rangle_T - \langle \Delta r \rangle_0 \approx \frac{3}{2} a_3 (\langle \Delta z^2 \rangle_T - \langle \Delta z^2 \rangle_0), \qquad (12.3$$

where a constant a_3(~ 2 Å^{-1}) is assumed to be analogous to the Morse parameter a for a diatomic molecule.[24,34,35] For evaluating $r_\alpha{}^0$ from $r_g{}^0$, Eq. (12) can be applied with the quadratic averages taken at zero Kelvin,

$$r_g{}^0 - r_\alpha{}^0 \approx \frac{\langle \Delta x^2 \rangle_0 + \langle \Delta y^2 \rangle_0}{2 r_e}. \qquad (12.3$$

As for the temperature dependence of the angle parameters, which may depend sensitively on the anharmonicity of bending displacements, there is much to be investigated in the future. Existing experimental evidence (in particular, accurate ϕ_e and ϕ_z angles determined by spectroscopy) suggests that ϕ_α and $\phi_\alpha{}^0$ should not differ by more than several minutes (at least for angles not involving hydrogen or deuterium atoms).[47,49] In summary, Fig. 5 shows the diagram of extrapolation.

In many circumstances, electron diffraction has proved to be a useful method for accurate determination of molecular geometry. However, the following are the main limitations.

(i) Only the distance parameters averaged for thermal equilibrium, instead of those for individual quantum states, can be obtained. Therefore, the parameters derived from electron diffraction are subject to various vibrational effects (Table V).

(ii) Even with extreme caution, it is generally hard to eliminate all the systematic errors in the measurement of electron scattering intensity, cf. Chapter 10. A similar feature is present

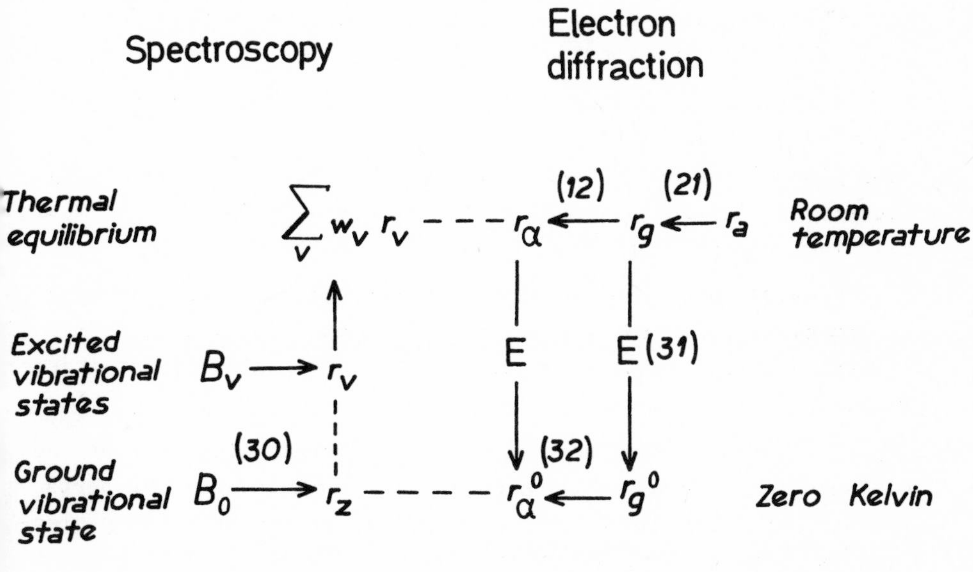

Fig. 12-5. A diagram showing conversions among electron diffraction parameters and those from rotation and rotation-vibration spectroscopy. E indicates extrapolation. The numbers refer to the equations in the text.

Table 12-V. Vibrational effects on internuclear distances.

	r_g	r_α	$r_\alpha^0 \approx r_z$	r_e
Shrinkage effect[a]	present	absent	absent	absent
Temperature effect[b]	present	present	absent	absent
Isotope effect[c]	present	present	present	absent

[a] Section 2.2.3.
[b] Section 2.2.2; Refs. 30-32.
[c] Refs. 15, 24, 67, 89.

in any spectroscopic experiment where a quantitative measurement of intensity (the vertical scale) is usually much more difficult than that of frequency (the horizontal scale). Any accident in the experiment or analysis can cause a serious (very often undiscovered) systematic error in the geometrical parameters derived from a diffraction experiment.

(iii) The resolving power of internuclear distances is only about 0.1 Å. This limit originates mainly from the zero-point vibration of the nuclei and cannot be controlled experimentally. Closely spaced, inequivalent distances are measured only as their weighted average values.

(iv) When the molecule under study has elements of very different atomic numbers, the parameters related to lighter atoms (weaker electron scatterers) may be very uncertain, if not indeterminate.

(v) Besides the above errors, the overall dimension of the distance parameters may be uncertain (by as much as a few parts per thousand) because of an error in the scale factor (the electron wavelength times the camera length).

12.3.3. Combined Use of Spectroscopic and Diffraction Methods

In view of the merits and demerits of the spectroscopic and diffraction methods mentioned above, it is worth while to investigate, when they are simultaneously available, whether the structure can be determined with more accuracy if rotational constants and diffraction intensities are used together than if they are used separately. This procedure may be particularly useful to resolve closely-spaced internuclear distances, so that one of the demerits of electron diffraction may be covered.

The method of least squares may be applied to this problem by treating spectroscopic and diffraction measurements as independent observables.[90,91] Acrolein,[67] vinylacetylene,[92] acrylonitrile,[69] $XeOF_4$,[93] and t-butyl chloride[71,72] are typical examples. By this method, it is possible to determine uniquely most of the hydrogen parameters, which cannot usually be resolved by the diffraction method alone. In many cases the r_g or r_z parameters determined in this way differ from the corresponding r_s parameters by more than the uncertainties in the former parameters (Table III). However, the following problems should be taken into account in the above procedure so as not to introduce additional systematic errors.

(i) The diffraction intensity must be free from various systematic errors of experimental and theoretical origin;[94] cf. Chapter 10. In particular, the scale factor must be sufficiently correct; otherwise, it should be calibrated against the rotational constants. A useful check may be made if a molecule of known structure is analysed simultaneously by the same procedure.

(ii) The structural parameters to be determined must be made consistent among one another by making proper corrections as treated in Section 12.2.

(iii) Even if rotational constants for more than one nuclide are available, it is not always advantageous to take them all in the analysis as independent observables. Only those constants which can improve the accuracy of the r_z structure, despite the uncertainty in the isotope effect [see Sections 12.2.3.2 and 12.3.1.3], can be used for this purpose.[67]

(iv) A question arises as to how relative weights should be assigned on the diffraction and spectroscopic measurements. It seems permissible to adjust the weights so as to make the residuals of the standard errors of the measurements comparable with their inherent uncertainties.[91]

(v) If the molecule has a large-amplitude vibration, special caution is necessary to estimate the vibrational corrections in regard to that degree of freedom.[86,95] For instance, the effective moments of inertia for a molecule with internal rotation have recently been formulated.[96-98]

Examples of the critical comparisons and combined use of spectroscopic and diffraction measurements are described in a separate review article.[99]

REFERENCES

1 D.R.Lide,Jr., Tetrahedron 17, 125 (1962).
2 B.P.Stoicheff, Tetrahedron 17, 135 (1962).
3 R.S.Mulliken, Tetrahedron 17, 247 (1962).
4 L.C.Allen, Ann.Rev.Phys.Chem. 20, 315 (1970).
5 M.Born and J.R.Oppenheimer, Ann.Physik 84, 457 (1927).
6 P.R.Bunker, J.Mol.Spectry. 36, 306 (1970).
7 G.Herzberg: Molecular Spectra and Molecular Structure. Vol. I. Spectra of Diatomic Molecules, 2. Ed., D. van Nostrand, Princeton, N.J. 1950.
8 J.T.Vanderslice, E.A.Mason, W.G.Maisch, and E.R.Lippincott, J.Mol.Spectry. 3, 17 (1951); erratum: ibid. 5, 83 (1960).
9 D.Steele, E.R.Lippincott, and J.T.Vanderslice, Revs. Modern Phys. 34, 239 (1962).
10 C.Eckart, Phys.Rev. 47, 552 (1935).
11 V.W.Laurie, Accts.Chem.Res. 3, 331 (1970).
12 H.D.Rudolph, Ann.Rev.Phys.Chem. 21, 73 (1971).
13 Y.Morino, Pure Appl.Chem. 18, 323 (1969).
14 Y.Morino and E.Hirota, Ann.Rev.Phys.Chem. 20, 139 (1970).
15 K.Kuchitsu, J.Chem.Phys. 49, 4456 (1968).
16 H.H.Nielsen, Revs. Modern Phys. 23, 90 (1951).
17 S.J.Cyvin: Molecular Vibrations and Mean Square Amplitudes, Universitetsforlaget, Oslo, and Elsevier, Amsterdam, 1968.
18 T.Oka and Y.Morino, J.Mol.Spectry. 6, 472 (1961).
19 J.H.Meal and S.R.Polo, J.Chem.Phys. 24, 1119 (1956).

20 T.Oka, J.Phys.Soc.Japan 15, 2274 (1960).
21 D.R.Herschbach and V.W.Laurie, J.Chem.Phys. 37, 1668 (1962).
22 V.W.Laurie and D.R.Herschbach, J.Chem.Phys. 37, 1687 (1962).
23 Y.Morino, K.Kuchitsu, and T.Oka, J.Chem.Phys. 36, 1108 (1962).
24 K.Kuchitsu and L.S.Bartell, J.Chem.Phys. 36, 2460 (1962).
25 M.Toyama, T.Oka, and Y.Morino, J.Mol.Spectry. 13, 193 (1964).
26 L.S.Bartell, J.Chem.Phys. 23, 1219 (1955).
27 K.Kuchitsu, Bull.Chem.Soc.Japan 40, 498 (1967).
28 K.Kuchitsu, Bull.Chem.Soc.Japan 40, 505 (1967).
29 M.Iwasaki and K.Hedberg, J.Chem.Phys. 36, 2961 (1962).
30 K.Hedberg and M.Iwasaki, J.Chem.Phys. 36, 589 (1962).
31 T.Ukaji and K.Kuchitsu, Bull.Chem.Soc.Japan 39, 2153 (1966).
32 A.Almenningen, S.P.Arnesen, O.Bastiansen, H.M.Seip, and R.Seip, Chem.Phys.Letters 1, 569 (1968).
33 D.R.Herschbach and V.W.Laurie, J.Chem.Phys. 35, 458 (1961).
34 K.Kuchitsu and Y.Morino, Bull.Chem.Soc.Japan 38, 805 (1965).
35 K.Kuchitsu and Y.Morino, Bull.Chem.Soc.Japan 38, 814 (1965).
36 L.S.Bartell, J.Chem.Phys. 38, 1827 (1963).
37 Y.Morino, J.Nakamura, and P.W.Moore, J.Chem.Phys. 36, 1050 (1962).
38 Y.Morino, S.J.Cyvin, K.Kuchitsu, and T.Iijima, J.Chem.Phys. 36, 1109 (1962).
39 O.Bastiansen and M.Trætteberg, Acta Cryst. 13, 1108 (1960).
40 M.Tanimoto, K.Kuchitsu, and Y.Morino, Bull.Chem.Soc.Japan 43, 2776 (1970).
41 A.Clark and H.M.Seip, Chem.Phys.Letters 6, 452 (1970).
42 K.Kuchitsu and S.Konaka, J.Chem.Phys. 45, 4342 (1966).
43 R.A.Bonham and L.S.Bartell, J.Am.Chem.Soc. 81, 3491 (1959).
44 K.Kuchitsu, Bull.Chem.Soc.Japan 32, 748 (1959).
45 J.Karle and H.Hauptman, J.Chem.Phys. 18, 875 (1950).
46 L.S.Bartell and D.A.Kohl, J.Chem.Phys. 39, 3097 (1963).
47 K.Kuchitsu, Bull.Chem.Soc.Japan 44, 96 (1971).
48 K.Kuchitsu and Y.Morino, Spectrochim.Acta 22, 33 (1966).
49 Y.Morino, K.Kuchitsu, and T.Moritani, Inorg.Chem. 8, 867 (1969).
50 K.Kuchitsu, T.Shibata, A.Yokozeki, and C.Matsumura, Inorg.Chem. 10, 2584 (1971).
51 G.Herzberg: Molecular Spectra and Molecular Structure. Vol. II. Infrared and Raman Spectra of Polyatomic Molecules, D. van

Nostrand, Princeton, N.J. 1945.
52 S.G.W.Ginn, J.K.Kenny, and J.Overend, J.Chem.Phys. 48, 1571 (1968).
53 B.T.Darling and D.M.Dennison, Phys.Rev. 57, 128 (1940).
54 K.Kuchitsu and L.S.Bartell, J.Chem.Phys. 36, 2470 (1962).
55 D.Kivelson, E.B.Wilson, and D.R.Lide, J.Chem.Phys. 32, 205 (196
56 C.H.Townes and A.L.Schawlow: Microwave Spectroscopy, McGraw-Hil New York 1955; p. 40.
57 J.Kraitchman, Am.J.Phys. 21, 17 (1953).
58 C.C.Costain, J.Chem.Phys. 29, 864 (1958).
59 C.C.Costain, Trans.Am.Cryst.Assoc. 2, 157 (1966).
60 L.Pierce, J.Mol.Spectry. 3, 575 (1959).
61 J.Casado, L.Nygaard, and G.O.Sørensen, J.Mol.Structure 8, 211 (1971).
62 R.Nelson and L.Pierce, J.Mol.Spectry. 18, 344 (1965).
63 T.Iijima, Bull.Chem.Soc.Japan 43, 1049 (1970).
64 M.Sinnott, J.Chem.Phys. 34, 851 (1961).
65 S.Tsuchiya and M.Kimura, Bull.Chem.Soc.Japan 45, 736 (1972).
66 E.A.Cherniak and C.C.Costain, J.Chem.Phys. 45, 104 (1966).
67 K.Kuchitsu, T.Fukuyama, and Y.Morino, J.Mol.Structure 4, 41 (1969).
68 C.C.Costain and B.P.Stoicheff, J.Chem.Phys. 30, 777 (1959).
69 T.Fukuyama and K.Kuchitsu, J.Mol.Structure 5, 131 (1970).
70 D.R.Lide,Jr. and M.Jen, J.Chem.Phys. 38, 1504 (1963).
71 R.L.Hilderbrandt and J.D.Wieser, J.Chem.Phys. 55, 4648 (1971).
72 R.L.Hilderbrandt and J.D.Wieser, J.Chem.Phys. 56, 1143 (1972).
73 L.H.Scharpen and V.W.Laurie, J.Chem.Phys. 39, 1732 (1963).
74 I.Tokue, T.Fukuyama, and K.Kuchitsu, to be published.
75 D.R.Lide,Jr., J.Chem.Phys. 33, 1514 (1960).
76 T.Iijima, Bull.Chem.Soc.Japan (in press).
77 D.R.Lide,Jr. and D.Christensen, J.Chem.Phys. 35, 1374 (1961).
78 I.Tokue, T.Fukuyama, and K.Kuchitsu, to be published.
79 C.C.Costain and J.R.Morton, J.Chem.Phys. 31, 389 (1959).
80 M.Sugié, T.Fukuyama, and K.Kuchitsu, J.Mol.Structure (in press)
81 K.Kuchitsu and L.S.Bartell, J.Chem.Phys. 35, 1945 (1961).
82 K.Hedberg and M.Iwasaki, Acta Cryst. 17, 529 (1964).
83 Y.Morino, K.Kuchitsu, and Y.Murata, Acta Cryst. 18, 549 (1965).

84 H.M.Seip, Acta Chem.Scand. 19, 1955 (1965).
85 S.Saito, J.Mol.Spectry. 30, 1 (1969).
86 J.T.Hougen, P.R.Bunker, and J.W.C.Johns, J.Mol.Spectry. 34, 136 (1970).
87 Y.Morino, K.Kuchitsu, and S.Yamamoto, Spectrochim.Acta 24A, 335 (1968).
88 W.Lafferty, D.R.Lide, and R.A.Toth, J.Chem.Phys. 43, 2063 (1965).
89 K.Kuchitsu, J.Chem.Phys. 44, 906 (1966).
90 G.Dallinga and L.H.Toneman, J.Mol.Structure 1, 11 (1967-68).
91 K.Kuchitsu, T.Fukuyama, and Y.Morino, J.Mol.Structure 1, 463 (1967-68).
92 T.Fukuyama, K.Kuchitsu, and Y.Morino, Bull.Chem.Soc.Japan 42, 379 (1969).
93 E.J.Jacob, H.B.Thompson, and L.S.Bartell, J.Mol.Structure 8, 383 (1971).
94 R.L.Hilderbrandt and R.A.Bonham, Ann.Rev.Phys.Chem. 22, 279 (1971).
95 R.Meyer and E.B.Wilson, J.Chem.Phys. 53, 3969 (1970).
96 P.R.Bunker, J.Chem.Phys. 47, 718 (1967).
97 P.R.Bunker, J.Chem.Phys. 48, 2832 (1968).
98 T.Iijima and S.Tsuchiya, J.Mol.Spectry. (in press).
99 K.Kuchitsu, MTP International Review of Science, Physical Chemistry Series One, Vol. 2, Molecular Structure and Properties (Ed. G.Allen), Chapter 6, Medical and Technical Publ.Co., Oxford 1972.

CHAPTER 13

The Material Point Method in the Interpretation of Electron Diffraction Data

N. M. EGOROVA and N. G. RAMBIDI

An adaptation of the global method of the search for minima, i.e. the material point method, to the electron diffraction determination of the molecular structure parameters is discussed. The method has been tested on a theoretical model and used for interpretation of electron diffraction patterns of vaporous molybdenum trioxide. It is shown that the molecular configuration of Mo_3O_9 consists of three tetrahedral MoO_4 groups, connected by oxygen atoms in such a manner that a flat ring of six alternate molybdenum and oxygen atoms is formed.

ИСПОЛЬЗОВАНИЕ МЕТОДА МАТЕРИАЛЬНОЙ ТОЧКИ ДЛЯ ИНТЕРПРЕТАЦИИ ДАННЫХ ЭЛЕКТРОНОГРАФИЧЕСКОГО ЭКСПЕРИМЕНТА

Н. М. Егорова и Н. Г. Рамбиди

Рассмотрено применение глобального метода минимизации - метода материальной точки - для электронографического определения структурных параметров молекул. Метод опробован на теоретической модели и применен к расшифровке электронограмм парообразной трехокиси молибдена. Показано, что конфигурация молекулы Mo_3O_9 представляет собой три тетраэдрические группировки MoO_4, соединенные атомами кислорода так, что образуется плоское кольцо чередующихся атомов молибдена и кислорода.

13.1. Introduction

In spite of the great success in the development of the theory and mathematics of gas electron diffraction, the direct interpretation of the diffraction pattern from the viewpoint of the structure of the molecule being studied remains, in general, a difficult task. This is caused both by a great variety of the criteria for comparison of experimental curves with theoretical analogues, and by the complexity of calculation methods based on these criteria. The most widely used criterion at present involves the requirement of the minimum of a quadratic functional[1] characterizing the root-mean-square degree of approximation of the experimental molecular intensity curve or its Fourier transform by means of appropriate theoretical expressions. The versions of the method of the search for minima[2-4] based on this criterion make it possible to find reliable values of the structure parameters as well as to evaluate their errors formally. Nevertheless, the use of the methods of the search for minima for solving problems of gas electron diffraction involves a number of peculiarities which until now have been given but little attention in the literature.

13.2. Preliminary Theoretical Developments

In practice the determination of the parameters of a molecular model reduces to the solution of an over-determined system of the transcendental equations

$$\phi_{\text{theor}}(x_i, \theta_k) = \phi_{\text{exp}}(x_i);$$
$$i = 1,2,\ldots,n; \quad k = 1,2,\ldots,m; \quad n > m, \tag{13.1}$$

which in general has not a unique solution. For the sake of convenience we shall consider the function $\phi(x, \theta_k)$ without specifying whether it describes the molecular intensity curve or its Fourier transform.

Firstly we shall consider the case when the system has a unique solution. If the experimental values of the function $\phi^{(o)}_{\exp}(x_i)$ are obtained without any error and the system

$$\phi_{\text{theor}}(x_i, \theta_k) = \phi^{(o)}_{\exp}(x_i) \quad (13.2$$

has the solution θ_{ok}, then it is clear that

$$\phi_{\text{theor}}(x_i, \theta_{ok}) = \phi^{(o)}_{\exp}(x_i) \quad (13.3$$

when there is no theoretical error. However, the values of $\phi_{\exp}(x_i$ found experimentally have some errors:

$$\phi_{\exp}(x_i) = \phi^{(o)}_{\exp}(x_i) + \delta\phi(x_i). \quad (13.4$$

That may result in the system (1) proving incompatible in the general case, i.e. it will have no accurate solution.[5,6] Therefore, the problem is to determine the values of the parameters θ_k in such a manner that, in a sense, all the equations of the system will be satisfied with the smallest error. This problem may be formulated as a search for a solution leading to minimum discrepancy

$$\Delta\phi(x_i) = ||\phi_{\text{theor}}(x_i)| - |\phi_{\exp}(x_i)||. \quad (13.5$$

Usually when solving the problem under consideration, use is made of the 'maximum probability' principle.[7] In this case, when assuming that the normal law of error distribution is acting, it is possible to show that the probability function reaches its maximum value when the values of the parameters, $\bar{\theta}_k$, are consistent with the minimum of the functional

$$\Phi(\theta_1, \ldots, \theta_m) = \sum_i w_i [|\phi_{\text{theor}}(x_i, \theta_k)| - |\phi_{\exp}(x_i)|]^2, \quad (13.6)$$

where $w_i = 1/(2\sigma_i)$, and σ_i is the root-mean-square deviation of the function $\phi_{\exp}(x_i)$. Thus, the solution of the problem under consideration reduces to find the values of the parameters $\bar{\theta}_k$, which cause the sum of the squares of discrepancies to be a minimum. In contrast to the accurate solution of the system (2) we shall refer to these $\bar{\theta}_k$ values as the 'average solution'. It is clear that these values differ from the accurate solutions of the system;

however, in the range where the error tends to zero they coincide with the accurate solution.

It is easy to show that in the case where there is no experimental error the search for the accurate solution of the system (2) is equivalent to the search for the minimum of the quadratic functional corresponding to its zero value (hereafter called 'zero minimum of the functional'). Let θ_{ok} be the solution of the system (2). Then by virtue of the relations (3)

$$\Phi \equiv 0. \tag{13.7}$$

Vice versa, if the condition (7) is true for a point θ_{ok}, then the relations (3) must be true as well.

If the system of the equations is linear with respect to the parameters to be determined, it is evident that the zero minimum corresponding to θ_{ok} is the only possible one. In the case where the equations (2) are transcendental (and that is just the case often encountered when interpreting the electron diffraction patterns of molecules), the situation becomes somewhat more complex. In order to understand it we shall proceed from the necessary condition of extremum for Φ,

$$\frac{\partial \Phi}{\partial \theta_k} = 0 \; ; \quad k = 1, 2, \ldots, m \tag{13.8}$$

with

$$\begin{vmatrix} \dfrac{\partial^2 \Phi}{\partial \theta_1^2} & \dfrac{\partial^2 \Phi}{\partial \theta_1 \partial \theta_2} & \cdots & \dfrac{\partial^2 \Phi}{\partial \theta_1 \partial \theta_m} \\ \cdot & \cdot & \cdots & \cdot \\ \cdot & \cdot & \cdots & \cdot \\ \dfrac{\partial^2 \Phi}{\partial \theta_m \partial \theta_1} & \dfrac{\partial^2 \Phi}{\partial \theta_m \partial \theta_2} & \cdots & \dfrac{\partial^2 \Phi}{\partial \theta_m^2} \end{vmatrix} > 0.$$

It should be noted immediately that this condition does not specify the number of extrema which exist on the surface of the functional, which of them will be obtained, and whether they correspond to the solution of the system (2). In fact the application of the condition (8) results in a system of normal equations for the determination of the parameters of θ_k:

$$\frac{\partial \Phi}{\partial \theta_k} =$$

$$\sum_i w_i \left[|\phi_{\mathrm{theor}}(x_i, \theta_k)| - |\phi_{\mathrm{exp}}(x_i)| \right] \frac{\partial |\phi_{\mathrm{theor}}(x_i, \theta_k)|}{\partial \theta_k} = 0. \quad (13.9$$

It is easy to see that two characteristic cases of solutions of the system (9) are possible.

(a) Such values of the parameters $\theta_k = \theta_{ok}$ are obtained as to make

$$\Delta\phi(x_i) = \| \phi_{\mathrm{theor}}(x_i, \theta_{ok})| - |\phi_{\mathrm{exp}}(x_i)\| = 0, \quad (13.1$$

i.e. $|\phi_{\mathrm{theor}}(x_i, \theta_{ok})| \equiv |\phi_{\mathrm{exp}}(x_i)|$ and $\Phi \equiv 0$. This means that the set of parameters $\theta_k = \theta_{ok}$ which turns the discrepancies into zero is the solution of the system (1), and the zero minimum of the functional is consistent with the system of equations.

(b) Such a solution of the system (9) is possible, which makes the linear combination of the partial derivatives $\partial\phi_{\mathrm{theor}}(x_i, \theta_k)/\partial\theta_k$ equal to zero, and not the discrepancies $\Delta\phi(x_i)$. This solution agrees with minima of the functional, which hereafter will be called local. They are in disagreement with the accurate solution of the initial system since the discrepancies $\Delta\phi(x_i)$ are different from zero. Consequently these minima should lie above the 'zero minimum', i.e. in all the points of the local minima $\Phi > 0$.

13.3. Preliminary Discussion

If there are no errors in the measurement of $\phi_{\mathrm{exp}}(x_i)$ and one has the case of a transcendental system with a unique solution, there is in general observed a plurality of minima on the surface of a corresponding functional. Only one of these, namely the 'zero minimum', is consistent with the solution of the initial system of equations.

When the experimental function has been obtained with some error there are no such values of θ_k that would simultaneously make all the discrepancies equal to zero, and a zero minimum is not to be found on the surface of the functional. In this case the system that is linear relative to θ_k has the only minimum which agrees with the 'average solution' $\bar{\theta}_k$.

In the case of a transcendental system it is possible by using the condition (8) to obtain only such minima of the functional which correspond to the parameters θ_k that turn the linear combination of derivatives into zero. In this sense all these minima may be considered as local. The majority of them may immediately be neglected if at these points the value of the functional exceeds the level permitted from the viewpoint of the accuracy of the experiment. The deepest minimum of the functional (hereafter called 'global minimum') is in agreement with the approximation that is the best from the viewpoint of the principle of maximum probability. Therefore it is this minimum that should be considered as the desired 'average solution' of the system (1). In practice the situation may be additionally complicated by several features: (a) Besides the global minimum some other local minima may turn out to be in agreement with the accuracy of the experimental values of the function $\phi_{exp}(x_i)$. (b) The system (1), like any system of transcendental equations, may happen to have more than one real solution. In this case, with a given accuracy of the experiment, the method of the search for minima does not make it possible to choose between several equivalent 'average solutions'; in order to determine the structure of the molecule it is therefore necessary to use some additional information.

The problem of interpreting the molecular intensity curve or its Fourier transforms from the viewpoint of the structure of a molecule being studied should consequently be reduced to an examination of the whole surface of the functional to find the global minimum. It follows from this that the local methods of the search for minima, such as the method of the least squares, the method of the quickest descent, etc.,[2-4] which are widely used at present, should only be used in cases when the initial approximation falls within the boundaries of the global minimum, i.e. for refining the structure. In order to find the structure, methods are required which allow an examination of the whole surface of the functional. One of such methods is the material point method developed by

Shedrin;[8] cf. also B.M.Shedrin, thesis, MGU, Moscow 1967.

13.4. Material Point Method

The material point method is based on the following physical analogy. Assume a material point moving upon a surface. Under the influence of gravity the point tends to get into the region of the surface minimum, which corresponds to the minimum of potential energy. The trajectory of the point's movement should be the geodesic curve and, depending on the initial velocities and the character of the surface, the point will move either through maxima and minima, or within a sufficiently deep minimum.

Let the equation of the surface have the form

$$\phi(y, \theta_1, \ldots, \theta_m) = y - \Phi(\theta_1, \ldots, \theta_m) = 0. \quad (13.11)$$

The point moves under the action of a force of gravity directed along the axis y. Then the system of the differential equations describing the movement of the point is:

$$\frac{d^2\theta_i}{dt^2} = - \frac{g + \sum_{p=1}^{m} \sum_{q=1}^{m} \frac{\partial^2 \Phi}{\partial\theta_p \partial\theta_q} \frac{d\theta_p}{dt} \frac{d\theta_q}{dt}}{1 + \sum_{l=1}^{m} \left(\frac{\partial \Phi}{\partial \theta_l}\right)^2} \frac{\partial \Phi}{\partial \theta_i} ; \quad i = 1, 2, \ldots, m. \quad (13.12)$$

To solve this system a difference diagram of the first order has been chosen,

$$\frac{d\tilde{\theta}_i}{dt} = \frac{\theta_i^{(k)} - \theta_i^{(k-1)}}{h_k},$$

$$\frac{d^2\tilde{\theta}_i}{dt^2} = \left(\frac{\theta_i^{(k+1)} - \theta_i^{(k)}}{h_{k+1}} - \frac{\theta_i^{(k)} - \theta_i^{(k-1)}}{h_k}\right) \frac{1}{h_{k+1}}, \quad (13.13)$$

where $\left.\frac{d\tilde{\theta}_i}{dt}\right|_k$ and $\left.\frac{d^2\tilde{\theta}_i}{dt^2}\right|_k$ are approximated values of the first and second derivatives, respectively, for the i-th variable at the k-th point. $h_k = t_k - t_{k-1}$, and t_k is an independent variable which plays the role of time. Thus the $(k+1)$-th approximation is found in the following way.

$$\theta_i^{(k+1)} = \theta_i^{(k)} + \frac{h_{k+1}}{h_k}\left(\theta_i^{(k)} - \theta_i^{(k-1)}\right) -$$

$$\left.\frac{\partial\Phi}{\partial\theta_i}\right|_k \frac{g\, h_{k+1}^2 + \displaystyle\sum_{p=1}^{m}\sum_{q=1}^{m} \left.\frac{\partial^2\phi}{\partial\theta_p\partial\theta_q}\right|_k \cdot \frac{h_{k+1}^2}{h_k^2}\left(\theta_p^{(k)} - \theta_p^{(k-1)}\right)\left(\theta_q^{(k)} - \theta_q^{(k-1)}\right)}{1 + \displaystyle\sum_{s=1}^{m}\left(\left.\frac{\partial\Phi}{\partial\theta_s}\right|_k\right)^2} . \tag{13.14}$$

In this case the initial deviations are found from the formulae

$$\theta_i^{(1)} = \theta_i^{(0)} - h_0\left[\frac{\nu_1}{\sqrt{G^2 + \displaystyle\sum_{q=1}^{m}\left(\left.\frac{\partial\Phi}{\partial\theta_q}\right|_0\right)^2}}\left.\frac{\partial\Phi}{\partial\theta_i}\right|_0 + \nu_2 \operatorname{sgn}\left.\frac{\partial\Phi}{\partial\theta_i}\right|_0\right], \tag{13.15}$$

where

$$\operatorname{sgn}\left.\frac{\partial\Phi}{\partial\theta_i}\right|_0 = \begin{cases} +1, \text{ if } \left.\dfrac{\partial\Phi}{\partial\theta_i}\right|_0 \geq 0 \\ -1, \text{ if } \left.\dfrac{\partial\Phi}{\partial\theta_i}\right|_0 < 0 , \end{cases}$$

and ν_1, ν_2 are some parameters determining the initial deviation. For further information on selection of the parameters G, g, ν_1, ν_2 of the method, see Shedrin.[8] This author has pointed out the following advantages of the material point method.

(a) The movement is effected in the direction of the quickest descent with $\frac{\partial\theta_i}{\partial t}$ being small; i.e. it comprises gradient methods with a higher convergence.

(b) The solution is achieved by means of a continuous function, and consequently there is an opportunity to examine the whole region of variation of initial parameters in a given direction.

(c) By presetting several initial points we can examine a large region of variation of parameters and, after having obtained information on the distribution of minima, form a judgement on the distribution of the global minimum.

(d) In the course of movement along the surface use is made of both separate characteristics of the surface and of all the information about the geometrical surface at the 'preceding moment of time' (acceleration in the course of descent and retardation in the course of ascent).

(e) If the surface being studied is of the ravine type, the movement takes place along the ravine due to the force of inertia of the system. Indeed, the greater the inertia of the system, the greater the deviation of the trajectory of movement from the direction gradΦ and its adaptation in the direction of the ravine.

13.5. Applications of the Material Point Method

13.5.1. Introduction

In the present paper the material point method has been used to determine the parameters of the modified radial distribution function

$$f_{\text{theor}}(r) = k \sum_{n=1}^{N} k_n \frac{c_n Z_{p(n)} Z_{q(n)}}{r_n \sqrt{2b + l_n^2}} \exp\left[- \frac{(r_n - r)^2}{2(2b + l_n^2)} \right] , \qquad (13.1$$

in which it is possible to find the values r_n, l_n, and k_n. The value of k is calculated during every step of the search for minima according to the formula:

$$k = \sum_{j=1}^{N} f_{\text{theor}}^{j} f_{\text{exp}}^{j} \Big/ \sum_{j=1}^{N} (f_{\text{theor}}^{j})^2 . \qquad (13.1$$

If necessary, the connections between separate distances between nuclei may be written down by formulae of an arbitrary type.

13.5.2. Theoretical Model

Initially, the method was tested on a theoretical model, in which case $f_{exp}(r_k)$ was substituted with the theoretical curve (16) having the parameters of Table I. The values of r_n were sought for at fixed values of k_n and l_n, and the values of r_n and l_n at fixed values of k_n. The possibilities of the use of the method for the case of simultaneous determination of r_n and l_n are illustrated by Fig. 1 and Table II. Fig. 1 clearly shows two main peculiar features of the functional being studied: a great number of local minima and a relatively small 'width' of the global minimum. Generally speaking the latter feature which considerably hampers the search for the global minimum is in agreement with the experiment by Hedberg and Iwasaki,[4] who used the local method of the search for minima, i.e. the method of the least squares. The information on separate local and global minima is given in Table II. This table shows inter alia initial values of the parameters (the first line of the table) and the precise values corresponding to the theoretical function $f(r)$ (the last line). It should be noted in this connection that the material point method makes it possible to find the global minimum in case the initial approximation differs by $\pm$ 0.05 Å with respect to r_n and by $\pm$ 0.03 Å with respect to l_n. The time of the search for the global minimum for this particular case by means of the BESM-4-type electronic computer (the average speed is 18 000 operations per second) is about 15 min.

Table 13-I. Parameters used in the theoretical model of Section 13.5.2.

n	k_n	c_n	r_n	l_n	$Z_{p(n)}$	$Z_{q(n)}$
1	1	2	2.00	0.08	10	10
2	1	1	3.00	0.12	10	10
3	1	1	3.40	0.10	15	10
			a = 0.005			

Fig. 13-1. Behaviour of the functional Φ in the course of finding the parameters of the theoretical model. The upper curve shows $f(r)$ for this model with $r_1 = 2.00$, $r_2 = 3.00$, $r_3 = 3.40$, $l_1 = 0.08$, $l_2 = 0.12$, and $l_3 = 0.10$ (Å). Upper curve (a) and lower curve (b) for Φ pertain to (a) determination of r_{ij} and (b) simultaneous determination of r_{ij} and l_{ij}, respectively. The indicated minima:

	Φ	r_1	r_2	r_3	l_1	l_2	l_3
a_1	0.11×10^5	2.00	2.96	3.38	-	-	-
a_2	0.64×10^4	2.00	2.96	3.39	-	-	-
a_3	28.66	2.00	3.00	3.40	-	-	-
b_1	0.53×10^4	2.00	2.97	3.38	0.079	0.122	0.102
b_2	46.85	2.00	3.00	3.40	0.080	0.119	0.997

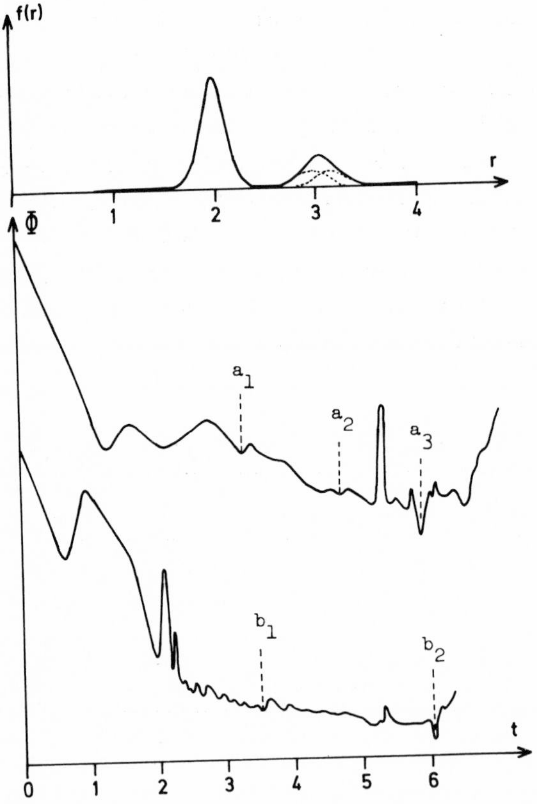

Table 13-II. Behaviour of the functional Φ and parameters (r and l in Å) for the theoretical model when using the material point method.

Φ	r_1	r_2	r_3	l_1	l_2	l_3
0.2957×10^6	2.0500	2.9500	3.3500	0.0600	0.1500	0.0700
0.2956×10^6	2.0499	2.9500	3.3500	0.0600	0.1499	0.0700
0.2952×10^6	2.0499	2.9500	3.3500	0.0601	0.1499	0.0700
0.2789×10^6	2.0493	2.9500	3.3500	0.0636	0.1498	0.0711
0.2267×10^6	2.0465	2.9500	3.3505	0.0765	0.1491	0.0755
0.9682×10^5	*2.0264*	*2.9505*	*3.3636*	*0.0999*	*0.1431*	*0.1050*
0.1739×10^6	2.0409	2.9502	3.3513	0.0920	0.1480	0.0840
0.1526×10^6	2.0376	2.9502	3.3518	0.0950	0.1470	0.0890
0.1087×10^6	2.0310	2.9510	3.3530	0.0920	0.1447	0.0981
0.1575×10^5	1.9930	2.9560	3.7710	0.0804	0.1280	0.1120
0.1552×10^5	*0.9998*	*0.9560*	*3.3707*	*0.0799*	*0.1289*	*0.1099*
0.1699×10^5	1.9968	2.9530	3.3630	0.0800	0.1330	0.1160
0.1580×10^5	1.9983	2.9534	3.3640	0.0798	0.1330	0.1140
0.8921×10^4	0.9945	2.9700	3.3970	0.0806	0.1103	0.8971
0.5300×10^4	*2.0000*	*2.9700*	*3.3800*	*0.0790*	*0.1220*	*0.1020*
0.6053×10^4	1.9999	2.9644	3.3826	0.0808	0.1233	0.1051
0.5309×10^4	1.9999	2.9662	3.3845	0.0789	0.1228	0.1045
0.4823×10^4	*1.9999*	*2.9677*	*3.3857*	*0.0795*	*0.1226*	*0.1041*
0.4830×10^4	1.9999	2.9674	3.3857	0.0795	0.1227	0.1032
0.2782×10^4	2.0000	2.9750	3.3890	0.0801	0.1230	0.1045
0.1013×10^4	*2.0000*	*2.9852*	*3.3950*	*0.0795*	*0.1230*	*0.0997*
0.1781×10^5	2.0003	2.9923	3.3960	0.0986	0.1180	0.1086
0.1181×10^3	2.0002	2.9997	3.3980	0.0788	0.1192	0.9980
0.4682×10^2	*2.0000*	*3.0000*	*3.4000*	*0.0800*	*0.1190*	*0.9970*
	2.0000	3.0000	3.4000	0.0800	0.1200	0.1000

13.5.3. Molybdenum Trioxide

The material point method was also used for the interpretation of the electron diffraction patterns of molybdenum trioxide vapour. The electron diffraction patterns were obtained with the aid of an electron diffraction unit for studying molecules of refractory compounds.[9] Use was made of an evaporator and of an ampoule containing the substance, which was subjected to heating by electron bombardment. The temperature of the ampoule was about 1000 °C. The main parameters of the electron diffraction unit were

(a) Electron-accelerating voltage: 60 kV.

(b) Distance from the centre of the nozzle of the evaporator ampoule to the plane of the photographic plate: 212 mm and 340 mm.

(c) Maximum radius of the r^3 sector: 50 mm.

The diffraction pattern was obtained in the region s = 3.6 – 20.6 $Å^{-1}$. The microphotograms were treated with the aid of the programme of the primary processing of experimental data.[10] In accordance with the results of the mass-spectrometric investigations[11] the electron diffraction patterns were interpreted under

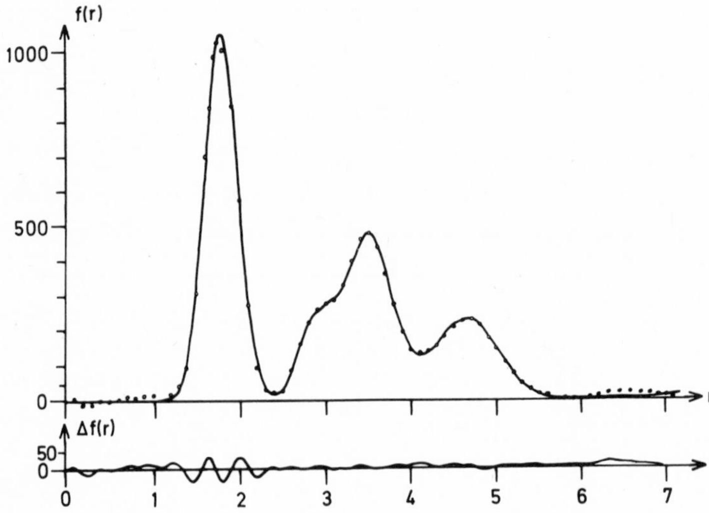

Fig. 13-2. Theoretical (solid line) and experimental modified radial distribution curves for the Mo_3O_9 molecule, and the difference function $\Delta f(r) = f_{exp}(r) - f_{theor}(r)$.

the assumption that the main component of the vapour of molybdenum trioxide are trimer molecules of Mo_3O_9.

Fig. 2 shows the modified radial distribution curve for vaporous molybdenum trioxide. The extrapolation of the molecular intensity curve into the initial sphere of the reverse space, which is necessary for the calculation of this function, was found by the method of subsequent approximations with the use of Fourier transforms for the $rD(r/s_{min}, s_{max})$ (for details see Ref. 12). A preliminary analysis of the modified radial distribution curve has shown that it is best satisfied with the cyclic model of the Mo_3O_9 molecule, in which three tetrahedral groups of MoO_4 are interconnected by their tops in such a manner that a planar six-membered ring of alternate atoms of molybdenum and oxygen is formed (Fig. 3). The preliminary analysis gave the initial values of the parameters which were used as the zero approximation for the material point method. The parameters were determined with the aid of the material point method in two stages.

In the first stage the atoms of molybdenum and oxygen of the hexagonal ring were assumed to lie in one plane. This configuration of the molecule belongs to the symmetry group D_{3h}. The result of this analysis gave the following parameters:

$$r(\text{Mo=O}) = 1.73\ \text{Å},$$
$$r(\text{Mo-O}) = 1.87\ \text{Å},$$
$$\angle(\text{O=Mo=O}) = 105^\circ,$$
$$\angle(\text{O-Mo-O}) = 103^\circ.$$

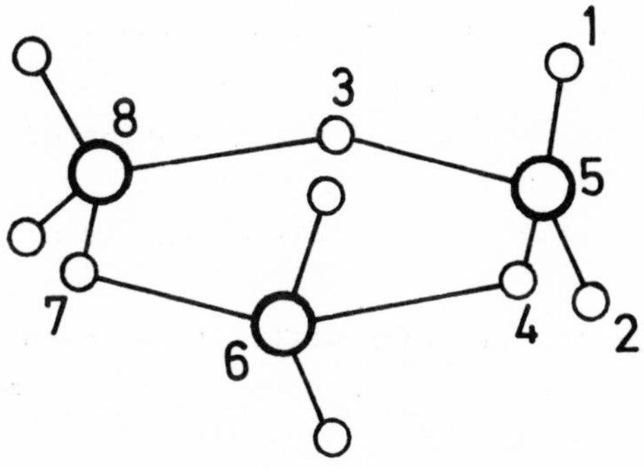

Fig. 13-3. Model of the Mo_3O_9 molecule.

The theoretical curve of $f(r)$ calculated by using these values of the parameters (Fig. 2) is in good agreement with the experimental modified radial distribution curve. It should also be noted that the values of k_n thus obtained approach unity. This apparently testifies to the fact that the chosen model is correct and that the extrapolation of $sM(s)$ into the initial sphere of the reverse space does not contain any considerable errors.

In the second stage the values of the parameters obtained in the first step were used as a zero approximation in the course of calculations with the assumption of a nonplanar structure of the six-membered ring of molybdenum and oxygen atoms. In this analysis no minimum was obtained that would be deeper than the global minimum found with the assumption of the planar structure of the ring.

In spite of good agreement of the theoretical and experimental modified radial distribution curves the correspondence between molecular intensity curves was excellent in the s = 3.6 - 13.0 Å^{-1} region, but rather poor at s > 13.0 Å^{-1}. An analysis of the experimental molecular intensity curve using the material point method gave us the opportunity to determine the structural parameters more precisely. Thus, the molecule of Mo_3O_9 agrees with the planar cyclic structure having the following parameters:

$$r(\text{Mo=O}) = 1.67 \pm 0.01 \text{ A},$$

$$r(\text{Mo-O}) = 1.89 \pm 0.01 \text{ Å},$$

$$\angle(\text{O=Mo=O}) = 106 \pm 2^{\circ},$$

$$\angle(\text{O-Mo-O}) = 107 \pm 2^{\circ}.$$

It should be noted that the values of these parameters naturally are mean values with respect to the whole complex of vibration and rotation levels (cf. N.G.Rambidi, thesis, MGU, Moscow 1970). Therefore the conclusion of the planarity of the six-membered ring cannot be related to the equilibrium geometrical configuration of the Mo_3O_9 molecule. In order to solve this problem definitely it would be necessary to carry out a simultaneous analysis of electron diffraction and spectroscopic data.

A detailed description of the electron diffraction investigation of vaporous molybdenum trioxide, performed together with E.Z. Zasorin and S.A. Komarov, will be published elsewhere.

REFERENCES

1 A.D.Booth, Quart.J.Mech.Appl.Math. 2, 4 (1949).
2 R.A.Bonham and L.S.Bartell, J.Chem.Phys. 31, 702 (1959).
3 O.Bastiansen, L.Hedberg, and K.Hedberg, J.Chem.Phys. 27, 1311

(1956).
4 K.Hedberg and M.Iwasaki, Acta Cryst. 17, 529 (1964).
5 I.S.Beresin and N.P.Zhidkov: Metody vychislenii (Methods of calculations), Vol. 2, Fizmatgiz, Moscow 1959.
6 C.Lanczos: Applied Analysis, Prentice-Hall, Englewood Cliffs (N.J.) 1956.
7 R.A.Fisher, Messenger Math. 41, 155 (1912).
8 B.M.Shedrin, N.V.Belov, and N.P.Zhidkov, Dokl.Akad.Nauk SSSR 170, 1070 (1966).
9 P.A.Akishin, M.I.Vinogradov, K.D.Danilov, N.P.Levkin, E.I. Martinson, N.G.Rambidi, and V.P.Spiridonov, Pribory i tekhnika eksperimenta (Instruments and experimental techniques), 3, 70 (1958).
10 N.G.Rambidi, E.Z.Zasorin, E.P.Vedeneev, Yu.S.Ezhov, N.M.Egorova, L.I.Ermolaeva, V.S.Vinogradov, and M.G.Anashkin, Zh.Strukt.Khim. 11, 361 (1970).
11 J.Berkowitz, M.G.Inghram, and W.Chupka, J.Chem.Phys. 26, 842 (1957).
12 N.G.Rambidi and E.Z.Zasorin, Zh.Strukt.Khim. 7, 483 (1966).

CHAPTER 14

Vibrational Analyses and Mean Amplitudes for Some Simple Molecules - Part I: Planar XYZUV (C_s) Model with Application to Nitric Acid

B. VIZI, B. N. CYVIN, and S. J. CYVIN

Mean amplitudes of vibration calculated from spectroscopic data for HNO_3 are given. Internal coordinates for the appropriate molecular model (also applicable to HCOOH) are included.

14.1. Planar XYZUV (C_s) Model

The considered model is shown in Fig. 1; it applies to formic acid,[1,2] nitric acid,[3,4] and other important molecules. A short account on this model is included here mainly because of the subsequent treatment of the formic acid molecule (Part II).

The valence coordinates are already symmetry coordinates in the present case. The selection which is shown in Fig. 1 may serve as a complete set of independent internal coordinates. Specifically the symmetry coordinates $S_i(A')$; $i = 1,\ldots,7$, and $S_j(A'')$; $j = 1,2$, as chosen here, are: (A') r_1, r_2, d, t, $(R_1D)^{\frac{1}{2}}\beta_1$, $(R_2D)^{\frac{1}{2}}\beta_2$, $(DT)^{\frac{1}{2}}\varphi$, and (A'') $[R_1(R_2D)^{\frac{1}{2}}]^{\frac{1}{2}}\gamma$, $(R_2T)^{\frac{1}{2}}\tau$; respectively.

14.2. Mean Amplitudes for Nitric Acid

A harmonic force field for nitric acid is to be published elsewhere. The force constants were used to calculate the mean amplitudes of vibration for $H^{14}NO_3$. The results are shown in Table

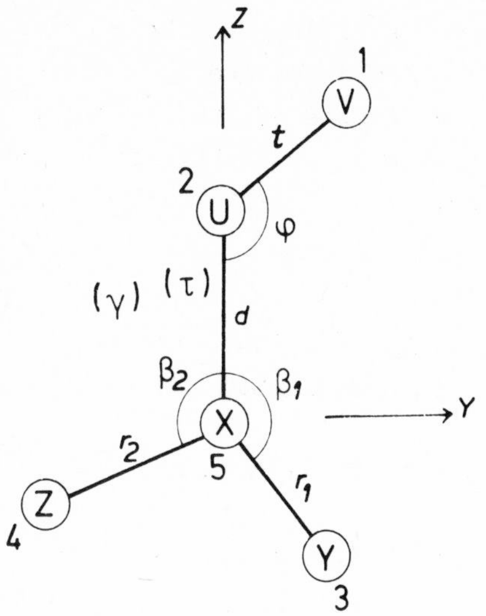

Fig. 14-1. Planar XYZUV molecular model; symmetry C_s. Equilibrium parameters: R_1(X-Y), R_2(X-Z), D(X-U), T(U-V), B_1(∠ UXY), B_2(∠ UXZ), Φ(∠ XUV). The parenthesized symbols indicate out-of-plane coordinates, viz. out-of-plane bending (γ) for the (2,4,5,3) atoms, where 3 is the end atom, and torsion (τ) for 1-2-5-4.

Table 14-I. Calculated mean amplitudes for $H^{14}NO_3$; Å units.

Distance	Atom pair[a]	(Equil.)[b]	T = 0	298 K
O-H	1-2	(0.964)	0.071	0.071
N=O	3-5	(1.211)	0.039	0.039
N=O'	4-5	(1.199)	0.039	0.039
N-<u>O</u>	2-5	(1.406)	0.049	0.050
N•••H	1-5	(1.865)	0.105	0.105
O•••O'	3-4	(2.187)	0.048	0.049
O•••<u>O</u>	2-3	(2.220)	0.054	0.057
O'•••<u>O</u>	2-4	(2.186)	0.054	0.056
O•••H	1-3	(2.143)	0.128	0.131
O'•••H	1-4	(2.923)	0.096	0.096

[a] According to the numbering of atoms in Fig. 1.

[b] Interatomic separation (in Å) calculated from the structural data adopted as equilibrium parameters.

The complete set of mean amplitudes for all distance types i nitric acid has not been reported previously. From the fragmentary data[5] of mean amplitudes for HNO_3 we have deduced l(N-O) = 0.0500 in consistence with our value. Also the results of Müller et al.[6] for HNO_3, in which the OH group is considered as one particle, sh good agreement with the present values.

REFERENCES

1 R.C.Millikan and K.S.Pitzer, J.Chem.Phys. 27, 1305 (1957).

2 S.J.Cyvin, I.Alfheim, and G.Hagen, Acta Chem.Scand. 24, 3038 (1970).

3 G.E.McGraw, D.L.Bernitt, and I.C.Hisatsune, J.Chem.Phys. 42, 237 (1965).

4 A.Palm, A.Castelli, and C.Alexander,Jr., Spectrochim.Acta 24A, 1658 (1968).

5 N. Rajeswara Rao and S.Sakku, Indian J. Pure Appl. Phys. 6, 4 (1968).

6 A.Müller, B.Krebs, A.Fadini, O.Glemser, S.J.Cyvin, J.Brunvoll, B.N.Cyvin, I.Elvebredd, G.Hagen, and B.Vizi, Z.Naturforschg. 23a 1656 (1968).

Vibrational Analyses and Mean Amplitudes for Some Simple Molecules - Part II: Formic Acid Monomer

I. ALFHEIM, S. J. CYVIN, G. HAGEN, and T. MOTZFELDT

Calculated mean amplitudes of vibration for HCOOH from spectroscopic data are compared with electron-diffraction values. The method of specific imposition of a potential parameter (h) was applied, where h corresponds to the hydrogen bond stretching.

14.3. Introduction to Part II

Modern normal coordinate analyses of formic acid have been performed by several investigators,[1-5] the latter one by Susi and Scherer[5] being the most complete one. The experimental frequency data[1,6] applied in that work[5] are preceded by an assignment of Wilmshurst,[7] who has reviewed earlier spectroscopic investigations, not including a significant Russian work.[8] The structural data used in the present work were obtained by electron diffraction. The detailed report of that work[9] includes references to earlier structural investigations from infrared, microwave and electron diffraction.

14.4. Force Fields

Table II shows a harmonic force field for formic acid monomer. Its derivation is described elsewhere,[10] where it is referred to as $\mathbf{F}^{(1)}$. It reproduces exactly the observed vibrational frequencies[1,6,8] for HCOOH, and with satisfactory accuracy those of[1,6] DCOOD, DCOOH, and HCOOD. This force field was used to calculate the

Table 14-II. Force constants $\mathbf{F}^{(1)}$ (mdyne/Å) of formic acid monomer.*

11.05	-0.08	0.27	-0.02	0.01	-0.33	-0.04
	4.65	0.04	-0.00_2	0.01	-0.01	-0.01
in-plane		6.12	0.01	-0.06	0.10	-0.04
			7.11	0.00_1	0.01	-0.01
out-of-plane				1.17	-0.01	0.03
	0.28	0.04			0.90	0.01
		0.21				0.47

* For internal coordinates, see Part I.

mean amplitudes of vibration for HCOOH, which were compared with those from the electron diffraction experiment (see below).

The calculated value of l(O···H'), i.e. for the atom pair 1-3 (cf. Fig. 1), may be expected to be too large because the existence of a hydrogen bond has not been taken into account duri the derivation of $\mathbf{F}^{(1)}$. The calculated value at 448.15 K is 0.145 Å versus the estimate of 0.12 Å used in the electron diffraction investigation. Unfortunately it was not found possible t refine this value of l, so it remains very uncertain. Nevertheles it was believed that 0.12 Å is nearer to the correct value.

Consequently it was decided to investigate the effect on calculated mean amplitudes by explicit inclusion of a stretching coordinate, say ρ, which corresponds to the hydrogen bond. An in-plane force field may be constructed by adding one row and column to the $\mathbf{F}^{(1)}$ (Species A') block as indicated on Fig. 2. Hereby one nonvanishing constant, viz. h, is introduced into the potential function. The modified force field, say $\mathbf{F}_h$, is based on internal coordinates with one redundancy. In consequence several of the elements in $\mathbf{F}_h$ (including h) are not uniquely defined from a giv potential function. It would be meaningless to interpret h as a "force constant of the hydrogen bond". For the sake of security h is not referred to as a "force constant" at all, but is designate "potential parameter". Both positive and negative values of potential parameters are possible, but $h < 0$ is not expected to be physically real in the present case.

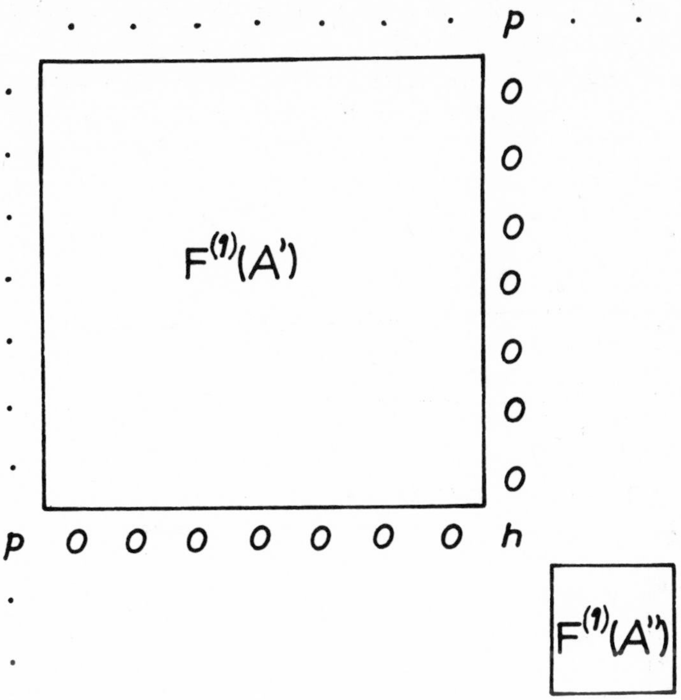

Fig. 14-2. Schematic representation of the harmonic force field $\mathbf{F}_h$.

14.5. Mean Amplitudes of Vibration

Series of mean amplitudes of vibration for HCOOH were calculated from force fields $\mathbf{F}_h$ with h ranging from -0.1 to 0.9 mdyne/Å. During these computations $\mathbf{F}_h$ was first converted to an $\mathbf{F}$ matrix in terms of the same standard coordinates which were used as the basis for $\mathbf{F}^{(1)}$. h = 0 corresponds to the original $\mathbf{F}^{(1)}$. The force fields of $\mathbf{F}_h$ for nonvanishing h are not exactly consistent with the observed frequencies for HCOOH. The force fields were adjusted to the observed frequencies (given by $\boldsymbol{\lambda}$) according to

$$\mathbf{GFL}_h = \mathbf{L}_h \lambda_h , \quad \mathbf{F}_a = \mathbf{K}_h' \lambda \mathbf{K}_h ,$$

where $L_h = K_h^{-1}$ and λ_h are calculated from F_h, and F_a is the force constant matrix adjusted to observed frequencies.

Fig. 3 shows the variation of mean amplitudes with different values of h, as calculated from F_h (dotted curve) and from F_a

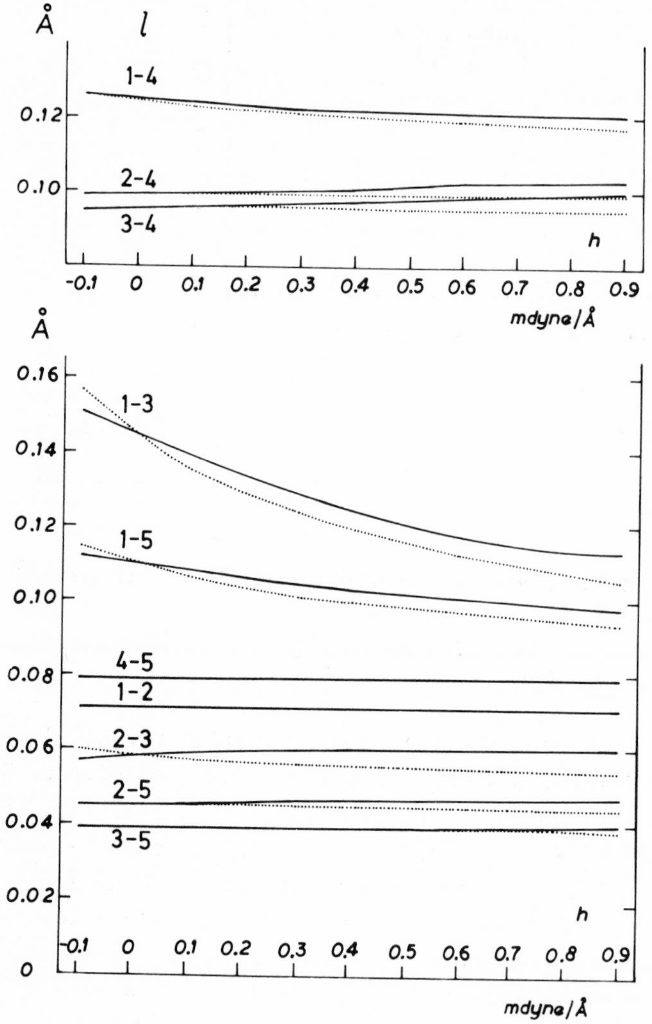

Fig. 14-3. Calculated mean amplitudes of vibration for HCOOH at 448.15 K from (a) F_h (dotted curves) and (b) the adjusted force field F_a (solid curves) for different values of the potential parameter h.

after adjusting to the observed frequencies (solid curve). The here applied procedure of calculating modified mean amplitudes is nothing else than an example of the specific imposition of a potential parameter (cf. Chapter 6). It is seen from Fig. 3 that

Table 14-III. Final values of calculated mean amplitudes for HCOOH at 448.15 K along with electron diffraction results; Å units.

Distance	Atom pair[a]	(Equil.)[b]	Calc.	Obs.
O-H'	1-2	(0.984)	0.071	0.068[c]
C-H	4-5	(1.106)	0.079	0.077[c]
C-O	2-5	(1.361)	0.047	0.042 (0.0014)
C-O	3-5	(1.217)	0.039	0.032 (0.0010)
C•••H'	1-5	(1.902)	0.101	0.09[c]
O•••$\underline{\underline{O}}$	2-3	(2.271)	0.061	0.054 (0.0018)
O•••H	2-4	(2.016)	0.102	0.09[c]
$\underline{\underline{O}}$•••H	3-4	(2.083)	0.099	0.09[c]
$\underline{\underline{O}}$•••H'	1-3	(2.325)	0.120	0.12[c]
H•••H'	1-4	(2.829)	0.122	-

[a] According to the numbering of atoms in Fig. 1.

[b] Interatomic separation (in Å) calculated from the structural data adopted as equilibrium parameters.

[c] Unrefined during the least-squares procedure of the electron-diffraction investigation. Standard deviation in parentheses.

Table 14-IV. Force constants $\mathbf{F}_a$ (mdyne/Å) obtained by adjusting $\mathbf{F}_h$ with h = 0.5 mdyne/Å.*

10.16	-0.12	0.16	0.07	0.16	-0.10	0.33
	4.64	-0.02	-0.00_5	0.04	-0.01	-0.01
in-plane		5.42	0.12	0.20	0.01	0.27
			7.11	0.11	-0.00_0	0.10
out-of-plane				1.17	-0.06	0.33
	0.28	0.04			0.79	-0.06
		0.21				0.76

*See footnote to Table II.

$l(O\cdots H')$ is the mean amplitude being most affected by a variati of h, as also was expected.

The value of h = 0.5 mdyne/Å leads to $l(O\cdots H')$ = 0.12 Å f the corresponding F_a, and coincides with the electron diffracti l value (cf. Table III). This force field (see Table IV) was con sidered as the final result of the normal coordinate analysis.[10] The cited paper[10] contains a table of the calculated mean amplitudes and perpendicular amplitude corrections for HCOOH at 0 K a 298.15 K from the final force field.

REFERENCES

1 T.Miyazawa and K.S.Pitzer, J.Chem.Phys. 30, 1076 (1959).
2 R.Blinc and D.Hadži, Spectrochim.Acta 15, 82 (1959).
3 K.Nakamoto and S.Kishida, J.Chem.Phys. 41, 1554 (1964).
4 W.V.F.Brooks and C.M.Haas, J.Phys.Chem. 71, 650 (1967).
5 H.Susi and J.R.Scherer, Spectrochim.Acta 25A, 1243 (1969).
6 R.C.Millikan and K.S.Pitzer, J.Chem.Phys. 27, 1305 (1957).
7 J.K.Wilmshurst, J.Chem.Phys. 25, 478 (1956).
8 L.M.Sverdlov, Izvest.Akad.Nauk SSSR, Ser.Khim. 17, 567 (1953)
9 A.Almenningen, O.Bastiansen, and T.Motzfeldt, Acta Chem.Scand 23, 2848 (1969).
10 S.J.Cyvin, I.Alfheim, and G.Hagen, Acta Chem.Scand. 24, 3038 (1970).

Vibrational Analyses and Mean Amplitudes for Some Simple Molecules - Part III: Carbonyl Cyanide

V. UNNIKRISHNAN NAYAR, G. ARULDHAS, K. BABU JOSEPH,

and S.J. CYVIN

Symmetry coordinates are specified for the planar $WX(YZ)_2$ model with linear XYZ chains and of symmetry C_{2v}. The model is applied to carbonyl cyanide, $CO(CN)_2$, for which a harmonic force field is given. Calculated results of mean amplitudes and shrinkage effect are reported.

कारबोणिल सैनेड़ का प्रसामान्य निर्देशांक विश्लेषण और परिकलित कम्पविस्तार (आयाम) का वर्ग माध्य मूल

वी. उण्णिकृष्णन् नायर्, जी. अरुलदास
के. बाबुजोसफ़ और एस्. जे. सिविन्

रेखीय XYZ शृन्खलाओं के साथ C_{2v} सममितिवाली समतल-सम्बन्धी $WX(YZ)_2$ नमूने के सममिति-निर्देशांक निदिष्ट किये हुये हैं। इस नमूने को कारबोणिल सैनेड़, $CO(CN)_2$ अणु पर प्रयुक्त करके प्रसंवादी बल क्षेत्र निश्चित किया गया है। उस अणु के कम्पविस्तार का वर्ग माध्य मूल और संकुचन - प्रभाव भी निर्णय किये गये हैं।

കാർബോണിൽ സൈനേഡിന്റെ നോർമൽ നിർദ്ദേശാങ്കവിശകലനവും പരികലനംചെയ്ത ആയാമവർഗ മാധ്യമൂലവും

വി. ഉണ്ണികൃഷ്ണൻനായർ, ജി. അരുൾദാസ്, കെ. ബാബുജോസഫ് & എസ്. ജെ. സിവിൻ

XYZ രേഖീയ ശൃംഖലകളോട് കൂടിയ C_{2v} സമമിതിയുള്ള സമപ്രതലീയമായ $WX(YZ)_2$ മാതൃകയുടെ സമമിതി നിർദ്ദേശാങ്കങ്ങളും നിർണ്ണയിക്കപ്പെട്ടിരിക്കുന്നു. ഈ മാതൃക കാർബോണിൽ സൈനേഡ് തന്മാത്രയിൽ പ്രയോഗിച്ച് അതിന്റെ ഹാർമോണിക ബലക്ഷേത്രം കണ്ടുപിടിച്ചിട്ടുണ്ട്. തന്മാത്രയുടെ ആയാമ വർഗ മാധ്യമൂലങ്ങളും സങ്കോചന പ്രഭാവവും പരികലനംചെയ്ത് കൊടുത്തിരിക്കുന്നു.

14.6. Introduction to Part III

Mean amplitudes of vibration have been calculated for many linear organic molecules with the C≡N bond (cf. Chapter 22). Also molecules with the C=O bond have been studied extensively.[1] An extensive survey of all kinds of molecules with C≡N bonds is still to be awaited. The CN triple bond is especially mentioned in Sectio 12.13 of Cyvin's book.[2] In the present chapter a normal coordinate analysis with calculated mean amplitudes is reported for a small molecule containing both the carbonyl bond, C=O, and two cyano groups, C≡N: carbonyl cyanide.

14.7. Molecular Model and Symmetry Coordinates

Fig. 4 shows the adopted planar $WX(YZ)_2$ model of symmetry C_{2v} and with linear XYZ chains. The applied valence coordinates are specified in the legend of the figure. Symmetry coordinates:

$$S_1(A_1) = 2^{-\frac{1}{2}}(r_1 + r_2), \quad S_2(A_1) = 2^{-\frac{1}{2}}(d_1 + d_2), \quad S_3(A_1) = t,$$

$$S_4(A_1) = (RT/2)^{\frac{1}{2}}(\beta_1 + \beta_2), \quad S_5(A_1) = (RD/2)^{\frac{1}{2}}(\varphi_1 + \varphi_2);$$

$$S(A_2) = (RD/2)^{\frac{1}{2}}(\theta_1 - \theta_2);$$

$$S_1(B_1) = (RD/2)^{\frac{1}{2}}(\theta_1 + \theta_2), \quad S_2(B_1) = (RT)^{\frac{1}{2}}\gamma;$$

$$S_1(B_2) = 2^{-\frac{1}{2}}(r_1 - r_2), \quad S_2(B_2) = 2^{-\frac{1}{2}}(d_1 - d_2),$$

$$S_3(B_2) = (RT/2)^{\frac{1}{2}}(\beta_1 - \beta_2), \quad S_4(B_2) = (RD/2)^{\frac{1}{2}}(\varphi_1 - \varphi_2).$$

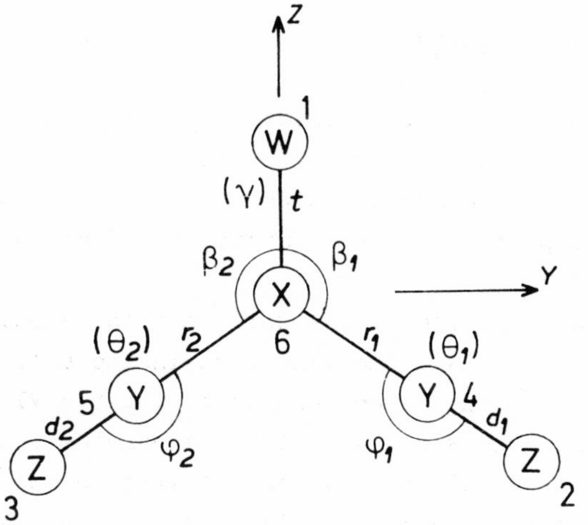

Fig. 14-4. Planar $WX(YZ)_2$ model with linear XYZ chains; symmetry C_{2v}. The applied valence coordinates are indicated. Parenthesized symbols designate the out-of-plane coordinates: linear bendings θ_1 and θ_2, and the out-of-plane WXY_2 bending γ. Capital letters R, D, and T are used to denote the equilibrium distances of X-Y, Y-Z, and W-X, respectively. One more parameter is needed to define the equilibrium structure, say $2A$, the YXY angle.

14.8. Structural and Spectral Data

The structural parameters applied here are[3] R(C-C) = 1.460 Å, D(C≡N) = 1.156 Å, T(C=O) = 1.220 Å, and $2A$(CCC) = 120°. Infrared and Raman spectra of carbonyl cyanide were investigated by Tramer and Wierzchowski,[4,5] and were used along with some additional measurements to produce a complete vibrational assignment;[6] cf. Table V. The table includes another independently published assignment from new infrared and Raman data by Bates and Smith.[7] These authors admit that the assignments of $\nu_4(A_1)$, $\nu_7(B_1)$, and $\nu_{11}(B_2)$ are especially uncertain. For ν_4 they have observed a wave number of 526 cm^{-1} in Raman liquid. Their ν_7 = 557 cm^{-1} might possibly be assigned to $\nu_{11}(B_2)$. Anyhow it seems clear that the problem of assignment is not yet definitely solved; $\nu_6(A_2)$ should perhaps also be regarded as uncertain.

14.9. Computational Results

Different harmonic force field were developed in consistence with the two vibrational assignments shown in Table V. Table VI shows the final force constants which are consistent with the late assignment.[7] They pertain to the symmetry coordinates specified above (Section 14.7).

Two alternative force fields were used to calculate the mean amplitudes of vibration and the linear shrinkage effect.[2] The results obtained from the force constants of Table VI are shown in Table VII, and are consistent with the frequencies from Ref. 7. Th calculations using the alternative set of frequencies[6] gave simila results; parenthesized figures are included in Table VII in the cases where the deviations in mean amplitudes exceed 0.005 Å at 298 K and for the shrinkage effect. As to the orders of magnitude the present results are well comparable with related quantities in sulphur cyanide;[1,8] cf. Table VII. The present result for l(C=O) is consistent with the characteristic magnitudes found for C=O in the aldehyde, ketone, and carboxyl groups.[1]

Table 14-V. Fundamental frequencies (in cm^{-1}) for carbonyl cyanide.*

Species	No.	Ref. 6	Ref. 7
A_1	1	2242 (g)	2242 (g)
	2	1714 (g)	1712 (g)
	3	712 (g)	715 (g)
	4	520 (l)	495 (g)
	5	142 (l)	127 (g)
A_2	6	370 (ov)	298 (l)
B_1	7	475 (g,l)	557 (g)
	8	255 (l)	208 (g)
B_2	9	2222 (g)	2234 (g)
	10	1115 (g)	1131 (g)
	11	567 (g)	456 (calc)
	12	306 (l)	244 (g)

*(g) = infrared gas, (l) = Raman liquid, (ov) = from overtone frequencies, (calc) = calculated.

Table 14-VI. Symmetry force constants (mdyne/Å) for carbonyl cyanide with frequencies from Ref. 7.

Species A_1					A_2
5.314	0.761	1.041	0.053	-0.003	0.122
	18.306	0.135	0.030	-0.001	
		11.363	0.016	0.023	
			0.709	0.045	
				0.181	

B_1		Species B_2			
0.159	-0.008	3.49	0.449	-0.080	0.017
	0.121		18.329	-0.039	0.010
				0.406	0.026
					0.142

Table 14-VII. Mean amplitudes of vibration (l) and shrinkage effect (δ) for carbonyl cyanide, $CO(CN)_2$. Data for $S(CN)_2$ are included for comparison; Å units.

	$CO(CN)_2$ [a]		$S(CN)_2$ [b]
	T = 0	298 K	298 K
l(C=O)	0.038	0.038	
l(C–C)	0.048	0.049	0.0455 (S–C)
l(C≡N)	0.034	0.034	0.0349
l(C•••N) [c]	0.050	0.051	0.0484 (S•••N)
l(C•••C)	0.064	0.080	0.0987
l(C•••N) [d]	0.075 (0.071)	0.109 (0.099)	0.1260
l(N•••N)	0.094 (0.089)	0.162 (0.147)	
l(O•••C)	0.059 (0.056)	0.067 (0.061)	
l(O•••N)	0.069 (0.064)	0.092 (0.080)	
δ(C•••N) [c]	0.008_2 (0.007_3)	0.013_0 (0.010_5)	0.0116 (S•••N)

[a] Frequencies from Ref. 7; parenthesized values with frequencies from Ref. 6.
[b] From Ref. 8.
[c] Linear.
[d] Nonlinear.

REFERENCES

1 S.J.Cyvin and B.Vizi, Acta Chim.Hung. 70, 55 (1971).
2 S.J.Cyvin: Molecular Vibrations and Mean Square Amplitudes, Universitetsforlaget, Oslo, and Elsevier, Amsterdam, 1968.
3 J.F.Westerkamp, Bol.Acad.Nac.Cience Argentina 42, 191 (1961).
4 A.Tramer and K.L.Wierzchowski, Bull.Acad.Polon.Sci. Cl.III 5, 411 (1957).
5 A.Tramer and K.L.Wierzchowski, Bull.Acad.Polon.Sci. Cl.III 5, 417 (1957).
6 J.Prochorow, A.Tramer, and K.L.Wierzchowski, J.Mol.Spectry. 19, 45 (1966).
7 J.B.Bates and W.H.Smith, Spectrochim.Acta 26A, 455 (1970).
8 J.Brunvoll, Acta Chem.Scand. 21, 820 (1967).

CHAPTER 15

Studies of Some Alkali Fluorides - Part I: Planar Cyclic X_2YZ Molecular Model

B. N. CYVIN and S. J. CYVIN

Symmetry coordinates of molecular vibrations are specified for the planar cyclic X_2YZ model of C_{2v} symmetry. The $\mathbf{G}$, $\mathbf{C}^{\alpha}$, and $\mathbf{T}_{\alpha\beta,S}$ matrix elements are given.

The considered model is shown in Fig. 1, which includes the applied notation for valence coordinates of molecular vibrations. The main purpose of this chapter part is to specify a suitable set of symmetry coordinates to be used in an analysis of $LiNaF_2$ (see subsequent parts).

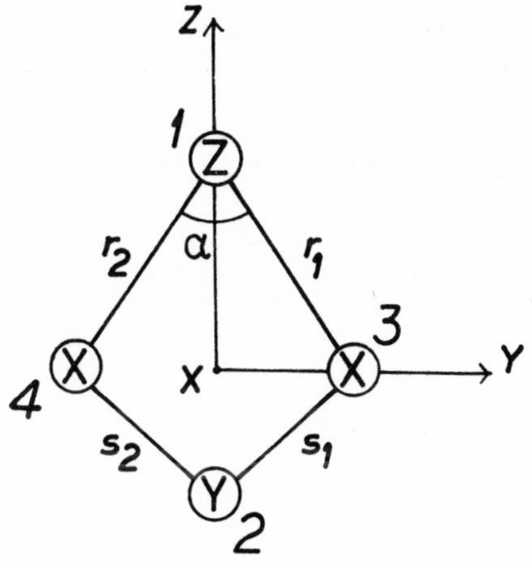

Fig. 15-1. Planar cyclic X_2YZ molecular model; symmetry C_{2v}. The equilibrium structure is determined by three parameters, e.g. R, S, $2A$, denoting the X-Z distance, X-Y distance, and the XZX angle, respectively. When the XYX angle is denoted $2B$, one has $\sin B = (R/S)\sin A$.

Table 15-I. **G** matrix for the planar cyclic X_2YZ molecular model.

G	$S_1(A_1)$	$S_2(A_1)$	$S_3(A_1)$
$S_1(A_1)$	$\mu_X+2\mu_Z\cos^2 A$	$-\mu_X\cos(A+B)$	$-2^{\frac{1}{2}}\mu_Z\sin 2A$
$S_2(A_1)$		$\mu_X+2\mu_Y\cos^2 B$	$2^{\frac{1}{2}}\mu_X\sin(A+B)$
$S_3(A_1)$			$2(\mu_X+2\mu_Z\sin^2 A)$
	$S(B_1)$	$S_1(B_2)$	$S_2(B_2)$
$S(B_1)$	$2\mu_X+\mu_Y+\mu_Z$		
$S_1(B_2)$		$\mu_X+2\mu_Z\sin^2 A$	$-\mu_X\cos(A+B)$
$S_2(B_2)$			$\mu_X+2\mu_Y\sin^2 B$

Table 15-II. $\mathbf{C}^{\alpha}$ matrix elements for the planar cyclic X_2YZ model: $\mathbf{C}^{y}(A_1\times B_1)$, $\mathbf{C}^{x}(A_1\times B_2)$, and $\mathbf{C}^{z}(B_1\times B_2)$.

$\mathbf{C}^{\alpha}$	$S(B_1)$	$S_1(B_2)$	$S_2(B_2)$
$S_1(A_1)$	$2^{\frac{1}{2}}(\mu_X+\mu_Z)\cos A$	$\mu_Z\sin 2A$	$\mu_X\sin(A+B)$
$S_2(A_1)$	$-2^{\frac{1}{2}}(\mu_X+\mu_Y)\cos B$	$-\mu_X\sin(A+B)$	$-\mu_Y\sin 2B$
$S_3(A_1)$	$-2(\mu_X+\mu_Z)\sin A$	$-2^{\frac{1}{2}}(\mu_X+2\mu_Z\sin^2 A)$	$2^{\frac{1}{2}}\mu_X\cos(A+B)$
$S(B_1)$		$-2^{\frac{1}{2}}(\mu_X+\mu_Z)\sin A$	$-2^{\frac{1}{2}}(\mu_X+\mu_Y)\sin B$

The following set of symmetry coordinates has been constructed.

$$S_1(A_1) = 2^{-\frac{1}{2}}(r_1 + r_2),\quad S_2(A_1) = 2^{-\frac{1}{2}}(s_1 + s_2),\quad S_3(A_1) = R\alpha\ ;$$

$$S(B_1) = x_1 + x_2 - x_3 - x_4\ ;$$

$$S_1(B_2) = 2^{-\frac{1}{2}}(r_1 - r_2),\quad S_2(B_2) = 2^{-\frac{1}{2}}(s_1 - s_2).$$

Table 15-III. $T^{(i)}_{\alpha\beta,S}$ elements for the planar cyclic X_2YZ model.

$T_{\alpha\alpha,S}$	xx	yy	zz
$S_1(A_1)$	$8^{\frac{1}{2}}R$	$8^{\frac{1}{2}}R\cos^2 A$	$8^{\frac{1}{2}}R\sin^2 A$
$S_2(A_1)$	$8^{\frac{1}{2}}S$	$8^{\frac{1}{2}}S\cos^2 B$	$8^{\frac{1}{2}}S\sin^2 B$
$S_3(A_1)$	0	$-2R\sin 2A$	$2R\sin 2A$
$T_{\alpha\beta,S}$	yz		zx
$S_1(B_2)$	$-2^{\frac{1}{2}}R\sin 2A$	$S(B_1)$	$R\cos A-S\cos B$
$S_2(B_2)$	$2^{\frac{1}{2}}S\sin 2B$		

The out-of-plane (B_1) coordinate is given in terms of cartesian displacements. The set is complete, as the six coordinates are independent and contain no redundancies. Notice that three bendings in the ring were simply left out from the valence coordinates used for constructing the symmetry coordinates. This was the easy way of removing redundancies, which normally would appear in cyclic structures. The principles of this procedure for avoiding redundancies is discussed elsewhere.[1]

The **G**, **C**$^{\alpha}$, and $\mathbf{T}_{\alpha\beta,S}$ matrix elements in terms of the present symmetry coordinates are given in Tables I, II, and III, respectively. For a short account on these matrices, including their definitions, see Chapter 1.

REFERENCE

1 S.J.Cyvin, Acta Chem.Scand. 20, 2616 (1966).

-- Part II: Infrared Spectra of $^6LiNaF_2$, $^7LiNaF_2$, and Normal Coordinate Analysis of Planar Cyclic $LiNaF_2$, Li_2F_2, and Na_2F_2

A. SNELSON, B. N. CYVIN, and S. J. CYVIN

Spectra of the vapors over mixtures of LiF + NaF at 900 - 1000 °C have been recorded in the spectral region 4000 - 190 cm^{-1} using the matrix isolation technique. Five of the six infrared active frequencies of $LiNaF_2$ have been observed and assigned on the basis of a planar C_{2v} structure. Harmonic force constants are given for Li_2F_2, Na_2F_2, and $LiNaF_2$.

15.1. Experimental

The vapor species of the alkali metal fluorides only exist at relatively high temperatures. Conventional infrared techniques under these conditions are extremely difficult and for this reason the matrix isolation technique was used to investigate the $LiNaF_2$ molecule. The matrix isolation cryostat and molecular beam furnace used in the study have been described previously.[1] Neon was used as the matrix gas and liquid helium as the coolant. The Knudsen cell was made of graphite, and the $LiNaF_2$ vapor species formed at 900 - 1000 °C. Spectra were obtained in the region 4000 - 190 cm^{-1} from mixtures of ^6LiF + NaF, ^7LiF + NaF and ^6LiF + ^7LiF + NaF at various molar ratios. A typical spectrum is shown in Fig. 2.

15.2. Spectral and Structural Data

15.2.1. Interpretation of the Spectrum

Comparison of the spectra obtained from vaporizing a mixture

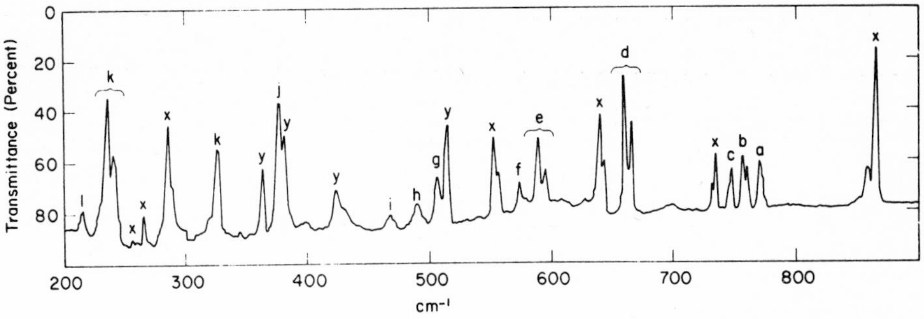

Fig. 15-2. Infrared spectrum of the vapor, species over an equimolar mixture of ^{7}LiF and NaF in a neon matrix. Deposition time, 6 hrs at 1000 °C. Matrix dilution approximately 7000:1. Absorption bands designated by x and y are due to species containing only pure ^{7}LiF or NaF respectively. All other absorption bands are due to species containing ^{7}LiF + NaF.

of LiF + NaF, with those obtained from pure LiF [2] and NaF [3] separately, easily allow identification of those absorption bands which are due to species containing LiF + NaF. In the spectra shown in Fig. 2 there are twelve such absorption bands. Relative intensity measurements show that five of these bands may be assigned to the same species. These five bands moreover, are of similar intensity as the cyclic dimer bands of Li_2F_2 and Na_2F_2, whilst the remaining seven are quite weak. Mass spectrographic[4] studies on the LiF + NaF system indicate that the major gas phase species are LiF, Li_2F_2, NaF, Na_2F_2 and $LiNaF_2$, all in about equal proportion. Thus it appears reasonable to assign the five absorption bands to the species $LiNaF_2$. By analogy with Li_2F_2 and Na_2F_2 the mixed dimer is assumed to be planar cyclic with C_{2v} symmetry. On this assumption, the Teller-Redlich product rule has been applied to the frequencies observed for the ^{6}Li and ^{7}Li substituted species. This is shown in Table IV where the agreement between the observed and calculated shifts are found to lie within the limits of the experimental determination of ± 1 cm^{-1}.

Table 15-IV. Spectral frequencies assigned to $LiNaF_2$.*

cm^{-1}	Assignment
700	(^{6}Li) A_1
660	(^{7}Li) A_1
621	(^{6}Li) B_2
589	(^{7}Li) B_2
377	(^{6}Li) A_1
376	(^{7}Li) A_1
327	(^{6}Li) B_2
326	(^{7}Li) B_2
249	(^{6}Li) B_1 (out-of-plane)
238	(^{7}Li) B_1 (out-of-plane)

* Theoretical frequency ratios (^{6}Li/^{7}Li) from the Teller-Redlich product rule with parenthesized observed values: A_1 1.072, B_2 1.064 (1.058), B_1 1.040 (1.046).

15.2.2. Experimental Data and Assignments of Vibrational Frequencies in Li_2F_2 and Na_2F_2

The molecular structure of $Li_2F_2(g)$ has been determined as D_{2h} by Büchler et al.[5] using electric deflection techniques. This structure is also consistent with that determined by electron diffraction techniques.[6] The latter technique has also been used on $Na_2F_2(g)$ and the same D_{2h} structure is indicated.[7,8] Berkowitz[9,10] has performed an extensive analysis of the alkali halide dimers, based on an ionic model for these molecules. His calculated frequencies have not been confirmed by later experimental studies, the calculated values in general being too low. Klemperer and Norris[11] reported frequencies of 640 ± 15 and 460 ± 15 cm^{-1} for the in-plane vibrations of Li_2F_2 observed in a high temperature infrared gas phase study. The infrared spectra of Li_2F_2[1,2,12-16] and Na_2F_2[3] have been studied extensively by the matrix isolation technique. For lithium fluoride[2] the use of isotopes has allowed an unambiguous assignment for the three infrared active frequencies belonging to the symmetry species B_{1u}, B_{2u} and B_{3u} at 678, 585 and 303 cm^{-1} for $^6Li_2F_2$, and 641, 553, and 287 cm^{-1} for $^7Li_2F_2$. For $^6Li^7LiF_2$ the corresponding observed wave numbers are 662, 565, and 296 cm^{-1}. For Na_2F_2, the B_{1u} and B_{2u} species were assigned at 380 and 363 cm^{-1} respectively, and the out-of-plane, B_{3u} bending mode at < 190 cm^{-1}. This latter number represented the long wavelength limit of the investigation.

15.2.3. Structural Data

The here adopted data as equilibrium structure parameters for Li_2F_2 and Na_2F_2 are[17] R(Li-F) = 1.68 Å, $2A$(FLiF) = 107°; and R(Na-F) = 2.02 Å, $2A$(FNaF) = 96°. From these parameters the distances on Fig. 3 were derived, from which $\frac{1}{2}$(Li···Li) and $\frac{1}{2}$(Na···Na) were transferred to the $LiNaF_2$ molecule. The F···F distance in this molecule was taken as the average of the corresponding distances in Li_2F_2 and Na_2F_2. The present structural data (for Li_2F_2 and Na_2F_2) are fairly well consistent with those from Berkowitz,[9] which actually were used in a preliminary normal coordinate analysis of the present work.

15.3. Theoretical Potential Functions

15.3.1. Planar Cyclic X_2YZ Model

The adopted molecular model for $LiNaF_2$ is treated in Part I,

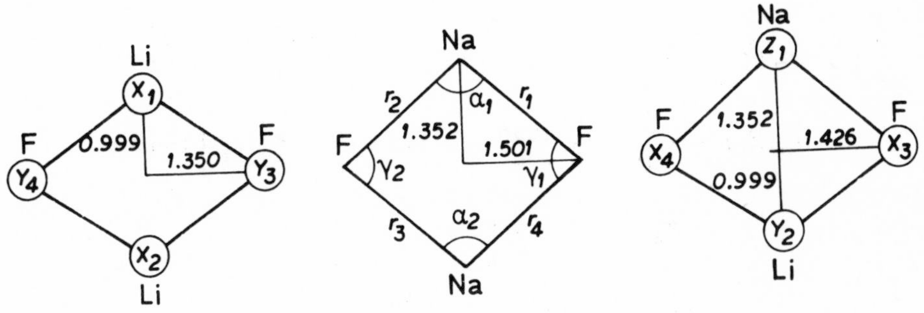

Fig. 15-3. Adopted equilibrium parameters for Li_2F_2, Na_2F_2, and $LiNaF_2$. In-plane valence coordinates for the X_2Y_2 (D_{2h}) model are indicated.

where a complete set of independent symmetry coordinates is given. Arbitrary valence coordinates have been omitted in constructing the symmetry coordinates. For this reason the symmetry coordinates are not believed to be suitable for setting up an initial force field. Especially there is no reason to believe that a diagonal **F** matrix in terms of these coordinates would be a good approximation. Therefore an augmented set of in-plane valence coordinates was considered. In addition to the coordinates shown in Fig. 1 the XYX bending (β) and two YXZ bendings, viz. γ_1 (1-3-2) and γ_2 (1-4-2), are included. An initial approximate in-plane valence force field was assumed, and is represented by a five-constant potential function. The case is simple enough to be worked out algebraically. Using the method of a transformation matrix $\boldsymbol{T}$ (cf. Chapter 6) we arrived at the following expressions for the symmetry force constants.

$$F_1(A_1) = f_r + (2f_\beta + \frac{R}{S} f_\gamma) \frac{\sin^2 A}{\cos^2 B} , \qquad (15.1)$$

$$F_2(A_1) = f_S + (2f_\beta + \frac{R}{S} f_\gamma) \tan^2 B , \qquad (15.2)$$

$$F_3(A_1) = f_\alpha + f_\beta \frac{\cos^2 A}{\cos^2 B} + \frac{1}{2} \left[(\frac{S}{R})^{\frac{1}{2}} + (\frac{R}{S})^{\frac{1}{2}} \frac{\cos A}{\cos B} \right]^2 f_\gamma , \qquad (15.3)$$

$$F_{12}(A_1) = -(2f_\beta + \frac{R}{S} f_\gamma) \frac{\sin A}{\cos B} \tan B , \qquad (15.4)$$

$$F_{13}(A_1) = \left[2^{\frac{1}{2}} f_\beta \frac{\cos A}{\cos B} + 2^{-\frac{1}{2}} (1 + \frac{R}{S} \frac{\cos A}{\cos B}) f_\gamma \right] \frac{\sin A}{\cos B} , \qquad (15.5)$$

$$F_{23}(A_1) = - \left[2^{\frac{1}{2}} f_\beta \frac{\cos A}{\cos B} + 2^{-\frac{1}{2}} (1 + \frac{R}{S} \frac{\cos A}{\cos B}) f_\gamma \right] \tan B ; \qquad (15.6)$$

$$F_1(B_2) = f_r + \frac{S}{R} f_\gamma \cot^2 A , \qquad (15.7)$$

$$F_2(B_2) = f_S + \frac{R}{S} f_\gamma \cot^2 B , \qquad (15.8)$$

$$F_{12}(B_2) = f_\gamma \cot A \cot B . \qquad (15.9)$$

15.3.2. Planar Cyclic X_2Y_2 Model

The planar cyclic molecular model of symmetry D_{2h} is treated elsewhere.[18] For the sake of convenience we repeat here the adopted symmetry coordinates; for valence coordinates, see Fig. 3. $S_1(A_g) = \frac{1}{2}(r_1 + r_2 + r_3 + r_4)$, $S_2(A_g) = 2^{-\frac{1}{2}}R(\alpha_1 + \alpha_2)$; $S(B_{3g}) = \frac{1}{2}(r_1 - r_2 + r_3 - r_4)$; $S(B_{1u}) = \frac{1}{2}(r_1 + r_2 - r_3 - r_4)$; $S(B_{2u}) = \frac{1}{2}(r_1 - r_2 - r_3 + r_4)$; $S(B_{3u}$; in terms of out-of-plane cartesian displacements) $= x_1 + x_2 - x_3 - x_4$. An approximate potential function was set up, following the same principles as outlined for the X_2YZ model in the preceding paragraph. In the present case the in-plane part contains three force constants, viz. f_r, f_α, and f_γ, and their connections with the six in-plane symmetry force constants are given in the following.

$$F_1(A_g) = f_r, \quad F_2(A_g) = f_\alpha + f_\gamma, \quad F_{12}(A_g) = 0 ; \tag{15.10}$$

$$F(B_{3g}) = f_r ; \tag{15.11}$$

$$F(B_{1u}) = f_r + 2f_\alpha \tan^2 A ; \tag{15.12}$$

$$F(B_{2u}) = f_r + 2f_\gamma \cot^2 A . \tag{15.13}$$

15.4. Normal Coordinate Analysis

15.4.1. $LiNaF_2$

After several trials the following values (all in mdyne/Å) were found, which reproduce satisfactorily the observed frequencies in $LiNaF_2$. $f_r(NaF) = 0.53$, $f_s(LiF) = 0.73$, $f_\alpha(FNaF) = 0.16$, $f_\beta(FLiF) = 0.30$, and $f_\gamma(NaFLi) = 0.18$. Table V shows the observed frequencies (cf. Section 15.2) along with the calculated values from the above approximate force field. The force constants were converted to those of the symmetrized **F** matrix, which was refined in order to give a still better fit to the observed frequencies. The final symmetry force field (including the out-of-plane force constant) is shown in Table VI. It was adjusted to fit exactly the observed frequencies for $^7LiNaF_2$ including the uncertain value of

Table 15-V. Observed and approximately calculated in-plane frequencies (cm^{-1}) for $^{6}LiNaF_2$ and $^{7}LiNaF_2$.

	$^{6}LiNaF_2$ Approx. calc.	$^{6}LiNaF_2$ Obs.	$^{7}LiNaF_2$ Approx. calc.	$^{7}LiNaF_2$ Obs.
Species A_1	707	700	662	660
	354	377	354	376
	332	-	331	(321?)
Species B_2	632	621	595	589
	326	327	325	326

321 cm^{-1} obtained from a shoulder. The calculated frequencies for $^{6}LiNaF_2$, viz.(in cm^{-1}) 705, 376, 322, 248 (B_1), 625, and 327 agree satisfactorily with the observed values (cf. Tables V and VI).

15.4.2. Li_2F_2

For $^{7}Li_2F_2$ the value of f_r(LiF) = 0.73 mdyne/Å (cf. Section 15.4.1) was tentatively used. Then it was possible to calculate f_α = 0.28 mdyne/Å and f_γ = 0.00 mdyne/Å from the observed B_{1u} and B_{2u} frequencies using Eq. (12) and (13), respectively. The value of f_α exhibits pleasingly good agreement with f(FLiF) = 0.30 mdyne/Å from the mixed dihalide. In these preliminary calculations the structural data from Berkowitz[1] were applied. The final force field as shown in Table VII reproduces exactly the observed B_{1u}, B_{2u}, and B_{3u} frequencies for $^{7}Li_2F_2$ with the new parameters[17] (cf. Fig. 3). The calculated A_g and B_{3g} frequencies (in cm^{-1}) for $^{7}Li_2F_2$ are

Table 15-VI. Final symmetry force constants (mdyne/Å) for $LiNaF_2$.

Species A_1			B_1	B_2	
1.826	-1.450	1.075	0.1145	0.675	0.115
	2.354	-1.147			0.812
		1.088			

Table 15-VII. Symmetry force constants (mdyne/Å) for Li_2F_2 and Na_2F_2.

Li_2F_2	Species A_g		B_{3g}	B_{1u}	B_{2u}	B_{3u}
	0.73	0.00	0.73	1.752	0.714	0.124
		0.28				
Na_2F_2	Species A_g		B_{3g}	B_{1u}	B_{2u}	B_{3u}
	0.53	0.00	0.53	0.987	0.731	0.11
		0.30				

(A_g) 551, 342, and (B_{3g}) 524. For $^6Li_2F_2$ the calculated frequencies from the final force field are (A_g) 594, 342, (B_{3g}) 559, (B_{1u}) 679, (B_{2u}) 586, and (B_{3u}) 304 cm^{-1}. For $^6Li^7LiF_2$ they are: (A_1) 664, 568, 342, (B_1) 296, and (B_2) 577 and 533 cm^{-1}. According to the here chosen orientation of cartesian axes the correlation between symmetry species of C_{2v} and (D_{2h}) are: A_1 (A_g + B_{1u}), B_1 (B_{3u}), and B_2 (B_{3g} + B_{2u}). Hence the observed frequencies of 662 and 296 cm^{-1} are easily assigned to the highest A_1 and the B_1 frequency, respectively. One of the B_2 frequencies would be expected to be the third one with appreciable intensity, while the remaining three frequencies are presumably too weak to allow detection. The present calculations suggest that the position of this frequency (observed value 565 cm^{-1}) may have been obscured by virtue of its proximity to one of the A_1 frequencies.

15.4.3. Na_2F_2

The same procedure as outlined for Li_2F_2 above was used for Na_2F_2. f_r(NaF) = 0.53 mdyne/Å gave f_α = 0.19 mdyne/Å and f_γ = 0.11 mdyne/Å. Here again the agreement between f_α and f(FNaF) = 0.16 mdyne/Å is a pleasing feature. f_γ on the other hand shows unsystematic variation on comparison between the different molecules. The final force constants are shown in Table VII. They reproduce the observed B_{1u} and B_{2u} frequencies. The calculated frequencies (in cm^{-1}) for the other species are: (A_g) 320, 287, (B_{3g}) 293, and (B_{3u}) 190. The out-of-plane (B_{3u}) force constant was estimated by rough extrapolation from the out-of-plane force constants in Li_2F_2 and $LiNaF_2$. The conclusion of $\nu(B_{3u}) < 190\ \mathrm{cm}^{-1}$ (cf. Section 15.2.2) would indicate $F(B_{3u}) < 0.11$ mdyne/Å.

REFERENCES

1 A.Snelson, J.Phys.Chem. 73, 1919 (1969).
2 A.Snelson, J.Chem.Phys. 46, 3652 (1967).
3 A.Snelson, J.Phys.Chem. 67, 882 (1963).
4 R.F.Porter and R.C.Schoonmaker, J.Chem.Phys. 29, 1070 (1958).
5 A.Büchler, J.L.Stauffer, and W.A.Klemperer, J.Am.Chem.Soc. 86, 4544 (1964).
6 P.A.Akishin, L.N.Gorokhov, and L.N.Sidorov, Russ.J.Phys.Chem. (English translation) 33, 648 (1959).
7 P.A.Akishin and N.G.Rambidi, Zh.neorg.Khim. 5, 23 (1960).
8 P.A.Akishin and N.G.Rambidi, Z.phys.Chem. 213, 111 (1960).
9 J.Berkowitz, J.Chem.Phys. 29, 1386 (1958).
10 J.Berkowitz, J.Chem.Phys. 32, 1519 (1960).
11 W.A.Klemperer and W.G.Norris, J.Chem.Phys. 34, 1071 (1961).
12 M.J.Linevsky, J.Chem.Phys. 38, 658 (1963).
13 A.Snelson and K.S.Pitzer, J.Phys.Chem. 67, 882 (1963).
14 S.Schlick and O.Schnepp, J.Chem.Phys. 41, 463 (1964).
15 R.L.Redington, J.Chem.Phys. 44, 1238 (1964).
16 S.Abramowitz, N.Acquista, and I.W.Levin, J.Res.Nat.Bur.Stand. 72A, 487 (1968).
17 K.S.Krasnov, V.S.Timoshinin, T.G.Danilova, and S.V.Khandozhko: Molekulyarnye postoyannye neorganicheskikh soedineii (Molecular constants of inorganic compounds), Izd. Khimiya, Leningrad 1968
18 S.J.Cyvin, J.Brunvoll, B.N.Cyvin, I.Elvebredd, and G.Hagen, Mol.Phys. 14, 43 (1968).

-- Part III: Mean Amplitudes for Lithium and Sodium Fluorides

S. J. CYVIN and B. N. CYVIN

Calculated mean amplitudes are reported for the LiF, NaF, Li_2F_2, Na_2F_2, and $LiNaF_2$ molecules. Classical limits and low-temperature anomalies are discussed.

15.5. Calculated Mean Amplitudes

Mean amplitudes of vibration were calculated for lithium and sodium fluoride monomers and dimers, and for lithium-sodium difluoride.

The vibrational frequency data for $^{6}Li^{19}F$ (964.07 cm^{-1}), $^{7}Li^{19}F$ (910.34 cm^{-1}), and $^{23}Na^{19}F$ (536.1 cm^{-1}) were taken from the quotations of Krasnov et al.[1] The appropriate force constants were found to be 2.502 mdyne/Å for LiF, and 1.761 mdyne/Å for NaF; and the resulting mean amplitudes of vibration at different temperatures are given in Table VIII.

The developed harmonic force fields for Li_2F_2 and Na_2F_2 (see Part II) were used to calculate the mean amplitudes of vibration

Table 15-VIII. Mean amplitudes of vibration (Å units) for LiF and NaF.

Temp. (K)	$^{6}Li^{19}F$	$^{7}Li^{19}F$	$^{23}Na^{19}F$
0	0.0619	0.0601	0.0550
298.15	0.0625	0.0609	0.0593
500	0.0659	0.0647	0.0683
1000	0.0799	0.0793	0.0907
1500	0.0941	0.0938	0.1096
2000	0.1071	0.1069	0.1260

Table 15-IX. Mean amplitudes of vibration (Å units) for Li_2F_2 molecules.

Distance	Temp. (K)	$^6Li_2F_2$	$^6Li^7LiF_2$		$^7Li_2F_2$
Li-F		^{6}Li-F	^{6}Li-F	^{7}Li-F	^{7}Li-F
	0	0.080	0.080	0.078	0.078
	298.15	0.087	0.087	0.085	0.085
	500	0.099	0.099	0.098	0.098
	1000	0.131	0.131	0.131	0.131
	1500	0.158	0.158	0.158	0.158
	2000	0.181_5	0.181_5	0.181_4	0.181_4
Li···Li		^{6}Li···^{6}Li	^{6}Li···^{7}Li		^{7}Li···^{7}Li
	0	0.098	0.096		0.094
	298.15	0.104	0.102		0.101
	500	0.118	0.117		0.116
	1000	0.155	0.154		0.154
	1500	0.187	0.186		0.186
	2000	0.214	0.214		0.214
F···F	0	0.072	0.072		0.072
	298.15	0.087	0.087		0.087
	500	0.106	0.106		0.106
	1000	0.146	0.146		0.146
	1500	0.178	0.178		0.178
	2000	0.205	0.205		0.205

for the molecules in question. The results for three isotopic mole cules of Li_2F_2 are shown in Table IX, and those for Na_2F_2 in Table

The calculated mean amplitudes for $^6LiNaF_2$ and $^7LiNaF_2$ (for force constants, see Part II) are given in Table XI.

As expected the mean amplitude for a bridged alkali-fluorine distance is found to be larger than the corresponding distance in the monomer. Otherwise the mean amplitudes of corresponding distances in the different (four-atomic) molecules are seen to be fairly characteristic. The unsystematic variation of the F···F mea amplitude through the series Li_2F_2-$LiNaF_2$-Na_2F_2 is not necessaril

Table 15-X. Mean amplitudes of vibration (Å units) for Na_2F_2.

Distance	Temperature (K)					
	0	298.15	500	1000	1500	2000
Na-F	0.070	0.087	0.107	0.147	0.179	0.207
Na···Na	0.071	0.092	0.113	0.157	0.192	0.221
F···F	0.075	0.093	0.115	0.158	0.193	0.223

real. In general the temperature influence on the mean amplitudes is found to be important, both quantitatively and qualitatively. For some additional discussions of the results, see Section 15.8.

15.6. Classical Limits and Low-Temperature Anomalies

Some regularities can be assumed for force constants and compliants in related molecules. This knowledge can be used to deduce certain regularities also for the classical limits of mean-square amplitudes, and should consequently hold for the rigorously

Table 15-XI. Mean amplitudes of vibration (Å units) for $LiNaF_2$ molecules.

Distance	Temp.(K)	$^6LiNaF_2$	$^7LiNaF_2$	Distance	Temp.(K)	$^6LiNaF_2$	$^7LiNaF_2$
Li-F		^{6}Li-F	^{7}Li-F	Li···Na		^{6}Li···Na	^{7}Li···Na
	0	0.079	0.077		0	0.082	0.080
	298.15	0.085	0.084		298.15	0.092	0.091
	500	0.097	0.096		500	0.108	0.108
	1000	0.128	0.128		1000	0.145	0.145
	1500	0.154	0.154		1500	0.176	0.176
	2000	0.177	0.177		2000	0.202	0.202
Na-F	0	0.070	0.070	F···F	0	0.070	0.070
	298.15	0.086	0.086		298.15	0.083	0.083
	500	0.105	0.105		500	0.100	0.100
	1000	0.145	0.145		1000	0.137	0.137
	1500	0.177	0.177		1500	0.167	0.167
	2000	0.204	0.204		2000	0.193	0.193

calculated mean-square amplitudes at sufficiently high temperatures. Cyvin[2] has pointed out some deviations from such regularities, and referred to them as low-temperature anomalies. They arise when effects of atomic masses predominate the force-constant effects, and may be understood from the following expansion of a mean-square amplitude,[2]

$$l_{ij}^2 = kTn_{ij} + \frac{h^2}{48\pi^2 kT}(\mu_i + \mu_j) - \ldots , \qquad (15.14)$$

where n_{ij} is the compliant corresponding to the atom pair ij; μ_i and μ_j are the inverse masses of the i and j atoms, respectively. The first term of the expansion represents the classical limit:

$$(l_{ij}^{cl})^2 = kTn_{ij} . \qquad (15.15)$$

The cited paper[2] contains examples of low-temperature anomalies for the series of boron trihalides. In the next section some examples are pointed out within the here studied series of molecules.

15.7. Examples of Low-Temperature Anomalies

The calculated results show the absence of any significant secondary isotope effect; cf. F···F in Table IX, and Na-F and F···F in Table XI. Also the primary isotope effects are small concerning the ^{6}Li and ^{7}Li substitutions, and they become quite insignificant at high temperatures (cf. Tables VIII, IX, and XI). These features are consistent with the statement that all isotope effects on mean amplitudes are to be regarded as low-temperature anomalies.[2] This statement holds under the usual assumption of identical force fields for isotopic molecules.

In the monomer molecules $f_{LiF} > f_{NaF}$ (cf. Section 15.5), and consequently $n_{LiF} < n_{NaF}$ for the compliants, which also holds for the two dimer molecules. Hence for the classical limits of mean amplitudes

$$l_{LiF}^{cl} < l_{NaF}^{cl} . \qquad (15.16)$$

This inequality also holds for the rigorous mean amplitudes at high temperatures. The inequality is reversed as an effect of the low-

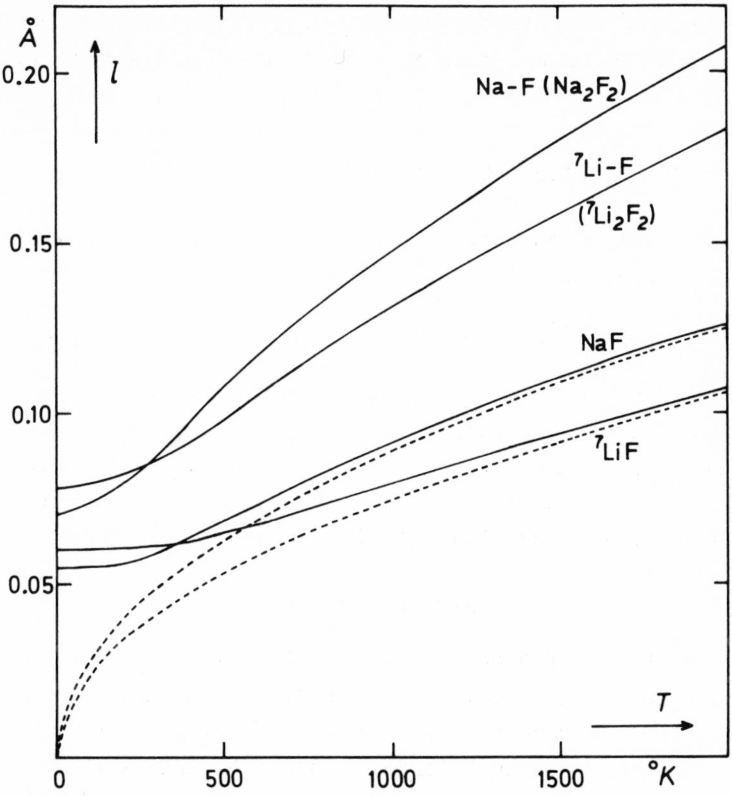

Fig. 15-4. Mean amplitudes of vibration at varying temperature for Li-F and Na-F distances in ^7LiF, NaF, $^7Li_2F_2$, and Na_2F_2. Classical limits of mean amplitudes for the monomeric molecules are shown as broken curves, which are parabola branches.

temperature anomaly at $< \sim 400$ K for LiF/NaF (monomer) and $< \sim 300$ K for Li_2F_2/Na_2F_2; cf. Fig. 4. Other low-temperature anomalies of a similar kind are found. $l_{Na \cdots Na}$ being smaller than $l_{Li \cdots Na}$ and $l_{Li \cdots Li}$ at low temperatures becomes the largest one of these l values above about 1000 K.

Here we find also examples of molecules, viz. Li_2F_2 and $LiNaF_2$, in which the calculated mean amplitude for a bonded distance at low temperatures is larger than the mean amplitude of a nonbonded distance. This is the case for Li-F as the bonded distance (though through bridged fluorine) versus F···F as the nonbonded one, at temperatures <~298 K. The same effect was described for boron tri iodide.[2]

15.8. Comparison with Electron Diffraction Results

In the high-temperature gas electron-diffraction work on lithium fluoride dimer[3,4] the conclusion of

$$l_{F\cdots F}/l_{Li-F} \approx 3 \quad (15.17$$

is made. It is based on l_{Li-F} = 0.07 Å (assumed), which according to the present calculations seems to be too low. The here calculated ratio of $l_{F\cdots F}/l_{Li-F}$ varies from 0.9 at absolute zero to 1.1 at 2000 K. The reason for this discrepancy between electron-diffraction and spectroscopical results is not clear.

REFERENCES

1 K.S.Krasnov, V.S.Timoshinin, T.G.Danilova, and S.V.Khandozhko: Molekulyarnye postoyannye neorganicheskikh soedineii (Molecular constants of inorganic compounds), Izd. Khimiya, Leningrad 1968.
2 S.J.Cyvin, Kgl. Norske Videnskab. Selskabs Skrifter, No. 1 (1969
3 P.A.Akishin and N.G.Rambidi, Zh.neorg.Khim. 5, 23 (1960).
4 P.A.Akischin (Akishin) and N.G.Rambidi, Z.phys.Chem. 213, 111 (1960).

CHAPTER 16

Sandwich Compounds - Part I: Symmetry Coordinates of Molecular Vibrations for Trigonal Prism and Antiprism with Central Atom

J. BRUNVOLL, B. N. CYVIN, S. J. CYVIN, G. HAGEN, and LOTHAR SCHÄFER

The trigonal XY_6 prism (D_{3h}) and antiprism (D_{3d}) models are treated as simplified illustrations of a sandwich compound. Symmetry coordinates of molecular vibrations of standard types are specified. For the D_{3h} model also an alternative set of symmetry coordinates is given, based on an approach which tends to preserve the identity of the ligands (i.e. Y_3 triangles). **G** matrix elements are given for the eclipsed (D_{3h}) model in terms of the two symmetry coordinate sets.

To trigonale XY_6 *modellar, prisme* (D_{3h}) *og antiprisme* (D_{3d}), *blir brukte som enkle eksempel på stoff av* sandwich-*type. Symmetrikoordinatar av standardtypen for molekylvibrasjonar blir gjevne. For* D_{3h} *modellen blir det og laga eit anna symmetrikoordinatsett. Dette har ei form som er nærare knytt til vibrasjonane for dei individuelle ligandane* (Y_3 *gruppene). For begge dei to symmetrikoordinatsetta* (D_{3h} *modellen) blir* **G**-*matriseelementa gjevne.*

16.1. Introduction

In this chapter two XY_6 models of symmetries D_{3h} and D_{3d}

are treated. They are simplified models of a sandwich compound and were studied in connection with the study of dibenzene chromium; cf. Part II. The derivations of algebraic expressions for the XY_6 models are not as formidable as they would be for a complete twenty-five atomic dibenzene chromium model. Nevertheless the present treatment illustrates many of the interesting features encountered in the study of the dibenzene chromium molecule.

16.2. Tentatively Standardized Symmetry Coordinates

16.2.1. Molecular Models

Fig. 1 shows the eclipsed and staggered XY_6 molecular models of symmetries D_{3h} and D_{3d}, respectively. Structural parameters which define the equilibrium configuration: the X-Y distance R and the Y_1XY_2 angle, say $2A$. If the Y_1XY_4 angle in the

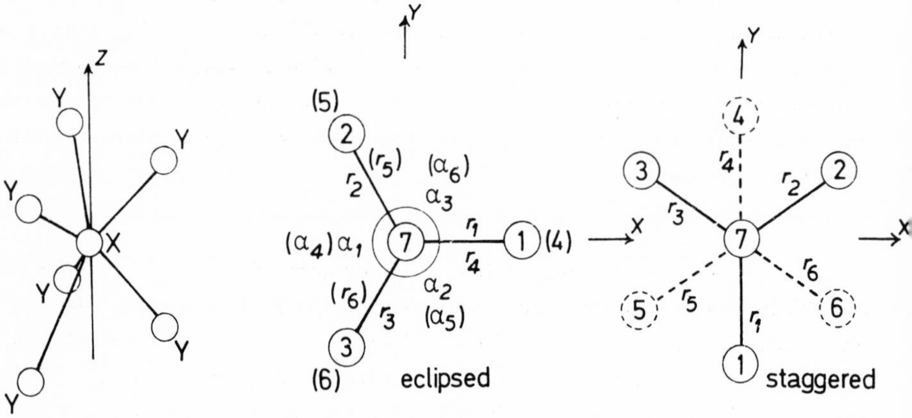

Fig. 16-1. The eclipsed (D_{3h}) and staggered (D_{3d}) XY_6 molecular models: trigonal prism and antiprism with central atom. The notation for X-Y stretchings and α-type YXY bendings are indicated. Notation for β-type bendings: $\beta_1(Y_1XY_4)$, $\beta_2(Y_2XY_5)$, $\beta_3(Y_3XY_6)$. The coordinate describing internal rotation in terms of cartesian displacements: $R\varphi = y_1 - \frac{1}{2}3^{\frac{1}{2}}x_2 - \frac{1}{2}y_2 + \frac{1}{2}3^{\frac{1}{2}}x_3 - \frac{1}{2}y_3 - y_4 + \frac{1}{2}3^{\frac{1}{2}}x_5 + \frac{1}{2}y_5 - \frac{1}{2}3^{\frac{1}{2}}x_6 + \frac{1}{2}y_6$.

eclipsed model is $2B$ one has

$$\sin A = \frac{1}{2} 3^{\frac{1}{2}} \cos B \,. \tag{16.1}$$

16.2.2. Symmetry Coordinates

In this section complete sets of symmetry coordinates for the models in question are specified. They were chosen in the style of tentatively standardized symmetry coordinates; cf. Chapter 20. In particular they are related to the coordinates of trigonal X_2Y_6 (ethane-like) models.[1] The distribution of normal modes of vibration among the symmetry species are found to be

$$\Gamma(D_{3h}) = 2A_1' + 3E' + A_1'' + 2A_2'' + 2E'' \tag{16.2}$$

and

$$\Gamma(D_{3d}) = 2A_{1g} + 2E_g + A_{1u} + 2A_{2u} + 3E_u \tag{16.3}$$

in the two cases of the XY_6 model. The applied valence coordinates are indicated in Fig. 1 and its legend, where also the coordinate of internal rotation is given in terms of cartesian displacements.

Symmetry coordinates for the trigonal prism with central atom:

$$S_1(A_1') = 6^{-\frac{1}{2}}(r_1 + r_2 + r_3 + r_4 + r_5 + r_6),$$

$$S_2(A_1') = \frac{R}{[6(2+k^2)]^{\frac{1}{2}}} [k(\alpha_1+\alpha_2+\alpha_3+\alpha_4+\alpha_5+\alpha_6)$$

$$- 2(\beta_1+\beta_2+\beta_3)]; \tag{16.4}$$

$$S_{1a}(E') = 12^{-\frac{1}{2}}(2r_1 - r_2 - r_3 + 2r_4 - r_5 - r_6),$$

$$S_{2a}(E') = 12^{-\frac{1}{2}}R(2\alpha_1 - \alpha_2 - \alpha_3 + 2\alpha_4 - \alpha_5 - \alpha_6),$$

$$S_{3a}(E') = 6^{-\frac{1}{2}}R(2\beta_1 - \beta_2 - \beta_3); \tag{16.5}$$

$$S_{1b}(E') = \frac{1}{2}(r_2 - r_3 + r_5 - r_6),$$

$$S_{2b}(E') = \frac{1}{2}R(\alpha_2 - \alpha_3 + \alpha_5 - \alpha_6),$$

$$S_{3b}(E') = 2^{-\frac{1}{2}}R(\beta_2-\beta_3); \quad (16.6)$$

$$S(A_1'') = R\varphi; \quad (16.7)$$

$$S_1(A_2'') = 6^{-\frac{1}{2}}(r_1+r_2+r_3-r_4-r_5-r_6),$$

$$S_2(A_2'') = 6^{-\frac{1}{2}}R(\alpha_1+\alpha_2+\alpha_3-\alpha_4-\alpha_5-\alpha_6); \quad (16.8)$$

$$S_{1a}(E'') = \frac{1}{2}(r_2-r_3-r_5+r_6),$$

$$S_{2a}(E'') = \frac{1}{2}R(\alpha_2-\alpha_3-\alpha_5+\alpha_6); \quad (16.9)$$

$$S_{1b}(E'') = 12^{-\frac{1}{2}}(2r_1-r_2-r_3-2r_4+r_5+r_6),$$

$$S_{2b}(E'') = 12^{-\frac{1}{2}}R(2\alpha_1-\alpha_2-\alpha_3-2\alpha_4+\alpha_5+\alpha_6). \quad (16.10)$$

In Eq. (4) the S_2 coordinate was produced on the basis of the redundancy condition

$$\alpha_1+\alpha_2+\alpha_3+\alpha_4+\alpha_5+\alpha_6 + k(\beta_1+\beta_2+\beta_3) = 0, \quad (16.11)$$

where

$$k = 3^{\frac{1}{2}} \frac{\sin B}{\cos A}. \quad (16.12)$$

Hence when $S(\alpha) = 6^{-\frac{1}{2}}R(\alpha_1+\alpha_2+\alpha_3+\alpha_4+\alpha_5+\alpha_6)$ and $S(\beta) = 3^{-\frac{1}{2}}R(\beta_1+\beta_2+\beta_3)$, the redundant zero coordinate and the chosen S_2 coordinate of Eq. (4) are given by the orthogonal transformation

$$\left[\frac{1}{2+k^2}\right]^{\frac{1}{2}}[2^{\frac{1}{2}}S(\alpha) + kS(\beta)] = 0,$$
$$\left[\frac{1}{2+k^2}\right]^{\frac{1}{2}}[kS(\alpha) - 2^{\frac{1}{2}}S(\beta)] = S_2(A_1'). \quad (16.13)$$

For the trigonal antiprism with central atom formally the same expressions for symmetry coordinates as given in Eqs. (4)-(10) for the prism can be adapted as far as the r and α type combinations are concerned. Then it is only necessary to observe the correlations between the appropriate species of the symmetry groups D_{3h} and D_{3d}, viz.[1]: A_1' - A_{1g}, E' - E_g, A_1'' - A_{1u}, A_2'' - A_{2u}, and E'' - E_u. The consideration of β's brings some interesting features, inasmuch as they have a different nature from the β(XXY) bendings in the ethane-type molecular models.[1] In the present case the corresponding equilibrium angles change during the internal rotation, and in the extreme case of the antiprism model the β's are linear bendings. Let them be defined in the sense that $\beta_1+\beta_2+\beta_3$ makes the central atom move upwards (along the Z axis). This (redundant) combination is not totally symmetric now, but belongs to the A_{2u} species.

Complete set of symmetry coordinates for the trigonal antiprism with central atom:

$$S_1(A_{1g}) = 6^{-\frac{1}{2}}(r_1+r_2+r_3+r_4+r_5+r_6),$$

$$S_2(A_{1g}) = 6^{-\frac{1}{2}}R(\alpha_1+\alpha_2+\alpha_3+\alpha_4+\alpha_5+\alpha_6); \qquad (16.14)$$

$$S_{1a}(E_g) = 12^{-\frac{1}{2}}(2r_1-r_2-r_3+2r_4-r_5-r_6),$$

$$S_{2a}(E_g) = 12^{-\frac{1}{2}}R(2\alpha_1-\alpha_2-\alpha_3+2\alpha_4-\alpha_5-\alpha_6); \qquad (16.15)$$

$$S_{1b}(E_g) = \frac{1}{2}(r_2-r_3+r_5-r_6),$$

$$S_{2b}(E_g) = \frac{1}{2}R(\alpha_2-\alpha_3+\alpha_5-\alpha_6); \qquad (16.16)$$

$$S(A_{1u}) = R\varphi; \qquad (16.17)$$

$$S_1(A_{2u}) = 6^{-\frac{1}{2}}(r_1+r_2+r_3-r_4-r_5-r_6),$$

$$S_2(A_{2u}) = \frac{R}{[6(2+k^2)]^{\frac{1}{2}}}[k(\alpha_1+\alpha_2+\alpha_3-\alpha_4-\alpha_5-\alpha_6)$$

$$+ 2(\beta_1+\beta_2+\beta_3)]; \qquad (16.18)$$

$$S_{1a}(E_u) = \frac{1}{2}(r_2-r_3-r_5+r_6),$$

$$S_{2a}(E_u) = \frac{1}{2}R(\alpha_2-\alpha_3-\alpha_5+\alpha_6),$$

$$S_{3a}(E_u) = 2^{-\frac{1}{2}}R(\beta_2-\beta_3); \qquad (16.19$$

$$S_{1b}(E_u) = 12^{-\frac{1}{2}}(-2r_1+r_2+r_3+2r_4-r_5-r_6),$$

$$S_{2b}(E_u) = 12^{-\frac{1}{2}}R(-2\alpha_1+\alpha_2+\alpha_3+2\alpha_4-\alpha_5-\alpha_6),$$

$$S_{3b}(E_u) = 6^{-\frac{1}{2}}R(-2\beta_1+\beta_2+\beta_3). \qquad (16.20$$

In order to follow the previously adopted conventions[1] the $S_{1b}(E_u)$ and $S_{2b}(E_u)$ coordinates were taken as the respective expressions for $S_{1b}(E'')$ and $S_{2b}(E'')$ from Eq. (10) with opposite sign. The coordinate of internal rotation (17) must be constructed properly in terms of cartesian displacements with coefficients different from those of the legend to Fig. 1.

16.3. Alternative Symmetry Coordinates for the Eclipsed Model

An alternative approach to the construction of symmetry coordinates is shown for the eclipsed XY_6 (D_{3h}) model in the following. This approach is believed to be very useful in treatments of sandwich compounds, and has been applied to dibenzene chromium; cf. Part II. By virtue of this analogy the Y_3 triangles are referred to as ligands in the present model; see Fig. 2.

The symmetric structure for the totality of the fifteen normal modes of vibration is given in Eq. (2). The symmetry coordinates are derived on the basis of a further classification of the vibrational modes.

(i) In-phase ligand vibrations. The symmetry coordinates for a regular triangle, viz. $S_1 = 3^{-\frac{1}{2}}(d_1+d_2+d_3)$, $S_{2a} = 6^{-\frac{1}{2}}(2d_1-d_2-d_3)$, and $S_{2b} = 2^{-\frac{1}{2}}(d_2-d_3)$, are also given elsewhere.[2] They have the symmetric structure of

$$\Gamma(\text{i}) = A_1' + E', \qquad (16.21)$$

which also is the symmetric structure of the in-phase ligand vibrations described by

$$\mathcal{S}_1(A_1') = 6^{-\frac{1}{2}}(d_1+d_2+d_3+d_4+d_5+d_6),$$

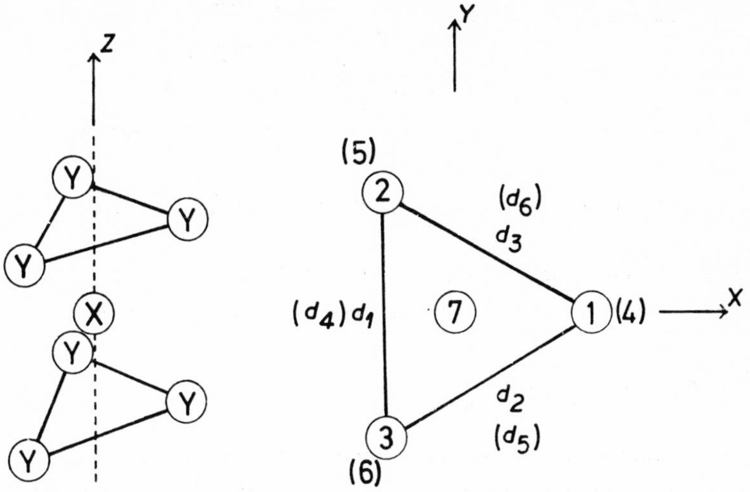

Fig. 16-2. The eclipsed (D_{3h}) XY_6 molecular model considered as a central atom (X) with two ligands (Y_3).

$$\mathcal{S}_{1a}(E') = 12^{-\frac{1}{2}}(2d_1 - d_2 - d_3 + 2d_4 - d_5 - d_6),$$

$$\mathcal{S}_{1b}(E') = \frac{1}{2}(d_2 - d_3 + d_5 - d_6). \tag{16.22}$$

(ii) Out-of-phase ligand vibrations. Table I shows the correlations between species for in-phase and out-of-phase vibrations. Hence there are three modes with the symmetric structure of

$$\Gamma(\text{ii}) = A_2'' + E'' , \tag{16.23}$$

in which the two triangles vibrate in opposite directions. These motions are described by

$$\mathcal{S}_1(A_2'') = 6^{-\frac{1}{2}}(d_1 + d_2 + d_3 - d_4 - d_5 - d_6),$$

$$\mathcal{S}_{1a}(E'') = \frac{1}{2}(d_2 - d_3 - d_5 + d_6),$$

Table 16-I. Correlations between symmetry species for in-phase and out-of-phase motions in D_{3h} structures.

$A_1' \leftrightarrow A_2''$
$A_2' \leftrightarrow A_1''$
$E' \leftrightarrow E''$

$$S_{1b}(E'') = 12^{-\frac{1}{2}}(2d_1 - d_2 - d_3 - 2d_4 + d_5 + d_6). \qquad (16.24)$$

(iii) Compensated in-phase translations. This type of motion has the symmetric structure of rigid translations of the molecule, viz.

$$\Gamma(\text{iii}) = E' + A_2'' , \qquad (16.25)$$

and may be described as parallel translations of the ligands compensated by the central atom, which moves in the opposite direction of the ligands. In terms of cartesian displacements (cf. Fig. 1):

$$S_{3a}(E') = x_1 + x_2 + x_3 + x_4 + x_5 + x_6 - 6x_7,$$

$$S_{3b}(E') = y_1 + y_2 + y_3 + y_4 + y_5 + y_6 - 6y_7,$$

$$S_2(A_2'') = z_1 + z_2 + z_3 + z_4 + z_5 + z_6 - 6z_7. \qquad (16.26)$$

(iv) Out-of-phase ligand librations. The symmetric structure of the species which correlate with the ones for rigid rotations according to Table I. It is found

$$\Gamma(\text{iv}) = E' + A_1'' , \qquad (16.27)$$

where the A_1'' mode represents the internal rotation involving the two ligands. The degenerate pair of E' modes was constructed from rigid rotations of the ligands about local X and Y axes situated in the respective ligand planes. Symmetry coordinates:

$$S_{2a}(E') = 12^{-\frac{1}{2}}(-2z_1 + z_2 + z_3 + 2z_4 - z_5 - z_6),$$

$$S_{2b}(E') = \frac{1}{2}(z_2 - z_3 - z_5 + z_6),$$

$$\mathcal{S}(A_1'') = S(A_1'') . \tag{16.28}$$

(v) Parallel out-of-phase ligand translation. The rigid translations of the two ligands moving in opposite directions parallel to the Z axis represent a genuine vibration of the whole molecule model. It belongs to the species

$$\Gamma(\text{v}) = A_1' , \tag{16.29}$$

and the corresponding symmetry coordinate reads

$$\mathcal{S}_2(A_1') = z_1 + z_2 + z_3 - z_4 - z_5 - z_6 . \tag{16.30}$$

(vi) Deformed out-of-phase ligand translations. The two remaining modes have the symmetry of out-of-phase rigid translations of the ligands in the direction of the X and Y axis, viz.

$$\Gamma(\text{vi}) = E'' . \tag{16.31}$$

This type of motion is not a genuine vibration for the whole model. It was found necessary to allow for ligand deformations under the appropriate vibrational motions. The easiest way of choosing symmetry coordinates in this case was to adopt some of those from the above treatment of Section 16.2.2. It was chosen

$$\mathcal{S}_{2a}(E'') = S_{1a}(E''), \qquad \mathcal{S}_{2b}(E'') = S_{1b}(E'') . \tag{16.32}$$

These coordinates are constructed from X-Y stretchings; see Eqs. (9) and (10). Nevertheless the motions of the central atom (X) cancel from the final combinations forming these symmetry coordinates.

16.4. G Matrices for the Eclipsed Model

Algebraic expressions were worked out for the **G** matrix elements of the treated D_{3h} model in terms of the two sets of symmetry coordinates. Table II shows the results on the basis of the coordinates of Section 16.2.2; cf. Eqs. (4)-(10). The elements in Table III are based on the alternative symmetry coordinates of

Table 16-II. **G** matrix elements for the eclipsed XY_6 model (D_{3h}) in standard symmetry coordinates (cf. Section 16.2.2).

		$S_1(A_1')$	$S_2(A_1')$
$S_1(A_1')$		μ_Y	0
$S_2(A_1')$			$\mu_Y(6-\sec^2 A)$
	$S_1(E')$	$S_2(E')$	$S_3(E')$
$S_1(E')$	$4\mu_X\sin^2 A+\mu_Y$	$4\mu_X\sin^2 A\tan A$	$-\frac{3}{2}2^{\frac{1}{2}}\mu_X\sin 2B$
$S_2(E')$		$4\mu_X\sin^2 A\tan^2 A + \mu_Y(1+\frac{1}{2}\sec^2 A)$	$-6^{\frac{1}{2}}(2\mu_X\sin^2 A-\frac{1}{2}\mu_Y)\sec A\sin B$
$S_3(E')$			$2(3\mu_X\sin^2 B+\mu_Y)$
	$S(A_1'')$	$S_1(A_2'')$	$S_2(A_2'')$
$S(A_1'')$	$6\mu_Y$		
$S_1(A_2'')$		$6\mu_X\sin^2 B+\mu_Y$	$-12\mu_X\tan A\sin^2 B$
$S_2(A_2'')$			$3(8\mu_X\sin^2 A+\mu_Y)\sec^2 A\sin^2 B$
		$S_1(E'')$	$S_2(E'')$
$S_1(E'')$		μ_Y	0
$S_2(E'')$			$\mu_Y(1+\frac{1}{2}\sec^2 A)$

Section 16.3. μ_X and μ_Y denote as usual the inverse masses of the X and Y atoms, respectively. The latter **G** matrix (Table III) is seen to contain more interaction terms equal to zero than the former one.

Table 16-III. **G** matrix elements for the eclipsed XY_6 model (D_{3h}) in alternative symmetry coordinates (cf. Section 16.3).

	$S_1(A_1')$	$S_2(A_1')$
$S_1(A_1')$	$3\mu_Y$	0
$S_2(A_1')$		$6\mu_Y$

	$S_1(E')$	$S_2(E')$	$S_3(E')$
$S_1(E')$	$\frac{3}{2}\mu_Y$	0	0
$S_2(E')$		μ_Y	0
$S_3(E')$			$6(6\mu_X+\mu_Y)$

	$S(A_1'')$	$S_1(A_2'')$	$S_2(A_2'')$
$S(A_1'')$	$6\mu_Y$		
$S_1(A_2'')$		$3\mu_Y$	0
$S_2(A_2'')$			$6(6\mu_X+\mu_Y)$

	$S_1(E'')$	$S_2(E'')$
$S_1(E'')$	$\frac{3}{2}\mu_Y$	$-\frac{1}{2}3^{\frac{1}{2}}\mu_Y\cos B$
$S_2(E'')$		μ_Y

REFERENCES

1 S.J.Cyvin, I.Elvebredd, J.Brunvoll, and G.Hagen, Acta Chem. Scand. 22, 1491 (1968).

2 S.J.Cyvin: Molecular Vibrations and Mean Square Amplitudes, Universitetsforlaget, Oslo, and Elsevier, Amsterdam, 1968.

Sandwich Compounds - Part II: Harmonic Force Field Analysis of the Complete Molecule of Dibenzene Chromium

LOTHAR SCHÄFER, J. BRUNVOLL, and S. J. CYVIN

An analysis of the molecular vibrations of dibenzene chromium is performed in which the sandwich compound is treated as a whole molecule, assuming an eclipsed D_{6h} model. The constructed symmetry coordinates tend to preserve the identity of the ligands. A complete harmonic force field is constructed. Some of the most interesting frequency shifts from free to complexed benzene are explained in a striking way by kinematic couplings. The calculated mean amplitudes show astonishingly good agreement with the electron diffraction values.

Die Resultate einer totalen Normalkoordinatenanalyse des Dibenzolchrom (D_{6h} Symetrie, Liganden eclipsed) werden mitgeteilt. Ein vollständiges harmonisches Kraftfeld wurde auf der Basis von Symetriekoordinaten berechnet, die die Identität der Liganden bewahren. Einige der auffälligsten Frequenzverschiebungen vom freien zum komplex gebundenen Benzol können, völlig unerwartet, durch kinematische Kopplung erklärt werden. Die errechneten, mittleren Schwingungsamplituden stimmen sehr genau mit den Elektronenbeugungswerten überein.

16.5. Introduction

Dibenzene chromium, $(C_6H_6)_2Cr$, has been the object of a lar number of structural investigations[1-16] including X-ray,[1-5] infrared,[6-11] Raman,[12] electron-diffraction,[13] thermodynamic,[14] and neutron diffraction[15,16] techniques. In the course of time a contr versy developed concerning the symmetry of benzene in this comple Some infrared,[6-9] X-ray,[1,2] and neutron diffraction[15,16] works favoured a threefold symmetry (D_{3d}); other infrared,[10-12] X-ray,[3-5] and thermodynamic[14] works, and to some degree the electron diffraction[13] study favoured a sixfold symmetry (D_{6h}) of benzene in $(C_6H_6)_2Cr$. The first experimental proof for D_{6h} was produced by the infrared spectrum of the complex in the vapour state.[11] Calculations of CC mean amplitudes from these frequencies combined with the electron diffraction study[13] supported the hexagonal structure.[17] The present vibrational analysis was partly inspired by the long time confusing situation. Furthermore we felt challenged by the vibrational problem of the total complex, having performed before[12] a normal coordinate analysis of the ligand considered as isolated from the rest of the molecule. In this analysis the frequency shifts from free benzene resulted in a force field for the ligand largely different from the generally accepted one for benzene. In the present work a normal coordinate analysis for $(C_6H_6)_2Cr$ treated as a twenty-five atomic molecule is performed for the first time (a preliminary communication is published elsewhere[18]).It is perhaps the first analysis of this kind for any sandwich compound. The main idea of the present work was to investigate the influence on the ligand frequencies from force constants associated with ligand-ligand and metal-ligand motions different from the internal vibrations of the ligands. It seemed to be of great interest to see whether these influences might explain some of the frequency shifts without changing the force field of the internal ligand vibrations from the one in free benzene.

16.6. Classification of Vibrational Modes

On the assumption of an eclipsed structure of D_{6h} symmetry the totality of the sixty-nine normal vibrations are distributed among the species according to[19]

$$\Gamma = 4A_{1g} + A_{2g} + 2B_{1g} + 4B_{2g} + 5E_{1g} + 6E_{2g}$$

$$+2A_{1u} + 4A_{2u} + 4B_{1u} + 2B_{2u} + 6E_{1u} + 6E_{2u} . \qquad (16.33)$$

The vibrations have been classified according to the scheme of Section 16.3, where it was applied to a simplified sandwich-type model. (i) In-phase ligand vibrations, which have the symmetric structure of free benzene:[20]

$$\Gamma(\text{i}) = 2A_{1g} + A_{2g} + 2B_{2g} + E_{1g} + 4E_{2g}$$

$$+ A_{2u} + 2B_{1u} + 2B_{2u} + 3E_{1u} + 2E_{2u} . \qquad (16.34)$$

(ii) Out-of-phase ligand vibrations, for which the symmetric structure is found from Eq. (34) with the aid of the correlation scheme[19] of Table IV;

$$\Gamma(\text{ii}) = 2A_{2u} + A_{1u} + 2B_{1u} + E_{1u} + 4E_{2u}$$

$$+ A_{1g} + 2B_{2g} + 2B_{1g} + 3E_{1g} + 2E_{2g} . \qquad (16.35)$$

(iii) Compensated in-phase ligand translations, which have the symmetric structure of the rigid translations for the whole molecule, viz.:

$$\Gamma(\text{iii}) = A_{2u} + E_{1u} . \qquad (16.36)$$

(iv) Out-of-phase ligand librations. The symmetric structure consists of the species which are correlated to the ones of the rigid rotations of the molecule;

$$\Gamma(\text{iv}) = A_{1u} + E_{1u} . \qquad (16.37)$$

The remaining three modes have the symmetry of out-of-phase ligand translations. This type of motion in the direction of the Z axis is a genuine vibration. It is classified as (v) parallel out-of-phase ligand translation, and has the symmetry of

$$\Gamma(\text{v}) = A_{1g} . \qquad (16.38)$$

Table 16-IV. Correlations between symmetry species for in-phase and out-of-phase motions in D_{6h} structures.

$A_{1g} \leftrightarrow A_{2u}$	$B_{2g} \leftrightarrow B_{1u}$
$A_{2g} \leftrightarrow A_{1u}$	$E_{1g} \leftrightarrow E_{1u}$
$B_{1g} \leftrightarrow B_{2u}$	$E_{2g} \leftrightarrow E_{2u}$

The out-of-phase ligand translations in the directions of X and Y are not genuine vibrations. In the appropriate vibrational modes the ligands are deformed, and we have classified this type as (vi) deformed out-of-phase ligand translations. The symmetry is

$$\Gamma(\text{vi}) = E_{1g} \,. \qquad (16.39)$$

A complete set of symmetry coordinates of molecular vibrations for dibenzene chromium was constructed in the following way. The (i) in-phase and (ii) out-of-phase ligand vibrations were taken as the symmetric and antisymmetric combinations of symmetry coordinates from free benzene[20] adapted to the two ligands; cf. also Section 16.3. The coordinates of types (iii), (iv), and (v) were constructed directly from cartesian displacements in the same way as shown in the mentioned section. Under the type (vi) we had to resort to using the metal-carbon stretchings, but the metal motions cancelled out in the finally constructed coordinates.

16.7. Normal Coordinate Analysis

16.7.1. Structural Data

In the present calculations the structural parameters from the gas electron-diffraction investigation of Haaland[13] were applied, viz. C-C = 1.423 Å, C-H = 1.090 Å, and a vertical ring to ring distance of 3.226 Å.

16.7.2. Approximate Force Field

An initial approximate harmonic force field was assumed to have a form as shown in Fig. 3. This figure also indicates the presence of kinematic coupling between different types of coordinates. Many of these couplings vanish on account of the assumed planarity of the ligands; pyramidal hexagonal ligands would have been possible without violating the D_{6h} symmetry. All force-constant blocks for in-phase and out-of-phase ligand vibrations were transferred from free benzene, assuming no interaction in the potential energy between the two ligands. These force constants reproduce the vibrational frequency assignment for free benzene from Brodersen and Langseth.[21] For dibenzene chromium the calcula-

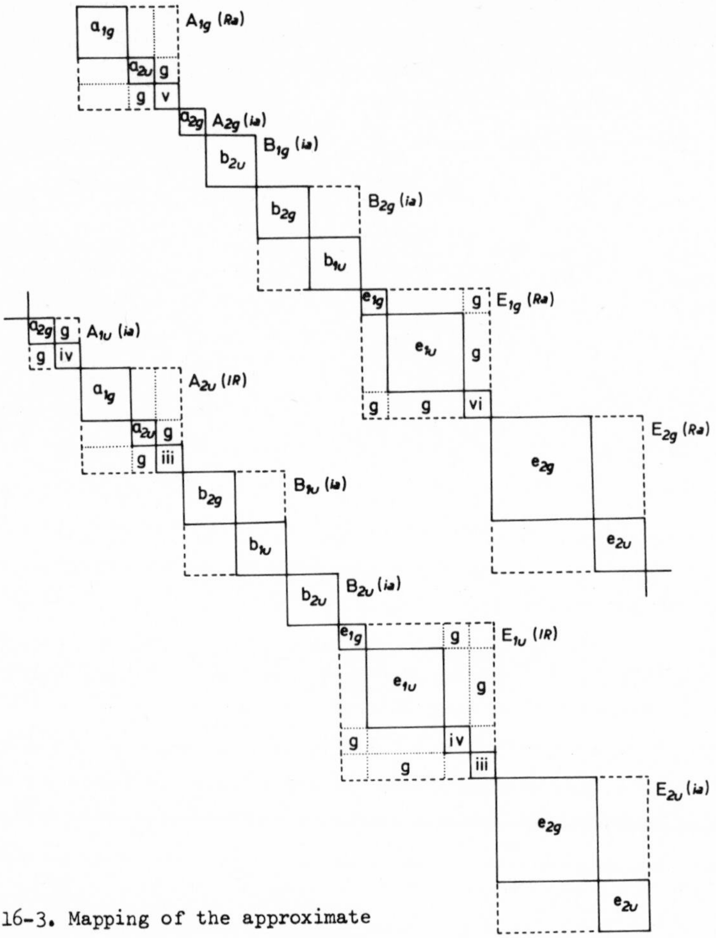

Fig. 16-3. Mapping of the approximate force field for dibenzene chromium type sandwich compounds. (Ra) = Raman active, (IR) = infrared active, (ia) = inactive. Species designations in small letters indicate the blocks from free benzene, and correspond to (i) in-phase and (ii) out-of-phase ligand vibrations; cf. Section 16.6. The other types of modes are indicated by roman numerals. Nonvanishing kinematic coupling between different blocks (of the G matrix) is indicated by 'g'. Where nothing is indicated this type of coupling is zero. In the approximate force field all interaction terms between full-drawn blocks were neglected.

ted frequencies of ligand vibrations from these approximate calculations are in the cases with absence of kinematic coupling identical to the corresponding ones in free benzene or almost identical; see Table V. The small discrepancies are explained by slightly different structure parameters used for the complexed and free benzene. In cases where kinematic coupling exists between the ligand vibrations and other types of modes the calculated frequencies for complexed benzene are seen to deviate more or less from those of free benzene; the largest shifts occur in species A_{1g} and A_{2u} for the frequency which corresponds to a_{2u} in free benzene. The vibrational frequencies for dibenzene chromium in the low-frequency region are known from far-infrared spectra,[6] and have been used in the present analysis. The calculated values in Table V correspond to the force constants of low-frequency modes as given in Table VI. The magnitudes of force constants depend on more or less arbitrary scaling factors used in defining the corresponding coordinates. In order to facilitate the interpretation of the force constant values in Table VI the corresponding numerical $\mathbf{G}$-matrix diagonal elements are included in the table.

The calculations give an astonishingly good support to an explanation of the frequency shifts from free to complexed benzene in terms of kinematic couplings involving ligand-ligand and metal-ligand motions. It is stressed that all the frequency shifts of the considered type as found in Table V arise from $\mathbf{G}$-matrix coupling terms between the different types of coordinates, apart from the small effects arising from differences in the applied structural parameters. The force field for the ligand vibrations is maintained the same as in free benzene. Thus it is found that the approximate force field of the form as shown in Fig. 3 is qualitatively compatible with the observed vibrational frequencies for dibenzene chromium. In particular it is exciting to notice the frequency shift of the a_{2u} frequency of free benzene. The calculated frequencies of species A_{1g} and A_{2u} in $(C_6H_6)_2Cr$ (see Table V) reflect the trends of the observed frequencies in a striking way and even exaggerate the frequency shifts quantitatively. In regard to the high-frequency region of CH stretching frequencies it is found that the present approach does not explain the observed frequency shifts. Having in mind the approximate separability of the CH stretching

Table 16-V. Observed and approximately calculated frequencies (cm^{-1}) for $(C_6H_6)_2Cr$ on the basis of D_{6h} symmetry. Free benzene frequencies are included for comparison.

Dibenzene chromium				Dibenzene chromium				Free benzene[d]	
		calc.	obs.			calc.	obs.		
A_{1g}	(i)[a]	3073	3053[b]	A_{2u}	(ii)	3073	3053[c]	a_{1g}	3073
	(i)	993	970[b]		(ii)	993	971[c]		993
	(ii)	880	791[b]		(i)	916	794[c]	a_{2u}	673
	(v)	254	277[b]		(iv)	439	490[c]		-
A_{2g}	(i)	1350	-	A_{1u}	(ii)	1375	-	a_{2g}	1350
					(iv)	152	152[d]		-
B_{1g}	(i)	1312	-	B_{2u}	(ii)	1312	1308[c]		-
	(i)	1151	-		(ii)	1151	1142[c]		-
B_{2g}	(ii)	3057	2855[b]	B_{1u}	(i)	3057	-	b_{1u}	3057
	(ii)	1010	-		(i)	1010	-		1010
	(i)	993	-		(ii)	993	-	b_{2g}	990
	(i)	709	-		(ii)	709	-		707
E_{1g}	(ii)	3065	2904[b]	E_{1u}	(i)	3075	2904[b]	e_{1u}	3064
	(ii)	1491	1430[b]		(i)	1506	1426[c]		1482
	(ii)	1054	999[b]		(i)	1052	999[c]		1037
	(i)	864	860[c]		(ii)	885	860[c]	e_{1g}	846
	(vi)	334	335[b]		(iii)	459	459[c]		-
					(iv)	170	171[d]		-
E_{2g}	(i)	3056	2955[b]	E_{2u}	(ii)	3056	-	e_{2g}	3056
	(i)	1600	1631[b]		(ii)	1600	-	e_{2g}	1599
	(i)	1185	1143[b]		(ii)	1185	-	e_{2g}	1178
	(ii)	970	910[b]		(i)	970	-	e_{2u}	967
	(i)	606	604[b]		(ii)	606	-	e_{2g}	606
	(ii)	399	409[b]		(i)	399	-	e_{2u}	398

[a] Parenthesized roman numerals according to the classification in Section 16.6.

[b] Solid state Raman; Ref. 12.

[c] Solid state infrared; quoted in Ref. 12.

[d] Solid state far infrared; Ref. 6.

Table 16-VI. Diagonal force constants (F in mdyne/Å) and inverse kinetic energy matrix elements (G in Amu^{-1}) for the low-frequency vibrations in dibenzene chromium.

Type[a]	Species	F	G
(iii)	A_{2u}	0.0141	23.978
(iii)	E_{1u}	0.00888	23.978
(iv)	A_{1u}	0.0513	0.7713
(iv)	E_{1u}	0.0680	0.7713
(v)	A_{1g}	0.0174	12.903
(vi)	E_{1g}	1.300	0.08331

[a] See footnote to Table V.

modes it seems inevitable to modify the corresponding force constants in complexed benzene accordingly in a similar way as was done in the treatment of the isolated ligand vibrations.[12]

16.7.3. Final Force Field

The harmonic force field was refined in the following way. For the lowest frequencies in species A_{1g} and A_{2u} an exact adjustment to the observed frequencies was achieved with the diagonal force constants shown in Table VI along with the a_{2u} force constant from benzene (0.1925 mdyne/Å), when the interaction force constants of 0.0112 mdyne/Å (in A_{1g}) and 0.01316 mdyne/Å (A_{2u}) were introduced. Force constants for CH stretchings were taken from the analysis of isolated ligands.[12] In a final run the produced force constants were adjusted to fit exactly all the observed frequencies which are shown in Table V. It was found reasonable to adopt the same force constants for the species B_{1g} , B_{1u} , and E_{2u} as in B_{2u}, B_{2g}, and E_{2g} , respectively. For A_{2g}, the highest frequency in A_{1u} , and the three lowest frequencies in B_{2g}, for which observed frequencies are not available, the calculated frequencies from the approximate force field (see Table V) were adopted.

16.8. Mean Amplitudes of Vibration

The final force field was used to calculate the mean ampli-

Table 16-VII. Mean amplitudes of vibration for dibenzene chromium from spectroscopic and electron diffraction data. Free benzene values are included for comparison; Å units.

Distance	(Equil.)	$(C_6H_6)_2Cr$ Spectroscopic 0 K	$(C_6H_6)_2Cr$ Spectroscopic 25 °C	$(C_6H_6)_2Cr$ Spectroscopic 180°C	E.D.[13] 180°C	C_6H_6 Spectr.[22] 25 °C
C–H	(1.090)	0.079	0.079	0.079	0.084	0.077
C–C	(1.423)	0.046	0.047	0.048	0.045	0.046
C···C	(2.465)	0.054	0.056	0.059	0.055	0.055
C···C	(2.846)	0.057	0.059	0.064	0.063	0.059
C···H	(2.183)	0.101	0.101	0.103	0.100	0.100
C···H	(3.452)	0.097	0.098	0.101	0.120	0.097
C···H	(3.936)	0.094	0.095	0.098	0.120	0.093
H···H	(2.513)	0.159	0.160	0.164		0.158
H···H	(4.353)	0.134	0.135	0.139		0.133
H···H	(5.026)	0.118	0.120	0.122		0.118
Cr–H	(2.986)	0.114	0.124	0.135	0.140	
Cr–C	(2.151)	0.059	0.073	0.084	0.070	
$C_1 \cdots C_1'$	(3.226)	0.095	0.134	0.160	0.140	
$C_1 \cdots C_2'$	(3.526)	0.089	0.121	0.144		
$C_1 \cdots C_3'$	(4.060)	0.078	0.097	0.113		
$C_1 \cdots C_4'$	(4.302)	0.072	0.084	0.096		
$C_1 \cdots H_1'$	(3.405)	0.168	0.203	0.231		
$C_1 \cdots H_2'$	(3.895)	0.157	0.186	0.210		
$C_1 \cdots H_3'$	(4.725)	0.138	0.155	0.171		
$C_1 \cdots H_4'$	(5.089)	0.128	0.138	0.151		
$H_1 \cdots H_1'$	(3.226)	0.226	0.267	0.302		
$H_1 \cdots H_2'$	(4.089)	0.207	0.241	0.269		
$H_1 \cdots H_3'$	(5.418)	0.174	0.192	0.210		
$H_1 \cdots H_4'$	(5.972)	0.156	0.165	0.177		

tudes of vibration for dibenzene chromium. The results are given in Table VII. Parenthesized values of interatomic separations (in Å) are included in order to facilitate the identification of the distance types. The calculated mean amplitudes in a ligand are seen to be very nearly equal to the corresponding ones in free benzene.[22] Mean amplitudes for dibenzene chromium were also calculated from the approximate force field (see Section 16.7.2). Those results did not deviate appreciably from the final ones.[18] The agreement with the reported mean amplitudes from electron diffraction[13] is astonishi‸gly good.

REFERENCES

1 F.Jellinek, Nature 187, 871 (1960).

2 F.Jellinek, J.Organometal.Chem. 1, 43 (1963).

3 F.A.Cotton, W.A.Dollase, and J.S.Wood, J.Am.Chem.Soc. 85, 1543 (1963).

4 J.A.Ibers, J.Chem.Phys. 40, 3129 (1964).

5 E.Keulen and F.Jellinek, J.Organometal.Chem. 5, 490 (1966).

6 H.P.Fritz and E.O.Fischer, J.Organometal.Chem. 7, 121 (1967).

7 H.P.Fritz and W.Lüttke, 5th International Conference on Coordination Chemistry, London 1959; Special Publication No. 13, Chem. Soc. London 1959, p. 123.

8 E.O.Fischer and H.P.Fritz, Advan.Inorg.Chem.Radiochem. 1, 55 (1959).

9 H.P.Fritz, Advan.Organometal.Chem. 1, 298 (1964).

10 H.P.Fritz, W.Lüttke, H.Stammreich, and R.Forneris, Chem.Ber. 92, 3246 (1959).

11 L.H.Ngai, F.E.Stafford, and L.Schäfer, J.Am.Chem.Soc. 91, 48 (1969).

12 L.Schäfer, J.F.Southern, and S.J.Cyvin, Spectrochim.Acta 27A, 1083 (1971).

13 A.Haaland, Acta Chem.Scand. 19, 41 (1965).

14 J.T.S.Andrews, E.F.Westrum, and N.Bjerrum, J.Organometal.Chem. 17, 293 (1969).

15 G.Albrecht, E.Förster, D.Sippel, F.Eichhorn, and E.Kurras, Z.Chem. 8, 311 (1968).

16 E.Förster, G.Albrecht, W.Dürselen, and E.Kurras, J.Organometal.Chem. 19, 215 (1969).

17 L.Schäfer, J.F.Southern, S.J.Cyvin, and J.Brunvoll, J.Organometal.Chem. 24, C13 (1970).

18 J.Brunvoll, S.J.Cyvin, and L.Schäfer, J.Organometal.Chem. 27, 69 (1971).

19 H.P.Fritz, W.Lüttke, H.Stammreich, and R.Forneris, Spectrochim.

Acta 17, 1068 (1961).

20 B.N.Cyvin, S.J.Cyvin, and A.Müller, Acta Chem.Scand. 23, 1352 (1969).

21 S.Brodersen and A.Langseth, Mat.Fys.Skr.Dan.Vid.Selsk. 1, No.1 (1956).

22 S.J.Cyvin: Molecular Vibrations and Mean Square Amplitudes, Universitetsforlaget, Oslo, and Elsevier, Amsterdam, 1968; p. 245.

CHAPTER 17

Molecular Vibrations and Atomic Mean-Square Amplitudes - Part I: Normal Coordinate Analysis of the Molecular Vibrations of Hexamethylene Tetramine

I. ELVEBREDD and S. J. CYVIN

A normal coordinate analysis was performed for the molecular vibrations of hexamethylene tetramine (HMT). A complete set of symmetry coordinates for the twenty-two atomic model of T_d symmetry is specified. The developed harmonic force field was used to calculate the following quantities for HMT and HMT-d_{12}: (a) mean amplitudes of vibration, (b) perpendicular amplitude correction coefficients, and (c) atomic mean-square amplitudes.

Det er utført en normalkoordinatanalyse for molekylvibrasjonene i heksametylentetramin (HMT). *Et fullstendig sett av symmetrikoordinater for den tjueto-atomige molekylmodellen av* T_d *symmetri er oppgitt. Det utviklede harmoniske kraftfelt ble benyttet til å beregne følgende størrelser for* HMT *og* HMT-d_{12}*: a) Midlere svingningsamplituder, b) korreksjonskoeffisienter av perpendikulære amplituder og c) atomære midlere kvadratiske amplituder.*

17.1. Introduction

Hexamethylene Tetramine (HMT) has been investigated by X-ray diffraction[1,2] from the molecular crystal, and the tetrahedral (T_d) symmetry for the $(CH_2)_6N_4$ units has been established. More accurate structure studies of HMT by X-ray[3] and neutron diffraction[4] have been performed, and joint refinements of the data from these methods have been discussed.[5] The accuracies of these measurements require the inclusion of internal motions in order to make the refinements more meaningful. The purpose of the present work is to make a vibrational analysis of the HMT molecule following the standard Wilson's **GF** matrix method[6] to develop a harmonic force field of molecular vibrations. Then it should be possible to derive the necessary quantities for the X-ray and neutron diffraction refinements.

Some Raman[7] and infrared[8] spectral studies of HMT have been reported. One of the cited papers[7] contains a vibrational analysis of a simplified model of the molecule, where the CH_2 groups are considered as single particles. Moreover this analysis was performed before the Wilson **GF** matrix method became the standard method. It seems to be of interest to give an account of the molecular vibrations of HMT as treated by modern methods, including a detailed specification of a complete set of symmetry coordinates. Such coordinates have been worked out in the present work, and for the first time the molecular vibrations of the complete twenty-two atomic model have been treated theoretically.

17.2. Molecular Model and Symmetry Considerations

The HMT molecule skeleton is shown in Fig. 1, where the numbering of C and N atoms and orientation of cartesian axes is indicated. The C atoms form an octahedron and are numbered in the same way as the case is for the octahedral Z_6 and XY_6 molecule models treated elsewhere.[9] The N atoms form a tetrahedron. The H atoms are included on the projections into the YZ and ZX planes drawn in Figures 2 and 3, respectively. The CH_2 planes are assumed to be perpendicular to the respective CN_2 planes at each carbon atom corner.

The equilibrium structure of the skeleton is determined by only two parameters, say the C-N distance, here denoted by S, and the CNC angle. To determine the structure of the whole molecule

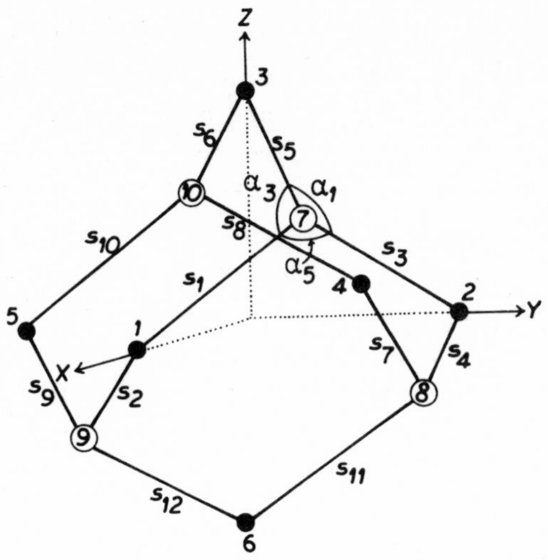

Fig. 17-1. The HMT skeleton model, showing the numbering of C and N atoms and orientation of cartesian axes. Also indicated are the C-N stretchings s_i (i = 1,2,...,12), while the twelve CNC bendings are exemplified by α_1, α_3 and α_5.

two additional parameters are needed, e.g. the C-H distance, here denoted by R, and the HCH angle.

The normal modes of vibration (twenty-four for the skeleton and sixty for the whole molecule) are distributed among the various species of the symmetry group T_d as

$$\Gamma_{\text{skeleton}} = 2A_1 + 2E + 2F_1 + 4F_2 \tag{17.1}$$

and

$$\Gamma_{\text{total}} = 4A_1 + A_2 + 5E + 6F_1 + 9F_2 \tag{17.2}$$

for the skeleton and whole molecule, respectively.

The models here considered are very convenient from the point of view of redundancy. In fact the problem of redundancies is simply resolved by a proper choice of valence coordinates; the twelve C-N stretchings (s_i) and twelve CNC bendings (α_i) together form a complete set of internal coordinates for the skeleton. Moreover these two sets of symmetrically equivalent coordinates have identical symmetric structures, viz.

$$\Gamma(s) = \Gamma(\alpha) = A_1 + E + F_1 + 2F_2. \tag{17.3}$$

These twenty-four coordinates may be used to construct a complete set of independent symmetry coordinates for the HMT skeleton model But it should be noticed that twelve NCN valence angle bendings exist, but do not appear explicitly in the symmetry coordinate expressions in our approach. The inclusion of the set of NCN bendings would introduce twelve redundancies, as these coordinates may all be expressed linearly in terms of the chosen twenty-four coordinates of the s and α types. The situation reminds one of the treatment of regular trigonal Z_3, tetrahedral Z_4 and octahedral Z_6 models.[9] In all these cases the Z-Z stretchings constitute

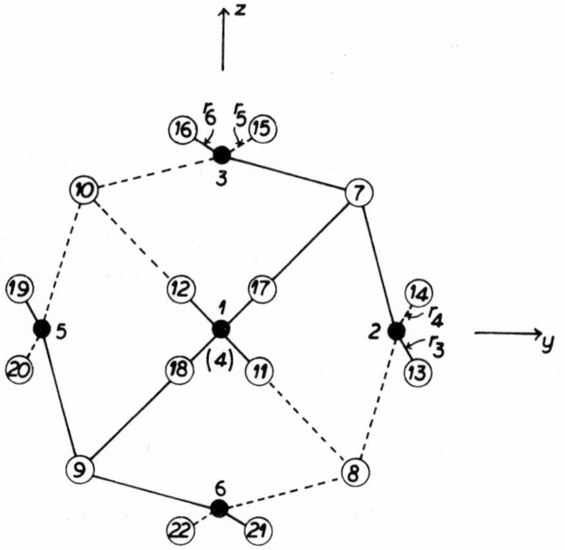

Fig. 17-2. The HMT molecule model: projection into the yz plane. Carbon atoms: black circles. The C-H stretchings (r_i; i = 1,2,...,12) are exemplified by r_3, r_4, r_5 and r_6.

complete sets of internal coordinates, and any inclusion of ZZZ angle bendings (two different sets in the octahedral Z_6 type) would introduce unnecessary redundancies. But the situation is not always as tractable as this, as was clearly demonstrated for the cubic Z_8 model.[9] In that case a complicated system of redundancies is inevitable if the vibrations are to be treated in terms of the stretchings and simple angle bendings.

When the H atoms are added to the skeleton of HMT the vibrations may still be treated without redundancies when the additional coordinates are taken as C-H stretchings (r_i) and NCH bendings (φ_{ij}) with the symmetric structures:

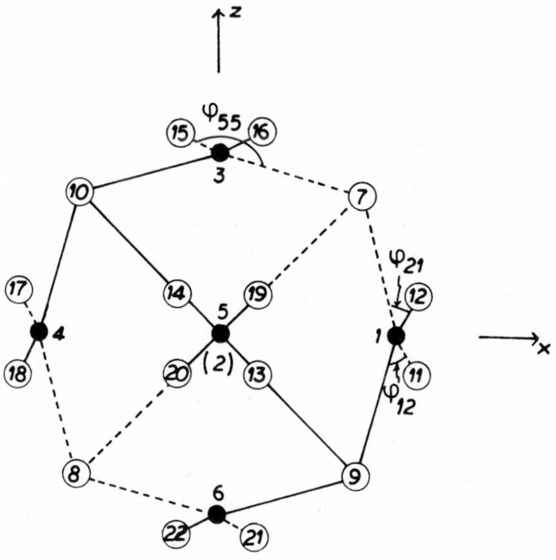

Fig. 17-3. The HMT molecule model: projection into the *ZX* plane. Carbon atoms: black circles. The NCH bendings (φ_{ij}; four at each C atom) are exemplified by φ_{12}, φ_{21} and φ_{55}.

$$\Gamma(r) = A_1 + E + F_1 + 2F_2 \tag{17.4}$$

identical with Eq. (3), and

$$\Gamma(\varphi) = A_1 + A_2 + 2E + 3F_1 + 3F_2 . \tag{17.5}$$

The set of eight HCH valence angle bendings may again be omitted as unnecessary redundancies.

17.3. Valence Coordinates and Intermediate Combinations of Them

The applied types of valence coordinates are the s and r stretchings along with α and φ bendings mentioned in the preceding section. Fig. 1 gives a precise definition of the individual s_i (i = 1,2,...,12) stretchings. Notice that s_1 and s_2 involve the motion of the carbon atom number 1, s_3 and s_4 are connected at C_2, s_5 and s_6 at C_3, etc. The system of numbering the α_i (i = 1,2,..., 12) bendings is that α_i comes opposite to s_i around an N atom. Thus, for instance, around atom number 7 one finds α_1, α_3 and α_5 (cf. Fig. 1). The symbols r_i (i = 1,2,...,12) identify the stretchings $C\text{-}H_{10+i}$. A few of these stretchings are indicated on Fig. 2. Finally the twenty-four NCH bendings are designated φ_{ii}, φ_{ij}, φ_{ji} and φ_{jj}, where (i,j) = (1,2),(3,4),...,(11,12), and φ_{mn} connects the two bonds whose stretchings are denoted r_m and s_n. A few of these bendings are indicated on Fig. 3.

The symmetric structures for the different sets of valence coordinates as given by Eqs. (3)-(5) display coefficients $n > 1$. Hence it must be possible to form intermediate combinations of the valence coordinates to make them separate into smaller sets of symmetrically equivalent coordinates.[6] This problem is readily solved for the s and α coordinates. For the former type we have formed the combinations

$$\begin{aligned} &s_1' = 2^{-\frac{1}{2}}(s_1 + s_2), \quad s_2' = 2^{-\frac{1}{2}}(s_3 + s_4), \quad s_3' = 2^{-\frac{1}{2}}(s_5 + s_6), \\ &s_4' = 2^{-\frac{1}{2}}(s_7 + s_8), \quad s_5' = 2^{-\frac{1}{2}}(s_9 + s_{10}), \quad s_6' = 2^{-\frac{1}{2}}(s_{11} + s_{12}); \end{aligned} \tag{17.6}$$

and

$$s_1''=2^{-\frac{1}{2}}(s_1 - s_2),\quad s_2''=2^{-\frac{1}{2}}(s_3 - s_4),\quad s_3''=2^{-\frac{1}{2}}(s_5 - s_6),$$
$$s_4''=2^{-\frac{1}{2}}(s_7 - s_8),\quad s_5''=2^{-\frac{1}{2}}(s_9 - s_{10}),\; s_6''=2^{-\frac{1}{2}}(s_{11} - s_{12}). \tag{17.7}$$

The bendings α_i transform exactly in the same way as the corresponding s_i. Hence a set of suitable combinations of α_i coordinates may be formed in the same way as specified in Eqs. (6) and (7), viz.

$$\alpha_k' = 2^{-\frac{1}{2}}(\alpha_{2k-1} + \alpha_{2k}) \tag{17.8}$$

and

$$\alpha_k'' = 2^{-\frac{1}{2}}(\alpha_{2k-1} - \alpha_{2k}) ; \tag{17.9}$$

($k = 1,2,\ldots,6$). Also the r stretchings have the same symmetry properties as s and α; cf. Eqs. (3) and (4). We have formed the following combinations.

$$r_1'=2^{-\frac{1}{2}}(r_7 + r_8),\quad r_2'=2^{-\frac{1}{2}}(r_9 + r_{10}),\quad r_3'=2^{-\frac{1}{2}}(r_{11} + r_{12}),$$
$$r_4'=2^{-\frac{1}{2}}(r_1 + r_2),\quad r_5'=2^{-\frac{1}{2}}(r_3 + r_4),\quad r_6'=2^{-\frac{1}{2}}(r_5 + r_6); \tag{17.10}$$

and

$$r_1''=2^{-\frac{1}{2}}(r_7 - r_8),\quad r_2''=2^{-\frac{1}{2}}(r_9 - r_{10}),\quad r_3''=2^{-\frac{1}{2}}(r_{11} - r_{12}),$$
$$r_4''=2^{-\frac{1}{2}}(r_1 - r_2),\quad r_5''=2^{-\frac{1}{2}}(r_3 - r_4),\quad r_6''=2^{-\frac{1}{2}}(r_5 - r_6). \tag{17.11}$$

Notice that r_7 and r_8 are situated opposite and in cis positions to s_1 and s_2, respectively; r_9 and r_{10} are opposite to s_3 and s_4, etc. This feature was utilized to establish the correlations between the different s_i', s_i'', r_i' and r_i'' combinations. For the φ coordinates a somewhat more complicated system of combinations is needed. In the first place we have formed:

$$\xi_1'= \tfrac{1}{2}(\varphi_{7\,7} + \varphi_{7\,8} + \varphi_{8\,7} + \varphi_{8\,8}),\quad \xi_2'= \tfrac{1}{2}(\varphi_{9\,9} + \varphi_{9\,10} + \varphi_{10\,9} + \varphi_{10\,10}),$$
$$\xi_3'= \tfrac{1}{2}(\varphi_{11\,11} + \varphi_{11\,12} + \varphi_{12\,11} + \varphi_{12\,12}),\quad \xi_4'= \tfrac{1}{2}(\varphi_{1\,1} + \varphi_{1\,2} + \varphi_{2\,1} + \varphi_{2\,2}),$$
$$\xi_5'= \tfrac{1}{2}(\varphi_{3\,3} + \varphi_{3\,4} + \varphi_{4\,3} + \varphi_{4\,4}),\quad \xi_6'= \tfrac{1}{2}(\varphi_{5\,5} + \varphi_{5\,6} + \varphi_{6\,5} + \varphi_{6\,6}); \tag{17.12}$$

and

$$\xi_1'' = \frac{1}{2}(\varphi_{7\,7} + \varphi_{7\,8} - \varphi_{8\,7} - \varphi_{8\,8}),\quad \xi_2'' = \frac{1}{2}(\varphi_{9\,9} + \varphi_{9\,10} - \varphi_{10\,9} - \varphi_{10\,10}),$$

$$\xi_3'' = \frac{1}{2}(\varphi_{11\,11} + \varphi_{11\,12} - \varphi_{12\,11} - \varphi_{12\,12}),\quad \xi_4'' = \frac{1}{2}(\varphi_{1\,1} + \varphi_{1\,2} - \varphi_{2\,1} - \varphi_{2\,2}),$$

$$\xi_5'' = \frac{1}{2}(\varphi_{3\,3} + \varphi_{3\,4} - \varphi_{4\,3} - \varphi_{4\,4}),\quad \xi_6'' = \frac{1}{2}(\varphi_{5\,5} + \varphi_{5\,6} - \varphi_{6\,5} - \varphi_{6\,6}). \tag{17.13}$$

The symmetric structures of the intermediate combinations given so far are all the same for the different kinds, viz.:

$$\Gamma(s') = \Gamma(\alpha') = \Gamma(r') = \Gamma(\xi') = A_1 + E + F_2 \tag{17.14}$$

and

$$\Gamma(s'') = \Gamma(\alpha'') = \Gamma(r'') = \Gamma(\xi'') = F_1 + F_2 . \tag{17.15}$$

To make the intermediate combinations of coordinates into a complete orthogonal transformation we need some more combinations of φ bendings, say:

$$\eta_1' = \frac{1}{2}(\varphi_{1\,1} - \varphi_{1\,2} + \varphi_{2\,1} - \varphi_{2\,2}),\quad \eta_2' = \frac{1}{2}(\varphi_{3\,3} - \varphi_{3\,4} + \varphi_{4\,3} - \varphi_{4\,4}),$$

$$\eta_3' = \frac{1}{2}(\varphi_{5\,5} - \varphi_{5\,6} + \varphi_{6\,5} - \varphi_{6\,6}),\quad \eta_4' = \frac{1}{2}(\varphi_{7\,7} - \varphi_{7\,8} + \varphi_{8\,7} - \varphi_{8\,8}),$$

$$\eta_5' = \frac{1}{2}(\varphi_{9\,9} - \varphi_{9\,10} + \varphi_{10\,9} - \varphi_{10\,10}),$$

$$\eta_6' = \frac{1}{2}(\varphi_{11\,11} - \varphi_{11\,12} + \varphi_{12\,11} - \varphi_{12\,12}); \tag{17.16}$$

and

$$\eta_1'' = \frac{1}{2}(\varphi_{1\,1} - \varphi_{1\,2} - \varphi_{2\,1} + \varphi_{2\,2}),\quad \eta_2'' = \frac{1}{2}(\varphi_{3\,3} - \varphi_{3\,4} - \varphi_{4\,3} + \varphi_{4\,4}),$$

$$\eta_3'' = \frac{1}{2}(\varphi_{5\,5} - \varphi_{5\,6} - \varphi_{6\,5} + \varphi_{6\,6}),\quad \eta_4'' = \frac{1}{2}(\varphi_{7\,7} - \varphi_{7\,8} - \varphi_{8\,7} + \varphi_{8\,8}),$$

$$\eta_5'' = \frac{1}{2}(\varphi_{9\,9} - \varphi_{9\,10} - \varphi_{10\,9} + \varphi_{10\,10}),$$

$$\eta_6'' = \frac{1}{2}(\varphi_{11\,11} - \varphi_{11\,12} - \varphi_{12\,11} + \varphi_{12\,12}). \tag{17.17}$$

The former set of these combinations, i.e. those of Eqs. (16) have the same symmetric structure as the s'' combinations given by Eq.(1

$$\Gamma(\eta') = F_1 + F_2 . \qquad (17.18)$$

The symmetric structure of the latter combinations (17) completes the whole scheme of the distribution of normal modes as given by Eq. (2). One has namely

$$\Gamma(\eta'') = A_2 + E + F_1 , \qquad (17.19)$$

and Eqs. (14), (15), (18) and (19) added properly together yield Eq. (2). Moreover the symmetric structures for the sets of intermediate combinations are seen to contain coefficients equal to unity exclusively.

17.4. Symmetry Coordinates

The symmetry coordinates of molecular vibrations were worked out as normalized combinations of the intermediate valence coordinate combinations of the preceding section.

Species A_1

$$S_1 = 6^{-\frac{1}{2}} \sum s_k' , \quad S_2 = 6^{-\frac{1}{2}} s \sum \alpha_k' ,$$
$$S_3 = 6^{-\frac{1}{2}} \sum r_k' , \quad S_4 = (RS/6)^{\frac{1}{2}} \sum \xi_k' , \qquad (17.20)$$

where all the summations are extended over $k = 1,2,\ldots,6$.

Species A_2

$$S = (RS/6)^{\frac{1}{2}}(-\eta_1''+\eta_2''+\eta_3''-\eta_4''+\eta_5''+\eta_6'') . \qquad (17.21)$$

Species E

$$S_{1a} = 12^{-\frac{1}{2}}(-s_1'-s_2'+2s_3'-s_4'-s_5'+2s_6') ,$$

$$S_{2a} = 12^{-\frac{1}{2}}(-\alpha_1'-\alpha_2'+2\alpha_3'-\alpha_4'-\alpha_5'+2\alpha_6') ,$$

$$S_{3a} = 12^{-\frac{1}{2}}(-r_1'-r_2'+2r_3'-r_4'-r_5'+2r_6') ,$$

$$S_{4a} = (RS/12)^{\frac{1}{2}}(-\xi_1'-\xi_2'+2\xi_3'-\xi_4'-\xi_5'+2\xi_6') ,$$

$$S_{5a} = \frac{1}{2}(RS)^{\frac{1}{2}}(\eta_1''+\eta_2''+\eta_4''+\eta_5'') ; \qquad (17.22a)$$

$$S_{1b}=\frac{1}{2}(s_1'-s_2'+s_4'-s_5'),\quad S_{2b}=\frac{1}{2}S(\alpha_1'-\alpha_2'+\alpha_4'-\alpha_5'),$$

$$S_{3b}=\frac{1}{2}(r_1'-r_2'+r_4'-r_5'),\quad S_{4b}=\frac{1}{2}(RS)^{\frac{1}{2}}(\xi_1'-\xi_2'+\xi_4'-\xi_5'),$$

$$S_{5b}=(RS/12)^{\frac{1}{2}}(\eta_1''-\eta_2''+2\eta_3''+\eta_4''-\eta_5''+2\eta_6''). \quad (17.22b)$$

Species F_1

$$S_{1a}=\frac{1}{2}(s_2''-s_3''+s_5''+s_6''),\quad S_{2a}=\frac{1}{2}S(\alpha_2''-\alpha_3''+\alpha_5''+\alpha_6''),$$

$$S_{3a}=\frac{1}{2}(r_2''-r_3''+r_5''+r_6''),\quad S_{4a}=\frac{1}{2}(RS)^{\frac{1}{2}}(\xi_2''-\xi_3''+\xi_5''+\xi_6''),$$

$$S_{5a}=\frac{1}{2}(RS)^{\frac{1}{2}}(\eta_2'-\eta_3'+\eta_5'+\eta_6'),\quad S_{6a}=(RS/2)^{\frac{1}{2}}(\eta_1''-\eta_4''); \quad (17.23a)$$

$$S_{1b}=\frac{1}{2}(s_1''-s_3''+s_4''-s_6''),\quad S_{2b}=\frac{1}{2}S(\alpha_1''-\alpha_3''+\alpha_4''-\alpha_6''),$$

$$S_{3b}=\frac{1}{2}(r_1''-r_3''+r_4''-r_6''),\quad S_{4b}=\frac{1}{2}(RS)^{\frac{1}{2}}(\xi_1''-\xi_3''+\xi_4''-\xi_6''),$$

$$S_{5b}=\frac{1}{2}(RS)^{\frac{1}{2}}(\eta_1'-\eta_3'+\eta_4'-\eta_6'),\quad S_{6b}=(RS/2)^{\frac{1}{2}}(\eta_2''-\eta_5''); \quad (17.23b)$$

$$S_{1c}=\frac{1}{2}(s_1''-s_2''-s_4''+s_5''),\quad S_{2c}=\frac{1}{2}S(\alpha_1''-\alpha_2''-\alpha_4''+\alpha_5''),$$

$$S_{3c}=\frac{1}{2}(r_1''-r_2''-r_4''+r_5''),\quad S_{4c}=\frac{1}{2}(RS)^{\frac{1}{2}}(\xi_1''-\xi_2''-\xi_4''+\xi_5''),$$

$$S_{5c}=\frac{1}{2}(RS)^{\frac{1}{2}}(\eta_1'-\eta_2'-\eta_4'+\eta_5'),\quad S_{6c}=(RS/2)^{\frac{1}{2}}(-\eta_3''+\eta_6''). \quad (17.23c)$$

Species F_2

$$S_{1a}=2^{-\frac{1}{2}}(s_1'-s_4'),\quad S_{2a}=\frac{1}{2}(s_2''+s_3''+s_5''-s_6''),$$

$$S_{3a}=2^{-\frac{1}{2}}S(\alpha_1'-\alpha_4'),\quad S_{4a}=\frac{1}{2}S(\alpha_2''+\alpha_3''+\alpha_5''-\alpha_6''),$$

$$S_{5a}=2^{-\frac{1}{2}}(r_1'-r_4'),\quad S_{6a}=\frac{1}{2}(r_2''+r_3''+r_5''-r_6''),$$

$$S_{7a}=(RS/2)^{\frac{1}{2}}(\xi_1'-\xi_4'),\quad S_{8a}=\frac{1}{2}(RS)^{\frac{1}{2}}(\xi_2''+\xi_3''+\xi_5''-\xi_6''),$$

$$S_{9a}=\frac{1}{2}(RS)^{\frac{1}{2}}(\eta_2'+\eta_3'+\eta_5'-\eta_6'); \quad (17.24a)$$

$$S_{1b}=2^{-\frac{1}{2}}(s_2'-s_5'),\quad S_{2b}=\frac{1}{2}(s_1''+s_3''+s_4''+s_6''),$$

$$S_{3b} = 2^{-\frac{1}{2}} S(\alpha_2' - \alpha_5'), \qquad S_{4b} = \frac{1}{2} S(\alpha_1'' + \alpha_3'' + \alpha_4'' + \alpha_6''),$$

$$S_{5b} = 2^{-\frac{1}{2}} (r_2' - r_5'), \qquad S_{6b} = \frac{1}{2}(r_1'' + r_3'' + r_4'' + r_6''),$$

$$S_{7b} = (RS/2)^{\frac{1}{2}} (\xi_2' - \xi_5'), \qquad S_{8b} = \frac{1}{2}(RS)^{\frac{1}{2}} (\xi_1'' + \xi_3'' + \xi_4'' + \xi_6''),$$

$$S_{9b} = \frac{1}{2}(RS)^{\frac{1}{2}} (\eta_1' + \eta_3' + \eta_4' + \eta_6'); \qquad (17.24b)$$

$$S_{1c} = 2^{-\frac{1}{2}} (s_3' - s_6'), \qquad S_{2c} = \frac{1}{2}(s_1'' + s_2'' - s_4'' - s_5''),$$

$$S_{3c} = 2^{-\frac{1}{2}} S(\alpha_3' - \alpha_6'), \qquad S_{4c} = \frac{1}{2} S(\alpha_1'' + \alpha_2'' - \alpha_4'' - \alpha_5''),$$

$$S_{5c} = 2^{-\frac{1}{2}} (r_3' - r_6'), \qquad S_{6c} = \frac{1}{2}(r_1'' + r_2'' - r_4'' - r_5''),$$

$$S_{7c} = (RS/2)^{\frac{1}{2}} (\xi_3' - \xi_6'), \qquad S_{8c} = \frac{1}{2}(RS)^{\frac{1}{2}} (\xi_1'' + \xi_2'' - \xi_4'' - \xi_5''),$$

$$S_{9c} = \frac{1}{2}(RS)^{\frac{1}{2}} (\eta_1' + \eta_2' - \eta_4' - \eta_5'). \qquad (17.24c)$$

17.5. Force Field

The initial force field was based on a five-constant potential function according to a valence-force approximation:

$$2V = \sum_k (f_s s_k^2 + S^2 f_\alpha \alpha_k^2 + f_r r_k^2)$$
$$+ S^2 \sum_l f_\beta \beta_l^2 + RS \sum_{i,j} f_\varphi \varphi_{ij}^2 . \qquad (17.25)$$

Here β_l ($l = 1,2,\ldots,6$) represent the NCN angle bendings, which have been omitted when constructing the symmetry coordinates. But their inclusion into the potential energy of the form (25) is supposed to be essential. It must be possible to express the β's as linear combinations of the symmetry coordinates of the present set because it is complete. The force field (25) when transformed to the present symmetry force constants is no longer represented by a diagonal **F** matrix, but significant interaction terms are introduced. Their magnitudes, as well as the magnitudes of some

Table 17-I. Symmetry force-constant matrix elements (mdyne/Å) and calculated frequencies (cm^{-1}). Figures in italics indicate observed values used in the calculations.

Species A_1:	*2933*	1127	912	697
1	5.23			
2	0.01	2.28		
3	0.01	0.00	4.83	
4	0.01	0.00	-0.00	0.57

Species A_2:	1232
1	0.45

Species E:	2883	*1354*	1144	787	*465*
1	7.02				
2	-0.69	0.89			
3	-0.00	-0.00	4.67		
4	-0.00	0.00	-0.00	0.55	
5	0.04	-0.00	0.00	0.00	0.43

Species F_1:	2942	1603	*1332*	1158	956	*320*
1	4.89					
2	0.01	0.66				
3	0.00	-0.00	4.57			
4	0.00	0.00	-0.00	0.56		
5	-0.01	0.00	0.00	0.00	0.47	
6	0.01	0.00	0.00	0.00	0.00	0.47

Species F_2:	*2948*	*2884*	*1438*	*1370*	*1240*	*1007*	*812*	*674*	*513*
1	8.32								
2	1.46	5.10							
3	0.44	0.51	1.21						
4	1.46	0.35	0.49	1.66					
5	-0.03	-0.02	0.01	-0.01	4.66				
6	-0.05	0.01	-0.00	-0.02	0.00	4.56			
7	0.07	0.06	-0.01	-0.01	-0.00	-0.00	0.64		
8	-0.11	0.02	-0.02	-0.06	0.01	0.01	0.00	0.69	
9	-0.01	0.12	-0.03	-0.04	0.00	-0.00	0.02	-0.00	0.40

of the diagonal terms could hardly be estimated in advance, if it was attempted to work with the symmetry **F** matrix from the beginning.

The final harmonic force field was produced after several steps of iteration utilizing the existing (incomplete) spectral assignment.[7,8] It seems not to be of interest to give here a detailed report on the various steps of these refinements. We only present the final force-constant matrix in the symmetrized form along with the corresponding calculated frequencies; cf. Table I. The frequencies in italics indicate observed values which have been used in the calculations and consequently are identical with the calculated ones. Two additional frequencies have been assigned[7] as fundamentals of species A_1, viz. 1041 and 782 cm^{-1}. These values have not been used in the present calculations because it is not clear where the magnitude of the remaining unobserved A_1 frequency is to be expected.

Table II shows the tentatively predicted frequencies for hexamethylene tetramine-d_{12} (HMT-d_{12}), as calculated from the present force field.

17.6. Mean Amplitudes of Vibration and Related Quantities

The developed harmonic force field was used to calculate the mean amplitudes of vibration[9] for HMT and HMT-d_{12}. The results are presented in Table III, which shows the values of l at absolute zero and 298 K for every type of bonded and nonbonded inter-

Table 17-II. Predicted fundamental frequencies (cm^{-1}) for hexamethylene tetramine-d_{12}.

Species A_1:	2145	1106	848	522		
A_2:	872					
E:	2108	1274	860	570	439	
F_1:	2206	1511	1033	866	767	265
F_2:	2234	2120	1373	1255	1118	
	796	629	577	411		

Table 17-III. Mean amplitudes of vibration (Å units) for hexamethylene tetramine (HMT) and HMT-d_{12}

Distance	(Equil.)	HMT		HMT-d_{12}	
		T = 0	298 K	T = 0	298 K
C-H	(1.097)	0.079	0.079	0.067	0.067
C-N	(1.465)	0.046	0.046	0.046	0.046
N•••N	(2.403)	0.054	0.055	0.053	0.055
$C_1\cdots C_2$	(2.365)	0.057	0.061	0.057	0.061
$C_1\cdots C_4$	(3.344)	0.062	0.067	0.062	0.066
C•••N	(2.793)	0.059	0.063	0.059	0.063
$C_1\cdots H_{13}$	(2.583)	0.144	0.152	0.124	0.137
$C_1\cdots H_{14}$	(3.315)	0.099	0.101	0.087	0.089
$C_1\cdots H_{17}$	(4.057)	0.126	0.132	0.109	0.119
$N_7\cdots H_{11}$	(2.080)	0.107	0.107	0.092	0.094
$N_7\cdots H_{17}$	(3.144)	0.146	0.156	0.125	0.142
$N_7\cdots H_{18}$	(3.777)	0.102	0.103	0.089	0.092
$H_{11}\cdots H_{12}$	(1.825)	0.147	0.149	0.124	0.130
$H_{11}\cdots H_{13}$	(2.313)	0.211	0.226	0.179	0.204
$H_{11}\cdots H_{14}$	(3.592)	0.162	0.169	0.138	0.149
$H_{11}\cdots H_{15}$	(4.138)	0.141	0.142	0.119	0.122
$H_{11}\cdots H_{17}$	(4.741)	0.172	0.180	0.145	0.160

atomic distances. In order to facilitate the identification of the various distance types the table includes the interatomic separations(R_e given in parentheses), as calculated from the adopted equilibrium parameters.

The mean amplitudes of vibration (l) would be of great interest in a modern gas electron diffraction investigation of the molecule. Other quantities which would be useful in the interpretation of such measurements are the perpendicular amplitude correction coefficients, also referred to as values of K. The calculated values of K for HMT and HMT-d_{12} are presented in Table IV.

Finally we have also calculated the atomic vibration mean

Table 17-IV. Perpendicular amplitude correction coefficients (Å units) for hexamethylene tetramine (HMT) and HMT-d_{12}

Distance	HMT		HMT-d_{12}	
	T = 0	298 K	T = 0	298 K
C-H	0.017	0.018	0.012	0.014
C-N	0.002	0.002	0.002	0.002
N•••N	0.001	0.001	0.001	0.001
$C_1 \cdots C_2$	0.001	0.002	0.001	0.002
$C_1 \cdots C_4$	0.001	0.001	0.001	0.001
C•••N	0.001	0.001	0.001	0.001
$C_1 \cdots H_{13}$	0.006	0.007	0.004	0.006
$C_1 \cdots H_{14}$	0.006	0.007	0.004	0.006
$C_1 \cdots H_{17}$	0.004	0.004	0.003	0.003
$N_7 \cdots H_{11}$	0.009	0.010	0.007	0.008
$N_7 \cdots H_{17}$	0.004	0.005	0.003	0.004
$N_7 \cdots H_{18}$	0.005	0.006	0.004	0.004
$H_{11} \cdots H_{12}$	0.018	0.019	0.012	0.015
$H_{11} \cdots H_{13}$	0.011	0.012	0.008	0.009
$H_{11} \cdots H_{14}$	0.009	0.010	0.006	0.008
$H_{11} \cdots H_{15}$	0.009	0.010	0.006	0.008
$H_{11} \cdots H_{17}$	0.006	0.006	0.004	0.005

square amplitudes for the various types of atoms. The results are shown in Table V, and are believed to be of interest in modern interpretation of accurate X-ray and neutron diffraction measurements.[3-5] The values in Table V pertain to the present numbering of atoms and also the here chosen orientations of the cartesian axes.

For more precise definitions of some of the quantities of the present section, see, e.g., Part II.

Table 17-V. Atomic vibration mean-square amplitudes (Å^2 units) for hexamethylene tetramine (HMT) and HMT-d_{12}.

	Temp.(K)	$\langle x^2 \rangle$	$\langle y^2 \rangle$	$\langle z^2 \rangle$	$\langle xy \rangle$	$\langle yz \rangle$	$\langle xz \rangle$
HMT							
C_1	0	0.0015	0.0014	0.0014	0.0000	-0.0002	0.0000
	298	0.0016	0.0017	0.0017	0.0000	-0.0003	0.0000
N_7	0	0.0011	0.0011	0.0011	0.0001	0.0001	0.0001
	298	0.0013	0.0013	0.0013	0.0001	0.0001	0.0001
H_{11}	0	0.0174	0.0136	0.0136	-0.0058	0.0005	0.0058
	298	0.0193	0.0150	0.0150	-0.0071	-0.0004	0.0071
HMT-d_{12}							
C_1	0	0.0015	0.0013	0.0013	0.0000	-0.0001	0.0000
	298	0.0016	0.0017	0.0017	0.0000	-0.0001	0.0000
N_7	0	0.0012	0.0012	0.0012	0.0001	0.0001	0.0001
	298	0.0014	0.0014	0.0014	0.0000	0.0000	0.0000
D_{11}	0	0.0122	0.0096	0.0096	-0.0040	0.0003	0.0040
	298	0.0151	0.0115	0.0115	-0.0058	-0.0006	0.0058

REFERENCES

1 R.G.Dickinson and A.L.Raymond, J.Am.Chem.Soc. 45, 22 (1922).

2 R.W.G.Wyckoff and R.B.Corey, Z.Krist. 89, 462 (1934).

3 L.N.Becka and D.W.J.Cruickshank, Proc.Roy.Soc.(London) A273, 435 (1963).

4 J.A.K.Duckworth, B.T.M.Willis, and G.S.Pawley, Acta Cryst. A26, 263 (1970).

5 J.A.K.Duckworth, B.T.M.Willis, and G.S.Pawley, Acta Cryst. A25, 482 (1969).

6 E.B.Wilson,Jr., J.C.Decius, and P.C.Cross: Molecular Vibrations, McGraw-Hill, New York 1955.

7 L.Couture-Mathieu, J.P.Mathieu, J.Cremer, and H.Polet, J.Chim. Phys. 48, 1 (1951).

8 A.Cheutin and J.P.Mathieu, J.Chim.Phys. 53, 106 (1956).

-- Part II: Atomic Vibration Mean-Square Amplitudes for Light and Heavy Naphthalene

S. J. CYVIN, B. N. CYVIN, G. HAGEN,
D. W. J. CRUICKSHANK, and G. S. PAWLEY

Calculated atomic vibration mean-square amplitudes for naphthalene at T=0 and 298.15 K are presented on a figure. The quantities are among the mean binary products of cartesian displacements, which are tabulated for $C_{10}H_8$ and $C_{10}D_8$ at 298.15 K. Also mentioned are the correlation coefficients, which are given numerically for the out-of-plane vibrations of $C_{10}H_8$ at 298.15 K.

Recently a spectroscopic analysis of the molecular vibrations of naphthalene has been taken up with the aim of calculating the mean amplitudes of vibration. Such quantities are of great interest in the interpretation of modern gas electron-diffraction measurements. In the first complete set of mean amplitudes for light and heavy naphthalene published[1,2] unfortunately some of the values were in error owing to reasons fully discussed elsewhere.[3] An improved set of mean amplitudes has now been published together with the same quantities for anthracene.[4] It is believed that the electron-diffraction studies of naphthalene and anthracene by Almenningen et al.[5] could be improved today using modern methods of data analysis, and in such studies the spectroscopic mean amplitudes would be extremely useful. Furthermore, if an electron-diffraction reinvestigation of naphthalene should be undertaken with the aim of determining the average structure,[6,7] (cf. also Chapter 12) the present authors can now supply the necessary set of perpendicular amplitude correction coefficients,[8] $K = (\langle \Delta x^2 \rangle + \langle \Delta y^2 \rangle)/(2R)$.

The current calculations of mean amplitudes of (interatomic) vibration are based on a harmonic force field of the molecule. Here we wish to point out that the atomic vibration mean-square amplitudes may be calculated as a matter of routine once the force field is established. In fact the atomic mean-square amplitudes are available by the same computer program which calculates the mean amplitudes of (interatomic) vibration at the Technical University of Trondheim (Norway). The atomic mean-square amplitudes are of great interest in precise crystal structure studies by X-ray and neutron diffraction, and we think the workers in these fields should be aware of the acessibility of these quantities.

The atomic vibration mean-square amplitudes are contained in the tensor

$$\begin{bmatrix} \langle x_i^2 \rangle & \langle x_i y_i \rangle & \langle x_i z_i \rangle \\ & \langle y_i^2 \rangle & \langle y_i z_i \rangle \\ & & \langle z_i^2 \rangle \end{bmatrix} \qquad (17.26)$$

for every atom i. Here x_i, y_i, and z_i designate the respective cartesian displacement coordinates. The quantities (26) are found among the mean binary products of cartesian displacements defined in Section 13.9.2 of Cyvin's book,[8] which should be consulted for a detailed description of the methods of calculation.

In the present calculations the force field for the in-plane vibrations is the same as that used in the revised calculations of mean amplitudes.[4] It fits exactly the observed frequencies given by Neto et al.,[9] and was derived with the aid of initial valence force constants from the same work.[9] The force field for the out-of-plane vibrations is not necessary for calculating the mean interatomic amplitudes. In the present calculations it was taken as the final force field in the previous analysis of naphthalene,[2] which had been adjusted to the observed frequencies by Krainov.[10] The out-of-plane force field is less reliable than the in-plane force field because of the greater uncertainties in the vibrational assignment for the out-of-plane modes. Hence the present results for the out-of-plane amplitudes should not be accepted without certain reservations.

The results of the calculations of the atomic mean-square amplitudes for naphthalene are presented on Fig. 4. Results at

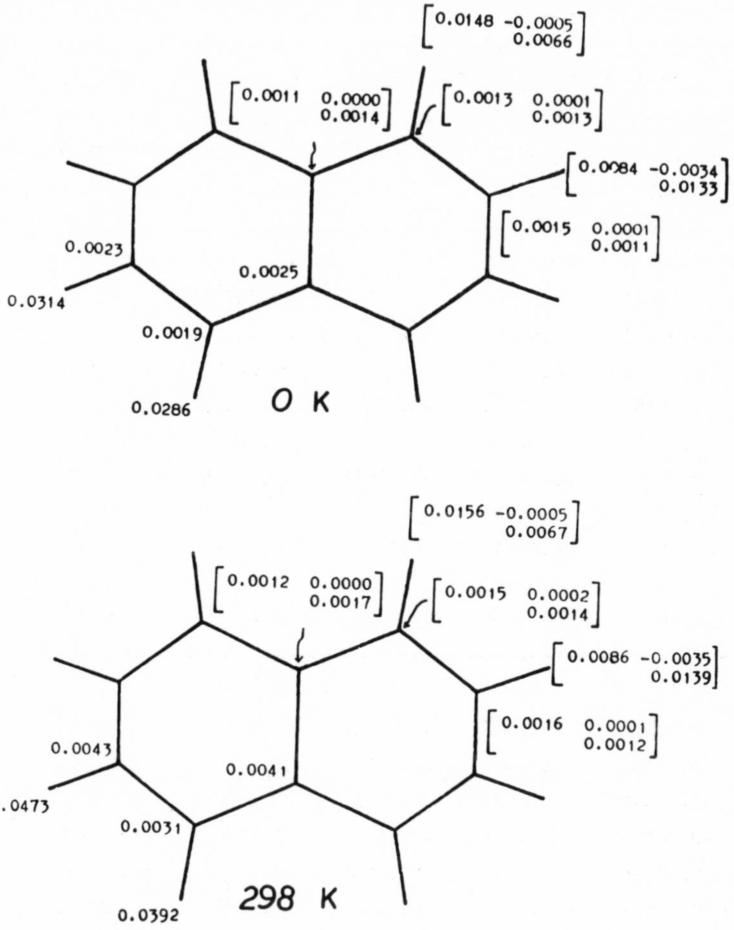

Fig. 17-4. Atomic vibration mean-square amplitudes in Å^2 for naphthalene at two temperatures. Lower-left part of each drawing gives the out-of-plane quantities, $\langle x_i^2 \rangle$. Upper-right parts give the in-plane quantities, viz.

$$\begin{bmatrix} \langle y_i^2 \rangle & \langle y_i z_i \rangle \\ & \langle z_i^2 \rangle \end{bmatrix} .$$

Table 17-VI. Mean binary products of cartesian displacements for naphthalene at 298.15 K; $Å^2$ units.

$\langle x_i x_j \rangle$	x_1	x_5	x_9	x_{11}	x_{15}
x_1	*0.00427**	-0.00011	-0.00221	0.00799	-0.00107
x_2	0.00038	-0.00215	-0.00221	0.00151	-0.00560
x_3	0.00313	0.00065	-0.00201	0.00728	0.00119
x_4	-0.00009	-0.00234	-0.00201	0.00023	-0.00589
x_5	-0.00011	*0.00305*	0.00062	-0.00145	0.00502
x_6	-0.00215	-0.00126	0.00062	-0.00477	-0.00255
x_7	0.00065	0.00148	0.00081	0.00164	0.00367
x_8	-0.00234	-0.00040	0.00081	-0.00578	-0.00006
x_9	-0.00221	0.00062	*0.00407*	-0.00517	0.00174
x_{10}	-0.00201	0.00081	0.00250	-0.00456	0.00210
x_{11}	0.00799	-0.00145	-0.00517	*0.04727*	-0.00660
x_{12}	0.00151	-0.00477	-0.00517	0.00475	-0.01348
x_{13}	0.00728	0.00164	-0.00456	0.01588	0.00258
x_{14}	0.00023	-0.00578	-0.00456	-0.00005	-0.01381
x_{15}	-0.00107	0.00502	0.00174	-0.00660	*0.03916*
x_{16}	-0.00560	-0.00255	0.00174	-0.01348	-0.00492
x_{17}	0.00119	0.00367	0.00210	0.00258	0.00987
x_{18}	-0.00589	-0.00006	0.00210	-0.01381	0.00438

$\langle y_i y_j \rangle$	y_1	y_5	y_9	y_{11}	y_{15}
y_1	*0.00161*	0.00038	-0.00004	0.00080	0.00030
y_2	-0.00033	-0.00032	-0.00004	-0.00030	-0.00054
y_3	-0.00042	-0.00028	-0.00027	-0.00030	-0.00002
y_4	-0.00003	-0.00026	-0.00027	-0.00019	-0.00013
y_5	0.00038	*0.00147*	0.00006	0.00050	0.00088
y_6	-0.00032	-0.00038	0.00006	-0.00041	-0.00093
y_7	-0.00028	-0.00003	-0.00023	-0.00012	0.00023
y_8	-0.00026	-0.00034	-0.00023	-0.00035	-0.00053
y_9	-0.00004	0.00006	*0.00117*	-0.00003	-0.00013
y_{10}	-0.00027	-0.00023	-0.00008	-0.00021	-0.00029
y_{11}	0.00080	0.00050	-0.00003	*0.00858*	0.00064
y_{12}	-0.00030	-0.00041	-0.00003	-0.00037	-0.00099
y_{13}	-0.00030	-0.00012	-0.00021	-0.00026	0.00014
y_{14}	-0.00019	-0.00035	-0.00021	-0.00024	-0.00026

Table 17-VI (Continued) -- naphthalene --

$\langle y_i y_j \rangle$	y_1	y_5	y_9	y_{11}	y_{15}
y_{15}	0.00030	0.00088	-0.00013	0.00064	*0.01556*
y_{16}	-0.00054	-0.00093	-0.00013	-0.00099	-0.00159
y_{17}	-0.00002	0.00023	-0.00029	0.00014	0.00076
y_{18}	-0.00013	-0.00053	-0.00029	-0.00026	-0.00044

$\langle z_i z_j \rangle$	z_1	z_5	z_9	z_{11}	z_{15}
z_1	*0.00120*	-0.00013	-0.00053	0.00091	-0.00016
z_2	0.00021	-0.00015	-0.00053	0.00063	-0.00014
z_3	0.00017	-0.00016	-0.00051	0.00036	-0.00016
z_4	0.00010	-0.00029	-0.00051	-0.00005	-0.00030
z_5	-0.00013	*0.00140*	0.00005	-0.00052	0.00103
z_6	-0.00015	-0.00022	0.00005	-0.00003	-0.00023
z_7	-0.00016	-0.00004	-0.00001	-0.00032	-0.00004
z_8	-0.00029	0.00038	-0.00001	-0.00035	0.00040
z_9	-0.00053	0.00005	*0.00168*	-0.00094	0.00000
z_{10}	-0.00051	-0.00001	0.00063	-0.00088	-0.00002
z_{11}	0.00091	-0.00052	-0.00094	*0.01388*	-0.00051
z_{12}	0.00063	-0.00003	-0.00094	0.00144	-0.00001
z_{13}	0.00036	-0.00032	-0.00088	0.00034	-0.00033
z_{14}	-0.00005	-0.00035	-0.00088	-0.00035	-0.00036
z_{15}	-0.00016	0.00103	0.00000	-0.00051	*0.00665*
z_{16}	-0.00014	-0.00023	0.00000	-0.00001	-0.00024
z_{17}	-0.00016	-0.00004	-0.00002	-0.00033	-0.00005
z_{18}	-0.00030	-0.00040	-0.00002	-0.00036	-0.00042

$\langle y_i z_j \rangle$	y_1	y_5	y_9	y_{11}	y_{15}
z_1	*0.00010*	-0.00006	0.00014	0.00006	-0.00023
z_2	-0.00020	-0.00044	-0.00014	-0.00044	-0.00080
z_3	-0.00024	-0.00039	-0.00014	-0.00035	-0.00069
z_4	0.00008	-0.00003	0.00014	0.00013	-0.00007
z_5	-0.00013	*0.00016*	0.00024	0.00009	-0.00053
z_6	-0.00015	-0.00020	-0.00024	-0.00022	0.00000
z_7	-0.00005	0.00003	0.00002	0.00004	0.00007
z_8	0.00028	0.00024	-0.00002	0.00031	0.00056
z_9	0.00016	0.00043	*0.00000*	0.00039	0.00112

Table 17-VI (Continued) -- naphthalene --

$\langle y_i z_j \rangle$	y_1	y_5	y_9	y_{11}	y_{15}
z_{10}	0.00014	0.00039	0.00000	0.00036	0.00086
z_{11}	0.00084	-0.00035	0.00011	*-0.00352*	-0.00075
z_{12}	-0.00014	-0.00060	-0.00011	-0.00061	-0.00158
z_{13}	-0.00045	-0.00070	-0.00026	-0.00046	-0.00095
z_{14}	-0.00017	-0.00017	0.00026	0.00005	-0.00029
z_{15}	-0.00012	0.00015	0.00023	0.00004	*-0.00055*
z_{16}	-0.00015	-0.00021	-0.00023	-0.00023	-0.00001
z_{17}	-0.00006	0.00002	0.00002	0.00004	0.00006
z_{18}	0.00028	0.00024	-0.00002	0.00030	0.00056

*Values for $i = j$ in *italics*.

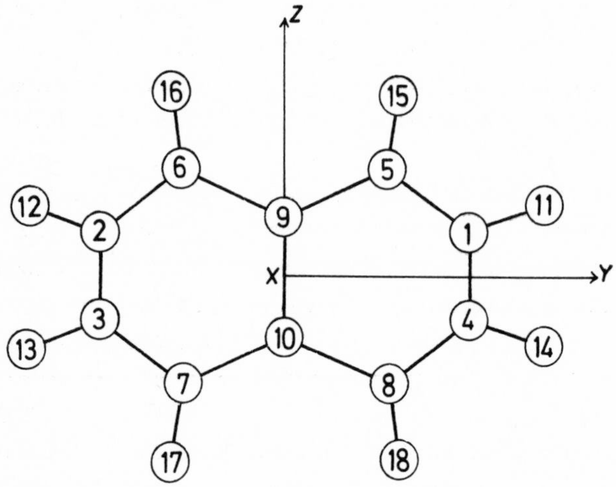

Fig. 17-5. Numbering of atoms in naphthalene. Symmetrically equivalent atoms are numbered consecutively within each set. In-plane cartesian axes are shown; X is perpendicular to the molecular plane.

Table 17-VII. Mean binary products of cartesian displacements for naphthalene-d_8 at 298.15 K; $Å^2$ units.

$\langle x_i x_j \rangle$	x_1	x_5	x_9	x_{11}	x_{15}
x_1	*0.00426**	-0.00002	-0.00215	0.00745	-0.00117
x_2	0.00026	-0.00218	-0.00215	0.00143	-0.00569
x_3	0.00297	0.00054	-0.00198	0.00724	0.00111
x_4	-0.00013	-0.00232	-0.00198	-0.00015	-0.00592
x_5	-0.00002	*0.00324*	0.00078	-0.00167	0.00489
x_6	-0.00218	-0.00117	0.00078	-0.00491	-0.00261
x_7	0.00054	0.00142	0.00089	0.00157	0.00368
x_8	-0.00232	-0.00036	0.00089	-0.00591	-0.00002
x_9	-0.00215	0.00078	*0.00427*	-0.00529	0.00178
x_{10}	-0.00198	0.00089	0.00265	-0.00464	0.00218
x_{11}	0.00745	-0.00167	-0.00529	*0.04073*	-0.00704
x_{12}	0.00143	-0.00491	-0.00529	0.00483	-0.01373
x_{13}	0.00724	0.00157	-0.00464	0.01624	0.00268
x_{14}	-0.00015	-0.00591	-0.00464	-0.00074	-0.01395
x_{15}	-0.00117	0.00489	0.00178	-0.00704	*0.03334*
x_{16}	-0.00569	-0.00261	0.00178	-0.01373	-0.00529
x_{17}	0.00111	0.00368	0.00218	0.00268	0.01010
x_{18}	-0.00592	-0.00002	0.00218	-0.01395	0.00445

$\langle y_i y_j \rangle$	y_1	y_2	y_5	y_{11}	y_{15}
y_1	*0.00161*	0.00039	-0.00003	0.00073	0.00027
y_2	-0.00031	-0.00030	-0.00003	-0.00031	-0.00057
y_3	-0.00041	-0.00028	-0.00027	-0.00030	-0.00007
y_4	-0.00004	-0.00026	-0.00027	-0.00020	-0.00018
y_5	0.00039	*0.00148*	0.00008	0.00045	0.00076
y_6	-0.00030	-0.00036	0.00008	-0.00043	-0.00096
y_7	-0.00028	-0.00003	-0.00023	-0.00011	0.00021
y_8	-0.00026	-0.00034	-0.00023	-0.00034	-0.00054
y_9	-0.00003	0.00008	*0.00119*	-0.00005	-0.00018
y_{10}	-0.00027	-0.00023	-0.00006	-0.00021	-0.00031
y_{11}	0.00073	0.00045	-0.00005	*0.00623*	0.00050
y_{12}	-0.00031	-0.00043	-0.00005	-0.00043	-0.00105
y_{13}	-0.00030	-0.00011	-0.00021	-0.00024	0.00012
y_{14}	-0.00020	-0.00034	-0.00021	-0.00025	-0.00029

Table 17-VII (Continued) -- naphthalene-d_8 --

$\langle y_i y_j \rangle$	y_1	y_5	y_9	y_{11}	y_{15}
y_{15}	0.00027	0.00076	-0.00018	0.00050	*0.01153*
y_{16}	-0.00057	-0.00096	-0.00018.	-0.00105	-0.00172
y_{17}	-0.00007	0.00021	-0.00031	0.00012	0.00075
y_{18}	-0.00018	-0.00054	-0.00031	-0.00029	-0.00049

$\langle z_i z_j \rangle$	z_1	z_5	z_9	z_{11}	z_{15}
z_1	*0.00121*	-0.00012	-0.00051	0.00072	-0.00016
z_2	0.00016	-0.00016	-0.00051	0.00058	-0.00016
z_3	0.00012	-0.00017	-0.00049	0.00033	-0.00018
z_4	0.00011	-0.00028	-0.00049	-0.00012	-0.00030
z_5	-0.00012	*0.00140*	0.00008	-0.00057	0.00094
z_6	-0.00016	-0.00020	0.00008	-0.00007	-0.00022
z_7	-0.00017	-0.00003	0.00002	-0.00034	-0.00004
z_8	-0.00028	-0.00036	0.00002	-0.00041	-0.00038
z_9	-0.00051	0.00008	*0.00173*	-0.00094	0.00002
z_{10}	-0.00049	0.00002	0.00067	-0.00089	0.00000
z_{11}	0.00072	-0.00057	-0.00094	*0.01019*	-0.00057
z_{12}	0.00058	-0.00007	-0.00094	0.00141	-0.00005
z_{13}	0.00033	-0.00034	-0.00089	0.00042	-0.00034
z_{14}	-0.00012	-0.00041	-0.00089	-0.00053	-0.00043
z_{15}	-0.00016	0.00094	0.00002	-0.00057	*0.00485*
z_{16}	-0.00016	-0.00022	0.00002	-0.00005	-0.00024
z_{17}	-0.00018	-0.00004	0.00000	-0.00034	-0.00005
z_{18}	-0.00030	-0.00038	0.00000	-0.00043	-0.00041

$\langle y_i z_j \rangle$	y_1	y_5	y_9	y_{11}	y_{15}
z_1	*0.00008*	-0.00006	0.00013	0.00010	-0.00019
z_2	-0.00018	-0.00041	-0.00013	-0.00042	-0.00078
z_3	-0.00022	-0.00036	-0.00013	-0.00034	-0.00067
z_4	0.00006	-0.00004	0.00013	0.00015	-0.00003
z_5	-0.00013	*0.00017*	0.00024	0.00012	-0.00042
z_6	-0.00014	-0.00019	-0.00024	-0.00020	0.00000
z_7	-0.00004	0.00005	0.00003	0.00005	0.00008
z_8	0.00027	0.00024	-0.00003	0.00033	0.00058
z_9	0.00016	0.00045	*0.00000*	0.00040	0.00109

Table 17-VII (Continued) -- naphthalene-d_8 --

$\langle y_i z_j \rangle$	y_1	y_5	y_9	y_{11}	y_{15}
z_{10}	0.00014	0.00040	0.00000	0.00037	0.00087
z_{11}	0.00078	-0.00027	0.00013	*-0.00252*	-0.00056
z_{12}	-0.00016	-0.00061	-0.00013	-0.00064	-0.00160
z_{13}	-0.00043	-0.00068	-0.00025	-0.00050	-0.00101
z_{14}	-0.00017	-0.00015	0.00025	0.00009	-0.00021
z_{15}	-0.00011	0.00016	0.00023	0.00009	*-0.00042*
z_{16}	-0.00015	-0.00020	-0.00023	-0.00021	-0.00002
z_{17}	-0.00004	0.00004	0.00002	0.00005	0.00006
z_{18}	0.00027	0.00024	-0.00002	0.00032	0.00058

*Values for $i = j$ in *italics*.

absolute zero and 298.15 K are given. The temperature-dependence is appreciable for the out-of-plane quantities; they increase about 50 per cent from 0 to 298 K. The effect is considerably smaller for the in-plane quantities. The secondary isotope effect on all carbon amplitudes is almost negligible.

In Tables VI and VII a more detailed account of the mean binary products of cartesian displacements at 298.15 K is given for naphthalene and naphthalene-d_8, respectively. The quantities in question are $\langle x_i x_j \rangle$, $\langle y_i y_j \rangle$, $\langle z_i z_j \rangle$, and $\langle y_i z_j \rangle$, where i runs through a representative set of five atoms, i.e. one from each set of symmetrically equivalent atoms. The index j runs through all atoms. All $\langle x_i y_j \rangle$ and $\langle x_i z_j \rangle$ quantities are zero to the first order approximation as a consequence of the planarity of the molecule. The atomic mean-square amplitudes (see above) are found as diagonal elements (i.e. $i = j$) in the tables. The numbering of atoms follows the conventions proposed in a paper on tentatively standardized symmetry coordinates,[11] and is shown in Fig. 5, which also defines the directions of cartesian coordinate axes.

Correlation coefficients are of prime interest to crystallographers, and are defined as scaled $\langle \alpha_i \alpha_j \rangle$ quantities ($\alpha = x,y,z$), viz. $\langle \alpha_i \alpha_j \rangle / (\langle \alpha_i^2 \rangle \langle \alpha_j^2 \rangle)^{\frac{1}{2}}$. Thus for $i = j$ the correlation coefficients are unity. Table VIII shows the calculated correlation coefficients for out-of-plane vibrations of naphthalene at 298.15

Table 17-VIII. Correlation coefficients for out-of-plane atomic vibrations of naphthalene at 298.15 K.

Atom No.	1	5	9	11	15
1	*1.000*	-0.030	-0.531	0.562	-0.083
2	0.090	-0.595	-0.531	0.106	-0.433
3	0.733	0.179	-0.481	0.513	0.092
4	-0.021	-0.648	-0.481	0.016	-0.456
5	-0.030	*1.000*	0.175	-0.121	0.459
6	-0.595	-0.414	0.175	-0.397	-0.233
7	0.180	0.483	0.230	0.137	0.336
8	-0.648	-0.129	0.230	-0.481	-0.005
9	-0.531	0.175	*1.000*	-0.373	0.138
10	-0.481	0.230	0.613	-0.328	0.167
11	0.562	-0.121	-0.373	*1.000*	-0.153
12	0.106	-0.397	-0.373	0.100	-0.313
13	0.513	0.137	-0.328	0.336	0.060
14	0.016	-0.481	-0.328	-0.001	-0.321
15	-0.083	0.459	0.138	-0.153	*1.000*
16	-0.433	-0.233	0.138	-0.313	-0.126
17	0.092	0.336	0.167	0.060	0.252
18	-0.456	-0.005	0.167	-0.321	0.112

K. The values are not much different for naphthalene-d_8. In general the coefficients are smaller for the in-plane motions.

Finally we wish to refer to an earlier paper[12] which mentioned the out-of-plane mean binary products of cartesian displacements for the naphthalene skeleton. The earlier results in part deviate largely from the present ones and must now be considered rather rudimentary. Comparison with the earlier work serves to illustrate the enormous progress made in this field during the last few years, mainly due to the accessibility of modern electronic computers.

REFERENCES

1 B.N.Cyvin, S.J.Cyvin, and G.Hagen, Chem.Phys.Letters 1, 211 (1967).
2 G.Hagen and S.J.Cyvin, J.Phys.Chem. 72, 1446 (1968).
3 S.J.Cyvin, B.N.Cyvin, and G.Hagen, Chem.Phys.Letters 2, 341 (1968).
4 B.N.Cyvin and S.J.Cyvin, J.Phys.Chem. 73, 1430 (1969).
5 A.Almenningen, O.Bastiansen, and F.Dyvik, Acta Cryst. 14, 1056 (1961).
6 Y.Morino, K.Kuchitsu, and T.Oka, J.Chem.Phys. 36, 1108 (1962).
7 K.Kuchitsu and S.Konaka, J.Chem.Phys. 45, 4342 (1966).
8 S.J.Cyvin: Molecular Vibrations and Mean Square Amplitudes, Universitetsforlaget, Oslo, and Elsevier, Amsterdam, 1968.
9 N.Neto, M.Scrocco, and S.Califano, Spectrochim.Acta 22, 1981 (1966).
10 E.P.Krainov, Opt.Spektroskopiya 16, 763 (1964).
11 G.Hagen and S.J.Cyvin, J.Phys.Chem. 72, 1451 (1968).
12 S.J.Cyvin and E.Meisingseth, Kgl. Norske Videnskab. Selskabs Skrifter, No. 2 (1962).

CHAPTER 18

Electron Diffraction Studies - Part I: Some Four-Membered Ring Systems Containing Silicon

L.V. VILKOV, V.S. MASTRYUKOV, V.D. OPPENHEIM, and N.A. TARASENKO

A gas-phase electron-diffraction investigation of some four-membered ring systems containing silicon has been carried out in an effort to obtain the following information: ring configuration, endo- and exocyclic internuclear distances, and the external valence angles. The conformational problems have been studied using the cyclobutane molecule as a model. The studied compounds: 1-silacyclobutane $SiH_2(CH_2)_3$, 1,1-dichloro-1-silacyclobutane $SiCl_2(CH_2)_3$, 1,1,3,3-tetrachloro-1,3-disilacyclobutane $SiCl_2(CH_2)_2SiCl_2$, and 4-sila-3,3-spiroheptane $(CH_2)_3Si(CH_2)_3$.

The four-membered rings are found to be significantly puckered in all studied molecules. The endocyclic internuclear distances are longer than normal values, while the exocyclic ones seem to be normal. Both external (ClSiCl) and internal (CSiC) valence angles are less than the tetrahedral angle in contrast to the data on cyclobutane. The structural parameters of the individual molecules are summarized in Table IV.

ЭЛЕКТРОНОГРАФИЧЕСКОЕ ИССЛЕДОВАНИЕ НЕКОТОРЫХ КРЕМНИЙ-СОДЕРЖАЩИХ ЧЕТЫРЕХЧЛЕННЫХ ЦИКЛИЧЕСКИХ МОЛЕКУЛ

Л. В. Вилков, В. С. Мастрюков, В. Д. Оппенгейм и Н. А. Тарасенко.

Проведено электронографическое исследование в газовой фазе некоторых четырехчленных циклических соединений, содержащих кремний, с целью получения информации о конфигурации кольца, эндо- и экзоциклических межъядерных расстояниях и внешних валентных углах. Конформационные проблемы изучались, используя в качестве модели молекулу циклобутана. Были исследованы следующие соединения : 1-силациклобутан $SiH_2(CH_2)_3$, 1,1-дихлор-1-силациклобутан $SiCl_2(CH_2)_3$, 1,1,3,3-тетрахлор-1,3-силациклобутан $SiCl_2(CH_2)_2SiCl_2$, 4-сила-3,3-спирогептан $(CH_2)_3Si(CH_2)_3$.

Найдено, что четырехчленные циклы во всех изученных молекулах являются неплоскими. Эндоциклические расстояния завышены по сравнению с их обычными значениями в открытых аналогах, а экзоциклические расстояния близки к нормальным. И внешний, ClSiCl, и внутренний, CSiC, валентные углы атома кремния меньше тетраэдрического, что противоречит данным по циклобутану. Структурные параметры всех изученных молекул приведены в Таблице IV.

18.1. General Structural Problems of Four-Membered Rings

The structure investigation of four-membered rings was started in the early forties. The chemistry of strained small ring systems has attracted much interest, and the number of papers in the field increased constantly during following decades. The four-membered ring molecules have in particular been studied in order to determine the influence of molecular strain on bond lengths, on bond angles, bond energies, and on reactivity. Although extensively studied, the stereochemistry of these compounds is far from being completely understood. Cyclobutane and its derivatives have lately been subject to intensive structure studies. Cyclobutane itself is by no means the simplest molecule of this class. On the contrary, the cyclobutane problem presents all the structural questions arising in an investigation of four-membered ring molecules. We shall take a closer look at these questions in order to specify the problem to be solved.

Ring Configuration. Two opposing effects are important for the configuration of four-membered rings. (a) The cyclobutane ring is said to be strained since the CCC bond angle is markedly different from the tetrahedral angle. The minimum of this strain is expected for a planar ring with all CCC angles equal to 90° (Bayer strain). (b) The repulsion between hydrogen atoms of adjacent methylene groups ought to be larger for a planar ring and should thus tend to pucker the ring (Pitzer strain).

The actual configuration of the ring may be considered as a result of a compromise between two opposing effects. Planar four-membered rings have been reported for trimethylene oxide,[1,2] cyclobutanone,[3] and β-propiolactone[4] in the gas phase. Nonplanar carbon rings have also been found in gas phase for cyclobutane[5-7] and several of its halogen derivatives.[8-11] Both planar and puckered cyclobutane rings are known in the solid state.[12] But it is necessary to emphasize that crystal forces may well influence the conformation of the carbon ring. Solid-state results on cyclobutane conformation are therefore not directly applicable to gases, liquids, or solutions. The nonplanarity of the four-membered ring is responsible for pseudo-axial-equatorial conformerism[8,13] whilst in the planar conformation all positions are equivalent.

Endocyclic Internuclear Distances. A considerable amount of structural evidence indicates that carbon-carbon single bonds of three-membered rings are shorter and those of four-membered rings longer than the ordinary $sp^3 - sp^3$ distance of 1.54 Å. The bond lengths reported for four-membered rings vary between 1.55 Å in methylenecyclobutane and 1.63 Å in octafluorocyclobutane.[11] The most accurate CC bond length determination in gas-phase for cyclo-

butane gives values 1.548 ± 0.003 Å [6] and 1.558 ± 0.001 Å.[14] X-ray
work also supports long CC bonds in four-membered rings.

Exocyclic Internuclear Distances. There is a large amount of
independent chemical and spectroscopical evidence of electron-
attracting to strained aliphatic ring systems. It is concluded that
the electron-attractive influence of the carbon ring is in the
sequence: cyclopropyl > cyclobutyl > cyclopentyl > cyclohexyl. In
general one may expect a similar change in exocyclic bond lengths
due to the different degree of 'ring-substituent' interaction. A
pronounced decrease of C-Hal bond length has been pointed out for
cyclopropane derivatives. It is also of interest, therefore, to
study this question in relation to four-membered ring systems.

External Valence Angles. It is well-known that the concept of
hybridization is in principle based upon values of valence angles.
In particular, the unusual value for the HCH angle in cyclopropane
forms the basis of Walsh's theory about the unsaturated nature of
cyclopropane.[15-17] The corresponding values of the HCH angle in
four-membered ring molecules are also of great interest.

The problems mentioned have all been considered in connection
with the silicon analogues of cyclobutane. For each compound we
attempted to solve the following specific problems.

(a) 1-Silacyclobutane, $\square$ SiH$_2$ (SCB): configuration of the
ring, endocyclic distances.

(b) 1,1-Dichloro-1-Silacyclobutane, $\square$ SiCl$_2$ (DCSB): configuration of the ring, Si-Cl bond distance, ClSiCl angle.

(c) 1,1,3,3-Tetrachloro-1,3-Disilacyclobutane, Cl$_2$Si $\square$ SiCl$_2$
(TCDB): the same as in the preceding case.

(d) 4-Sila-3,3-Spiroheptane, $\square$ Si $\square$ (SSH): the same as for
SCB.

18.2. Silacyclobutanes

Although silacyclobutane derivatives have been known since
1954,[18] extensive chemical studies were not started till ten years

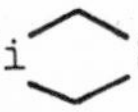

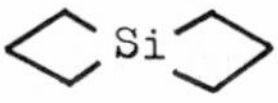

later.[19,20] The initial product in the synthesis of these compounds
is dichloroalkyl-γ-chloropropylsilane, which reacts according to
the following scheme.

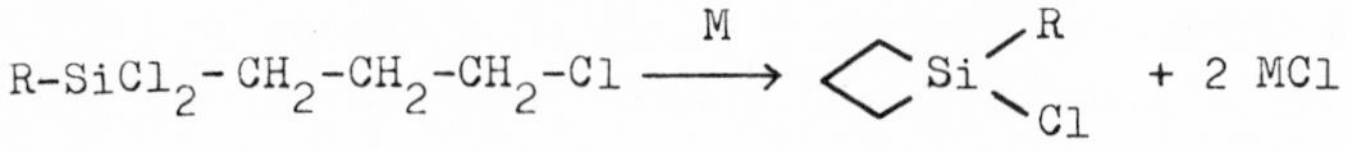

$$R\text{-SiCl}_2\text{-CH}_2\text{-CH}_2\text{-CH}_2\text{-Cl} \xrightarrow{\text{M}} \begin{array}{c} \diamondsuit \text{Si} \begin{array}{c} R \\ Cl \end{array} \end{array} + 2 \text{ MCl}$$

$$M = Na; \text{ Li}; \text{ Mg}$$

Silacyclobutanes react with many substances in a way characte
ristic only for this class of compounds.[20] They undergo many reactions such as polymerization,[21] telomerization,[22] and condensa-
tion.[23] A number of other characteristics have also been pointed
out.[24] In all the indicated reactions the Si-C bond seems to play
an important part exhibiting a higher degree of reactivity than the
corresponding bond in other organic compounds containing silicon.[20]

18.3. Experimental Part

All the samples of compounds investigated were generously
provided by Drs. V.M. Vdovin, P.L.Grinberg, and O.V. Kuzmin of the
Academy of Science, USSR, Moscow, and were used without further
purification.

Electron-diffraction photographs were taken at temperatures
ranging from 40° to $70\ ^\circ C$ with the diffraction unit at Moscow
State University using an r^3-sector and accelerating voltage of 55
kV. The exposure times were from 30 sec till 4 min for the 14-cm
and 26-cm camera distances. The observed data extended from $s = 3$
$Å^{-1}$ to s equal to about 29 $Å^{-1}$. The experimental procedure followed
that previously described.[25]

Intensity curves were determined by measuring absorbancies of
four selected plates for each camera distance. Measurements were
made with a modified MΦ-4 microphotometer. Corrections for non-
nuclear scattering and the failure of the Born approximation were
included only for TCDS. The value of b used in the damping function
$\exp(-bs^2)$, of the radial distribution curve (RDC) was 0.002.

The errors were estimated according to an earlier given formula.[26] The calculations were made with the "Setun'" digital computer.

18.4. Structure Analysis

18.4.1. 1-Silacyclobutane[27]

The experimental intensity function is shown in Fig. 1. Models with a planar four-membered ring were first considered. Calculations were made for a range of values of the CSiC angle in an attempt to reproduce the measured intensity function. In order to facilitate the analysis, the following assumptions were made: the angles HCH, HSiH are equal; the plane HCH is perpendicular to the ring plane and bisects the CCX angle (X = C or Si). The same is valid for the hydrogen atoms linked to the silicon atom. A possible wrong assign-

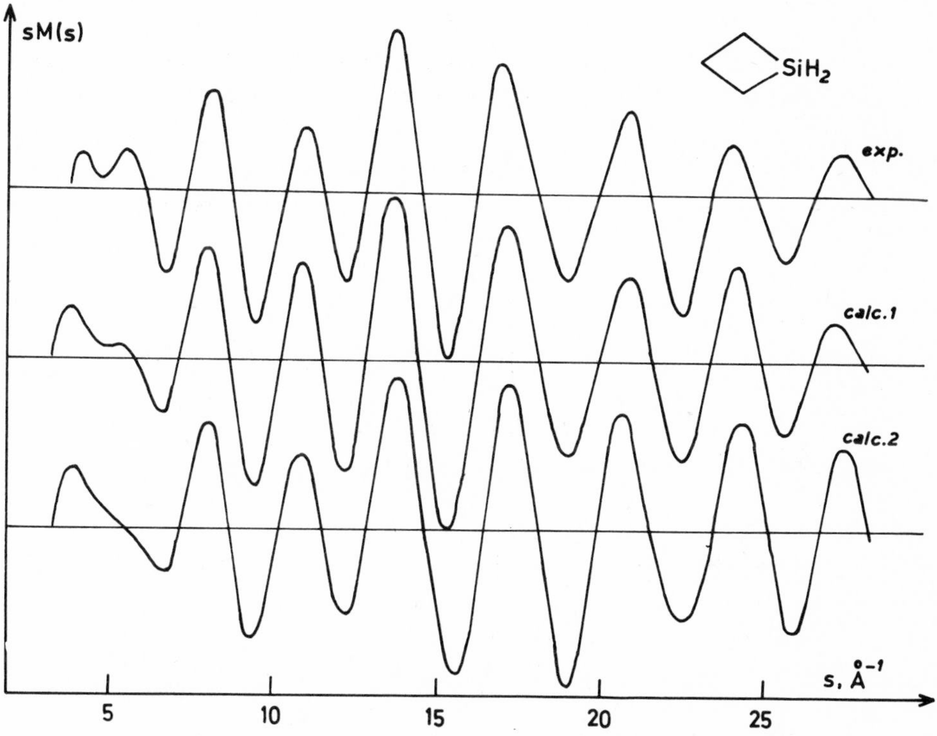

Fig. 18-1. Comparison of experimental and calculated *sM*(*s*) curves. Curve 1 corresponds to the final model. Curve 2 corresponds to the rejected planar ring model with the parameters taken from the experimental radial distribution curve.

ment of the hydrogen atom parameters does not influence the other parameters very much owing to the relatively small contribution of hydrogen atoms to the total intensity curve. The initial parameter were:

SiC	1.867 Å	HCH	110°
SiH	1.483 Å	HSiH	110°
CC	1.548 Å	CSiC	70°, 80°, 90°, 100°, 110°
CH	1.096 Å		

The best model was found to have the CSiC angle of 80°. When a reasonable range for the angle CSiC had been established, the othe parameters were refined from the experimental RDC (Fig. 2) calculated from the experimental intensity curve plus a theoretical curve for the inner unobserved s range.

The first peak in this curve corresponds to the C-H distance The second peak represents the contribution from the Si-H and C-C distances. The next major peak is due to the Si-C distance only, a the final peak is built up by Si···C, C···C, C···H, and Si···H di tance contributions. All the bonding parameters were taken directl

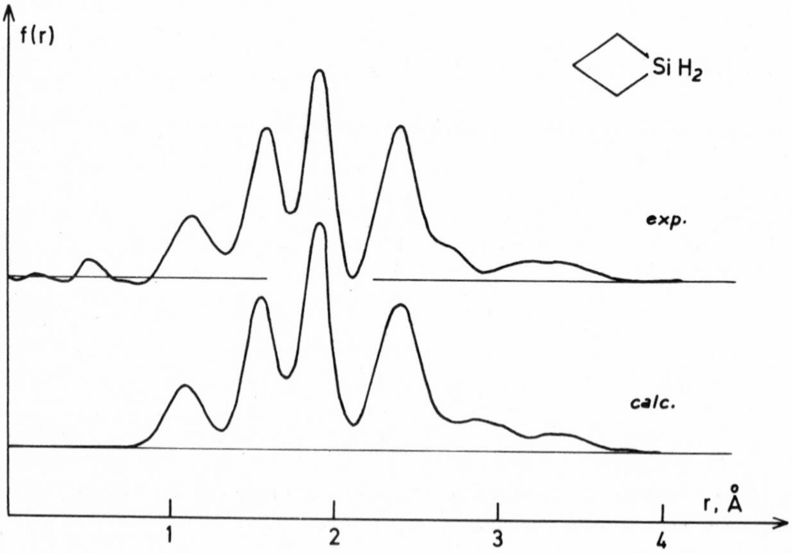

Fig. 18-2. Experimental and calculated radial distribution curves.

Table 18-I. Molecular parameters for the 1-silacyclobutane molecule.*

	r (Å) or angle	l (Å)
CH	1.13 ± 0.02	0.08 (assumed)
SiH	1.48_5± 0.03	0.05_3± 0.03[a]
CC	1.58_5± 0.01	0.04_7± 0.02
SiC	1.89_7± 0.01	0.04_5± 0.02[b]
CSiC	80° ± 2°	
α	30° ± 5°	

* Assumed HCH = 110° and HSiH = 110°.
[a] Δ = 0.010 Å. [b] Δ = 0.007 Å.

from the experimental RDC. These values were then applied as a correction to the previously assumed structure. It was immediately clear that a planar ring structure with these parameters could not be consistent with the experimental data (Fig. 1, curve 2). Accordingly a puckering of the ring was taken into account until a satisfactory agreement was reached. Structures with the dihedral angle α = 10°, 20°, 30°, and 40° (as defined in Fig. 3) were tested subsequently. The dihedral angle α is determined by the planes CCC and CSiC. It differs slightly (~1.5°) from the angle determined by two CCSi planes. Intensity and RD curves of the best structures with α = 30° are shown in Figs. 1 and 2. The final parameters are collected in Table I. Mean amplitudes (l_{ij}) for the SiH and SiC distances were obtained from the effective amplitudes (l_{ij}^{eff}) after the correction[28] $l_{ij} = l_{ij}^{eff} - \Delta$. The values for Δ are also listed in Table I. The shrinkage effect has not been taken into consideration. It is unlikely that the inclusion of shrinkage would change the results substantially.

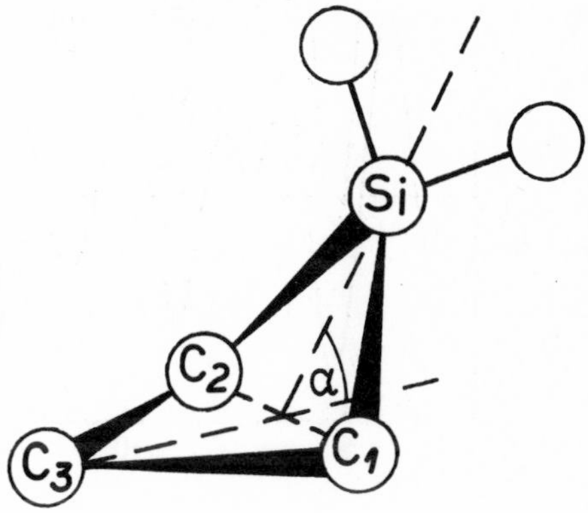

Fig. 18-3. Geometrical model of 1-silacyclobutane (SCB). Hydrogen atoms of the methylene groups are omitted.

18.4.2. 1,1-Dichloro-1-Silacyclobutane[27]

The experimental intensity function is shown in Fig. 4. In the analysis it was assumed that the main parameters of DCSB were the same as for SCB. The similarity of their structures is affirmed by the position of a peak which corresponds to the nonbonded Si···C and C···C distances: 2.40 Å in SCB and 2.39 Å in DCSB.

The experimental RDC contains two big peaks (Fig. 5). The first one is composed of the Si-C and Si-Cl distance contributions. The second one is more complex, containing Cl···Cl, C···Cl, and Cl···H distance contributions. The SiC and SiCl bond lengths and their mean amplitudes of vibration were determined by least-squares calculation on the experimental RDC. The principal parameters producing the second peak, namely the ClSiCl angle and the α angle, were obtained from an analysis of the experimental intensity function by trial and error. Structural results for DCSB are listed in Table II.

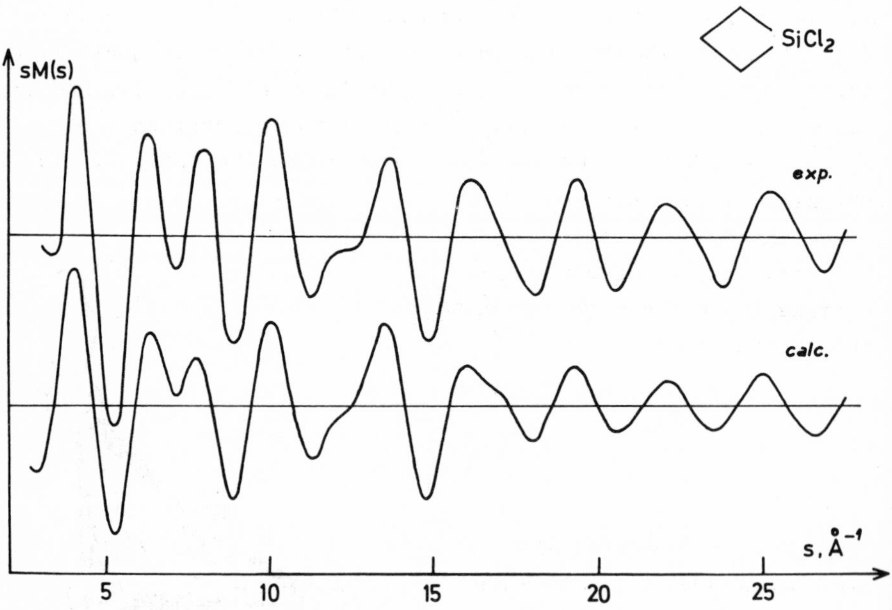

Fig. 18-4. Comparison of experimental and calculated *sM*(*s*) curves.

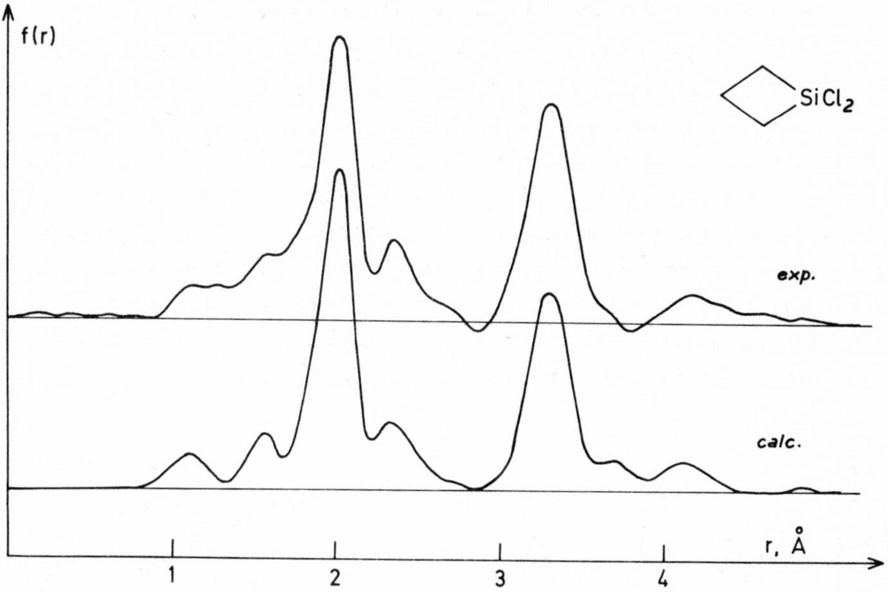

Fig. 18-5. Experimental and calculated radial distribution curves.

Table 18-II. Molecular parameters for 1,1-dichloro-1-silacyclobutane.*

	r (Å) or angle	l (Å)
CC	1.59 ± 0.03	0.06 (assumed)
SiC	1.88_2 ± 0.02	0.06_3 ± 0.03[a]
SiCl	2.050 ± 0.010	0.05 ± 0.01
ClSiCl	105° ± 1°	
CSiC	80° ± 2°	
α	30° ± 5°	

* Assumed: r(CH) = 1.13 Å, l(CH) = 0.08 Å, HCH = 110°.

[a] Δ = 0.007 Å.

18.4.3. 1,1,3,3-Tetrachloro-1,3-Disilacyclobutane[29]

The experimental intensity function is compared with the intensity function calculated for the final model in Fig. 6 (Curve 1). The experimental and theoretical RD curves are shown in Fig. 7. The procedure for obtaining the structural parameters was identical to that of DSCB. The SiC and SiCl bond lengths were deduced by fitting the experimental RDC with Gaussian peaks. The remaining parameters were derived from those of RDC and molecular intensity functions by trial and error. The range of the variation is shown in parentheses for each parameter: CSiC (75°-95°), ClSiCl (102°-107°), and α(0°-30°). The theoretical curves $sM(s)$ and $f(r)$ are found to be very

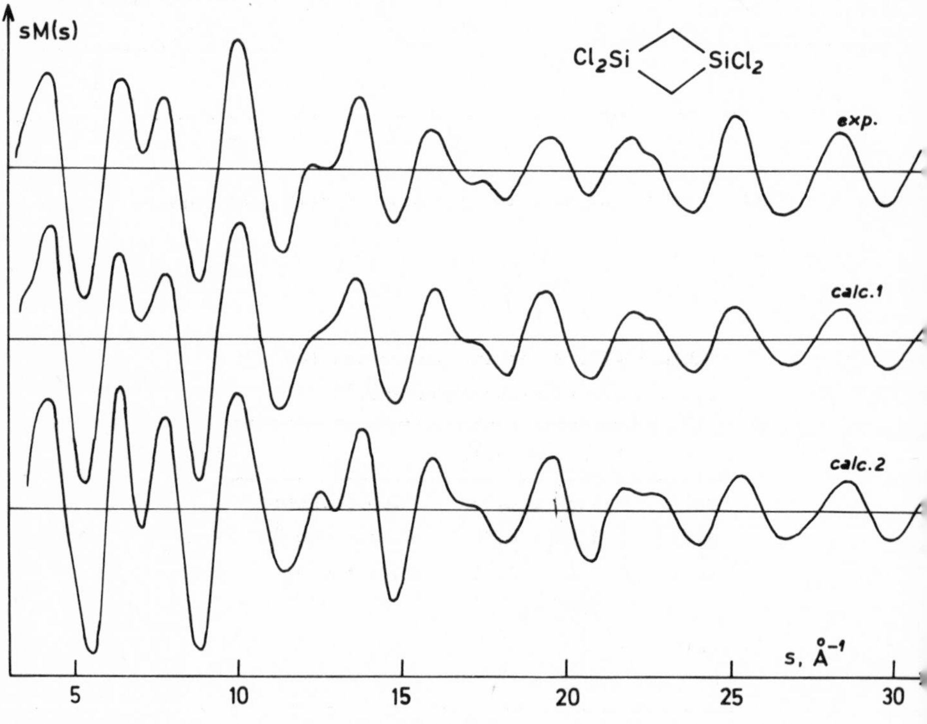

Fig. 18-6. Comparison of experimental and calculated $sM(s)$ curves. Curve 1 corresponds to the final model. Curve 2 corresponds to the rejected planar ring model with the parameters taken from the experimental radial distribution curve.

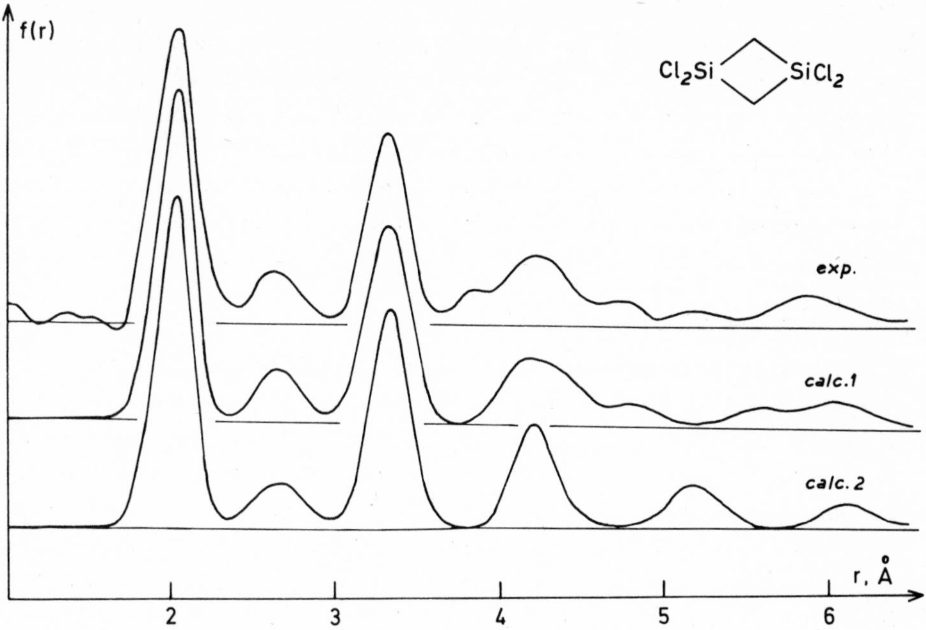

Fig. 18-7. Experimental and calculated radial distribution curves. Numbering of curves as in Fig. 6.

Table 18-III. Molecular parameters of 1,1,3,3-tetrachloro-1,3-disilacyclobutane.*

	r (Å) or angle	*l* (Å)
SiC	$1.89_5 \pm 0.01$	$0.05_3 \pm 0.01$[a]
SiCl	2.048 ± 0.005	$0.04_8 \pm 0.01$
ClSiCl	$104^\circ \pm 2^\circ$	
CSiC	$89^\circ \pm 1^\circ$	
α	$14^\circ \pm 3^\circ$	

* Assumed: *r*(CH) = 1.13 Å, *l*(CH) = 0.08 Å, HCH = 110°.

[a] Δ = 0.007 Å.

sensitive to small variations (1° - 2°) of these parameters. This feature facilitated an accurate determination of these parameters although the number of models analysed reached 35. As in the two preceding molecules the experimental results showed conclusively that the molecule ring is nonplanar. In Figs. 6 and 7 curve 2 corresponds to the planar ring model with main parameters taken from the experimental RDC which was rejected. The final molecular parameters are listed in Table III.

18.4.4. <u>4-Sila-3,3-Spiroheptane</u>

The experimental $sM(s)$ and $f(r)$ functions are compared in Figures 8 and 9 with their theoretical functions calculated for the parameters collected in Table IV. It is also seen that the experimental RDC of SSH has a great resemblance to that of SCB (Fig. 2) though there are two special features to be noted. A higher accurac in the determination of the CC internuclear distance from the singl peak position is due to the absence of SiH bonds in the SSH molecul In addition, a more exact determination of the CSiC bond angle in SSH is favoured by the symmetry of this molecule. The amplitudes o vibration are at present in the state of refinement.

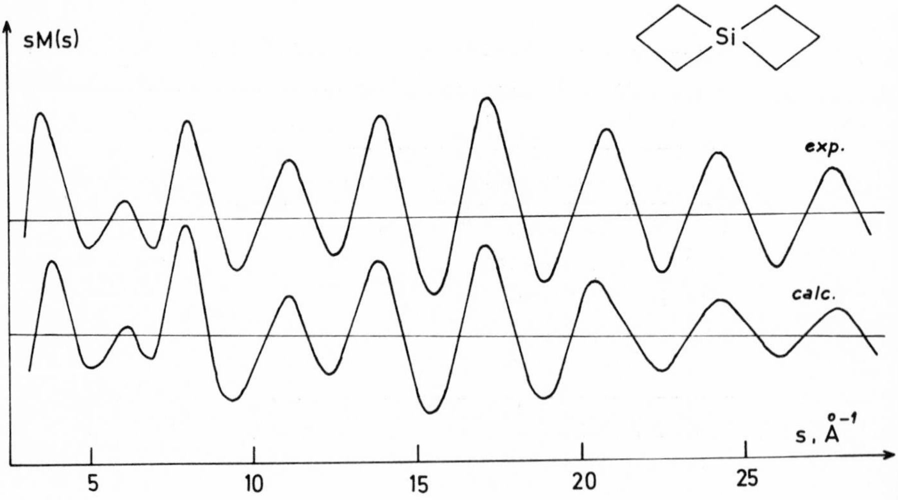

Fig. 18-8. Comparison of experimental and calculated $sM(s)$ curves.

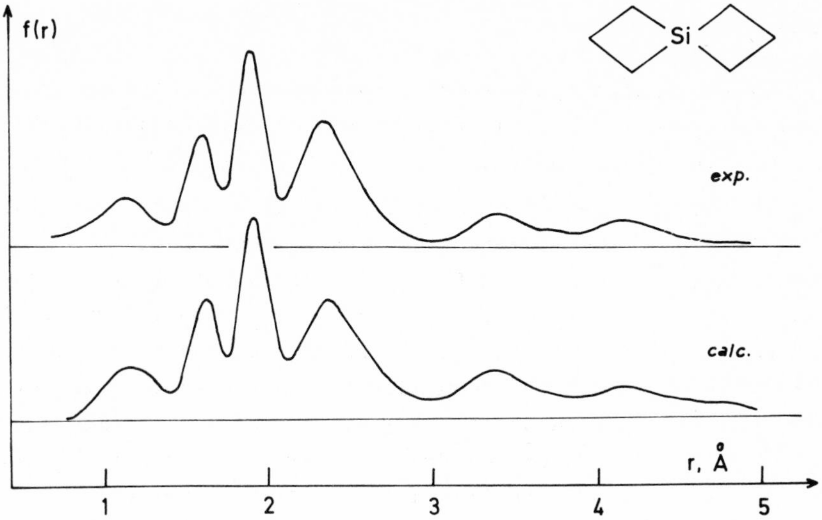

Fig. 18-9. Experimental and calculated radial distribution curves.

Table 18-IV. Molecular parameters (bond distances in Å) for the investigated molecules.

	SiH_2 SCB	$SiCl_2$ DCSB	Cl_2Si $SiCl_2$ TCDB	Si SSH
CH	1.13 ± 0.03	1.13 (assumed)	1.13 (assumed)	1.14 ± 0.02
SiC	$1.89_7 \pm 0.01$	$1.88_2 \pm 0.02$	$1.89_5 \pm 0.01$	1.90 ± 0.01
CC	$1.58_5 \pm 0.01$	1.59 ± 0.03	-	$1.59_6 \pm 0.01$
SiCl	-	2.050± 0.010	2.048± 0.005	-
CSiC	$80^\circ \pm 2^\circ$	$80^\circ \pm 2^\circ$	$89^\circ \pm 1^\circ$	$83^\circ \pm 0.5^\circ$
SiCC	$86^\circ \pm 2^\circ$	$86^\circ \pm 2^\circ$	-	$86^\circ 17' \pm 0.5^\circ$
SiCSi	-	-	$90^\circ 6' \pm 1^\circ$	-
CCC	$100^\circ 22' \pm 2^\circ$	$99^\circ 4' \pm 2^\circ$	-	$104^\circ 26' \pm 0.5^\circ$
ClSiCl	-	$105^\circ \pm 1^\circ$	$104^\circ \pm 2^\circ$	-
α	$30^\circ \pm 5^\circ$	$30^\circ \pm 5^\circ$	$14^\circ \pm 3^\circ$	$30^\circ \pm 2.5^\circ$

18.5. Discussion

All the principal parameters of the studied molecules are summarized in Table IV. In the following they will be discussed according to the scheme outlined in Section 18.1.

Ring Configuration. The general conclusion derived from this study is that the four-membered ring is nonplanar in all the inves gated molecules. It means that the forces for a planar cycle tendi to stagger the methylene and SiX_2 (X = H, Cl) groups exceed the fo ces tending to oppose a further decrease of the bond angles of the ring. The dihedral angle is presumably the parameter of greatest interest. The obtained value of $30^\circ \pm 5^\circ$ in SCB may be compared wi the value* of $35.9 \pm 2^\circ$ derived from the far-infrared investigatio and with the value of 28° estimated in a microwave research.[31] Th ring of SCB was also shown to be puckered by observing both axial and equatorial isomers.[32] Another recent spectroscopic investigati [V.D.Oppenheim, thesis, Academy of Sciences USSR, 1970] further co firms the ring nonplanarity of TCDB. The degree of puckering in SC is practically the same as in cyclobutane and its derivatives, for which the experimental data are available:** C_4H_8 $\alpha = 35^\circ$ [P.N. Skancke, thesis, NTH, Trondheim 1960] and $\alpha = 37^\circ$,[7] C_4H_7Br $\alpha = 29^\circ\ 22'$,[8] C_4H_7-C_4H_7 $\alpha = 33^\circ$.[33] This may seem interesting and unexpected. If one supposes a ring strain in cyclobutane equal or nearly equal to that of SCB one would expect the SCB ring to be le puckered due to release of the torsion which tends to keep the rin planar. The torsion contribution might be described using the etha potential which has a barrier to internal rotation of 2.9 kcal/mol combined with the torsion potential of methylsilane which has a ba rier of 1.7 kcal/mole.[36] However, the data of Table IV hardly justi such an approach. The nonplanarity of SCB remains about the same a in cyclobutane, although the Si-containing ring exhibits a pronoun inequality in the ring angles CSiC, SiCC, and CCC being 80°, 86°, and 100°, respectively. TCDB is the only molecule studied where al

* This value is based on an estimate of the reduced mass of this molecule.

** See also the excellent work of Bastiansen, Seip, and Boggs[34] for an illustration of conformational characteristics of four-membered rings.

the ring angles are very close to 90° as in cyclobutane, and it seems therefore reasonable to assume an equality of strain in these two molecules. The dihedral angle of 14° in TCDB may accordingly be rationalized in terms of a decrease in the torsional contribution.

This consideration demonstrates the strong relation between the degree of ring puckering and the values of the internal angles in four-membered rings. In this connection it may be of interest to discuss the ring configurations of three similar molecules I, II, and III:

$\alpha = 35^{\circ}$

I

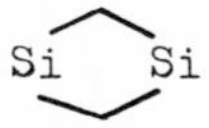

$\alpha = 14^{\circ}$

II

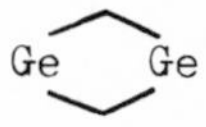

$\alpha = ?$

III

If the ring strains in I to III were equal, the ring III would be planar owing to the decrease of the torsional contribution. However, the infrared and Raman spectra for III indicate a strong violation of the rule of mutual exclusion. It means that this molecule has no center of symmetry, i.e. the ring is not planar [V.T.Aleksanyan, thesis, Academy of Sciences USSR, 1970]. Since the ring tends to become puckered as a consequence of a different strain distribution, one would be led to the conclusion that the ring III is rhomboid in shape.

In electron-diffraction studies it is extremely difficult to distinguish between (a) a really nonplanar ring (static nonplanarity) and (b) a planar ring with a large amplitude of out-of-plane bending (dynamic nonplanarity). This question in any particular case is probably best resolved by an investigation of the nature of the potential function which governs the ring-puckering vibration. A series of investigations along this line has been carried out using electron-diffraction techniques, though the significance and error limits of the determined parameters often remain questionable. The ring puckering motion has been favourably studied by microwave techniques[1] and also by far-infrared investigations.[30]

Endocyclic Internuclear Distances. For several decades the accepted reference distance for the CC single bond has been 1.544 Å, the distance occuring in diamond. However, the n-hydrocarbon CC distance of 1.533 Å seems to be a more suitable reference for saturat molecules.[37] Taking this into account, the distances of 1.558Å [14] a 1.555 Å [10] reported for the carbon ring seem to be longer than the 'standard' distance. Our distance of 1.59 ± 0.02 Å determined for three of the molecules is still longer. This result seems consiste with the value of 1.566 Å in cyclobutene[38] and of 1.583 Å in 3,4-d chlorocyclobutene-1.[39] The longest CC distance of 1.63 ± 0.02 Å ha been reported for perfluorocyclobutane.[11] The SiC distances (Table IV) also appear to be longer than the distance of 1.860 Å reported in dimethylsilane[40] and of 1.873 Å reported in disilylmethane.[41]

Summing up the results of this part one may say that the pre sent analysis provides a further example of a lengthening of the endocyclic bonds in four-membered ring systems in comparison with the corresponding bond distances in unstrained molecules.

Exocyclic Internuclear Distances. Accumulated evidence point to considerable delocalization of the electrons in small ring systems. Unsaturated side-chains seem to interact appreciably both wi three- and four-membered rings. The structural data available for exocyclic bond distances of cyclopropane derivatives agree well wi a π-character of cyclopropane (Table V). In contrast to this it is impossible to draw the same conclusion for cyclobutane derivatives In Table VI the C-Hal bond distances for some cyclobutane derivatives are collected; they are compared with selected bond lengths unsaturated molecules. It is not surprising that the SiCl bond dis tance of about 2.050 Å found in DCSB and TCDB agrees well with the value of 2.0479 Å reported for SiH_3Cl.[45] However, it seems a littl longer than the estimated value of 2.034 Å, which one would expect for the dichlorosilane molecule.[46] The 1.485 Å value found in this investigation for the SiH distance is in fair accordance with the previously reported valu in silyl groups.

Table 18-V. Comparison of CCl bond distances.

Molecule	C-Cl (Å)	Ref.
C_2H_5Cl	1.766 ± 0.010	42
$CCl_2=CCl_2$	1.718 ± 0.003	43
 CCl_2	1.734 ± 0.002	44

Table 18-VI. Comparison of C-Halogen bond distances.

Molecule	C-Hal (Å)	Ref.	Average value of r(C-Hal) in Å; Ref. 42
C_4F_8	1.332 ± 0.005	11	1.333 ± 0.005
C_4H_7Cl	1.775 ± 0.005	9	1.767 ± 0.005
$C_4H_6Br_2$	1.946	10	1.938 ± 0.005

External Valence Angles. There seems to exist conclusive evidence that the external valence angles in small ring derivatives are larger than the tetrahedral angle. For cyclopropane derivatives the external valence angles range from 114° to 118°; for cyclobutane derivatives they range from 110° to 114°. The ClSiCl angle value of about 105° seems somewhat unexpected, although it is well known that electronegative substituents tend to decrease the valence angles. A considerably smaller decrease is reported for perfluoro compounds. The FCF angle is reported to be 112.0° in C_3F_6 [J.F. Chiang and W.A.Bernett, 3rd Austin Symposium on Gas Phase Molecular Structure, 1970], and 108° in C_4F_8.[11]

On the basis of the present data one is led to the conclusion that the silicon atom in silacyclobutanes has a distorted tetrahedral surrounding with the valence angles CSiC of 84° ± 4° and ClSiCl of 104° ± 1°. This result is different from the features of CH_2Cl_2, where all valence angles were found to be about 112° (cited in Ref. 44).

Acknowledgments: This work was inspired by Dr. V.M.Vdovin, and we want to thank him for continuous interest. We are greatly indebted to Professor O. Bastiansen, who has read this manuscript and has made many valuable comments. We are grateful to Dr. W.C.Pringle,Jr., who has kindly made his results of a microwave investigation available prior to publication, and to Miss Yu.V.Baurova for having performed the calculations for SCB and DCSB.

REFERENCES

1 S.I.Chan, J.Zinn, J.Fernandez, and W.D.Gwinn, J.Chem.Phys. 33,

1643 (1960).
2 S.I.Chan, J.Zinn, and W.D.Gwinn, J.Chem.Phys. 34, 1319 (1964).
3 A.Bauder, F.Tank, and H.H.Günthard, Helv.Chim.Acta 46, 1453 (196?).
4 N.Kwak, J.H.Goldstein, and J.W.Simmons, J.Chem.Phys. 25, 1203 (1956).
5 J.D.Dunitz and V.Schomaker, J.Chem.Phys.20, 1703 (1952).
6 A.Almenningen, O.Bastiansen, and P.N.Skancke, Acta Chem.Scand. 15, 711 (1961).
7 D.A.Dows and N.Rich, J.Chem.Phys. 47, 333 (1967).
8 W.G.Rothschild and B.P.Dailey, J.Chem.Phys. 36, 2931 (1962).
9 H.Kim and W.D.Gwinn, J.Chem.Phys. 44, 865 (1966).
10 A.Almenningen, O.Bastiansen, and L.Walløe, Selected Topics in Structure Chemistry (Ed. P.Andersen, O.Bastiansen, and S.Furberg p. 91, Universitetsforlaget, Oslo 1967.
11 N.V.Alekseev, I.A.Ronova, and P.P.Barzdain, Zh.Strukt.Khim. 9, 1073 (1968).
12 E.Adman and T.N.Margulis, J.Am.Chem.Soc. 90, 4517 (1968).
13 K.B.Wiberg and G.M.Lampman, J.Am.Chem.Soc. 88, 4429 (1966).
14 R.C.Lord and B.P.Stoicheff, Can.J.Phys. 40, 725 (1962).
15 A.D.Walsh, Nature 159, 165 (1947).
16 A.D.Walsh, Nature 159, 712 (1947).
17 A.D.Walsh, Trans. Faraday Soc. 45, 179 (1949).
18 L.H.Sommer and G.A.Baum, J.Am.Chem.Soc. 76, 5002 (1954).
19 V.M.Vdovin, N.S.Nametkin, and P.L.Grinberg, Dokl.Akad.Nauk SSSR 150, 799 (1963).
20 N.S.Nametkin, V.M.Vdovin, and P.L.Grinberg, Dokl.Akad.Nauk SSSR 155, 849 (1964).
21 N.S.Nametkin, V.M.Vdovin, and V.I.Zavialov, Izv.Akad.Nauk SSSR Ser.Khim., 202 (1964).
22 N.S.Nametkin, V.M.Vdovin, and P.L.Grinberg, Izv.Akad.Nauk SSSR Ser.Khim., 1133 (1964).
23 N.S.Nametkin, L.E.Guselnikov, V.M.Vdovin, P.L.Grinberg, V.I. Zavialov, and V.D.Oppenheim, Dokl.Akad.Nauk SSSR 171, 630 (1966)
24 J.Laane, J.Am.Chem.Soc. 89, 1144 (1967).
25 L.V.Vilkov, V.S.Mastryukov, and P.A.Akishin, Zh.Strukt.Khim. 4, 323 (1964).

26 R.A.Bonham and L.S.Bartell, J.Chem.Phys. 31, 702 (1959).

27 L.V.Vilkov, V.S.Mastryukov, Yu.V.Baurova, V.M.Vdovin, and P.L. Grinberg, Dokl.Akad.Nauk SSSR 177, 1084 (1967).

28 R.A.Bonham and T.Ukaji, J.Chem.Phys. 36, 72 (1962).

29 L.V.Vilkov, M.M.Kusakov, N.S.Nametkin, and V.D.Oppenheim, Dokl. Akad.Nauk SSSR 183, 830 (1968).

30 J.Laane and R.C.Lord, J.Chem.Phys. 48, 1508 (1968).

31 W.C.Pringle,Jr., J.Chem.Phys. 54, 4979 (1971).

32 V.T.Aleksanyan, G.M.Kuzianz, V.M.Vdovin, P.L.Grinberg, and O.V. Kuzmin, Zh.Strukt.Khim. 10, 481 (1969).

33 O.Bastiansen and A. de Meijere, Angew.Chem. 78, 142 (1966).

34 O.Bastiansen, H.M.Seip, and J.E.Boggs, Perspectives in Structural Chemistry Vol. 4 (Ed. J.D.Dunitz and J.A.Ibers), p. 60, Wiley, New York 1971.

35 S.Weiss and G.E.Leroi, J.Chem.Phys. 48, 962 (1968).

36 R.W.Kilb and L.Pierce, J.Chem.Phys. 27, 108 (1957).

37 L.S.Bartell, J.Am.Chem.Soc. 81, 3497 (1959).

38 B.Bak, J.T.Led, L.Nygaard, J.Rastrup-Andersen, and G.O.Sørensen, J.Mol.Structure 3, 369 (1969).

39 O.Bastiansen and J.L.Derissen, Acta Chem.Scand. 20, 1089 (1966).

40 A.C.Bond and L.O.Brockway, J.Am.Chem.Soc. 76, 3312 (1954).

41 A.Almenningen, H.M.Seip, and R.Seip, Acta Chem.Scand. 24, 1697 (1970).

42 L.E.Sutton: Tables of Interatomic Distances and Configuration in Molecules and Ions, Supplement 1956-1959, London 1965.

43 T.G.Strand, Acta Chem.Scand. 21, 2111 (1967).

44 W.H.Flygare, A.Narath, and W.D.Gwinn, J.Chem.Phys. 36, 200 (1962).

45 B.Bak, J.Bruhn, and J.Rastrup-Andersen, Acta Chem.Scand. 8, 367 (1954).

46 L.V.Vilkov and V.S.Mastryukov, Dokl.Akad.Nauk SSSR 162, 1306 (1965).

Electron Diffraction Studies - Part II: 3-Chloro-1-Propanol

O. BASTIANSEN, J. BRUNVOLL, and I. HARGITTAI

An electron-diffraction study of 3-chloro-1-propanol has been carried out. Five different conformers have been considered, but only three of them have revealed their presence. Generally *gauche* positions seem to be preferred. The predominating conformer has the strainless *gauche-gauche* form. The possibility of intramolecular hydrogen bond formation has been discussed.

18.6. Introduction

Several rotational structures seem possible for 3-chloro-1-propanol in the gas phase. Probable directions of the bonds in the Cl-C_3-C_2-C_1-O chain are:

(1) The Cl-C_3 bond is <u>anti</u> to the C_1-C_2 bond, and the O-C_1 bond <u>anti</u> to the C_2-C_3 bond. (All <u>anti</u>)

(2) The Cl-C_3 bond is rotated with the C_2-C_3 bond as rotational axis in such a way that the Cl-C_3 bond is in <u>gauche</u> position to the C_1-C_2 bond. (Cl_gO_a)

(3) Cl-C_3 is <u>anti</u> to the C_1-C_2 bond, O-C_1 is <u>gauche</u> to the C_2-C bond. (Cl_aO_g)

(4) Cl-C_3 is <u>gauche</u> to the C_1-C_2 bond, O-C_1 <u>gauche</u> to the C_2-C_3 bond. The OH and the Cl are on different sides of the $C_1C_2C_3$ plane forming a long Cl···O distance. $(Cl_gO_g)_l$

(5) Cl-C_3 is <u>gauche</u> to C_1-C_2 and O-C_1 <u>gauche</u> to C_2-C_3. The OH and the Cl are on the same side of the $C_1C_2C_3$ plane, forming a short Cl···O distance. $(Cl_gO_g)_{sh}$

All these forms except the all <u>anti</u> exist in mirror image pairs.

In the above list of conformers the possibility of intramolecular hydrogen bonding has been ignored. Since intramolecular hydrogen bonding is known in ethylene chlorohydrin[1-3] it would be natural to consider this possibility also in 3-chloro-1-propanol.

In the present investigation the electron-diffraction method is used in an attempt to determine which conformers are present, and if possible also their relative abundance.

18.7. Experiment and Method of Calculation

The Oslo diffraction apparatus was used to obtain the intensity data. Four diffraction photographs were taken at each of the two distances between nozzle and photographic plate of approximately 48 cm and approximately 20 cm.

The data were treated in the usual way.[4,5] The experimental intensities were modified by $s/|f_C(s)|^2$, where $f_C(s)$ is the scattering amplitude for carbon. The experimental background was subtracted for each plate, and the average of the resulting intensity curve taken for each distance. A new intensity curve was found by combining the 48 cm and the 20 cm curve. This was transformed to a radial distribution curve, giving the first approximate molecular parameters. Preliminary theoretical intensity curves were calculated and were used to obtain a better background.

For a distance between atom i and atom j the contribution to the theoretical intensity curve was computed from*

$$I_{ij}(s)=\text{const}\cdot|f_i|\cdot|f_j|/|f_C|^2\cos(\eta_i-\eta_j)\exp(-\tfrac{1}{2}u_{ij}^2s^2)\sin(R_{ij}s)/R_{ij} \quad .$$

18.8. Determination of Molecular Structure

In the calculations it was assumed that the two C-C bonds are equal, and that all C-H bonds are equal. The assumptions for the bond angles were: $\angle\, \text{ClC}_3\text{H}_3 = \angle\, \text{ClC}_3\text{H}_3'$, $\angle\, \text{C}_i\text{C}_j\text{H}_j = \angle\, \text{C}_i\text{C}_j\text{H}_j'$, and

---

*The mean amplitudes of vibration are denoted u in this part (and in Chapter 11); the symbol corresponds to l used elsewhere in this book.

$\angle OC_1H_1 = \angle OC_1H_1'$. The same bond angles were used for all the forms (1) - (5) mentioned in the introduction. Shrinkage effects were not taken into consideration.

In addition to the bond distances, the bond angles, the torsional angles (rotational axes: the C_1-C_2, C_2-C_3 and C-O bonds) and amounts of the different forms were used as variables.

The part of the radial distribution curve with r-values shorter than 1.94 Å is built up by contributions from bond distances. The radial distribution curve from r = 0.82 Å to r = 1.94 Å wa transformed back to intensity form. A theoretical curve was fitted to this curve by the least-squares method, giving the following preliminary values for the bond distances and mean amplitudes of vibration:

$$R_{Cl-C} = 1.8082\ \text{Å}, \quad u_{Cl-C} = 0.0523\ \text{Å},$$
$$R_{C-C} = 1.5286\ \text{Å}, \quad u_{C-C} = 0.0470\ \text{Å},$$
$$R_{C-O} = 1.4189\ \text{Å}, \quad u_{C-O} = 0.0463\ \text{Å},$$
$$R_{C-H} = 1.1131\ \text{Å}, \quad u_{C-H} = 0.0724\ \text{Å},$$
$$R_{O-H} = 0.9688\ \text{Å}, \quad u_{O-H} = 0.0325\ \text{Å}.$$

The undamped radial distribution curve has a maximum at about 2.15 Å. The main contribution to this peak comes from H···C and H···O over one angle. An attempt was made to deduce these distances directly from the radial distribution curve by subtracting a preliminary theoretical curve built up by the other distances contri buting to the r range around 2.15 Å. But this approach failed.

The part of the radial distribution curve that did not contai contributions from bond distances was then converted to intensity form, and as many angles as possible determined by a least-squares refinement. The result was (standard deviations in parentheses): $\angle$ ClCC = 112.9° (0.25°), $\angle$ CCC = 113.4° (1.18°), $\angle$ CCO = 106.5° (0.43°), the Cl-C_3 bond for $(Cl_gO_g)_1$ is rotated 123.6° (1.38°) from the <u>anti</u> position, and the O-C_1 bond 104.3° (1.94°) from the <u>anti</u> position.

It was not possible to refine the mean amplitudes of vibration for the corresponding distances. The values used were obtained from approximate spectroscopic computations. Since no force constant determination exists for 3-chloro-1-propanol the necessary data wer obtained by combining information from spectroscopic studies of propanol[6] and ethylene chlorohydrin.[7]

The 48 cm intensity curve, the 20 cm curve, and the curve that was a combination of the two, were each analysed in the following way. The bond distances were determined by a least-squares refinement, keeping the other parameters constant. The bond distance u-values used were those given above except for u_{O-H} that seems unreasonably small. For the combined intensity curve least-squares refinements were further carried out varying both bond distances and u values. These refinements lead to reasonable values for u_{C-C}, u_{C-O}, and u_{C-Cl}, while u_{O-H} was still unreasonable (Table VII).

Keeping the refined bond distances and bond u-values constant (u_{O-H} put equal to the spectroscopical value of 0.07 Å) the valence angles and torsional angles were refined. This kind of refinement was carried out a series of times, systematically varying the amounts of the different conformers. At a certain state a new set of refinements were carried through, this time keeping all geometrical parameters and the u-values constant and refining the relative amounts of the conformers. This procedure was repeated several times. The final results are given in columns I, IV, and VII of Table VII.

Fig. 10 shows the theoretical and experimental intensity curves for the 48 cm and the 20 cm distance. The theoretical curves were calculated using the parameters of Table VII columns I and IV respectively. In Fig. 11 the combined intensity curve is compared to the theoretical curve of each of the three rotational forms that seem to be present (column VII). A comparison has also been made with the theoretical curve of the found conformational mixture. The corresponding radial distribution curves are given in Fig. 12.

The standard deviations of the final parameters were found allowing all the determined parameters to change simultaneously. The standard deviations for the amounts of the different forms were found in a separate run. The results obtained using diagonal weight matrix are given in columns II, V, and VIII of Table VII.

Correlation of the intensity values was taken into account in the standard deviation computations in a way given by Seip, Strand, and Stølevik.[8] This gave off-diagonal weight matrix elements p_2 = -0.618 and p_3 = 0.152 for the 48 cm curve. For the 20 cm curve the values obtained were p_2 = -0.495 and p_3 = 0.046. Standard deviations

Table 18-VII. Results of least-squares refinements for 3-chloro-1-propanol from electron diffraction data.*

	I	II	III	IV	V	VI	VII	VIII
R(Cl-C)	1.8016	0.0024	0.0031	1.8035	0.0021	0.0025	1.8062	0.0016
u(Cl-C)							0.0491	0.0019
R(C-C)	1.5224	0.0023	0.0029	1.5310	0.0020	0.0025	1.5306	0.0015
u(C-C)							0.0415	0.0023
R(C-O)	1.4280	0.0029	0.0036	1.4191	0.0027	0.0035	1.4215	0.0021
u(C-O)							0.0399	0.0032
R(O-H)	0.9509	0.0115	0.0152	0.9950	0.0168	0.0236	0.9756	0.0087
u(O-H)							0.0291	0.0113
R(C-H)	1.1107	0.0030	0.0040	1.1231	0.0042	0.0058	1.1219	0.0032
u(C-H)							0.0722	0.0040
∠ ClCC	112.9	0.43	0.56	112.9	0.70	0.65	112.4	0.25
∠ CCC	113.0	0.94	1.21	111.0	0.86	0.86	111.3	0.55
∠ CCO	106.4	0.37	0.49	106.0	0.51	0.55	106.1	0.32
(4) τ_{ClCCC}	120.0	3.1	3.9				120.8	1.6
(4) τ_{CCCO}	106.6	5.6	7.0				108.5	2.4
n_3	16.5	2.3	2.8				16.9	2.3
n_4	64.0	4.6	5.8				63.6	4.1
n_5	19.5	2.4	3.1				19.5	2.4

* Bond distances, mean amplitudes of vibration, and their standard deviations are given in Å units, angles in degrees, and amounts of different forms in % (n_3, n_4, n_5). The τ's give the torsional angles from *anti* form for $(Cl_gO_g)_1$. The values in different columns are:

I: 48 cm data, geometrical parameters and amounts of different forms.

II: 48 cm data, standard deviations with diagonal weight matrix.

III: 48 cm data, standard deviations, off-diagonal weight matrix elements $p_2 = -0.618$ and $p_3 = 0.152$.

IV: 20 cm data, geometrical parameters.

V: 20 cm data, standard deviations, diagonal weight matrix.

VI: 20 cm data, standard deviations, $p_2 = -0.495$, $p_3 = 0.046$.

VII: Parameters from combined intensity curve.

VIII: Standard deviations from combined curve, diagonal weight matrix.

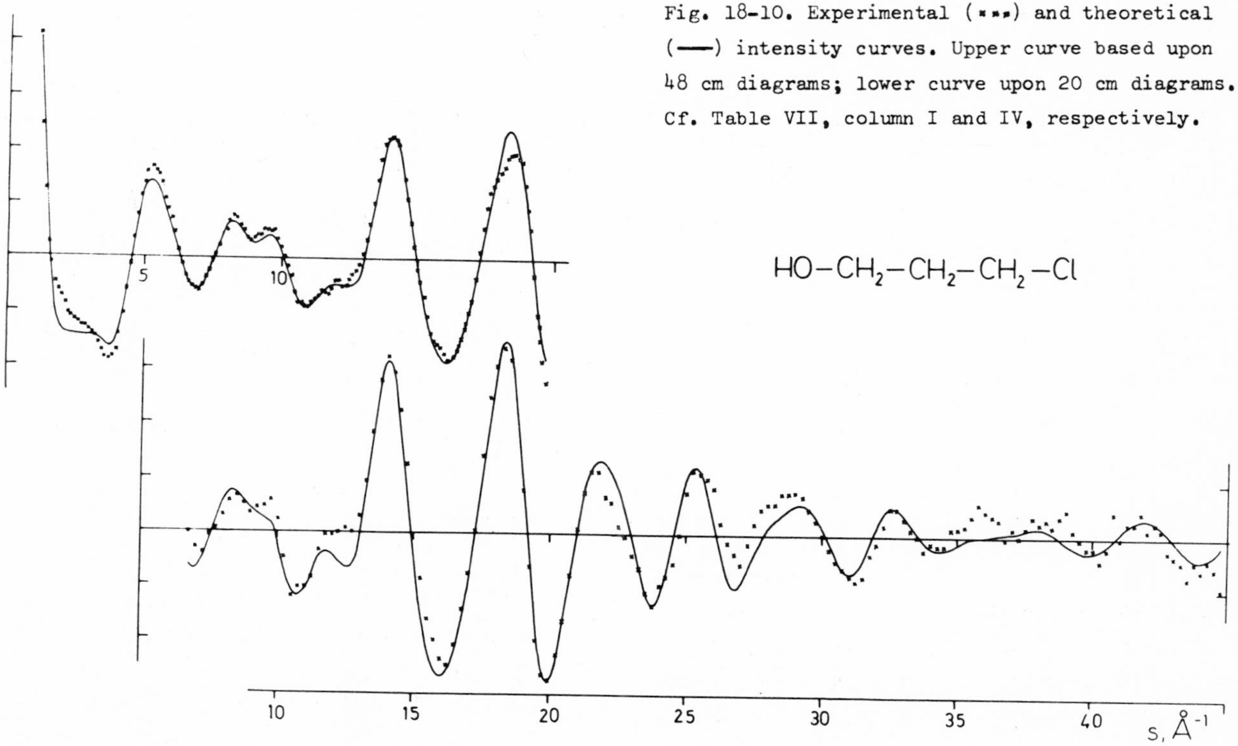

Fig. 18-10. Experimental (×××) and theoretical (——) intensity curves. Upper curve based upon 48 cm diagrams; lower curve upon 20 cm diagrams. Cf. Table VII, column I and IV, respectively.

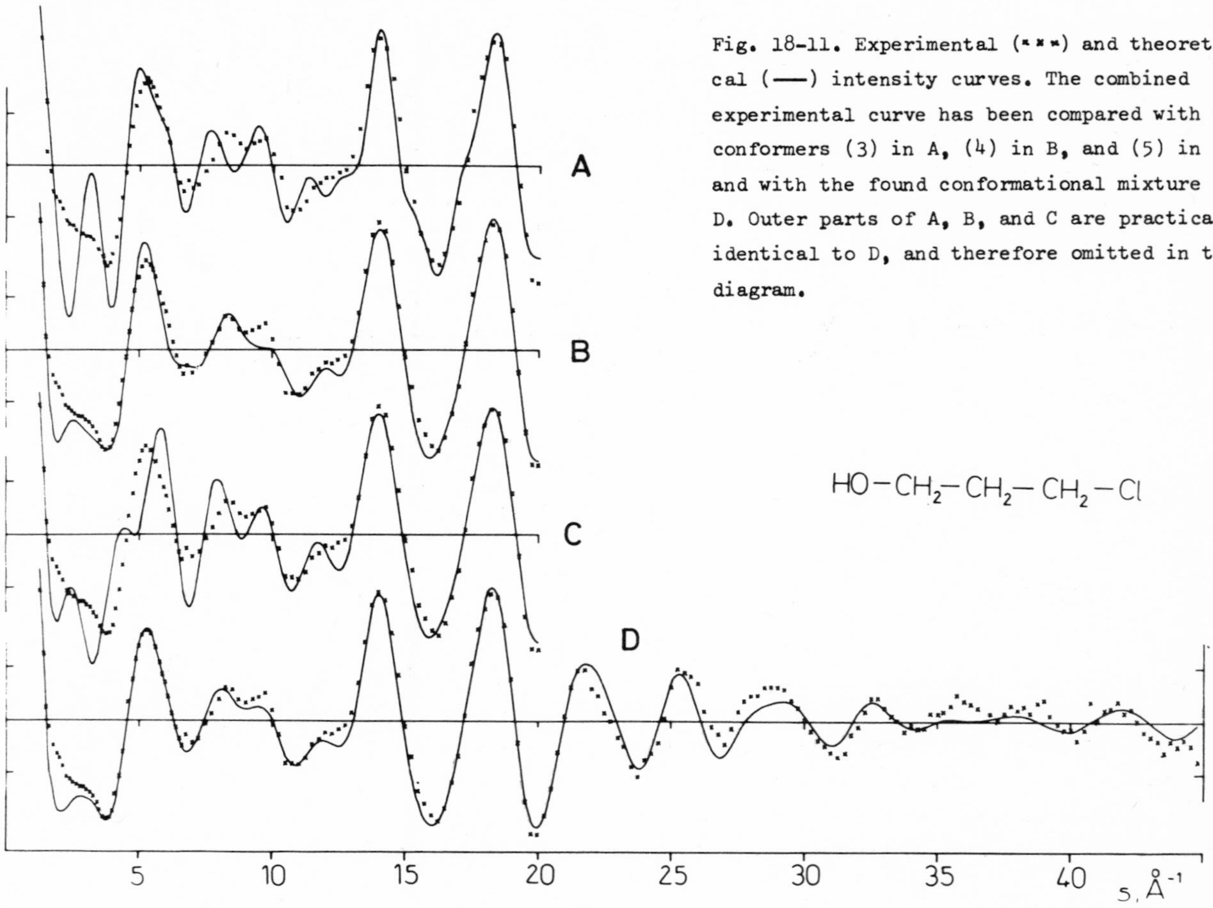

Fig. 18-11. Experimental (× × ×) and theoretical (——) intensity curves. The combined experimental curve has been compared with the conformers (3) in A, (4) in B, and (5) in C, and with the found conformational mixture in D. Outer parts of A, B, and C are practically identical to D, and therefore omitted in the diagram.

$HO-CH_2-CH_2-CH_2-Cl$

Fig. 18-12. Radial distribution curves corresponding to the intensity curves on Fig. 11. Damping factor k = 0.0015 $Å^2$. Inner parts of A, B, and C are practically identical to D, and therefore omitted in the diagram.

The full-drawn curve is experimental.

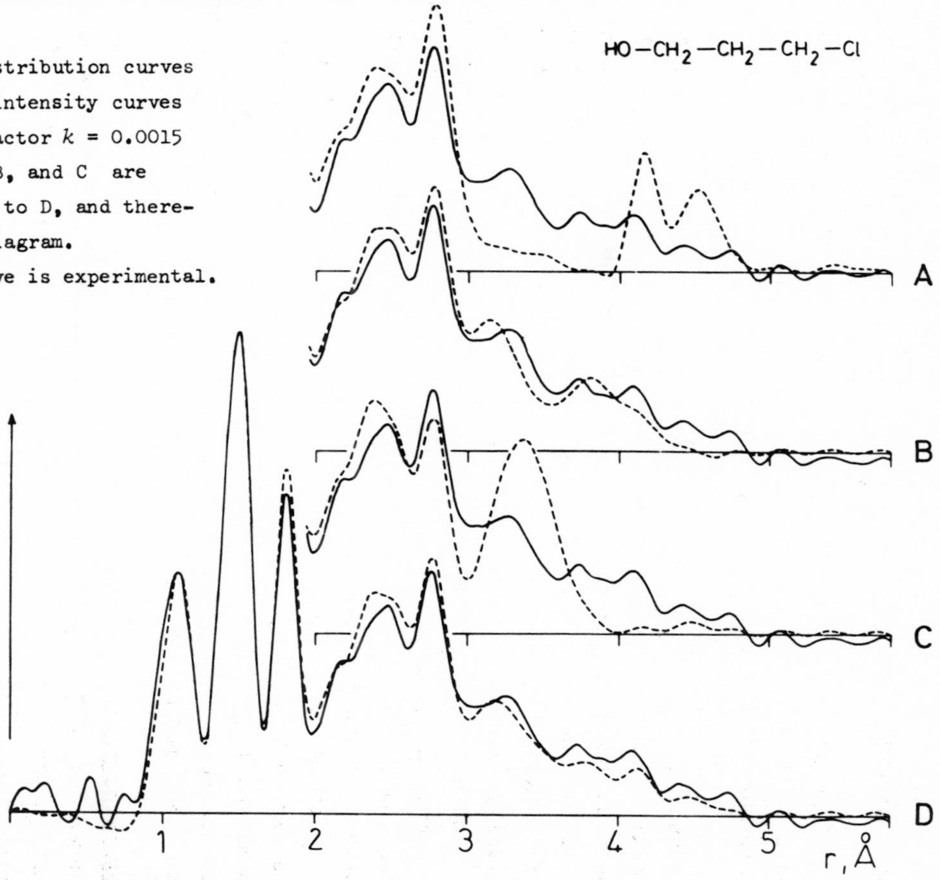

with these p-values included in the computations are given in columns III and VI of Table VII.

18.9. Discussion

The only conformer where a hydrogen bond between Cl and OH could be formed is conformer (5). The torsional angles for this conformer deviate considerably from the ideal 120° value. The 'ideal' form would be impossible for steric reasons. The O···Cl distance is found considerably longer than corresponding to the 'ideal' form, but the uncertainty for this distance is quite large. The best value (approximately 3.4 Å) is somewhat larger than the O···Cl distance found in the gauche conformer of ethylene chlorohydrin (approximately 3.15 Å). Whether a hydrogen bond formation actually takes place is perhaps an academic question. But in any case such a hydrogen bond does not lead to a predominating conformer. The energetic advantage of a hydrogen bond formation may have been compensated for by the considerable steric strain that conformer (5) has to suffer.

It is interesting to note that neither of the conformers with the oxygen bond in the anti position have been observed. The conformer having the chlorine atom in the anti position seems to contribute rather modestly, while the strainless gauche-gauche conformer (4) seems definitely to predominate.

Acknowledgments: The authors want to thank cand.real. Arne Almenningen for having taken the electron-diffraction diagrams. We also want to thank Professor Harold Hanson, the former Chairman of Department of Physics of The University of Texas at Austin. All the three authors enjoyed the best of working conditions and hospitality during a stay in Austin by invitation of Professor Hanson.

REFERENCES

1 O.Bastiansen: Om noen av de forhold som hindrer den fri dreibarhet om en enkeltbinding, A.Garnæs' boktr., Bergen 1948.
2 O.Bastiansen, Acta Chem.Scand. 3, 415 (1949).
3 A.Almenningen, O.Bastiansen, L.Fernholt, and K.Hedberg, Acta Chem.Scand. 25, 1946 (1971).

4 O.Bastiansen and P.N.Skancke, Advances Chem.Phys. 3, 323 (1960).

5 A.Almenningen, O.Bastiansen, A.Haaland, and H.M.Seip, Angew. Chem. 77, 877 (internat. Edit. 4, 819) (1965).

6 K.Fukushima and B.J.Zwolinski, J.Mol.Spectry. 26, 368 (1968).

7 E. Wyn-Jones and W.J. Orville-Thomas, J.Mol.Structure 1, 79 (1967-68).

8 H.M.Seip, T.G.Strand, and R.Stølevik, Chem.Phys.Letters 3, 617 (1969).

CHAPTER 19

Adamantane - Part I: Reinvestigation of the Molecular Structure by Electron Diffraction

ISTVÁN HARGITTAI and KENNETH HEDBERG

The results of a reinvestigation of the molecular structure of adamantane by electron diffraction are reported. They include r(C-C) 1.540 ± 0.002 Å, r(C-H) 1.112 ± 0.004 Å [both of them r_g(1) values], ∠C-C_{sec}-C 108.8 ± 1.0°, and ∠C-C_{ter}-C 109.8 ± 0.5° (the uncertainties shown are 2σ). The mean amplitude values obtained indicate a relatively stiff carbon skeleton.

19.1. Introduction

Adamantane, $C_{10}H_{16}$, has a carbon skeleton the shape of which may be imagined in terms of four fused cyclohexane-like carbon rin (Fig. 1); the configuration is the same as a portion of the diamond lattice. This unique structure has excited considerable interest. The inflexible character of the carbon skeleton and its cage-like form couple to produce unusual physical and chemical properties.[1,2]

Precise values for the bond distances and bond angles are of course important in the interpretations of the chemical properties and adamantane as a subject for study by gas electron diffraction is an especially favorable case because of the high symmetry of th molecule. Among the results to be expected is an accurate measure of the distance between a secondary and a tertiary carbon atom. An early electron diffraction study of the adamantane molecule geometry in the vapor phase was carried out by Nowacki and Hedberg[3]

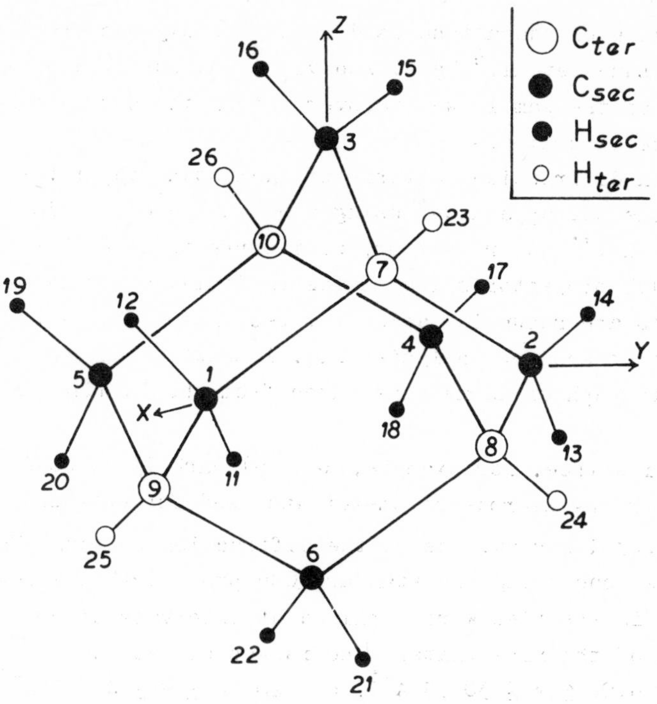

Fig. 19-1. Model of the adamantane molecule showing the atom numbering.

using the visual technique and resulted in the values 1.54 ± 0.01 Å for the C-C bond distance and 109.5 ± 1.5° for the C-C-C bond angles assuming T_d ($\bar{4}3m$) symmetry, C-H = 1.09 Å, and ∠HCH = 109°28'. It seemed probable that a reinvestigation of adamantane using modern diffraction techniques would lead to more accurate and more complete bond-distance and bond-angle determinations,[4,5] while at the same time yielding information about the amplitudes of vibration.* Accordingly, this work was initiated.

* A calculation of the mean amplitudes from spectroscopic data is presented in Part III. The final results from that calculation are included in Table II.

19.2. Experimental and Data Reduction

The sample of adamantane used in this study was prepared according to Landa et al.[6] by Z. Zubovicz[7] and purified by sublimation. Most of the sample was recovered from the nozzle cold trap after the experiments.

Electron-diffraction patterns of adamantane vapor were recorded with the OSU apparatus* using a nominal accelerating voltage of 44 kV,** an r^3 sector, beam currents of 0.25 - 0.30 µamp, and Kodak projector slide plates (medium). Other experimental conditions are shown in Table I. Three plates from each camera distance were chosen for analysis. Fig. 2 shows an electron-diffraction photograph of adamantane taken from the 'middle' camera distance.

A modified (see, for example, Ref. 8) Mark III C Joyce-Loebl Automatic Recording Microdensitometer was used for obtaining the optical density distributions of the diffraction pattern. The data reduction was done using the standard OSU procedure[8] and the total experimental intensities were obtained at intervals of $\Delta s = 0.25$ Å^{-1} for each of the nine plates. The ranges of data were $1.00 \leq s \leq 12.50$ Å^{-1}, $6.00 \leq s \leq 30.25$ Å^{-1}, and $20.50 \leq s \leq 45.75$ Å^{-1},

Table 19-I. Experimental conditions for the diffraction experiments.

	Nominal camera distance (cm)	Exposure (minutes)	Nozzle temperature (°C)
'long'	75	$\frac{3}{4} - 1\frac{1}{4}$	124 - 125
'middle'	30	1 - 3	126
'short'	12	5 - 6	135 - 141

* This apparatus will be described elsewhere.

** The electron wave length was calibrated in separate experiments against carbon dioxide using as the standard of length r_a(CO) = 1.1642 Å and r_a(O···O) = 2.3244 Å.

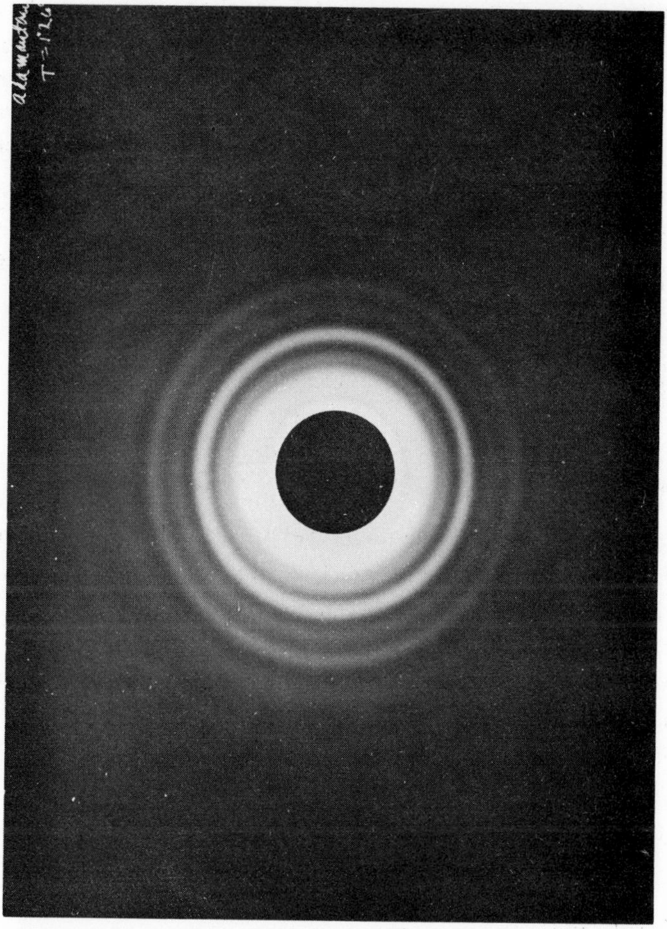

Fig. 19-2. Diffraction photograph (a negative print) of the scattering from adamantane taken at the 'middle' distance.

respectively [$s = (4\pi/\lambda)\sin\theta$ where λ is the electron wave length and 2θ is the scattering angle].

For drawing the first experimental backgrounds the theoretical molecular intensity curve calculated from the geometry deduced earlier[3] was especially useful. Backgrounds were fitted to the curves from the individual plates and during the structure refineme were slightly modified several times (at least twice for each indiv dual plate) in accordance with usual practice. After subtraction of the backgrounds the intensities were scaled and averaged to form a single composite curve. The data of the composite curve reflect the three camera distances as follows:

$1.00 \leq s \leq 6.25$ Å^{-1}	'long'
$6.50 \leq s \leq 12.25$ Å^{-1}	'long' and 'middle'
$12.50 \leq s \leq 21.75$ Å^{-1}	'middle'
$22.00 \leq s \leq 28.25$ Å^{-1}	'middle' and 'short'
$28.50 \leq s \leq 43.25$ Å^{-1}	'short'

The 'long' and 'middle' distances patterns in particular were of high quality. It was decided not to include the measurements beyond 43.25 Å^{-1} because of the obvious high noise level.

The experimental molecular intensities corresponding to the composite curve are shown in Fig. 3.* These values (as well as the corresponding theoretical ones from the final model) are in the form described by the equation

$$sI(s)=\sum_{i \neq j}\sum |f_i(s)||f_j(s)|\cos[\eta_i(s)-\eta_j(s)]\exp(-\tfrac{1}{2}l_{ij}^{\,2}s^2)\sin(r_{ij}s)/r_{ij} \tag{19.1}$$

where the $|f(s)|$ and $\eta(s)$ are the absolute values and phases of the complex electron scattering amplitudes, the r_{ij} are interatomic distances, and the l_{ij} are the corresponding mean amplitudes of vibration; the summation extends over all atomic pairs.

---

* It is very interesting to compare the experimental curve with that obtained by the visual technique. The agreement is excellent except for the shape of the maximum at about s = 9, which was discussed in detail in the account of the visual work.[3]

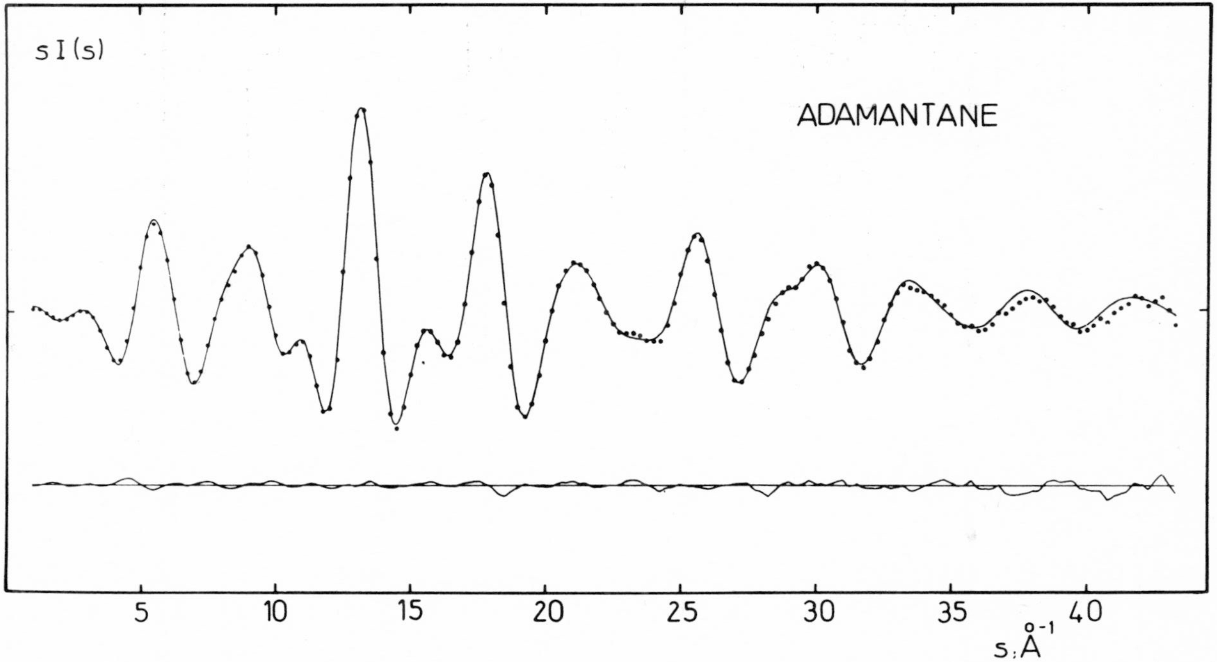

Fig. 19-3. Intensity curves for adamantane. The points represent the experimental curve and the solid line the theoretical curve. The difference (experimental minus theoretical) is shown below.

19.3. Structure Analysis

19.3.1. Experimental Radial Distribution

Radial distribution curves were calculated according to the formula

$$f(r) = \sum_{s=0.00}^{43.25} s\,I(s)\,\frac{Z_C^{\,2}}{|f_C(s)|^2}\exp(-a\,s^2)\sin(rs)\,\Delta s \qquad (19.2)$$

where Z_C and $|f_C(s)|$ refer to the carbon atom. For the experimental radial distribution curves either zeros or theoretical $sI(s)$ values were used for the unobserved range $0.25 \leq s \leq 0.75$. The experimental radial distribution curve shown in Fig. 4 was computed using $a = 0.001$ Å^2 and the values of the theoretical intensity curve of Fig. 3 for this unobserved range. For this and other calculations

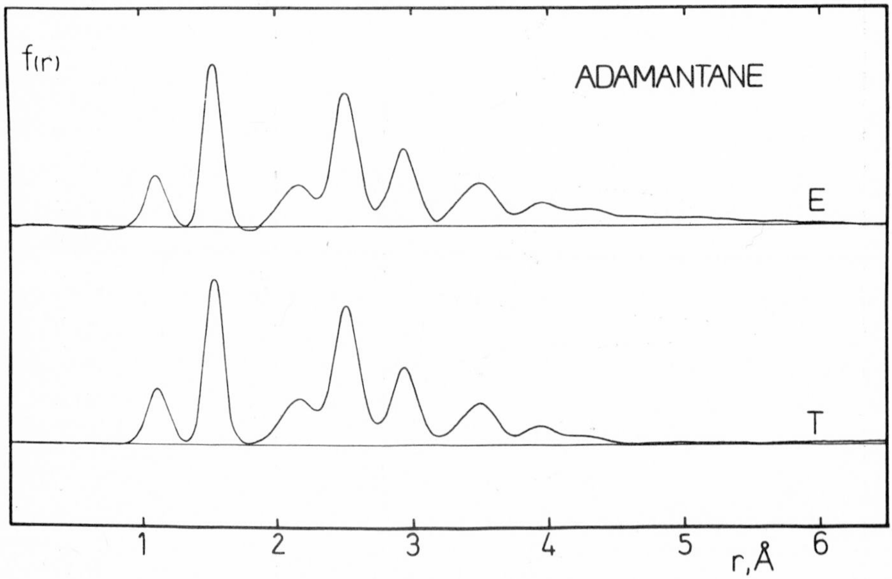

Fig. 19-4. Radial distribution curves. E is experimental and T theoretical. The curves correspond to the intensity curves of Fig. 3.

the partial waves electron scattering amplitudes and phases were interpolated for the 44 kV accelerating voltage from Cox and Bonham's tables[9] with L. Hedberg's program.*

Identification of the peaks of the experimental radial distribution curves was straightforward in terms of a molecule of T_d symmetry. The peak at 1.1 Å corresponds to the two different types of C-H bond distances which are unresolved. The peak at 1.5 Å is due to the C-C bond distance. The overlapping maxima in the region $1.8 \leq r \leq 4.5$ Å arise from non-bond interactions of the following types (a diagram of the structure is shown in Fig. 5 and the atom numbering in Fig. 1): 2.2 Å, $C_1 \cdots H_{23}$ and $C_7 \cdots H_{11}$; 2.5 Å, $C_1 \cdots C_2$ and $C_7 \cdots C_8$; 2.9 Å, $C_1 \cdots C_8$; 3.5 Å, $C_1 \cdots C_4$, $C_1 \cdots H_{14}$ and $C_7 \cdots H_{24}$; 3.9 Å, $C_1 \cdots H_{24}$ and $C_7 \cdots H_{18}$; and 4.3 Å, $C_1 \cdots H_{17}$. The distances $C_1 \cdots H_{13}$ and $C_7 \cdots H_{17}$ are found at about 2.7 Å and 3.3 Å respectively, and the H···H distances are distributed throughout the non-bond region.

19.3.2. The Geometry of the Model

Assuming T_d symmetry for the adamantane molecule and assuming that the two types of C-H bond distances are equal,** the molecular geometry can be described by four parameters. These were chosen to be the C-C bond distance (R_1), the C_7-C_1-C_9 bond angle (R_2), the C-H bond distance (R_3), and the H-C-H bond angle (R_4). There are twenty-two non-bond distances of types C...C, C···H, and H···H whose values are dependent on the values of these parameters.

It was convenient to consider the geometry of the adamantane molecule in terms of four imaginary cubes packed one inside the other.*** The tertiary carbon atoms (see Figures 1 and 5) occupy four of the tetrahedrally related corners of a cube with length of side 2TC, and thus

---

*For a description see Ref. 8.

**They are not equal, of course, but the refinement results showed the difference to be much too small to affect significantly the values of the other parameters.

***This description was applied to P_4O_{10} in Ref. 10.

$$TC = [R_1 \sin(R_2/2)]/\sqrt{2} .$$

The secondary carbon atoms are located at the centers of the six faces of a cube with length of side 2SC, so that

$$SC = TC + R_1 \cos(R_2/2) .$$

The hydrogen atoms adjoining the tertiary carbon atoms occupy four of the tetrahedrally related corners of a cube with length of side 2TH, and thus

$$TH = TC + R_3/\sqrt{3} .$$

Finally, the hydrogen atoms adjoining the secondary carbon atoms lie on the diagonals of the faces of a cube with length of side 2SH where

$$SH = SC + R_3 \cos(R_4/2) ,$$

and the distances of the hydrogen atoms from the center of the face is

$$Q = R_3 \sin(R_4/2) .$$

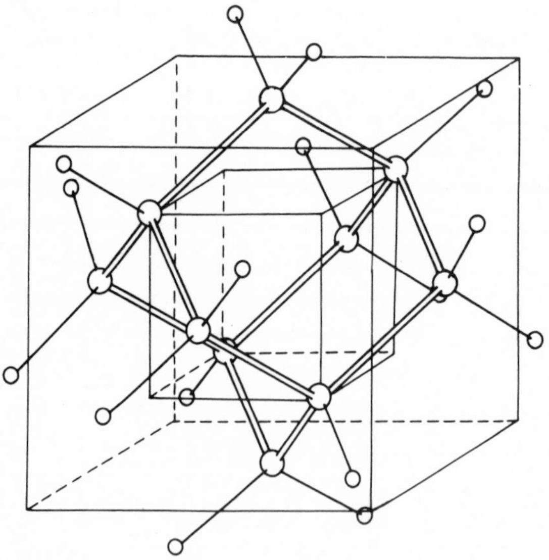

Fig. 19-5. Model of the adamantane molecule illustrating its symmetry.

The dependent distances were calculated from the Cartesian coordinates of the atoms and the partial derivatives $\partial d_d/\partial r_i$ were calculated numerically, where $i = 1,\ldots,4$ and $d = 5,\ldots,26$.

19.3.3. Trial Structures

The values 1.537 Å and 1.109 Å were inserted for R_1 and R_3 in all trial structures. The two types of C-H distances were assumed to be equal. That this assumption is reasonable to within experimental error is shown by the fact that it did not lead to larger-than-normal values for l(C-H) as would be expected if these distances were substantially different. The width and the height of the maximum at 2.52 Å do not indicate a large difference between the $C_7\cdots C_8$ and $C_1\cdots C_2$ distances, i.e. between the two types of C-C-C bond angles. Trial structures were formulated with C-C-C bond angles ranging from 107 to 112°. The H-C-H bond angle was given values equal or close to the tetrahedral value.

The vibration amplitudes for the trial structures were estimated mainly from those calculated for hexamethylene tetramine (Part I of Chapter 17).

19.3.4. Least Squares Refinement

The experimental radial distribution curves provided approximate values for the structural parameters of the molecule. Judged by the good agreement between the theoretical radial distribution curves corresponding to them and the experimental curve, the values were accurate enough to be refined by least squares. The least squares method was applied to the molecular intensities[11] in the form corresponding to Eq. (1). H.M. Seip's least squares program[12] was used with a diagonal weight matrix. For early refinements this was a unit matrix; in later refinements weights were given by the value of 0.2 for $s = 1.00$, 1.00 for the range $1.25 \leq s \leq 29.00$, and by $\exp[-0.0079(s - 29.25)^2]$ for the range $29.25 \leq s \leq 43.25$.

From the first refinements it was seen that the choice of trial structure did not significantly influence the results for the bond lengths (1.540 Å and 1.111 Å for C-C and C-H, respectively), but did have an effect on the refined values for the C-C-C bond angles. Refinements with starting values of $R_4 = 109.47°$ and $R_2 \leq$

109.47° resulted in values of 108.97° and 109.72° for the C_7-C_1-C_9 and C_1-C_7-C_2 angles, while starting values R_4 = 109.47° and R_2 > 109.47° gave 109.78° and 109.32° for these angles. The standard deviations for ∠ C_7-C_1-C_9 were 0.5 - 0.7° and for ∠ C_1-C_7-C_2, 0.3 - 0.4°. The H-C-H angle was found to be 111.3 - 111.7° with a standard deviation of about 1.3° in the different refinement schemes. In these cases all the independent geometrical parameters and the amplitudes for all the carbon-carbon and carbon-hydrogen distances were refined simultaneously; because of the small weight of the H···H distances, no attempts were made to refine the vibrational amplitudes associated with them.

The refinements just described were continued (i.e., some of the results were used as new starting values for parameters) using a 'block refinement' scheme. In the course of the continuation a slight change in the background was applied to the outer part of the 'middle' distance plates and to the inner part of the 'short' distance plates, which affected the intensity data in the range $21.25 \leq s \leq 33.75$. Groups or blocks of parameters were formed and only those parameters in the block were refined simultaneously. The blocks overlapped; that is, some parameters were included in more than one block. A further constraint was applied to the refinement of three pairs of amplitudes associated with geometrically almost equal distances (C_7···H_{11} and C_1···H_{23}, C_1···H_{14} and C_7···H_{24}, C_1···H_{24} and C_7···H_{18}): the assumption was made that the amplitudes for each of the members of a given pair were equal. Using starting values R_1 = 1.540 Å, R_3=1.111 Å, R_2 = 108.97° and R_4 = 111.65° (from the results of previous refinements), block refinements of the molecular geometry and the amplitudes $l(C_1 \cdots C_4)$, $l(C_7 \cdots H_{11})$ = $l(C_1 \cdots H_{23})$, $l(C_7 \cdots H_{17})$, and $l(C_1 \cdots H_{14})$ = $l(C_7 \cdots H_{24})$, led to an increased value for the H-C-H angle (116.5°) and an improvement of the agreement between the experimental and theoretical curves. Many of the results of this block refinement were next used as starting values for an additional unrestricted refinement of all important parameters. The results are given in Table II. The correlation matrix corresponding to this refinement is given in Table III.

The effect of different starting values on the refinement

Table 19-II. Structural results for adamantane: distances, $r_g(1)$, and vibrational amplitudes, l, in Å; angles in degrees. Parenthesized values were assumed.

Distance or angle	Multiplicity n_{ij}	r	σ_{LS}^r	$2\sigma_T^r$	l	σ_{LS}^l	$2\sigma_T^l$	l_{spec} [b]
C_1C_7	12	1.540	0.0003	0.002	0.052	0.0004	0.002	0.054
C_1H_{11}, C_7H_{23}	16	1.112	0.0014	0.004	0.061	0.0012	0.004	0.079
C_7C_8	6	2.503	0.0056	0.016	0.078	0.0067	0.018	0.059
C_1C_2	12	2.520	0.0030	0.009	0.066	0.0021	0.006	0.071
C_1C_8	12	2.946	0.0013	0.005	0.075	0.0011	0.004	0.069
C_1C_4	3	3.564	0.0042	0.012	0.080	0.0065	0.018	0.072
C_7H_{11}	24	2.156	0.0072	0.204	0.110	0.0229	0.064	0.114
C_1H_{23}	12	2.174	0.0025	0.007	0.110	0.0473	0.116	0.108
C_1H_{13}	24	2.697	0.0165	0.047	0.219	0.0177	0.050	0.166
C_7H_{17}	12	3.263	0.0188	0.053	0.166	0.0306	0.086	0.170
C_1H_{14}	24	3.471	0.0055	0.016	0.101	0.3772	1.066	0.108
C_7H_{24}	12	3.471	0.0057	0.017	0.101	0.7409	2.096	0.101
C_1H_{24}	12	3.951	0.0020	0.007	0.098	0.0300	0.088	0.108
C_1H_{17}	12	4.252	0.0147	0.042	0.117	0.0179	0.050	0.143
C_7H_{18}	12	3.923	0.0091	0.026	0.098	0.0300	0.084	0.109
$H_{11}H_{12}$	6	1.894	0.0223	0.063	(0.149)			0.142
$H_{11}H_{13}$	12	2.395	0.0357	0.101	(0.226)			0.244
$H_{11}H_{23}$	24	2.502	0.0026	0.008	(0.226)			0.160
$H_{11}H_{14}$	24	3.724	0.0139	0.039	(0.169)			0.180
$H_{11}H_{24}$	12	4.075	0.0204	0.058	(0.142)			0.206
$H_{11}H_{15}$	12	4.290	0.0139	0.040	(0.142)			0.149
$H_{23}H_{24}$	6	4.318	0.0061	0.018	(0.160)			0.138
$H_{11}H_{17}$	12	4.913	0.0294	0.083	(0.180)			0.190
$H_{11}H_{26}$	12	4.978	0.0079	0.023	(0.180)			0.131
C_7-C_1-C_9		108.8	0.37	1.0				
C_1-C_7-C_2		109.8	0.18	0.5				
H_{11}-C_1-H_{12}		116.9	2.14	6.0				
H_{11}-C_1-C_7 [a]		107.7	0.58	1.6				

[a] Values calculated from those for the other angles.

[b] Results at 400 K from spectroscopic calculations reported in Part III.

Table 19-III. Correlation matrix (×1000). The standard errors to be used with this table are the σ_{LS} of Table II.

	r(C–C)	∠CCH$_2$C	r(C–H)	∠HCH	l(C–C)	l(C–H)	$l(C_7C_8)$	$l(C_1C_2)$	$l(C_1C_8)$	$l(C_1C_4)$	$l(C_7H_{11})$
r(C–C)	1000	-256	-8	-156	-115	-54	-516	478	-41	-28	-143
∠CCH$_2$C		1000	-14	-215	73	33	641	-492	17	120	-174
r(C–H)			1000	126	14	3	-6	3	-22	-12	-51
∠HCH				1000	67	26	-5	47	7	-113	729
l(C–C)					1000	202	200	-51	227	56	104
l(C–H)						1000	85	-24	94	25	42
$l(C_7C_8)$							1000	-912	93	79	-94
$l(C_1C_2)$								1000	-28	-54	150
$l(C_1C_8)$									1000	38	20
$l(C_1C_4)$										1000	-73
$l(C_7H_{11})$											1000

	$l(C_1H_{23})$	$l(C_1H_{13})$	$l(C_7H_{17})$	$l(C_1H_{14})$	$l(C_7H_{24})$	$l(C_1H_{24})$	$l(C_1H_{17})$	$l(C_7H_{18})$	k
r(C–C)	171	167	-178	-72	72	76	6	-60	-161
∠CCH$_2$C	160	-269	54	27	-26	197	5	-211	108
r(C–H)	49	-45	-61	-87	87	63	-10	-75	10
∠HCH	-745	-15	343	107	-107	-403	-38	343	82
l(C–C)	-100	-27	-43	55	-54	1	11	8	706
l(C–H)	-40	-12	-17	25	-25	1	5	3	292
$l(C_7C_8)$	70	-127	101	52	-52	62	2	-75	284
$l(C_1C_2)$	-132	86	-89	-29	28	-56	0	66	-75

$l(C_1C_8)$	-17	136	135	6	-6	14	9	-15	321
$l(C_1C_4)$	74	-74	114	816	-816	139	30	-196	81
$l(C_7H_{11})$	-985	-91	252	103	-103	-311	-25	270	137
$l(C_1H_{23})$	1000	95	-263	-106	106	317	26	-274	-131
$l(C_1H_{13})$		1000	14	-76	75	-27	4	39	-38
$l(C_7H_{17})$			1000	61	-61	-94	0	21	-64
$l(C_1H_{14})$				1000	-1000	13	17	-53	77
$l(C_7H_{24})$					1000	-13	-17	52	-77
$l(C_1H_{24})$						1000	310	-915	7
$l(C_1H_{17})$							1000	-269	16
$l(C_7H_{18})$								1000	8
k									1000

results was tested in several ways. When the refinement described in the previous paragraph was repeated with $R_2 >$ 109.47° (e.g. 109.78°), the improvement stopped with $\angle C_7\text{-}C_1\text{-}C_9$ = 109.47°, $\angle C_1\text{-}C_7\text{-}C_2$ = 109.47°, $l(C_7\cdots C_8)$ = 0.0699 Å, $l(C_1\cdots C_2)$ = 0.0703 Å and $\angle$ H-C-H = 115.83°. By changing the values for $l(C_7\cdots C_8)$ and $l(C_1\cdots C_2)$ slightly, a continuation of the refinement gave nearly the same results as those in Table II. Because of the large correlation between the $C_7\cdots C_8$ and $C_1\cdots C_2$ distances and the associated amplitudes, a series of refinements was carried out using different trial values for the C-C-C and H-C-H angles and for the $C_7\cdots C_8$ and $C_1\cdots C_2$ amplitudes. The results were always close to those shown in Table II.

19.3.5. Final Results

Table II is based upon one of the refinements described above and we take it to be a fair representation of our results for the structure of adamantane. Theoretical intensity and radial distribution curves corresponding to the model defined by these results are shown in Figures 3 and 4; the agreement is seen to be very good except for the radial distribution curves in the region $r > 4.5$ Å. It would appear that this particular disagreement, which is quite negligible in terms of its effect on the determination of the important structural parameters, is due to small high-frequency errors in the low-angle region of the experimental intensity curve.

The standard deviations σ_{LS} shown in Table II are those from the least squares refinement already described using the diagonal weight matrix. To estimate the total standard error the values of σ_{LS} were multiplied by a factor of $\sqrt{2}$ to take account of possible correlation among the observations. These values were then combine with estimated systematic errors of $0.0005r$ and $0.02l$ in accordanc with the expressions

$$\sigma_r = [2\sigma_{LS}^2 + (0.0005r)^2]^{\frac{1}{2}}, \quad \sigma_l = [2\sigma_{LS} + (0.02l)^2]^{\frac{1}{2}}, \quad \sigma_\theta = \sqrt{2}\,\sigma_{LS}.$$

Justification for these formulas has been given by Hedberg and Iwasaki.[13] The total standard errors are also shown in Table II.

Inspection of the results from all refinements leads to a number of conclusions which are worth mention. First, the refined value of the C-C bond length was found to be largely independent o starting model and conditions of refinement, and accordingly may b regarded to be especially well determined. Second, the amplitude associated with the C-H bond has a rather smaller value than would be expected from other compounds (0.07 - 0.08 Å) despite the fact that the two types of C-H bonds were assumed to be equal in length which, were they actually significantly different, would tend to increase the apparent amplitude. Moreover, the value for this ampl tude seemed not to be appreciably affected by choice of starting model or conditions of refinement. We have no good explanation. Third, the strong correlations among the distances $C_1 \cdots C_2$ and $C_7 \cdots C_8$ and their associated amplitudes (Table III) makes any precise determination of the C-C-C bond angles impossible. Their values appear to be at most marginally different from the tetrahed value. Fourth, the 116.9° (6.0°) value determined for the H-C-H an seems implausibly large. The result for this angle is strongly affected by the refinement conditions (as is apparent from the preceding section) and, moreover, is strongly correlated with seve amplitude parameters: $l(C_7\cdots H_{11})$, $l(C_1\cdots H_{23})$, $l(C_7\cdots C_{17})$, $l(C_1\cdots H_{24})$ and $l(C_7\cdots H_{18})$. It seems quite likely that this angle is actually somewhat smaller than the tabulated value, and we do not weight our result heavily. Fifth, some of the total errors associated with amplitudes are so large as to suggest no determination at all [e.g. $l(C_1\cdots H_{14})$ and $l(C_7\cdots H_{24})$], even though the amplitude values themselves are plausible. These very large total errors doubtless have little quantitative significance; qualitatively they do indicate that no reliance can be placed on the amplitude values.

Our discussion of the third through fifth points above reflects a conservative interpretation of the significance of the results of Table II for those particular parameters. There is additional evidence that this view is justified: the improvement in the agreement between the observed and calculated intensity curves as a result of the slight background change mentioned in the preceding section was about the same as was obtained in the succeeding refinement series. Clearly the corresponding parameter changes are not to be given undue regard.

19.4. Discussion

The structure of adamantane from this investigation agrees well with that determined earlier.[3] The C-C bond length appears to be a bit different from that in diamond, but is quite comparable to that in other compounds as may be seen from Table IV. The average C-C-C bond angles are tetrahedral to within experimental error and

Table 19-IV. Comparison of some C-C bond lengths.

	r^*, Å	2σ, Å	Reference
Diamond	1.5445	0.0001	14
Adamantane	1.542	0.002	this work
Neo-pentane	1.541	0.002	15
Ethane	1.534	0.002	16

* The bond lengths for adamantane, neo-pentane and ethane are in terms of $r_g(0)$ values.

it seems unlikely that the H-C-H bond angle differs by much from this value.

The amplitude results for the C···C non-bond distances clearly support the expectation, based on the cage-like structure of the molecule, that the carbon skeleton is relatively stiff. These amplitude values for the long C···C distances (distances arising from third and fourth-neighbor atoms) are scarcely larger than the amplitude values for distances corresponding to second-neighbor atoms. Additional evidence of the stiffness of the carbon skeleton can be found in the magnitudes of the C-H amplitudes associated with long distances. As a group these amplitudes are smaller than would be expected were there large-amplitude vibrations of the carbon framework.

Acknowledgments: We thank the National Science Foundation for financial support under Grant GP 8453. We are grateful to Dr. Z. Zubovics for the sample of adamantane. One of us (I.H.) expresses his sincerest gratitude to Professor Harold P. Hanson (then Director of the Center for Structural Studies of the University of Texas) for an invitation to spend a year at Texas and for his interest in this work. Thanks are also due to Professor James E. Boggs (Austin) and Professor Sándor Lengyel (Budapest) for their interest and support.

REFERENCES

1 R.C.Fort,Jr. and P. von R. Schleyer, Chem.Rev. 64, 277 (1964).

2 R.C.Fort,Jr. and P. von R. Schleyer, Advances in Alicyclic Chemistry (Ed.: H.Hart and G.J.Karabatros), p. 283, Academic Press, New York 1966.

3 W.Nowacki and K.Hedberg, J.Am.Chem.Soc. 70, 1497 (1948).

4 O.Bastiansen, Structural Chemistry and Molecular Biology (Ed.: A.Rich and N.Davidson), p. 640, Freeman, San Francisco 1968.

5 S.J.Cyvin: Molecular Vibrations and Mean Square Amplitudes, Universitetsforlaget, Oslo, and Elsevier, Amsterdam, 1968.

6 S.Landa, S.Kriebel, and E.Konbloch, Chemické Listy 48, 61 (1954).

7 P.Sohár, Z.Zubovics, and Gy.Varsányi, Magyar Kémiai Folyóirat 75, 432 (1969).

8 G.Gundersen and K.Hedberg, J.Chem.Phys. 51, 2500 (1969).
9 H.L.Cox,Jr. and R.A.Bonham, J.Chem.Phys. 47, 2599 (1967).
10 B.Beagley, D.W.J.Cruickshank, T.G.Hewitt, and A.Haaland, Trans. Faraday Soc. 63, 836 (1967).
11 K.Hedberg and M.Iwasaki, Acta Cryst. 17, 529 (1964).
12 B.Andersen, H.M.Seip, T.G.Strand, and R.Stølevik, Acta Chem.Scand. 23, 3224 (1969).
13 K.Hedberg and M.Iwasaki, J.Chem.Phys. 36, 589 (1962).
14 K.Lonsdale, Phil.Trans.Roy.Soc. A240, 219 (1946).
15 B.Beagley, D.P.Brown, and J.J.Monaghan, J.Mol.Structure 4, 233 (1969).
16 L.S.Bartell and H.K.Higginbotham, J.Chem.Phys. 42, 851 (1965).

Adamantane - Part II: Spectroscopic Study

P. SOHÁR, Z. ZUBOVICS, and GY. VARSÁNYI

The types of normal vibrations in adamantane ($C_{10}H_{16}$) are described. Fundamental frequencies of infrared and Raman spectra are reported and assigned to the appropriate types and symmetry species.

Leirtuk az adamantán ($C_{10}H_{16}$) normálrezgéseinek tipusait. Közöltük az infravörös és Raman spektrum alapfrekvenciáit, és hozzárendeltük azokat a megfelelő rezgésmódokhoz és szimmetriatipusokhoz.

19.5. Classification of Normal Vibrations

Adamantane ($C_{10}H_{16}$) belongs to the symmetry group T_d , and being a molecule with twenty-six atoms it will show seventy-two normal vibrations.[1,2] They are distributed among the species as follows.[3-5]

$$\Gamma = 5a_1 + a_2 + 6e + 7f_1 + 11f_2, \qquad (19.3)$$

where as usual e and f designate doubly and triply degenerate species, respectively. Thirty-six of the 5 + 1 + 12 + 21 + 33 = 72 [cf. Eq.(3)] normal vibrations are localized to the methylene groups. Since the hydrogen atoms of each CH_2 group lie in a plane of symmetry, the distribution of the thirty-six vibrations among the species is

$$\Gamma_{methylene} = 2a_1 + a_2 + 3e + 4f_1 + 5f_2. \qquad (19.4)$$

With four methyne groups present twelve CH vibrations (four stretchings and eight bendings) are possible, and can be derived from the normal vibrations of a tetrahedral XY_4 (e.g. methane) molecule:

$$\Gamma_{methyne} = a_1 + e + f_1 + 2f_2. \quad (19.5)$$

The bending (δ CH) vibration of species f_1 can be derived from the nongenuine vibration (rotation) of CH_4.

Similarly to the CH vibrations the skeletal vibrations can be associated to the normal vibrations of XY_4 (e.g. methane) and octahedral XY_6 (e.g. sulphur hexafluoride) molecules. The secondary carbon atoms (C_s) are placed on the apices of an octahedron.

$$\Gamma_{skeleton} = 2a_1 + 2e + 2f_1 + 4f_2; \quad (19.6)$$

cf. also Eq. (17.1). In species a_1 the two possible skeletal vibrations are the combinations - in the same and in the opposite phase - of the ν_1 type vibrations of CH_4 and SF_6, respectively, producing one stretching and one bending vibration. The former is the breathing vibration of the ring system, and the latter is a skeletal deformation in the central direction. The two skeletal vibrations in species e can be regarded as the combinations - in the same and in the opposite phase - of the respective vibration types in SF_6 and CH_4. The former is a bending and the latter a stretching vibration. One of the two vibrations in species f_1 can be derived from the nongenuine vibration (rotation) of CH_4 (f_1) and SF_6 (f_{1g}). This is a bending vibration perpendicular to the planes $C_tC_sC_t$, where C_t denotes a tertiary C atom. The other f_1 vibration is the combination of the ν_6 vibration of SF_6 (f_{2u}) and of the rotation of CH_4, and has the character of an asymmetric skeletal stretching vibration. The direction of the degrees of freedom are: for C_t the direction of the C_3 axis, the direction perpendicular to this axis in one of the planes $C_tC_sC_t$, and the direction perpendicular to this plane; for C_s the direction of the C_2 axis, the direction perpendicular to this axis in the plane $C_tC_sC_t$, and the direction perpendicular to the latter plane. The four skeletal vibrations in species f_2 can be derived from the opposite phase combination of ν_3 and ν_4 vibrations of CH_4, and of the ν_3 , ν_4 and ν_5 vibrations of SF_6, respectively. It results in two skeletal stretching and two skeletal bending vibrations. The two stretching vibrations can be regarded as the opposite phase combinations of ν_3 and ν_4 of CH_4

with the ν_3 and ν_4 vibrations of SF_6, respectively. The bending vibrations result from the opposite phase combinations of the ν_3 and ν_4 vibrations of CH_4 with the ν_4 and ν_5 vibrations of SF_6, respectively. All the fluorine atoms in SF_6 participate in the ν_3 and ν_4 vibrations, whilst in the case of adamantane in the analogues to the ν_3 vibration only the four equatorial C_s atoms can vibrate, and in the vibrations resulting from the combinations of the ν_4 vibrations only the two axial C_s atoms can vibrate.

19.6. Infrared and Raman Spectrum

The infrared fundamental frequencies of species f_2 for adamantane measured by Mecke and Spiesecke[1] and by us, the frequencies of its Raman spectrum recorded by us on samples in CCl_4, and finally the Raman frequencies given in the literature[6] for solid samples, are collected in Table V, which also shows the assignments.

On the basis of frequency values alone the assignments of $\nu_{as}C_sH_2$ and νC_tH could very well be interchanged to conform the communication of Mecke and Spiesecke.[1] The νCH frequencies in simple paraffins appear at much lower wave numbers, generally between 2910 and 2880 cm^{-1},[7] while the $\nu_{as}CH_2$ bands have a range of appearance between 2940 and 2910 cm^{-1}. Intensities, Raman spectrum, and spectra of derivatives,[8] however, seem to support the assignment given here. There is also a deviation between the data for δCH found in the literature[1] and listed as present data in Table V. The study of spectra in KBr pellets reveals that there is no absorption (even at ten times higher concentrations) at 1155 cm^{-1} where a band is noted by other authors[1] as recorded in CCl_4 solutions. Neither does any absorption occur in the 1200 to 1100 cm^{-1} range of the Raman spectra of adamantane (in CCl_4 and in KBr pellets). In our infrared and Raman spectra, also in accordance with Raman data from literature, there is a weak band around 1320 cm^{-1}, and on the basis of its frequency and intensity its assignment to a δCH vibration is obvious. This assignment is well supported by the spectra of adamantane derivatives.[8] As to the third major revision with reference to data in the literature,[1] we propose a change of the frequency value from 476 cm^{-1} to 635 cm^{-1} for the skeletal bending vibration of f_2. No band was namely observed at 476 cm^{-1} even in spectra of samples in high concentrations. A band corresponding to the 635 cm^{-1} maximum also appears in the spectra of adamantane derivatives,[8] supporting the assignment made here, while in the range between 500

Table 19-V. Infrared and Raman fundamental frequencies (in cm^{-1}) for adamantane. vs = very strong, s = strong, m = medium, w = weak, vw = very weak.

Type	Infrared (solid)				Raman	
	Ref.1	Present	Species	in CCl_4[a]	Solid[6]	Species
νCH	2907	*2935*[*] s	f_2	*2932* s	2944	a_1 f_2
$\nu_{as}CH_2$	2933	*2910* vs	f_2	2910 s	2917	f_2
ν_sCH_2	2857	*2855* s	f_2	*2885* w	2895	e
				2865 m	2849	a_1 f_2
β_sCH_2				*1490* w		e
	1453	*1450* s	f_2	1450 w	1437	f_2
				1440		a_1
γ_sCH_2	1357	*1350* s	f_2			
δCH	1155	*1305* vw	f_2	*1320* w	1315	e f_2
$\gamma_{as}CH_2$				*1220* s	1223	e
νCC				1105	1099	f_2
	1101	*1100* s	f_2	*975* m	972	a_1
	966	*965* m[b]	f_2		*951*	e
$\beta_{as}CH_2$	799	*795* s	f_2			
δCC	476	*635* vw	f_2		*760*	a_1
	440	*440* w	f_2		443	f_2

[a] From present authors. On bands of wave numbers lower than 950 cm^{-1} the absorption of the solvent was superimposed.

[b] In spectra of samples in high concentration a weak band at 945 cm^{-1} appears, assigned to the split pair of the 965 cm^{-1} band of the degenerate νCC vibration.

[*] Values in italics were used in the normal coordinate analysis reported in Part III.

and 400 cm^{-1} only one band which corresponds to the adamantane 440 cm^{-1} band appears.

REFERENCES

1 R.Mecke and H.Spiesecke, Chem.Ber. 88, 1997 (1955).
2 R.Mecke and H.Spiesecke, Spectrochim.Acta 7, 387 (1956).
3 D.Chadwick, A.C.Legon, and D.J.Millen, J.Chem.Soc., 1116 (1968).
4 D.Sonland: Molecular Symmetry, D. van Nostrand, London 1965; p. 276.
5 G.Herzberg: Molecular Spectra and Molecular Structure. Vol.II. Infrared and Raman Spectra of Polyatomic Molecules, D. van Nostrand, New York 1945. Hungarian translation: Molekula-szinképek és molekula-szerkezet, Akadémiai Kiadó, Budapest 1959; p. 276.
6 R.C.Fort and P.R.Schleyer, Chem.Rev. 64, 277 (1964).
7 S.Holly and P.Sohár: Infravörös Spektroszkópia (Infrared spectroscopy), Müszaki Kiadó, Budapest 1968.
8 P.Sohár, Z.Zubovics, and Gy.Varsányi, Magyar Kémiai Folyóirat 75, 432 (1969).

Adamantane - Part III: Normal Coordinate Analysis and Mean Amplitudes

S. J. CYVIN, B. N. CYVIN, and J. BRUNVOLL

Mean amplitudes of vibration for adamantane ($C_{10}H_{16}$) were calculated from a developed harmonic force field for the molecule. The results are compared with electron diffraction data.

Adamantane (see the two preceding parts) has a structure which is derived by an extension of the model for hexamethylene tetramine (cf. Chapter 17, Part I). In the present analysis the described symmetry coordinates for the latter model were augmented and adapted to adamantane. Two new types of valence coordinates were introduced, viz. four C_t-H stretchings (t) and twelve C_sC_tH bendings (δ). The corresponding symmetric structures are

$$\Gamma(t) = A_1 + F_2 , \qquad (19.7)$$

$$\Gamma(\delta) = A_1 + E + F_1 + 2F_2 . \qquad (19.8)$$

This introduces redundancies belonging to $A_1 + F_2$. Therefore some appropriate combinations of the δ coordinates were omitted when constructing the symmetry coordinates. The t type symmetry coordinates are

$$S_5(A_1) = \tfrac{1}{2}(t_1 + t_2 + t_3 + t_4), \qquad (19.9)$$

$$S_{9a}(F_2) = \tfrac{1}{2}(t_1 - t_2 + t_3 - t_4), \qquad (19.10)$$

$$S_{9b}(F_2) = \tfrac{1}{2}(t_1 + t_2 - t_3 - t_4), \qquad (19.11)$$

$$S_{9c}(F_2) = \tfrac{1}{2}(t_1 - t_2 - t_3 + t_4). \qquad (19.12)$$

Table 19-VI. Observed and calculated vibrational frequencies (cm^{-1}) for adamantane.

Species	Observed		Calculated		
	Ref. 1	Part II	Ref. 1	Present approx.	Present final
A_1	-	2932	2909	2924	2932
	-	2865	2856	2842	2865
	-	1440	1453	1398	1440
	975	975	999	985	975
	759	760	746	696	760
A_2	-	-	1191	1493	1191
E	-	2885	2855	2843	2885
	-	1490	1460	1815	1490
	-	1320	1344	1510	1320
	1222	1220	1218	1089	1220
	-	951	912	931	951
	442	-	427	648	442
F_1	-	-	2926	2949	2926
	-	-	1371	1787	1371
	-	-	1328	1725	1328
	-	-	1136	1499	1136
	-	-	1088	1277	1088
	-	-	886	1045	886
	-	-	309	405	309
F_2	-	2935	2932	2961	2935
	-	2910	2904	2914	2910
	-	2855	2856	2842	2855
	1452	1450	1459	1836	1450
	1354	1350	1354	1765	1350
	1289	1305	1285	1378	1305
	1101	1100	1109	1174	1100
	966	965	971	966	965
	800	795	800	829	795
	632	635	649	820	635
	-	440	423	564	440

Here t_j (j = 1,...,4) involves the tertiary C atom number 6+j; cf. Fig. 1. Eight combinations of $(ST)^{\frac{1}{2}}\delta_i$ (i = 1,...,12), viz. $S_{6a}(E)$, $S_{6b}(E)$, $S_{7a}(F_1)$, $S_{7b}(F_1)$, $S_{7c}(F_1)$, $S_{11a}(F_2)$, $S_{11b}(F_2)$, and $S_{11c}(F_2)$, were constructed precisely as the corresponding combinations of s_i; see Chapter 17, Part I. In the F_2 species the s-combinations of S_1 were chosen. δ_i was taken as to involve the two atoms of the s_i stretching in addition to a tertiary hydrogen. S and T denote the equilibrium distances of C-C and C_t-H, respectively.

Preliminary force field calculations were performed on the basis of a valence-force approximation as in Eq. (17.25), but augmented with the appropriate terms containing f_t and f_δ for the new valence coordinates. This very simple, approximate force field gave generally good qualitative agreement with observed frequencies (cf. Table VI) when the following force constants were employed.[1] f_s = 4.337, f_α = 1.084, f_r = 4.554, f_φ = 0.656, f_t = 4.588, f_δ = 0.657, f_β = 1.130; all in mdyne/Å. The agreement is slightly better for the calculated frequencies by Snyder and Schachtschneider.[1] A final force field was adjusted to fit exactly the observed frequency values of Sohár et al. (see Table V; numbers in italics), along with the A_2 and F_1 frequencies (calculated) and the lowest E frequency (observed) from Ref. 1. The different sets of observed and calculated frequencies for adamantane are summarized in Table VI. The structural parameters from Part I were employed in the present analysis.

The final force field contains many interaction force constants and is not given here for the sake of brevity. It was used to calculate the mean amplitudes of vibration for all the twenty-five types of interatomic distances. The results at 400 K are reported in Table II. They display a very good agreement with the electron diffraction values shown in the same table when the appropriate limits of errors are taken into account.

REFERENCE

1 R.G.Snyder and J.H.Schachtschneider, Spectrochim.Acta 21, 169 (1965).

CHAPTER 20

Tentatively Standardized Symmetry Coordinates for Vibrations of Polyatomic Molecules: Survey

S. J. CYVIN

References are given to sixteen parts of an article series on tentatively standardized symmetry coordinates. One hundred and two molecular models of three through seven atoms are tabulated.

20.1. Introduction

The work of compiling useful arithmetical expressions for molecular models, initiated by Cyvin in Chapter 8 of Ref. 1, has been continued through a series of articles[2-17] with the same title as the present chapter. The main purpose of the article series was to specify sets of symmetry coordinates, which should be convenient as standard references when harmonic force fields are to be reported for molecules belonging to the various models. Particular emphasis has been laid on the orientation of cartesian axes, and especially the orientation of degenerate coordinates with respect to these axes.

20.2. Orientation of Cartesian Axes

When choosing the directions of cartesian axes (X, Y, Z) the recommendations of Mulliken's report[18] should preferably be followed. This implies (among other features) that (i) in C_s molecules the X axis should be perpendicular to the symmetry plane, (ii) in planar C_{2v} molecules X should be perpendicular to the molecular plane, and

(iii) the same in planar D_{2h} molecules. In the two latter cases (ii, iii) and in some others the species notation in classification of normal modes is influenced by the orientation of cartesian axes. Unfortunately in most infrared and Raman works the older conventions are still used. They follow the designations from Herzberg,[19] but violate Mulliken's recommendations in some respects. The water (H_2O; C_{2v}) and ethylene (C_2H_4; D_{2h}) molecules represent two very common and instructive examples. The species notation according to Herzberg[19] is consistent with the Y and Z axes perpendicular to the molecular planes in H_2O and C_2H_4, respectively. The X axis should be taken perpendicular to the molecular planes in both cases according to Mulliken.[18] It should be noted that the orientations of cartesian axes in the XY_2 and XY_2Z molecules of C_{2v} symmetry in Ref. 1 (Figures 8-2 and 14-7) follow the old conventions, but have later been changed[2] to conform Mulliken's recommendations.

20.3. Orientation of Degenerate Coordinates

Some conventions for orientation of degenerate symmetry coordinates have been reported in Ref. 1, and are also summarized in Ref. 2. For the sake of convenience they are repeated below.

(i)* In a symmetric top with a p-fold symmetry axis C_p ($p > 2$), the Z axis is oriented along C_p. Then a pair of degenerate coordinates (S_{ia}, S_{ib}) should conventionally fulfil the transformation

$$C_p \begin{bmatrix} S_{ia} \\ S_{ib} \end{bmatrix} = \begin{bmatrix} \cos\varphi & \sin\varphi \\ -\sin\varphi & \cos\varphi \end{bmatrix} \begin{bmatrix} S_{ia} \\ S_{ib} \end{bmatrix} \qquad (20.1)$$

* S.J. Daunt has pointed out to the writer (Private communication 1971 from Dr. Daunt, Queen's University, Kingston, Canada) that the E'' coordinates of the D_{3h} models treated in Ref. 8 do not obey the convention (i). The (S_a,S_b) coordinates transform like (α_{yz},α_{zx}) rather than (α_{zx},α_{yz}) as they should.[1] The imperfection is repaired on interchanging the a and b subscripts for the E'' coordinates. The same imperfection is present in the specified coordinates for the trigonal bipyramidal XY_3Z_2 model.[10] The simultaneous interchanging of all a and b coordinates does not affect the elements of **F** and **G** matrices.

as suggested by Boyd and Longuet-Higgins.[20] Here C_p is taken to rotate X towards Y by the angle $\varphi = 2\pi/p$.

(ii) In regard to a set of triply degenerate coordinates (S_{ia}, S_{ib}, S_{ic}) in a spherical top, we follow McDowell's[21] convention:

$$C_3(XYZ)\begin{bmatrix} S_{ia} \\ S_{ib} \\ S_{ic} \end{bmatrix} = \begin{bmatrix} 0 & 1 & 0 \\ 0 & 0 & 1 \\ 1 & 0 & 0 \end{bmatrix}\begin{bmatrix} S_{ia} \\ S_{ib} \\ S_{ic} \end{bmatrix}, \tag{20.2}$$

where $C_3(XYZ)$ lies in the XYZ-direction and rotates X into Y, Y into Z, and Z into X.

(iii) Doubly degenerate symmetry coordinates in a spherical top should be oriented (if possible) so as to fulfil the transformation property

$$C_3(XYZ)\begin{bmatrix} S_a \\ S_b \end{bmatrix} = \begin{bmatrix} -\frac{1}{2} & \frac{1}{2}3^{\frac{1}{2}} \\ -\frac{1}{2}3^{\frac{1}{2}} & -\frac{1}{2} \end{bmatrix}\begin{bmatrix} S_a \\ S_b \end{bmatrix}, \tag{20.3}$$

where $C_3(XYZ)$ has the same meaning as in Eq. (2).

20.4. Influence of Orientation on Coriolis Constants

Certain regularities among Coriolis coupling constants occur as a consequence of the chosen orientations of degenerate coordina[tes] with respect to the cartesian axes. These features have been treat[ed] in Ref. 1 and also summarized elsewhere.[2,3] Other papers[22-24] treating the problem have recently appeared. Here we repeat shortl[y] the most frequently encountered relations concerning Coriolis constants. They also imply certain conventions about notation. For further details some of the cited works[1-3] should be consulted.

(i) For a coupling of an $A\times E$ type in symmetric tops; cf. also Eq. (8.25) of Ref. 1:

$$\zeta_{it} = \zeta_{i\ ta}^{\ y} = \pm\,\zeta_{i\ tb}^{\ x}\,. \tag{20.4}$$

Here the two cases may be identified as (ia) and (ib), referring t[o] the + and - sign, respectively.

(ii) For the important $E\times E$ coupling with respect to the Z axis (along the unique symmetry axis, C_p) in symmetric tops; cf. also Eqs. (8.2),(8.3) of Ref. 1:

$$\zeta_i = \zeta_{ia}{}^{z}{}_{ib}\,, \quad \zeta_{ij} = \zeta_{ia}{}^{z}{}_{jb} = \zeta_{ja}{}^{z}{}_{ib}\,. \tag{20.5}$$

(iii) For $E\times F$ in a spherical top; cf. Eqs.(8.26),(8.27) of Ref. 1:

(*a*)

$$\zeta_{ia}{}^{x}{}_{ta} = -\zeta_{ia}{}^{y}{}_{tb} = -\frac{1}{2}\,3^{\frac{1}{2}}\,\zeta_{it}\,,$$

$$\zeta_{ib}{}^{x}{}_{ta} = \zeta_{ib}{}^{y}{}_{tb} = -\frac{1}{2}\,\zeta_{it}\,, \qquad \zeta_{ia}{}^{z}{}_{tc} = \zeta_{it}\,. \tag{20.6}$$

(*b*)

$$\zeta_{ia}{}^{x}{}_{ta} = \zeta_{ia}{}^{y}{}_{tb} = -\frac{1}{2}\,\zeta_{it}\,,$$

$$-\zeta_{ib}{}^{x}{}_{ta} = \zeta_{ib}{}^{y}{}_{tb} = -\frac{1}{2}\,3^{\frac{1}{2}}\,\zeta_{it}\,, \qquad \zeta_{ib}{}^{z}{}_{tc} = \zeta_{it}\,. \tag{20.7}$$

(iv) For the important $F\times F$ type in spherical tops; cf. also Eqs. (8.5),(8.6) of Ref. 1:

$$\zeta_i = \zeta_{ia}{}^{z}{}_{ib} = \zeta_{ib}{}^{x}{}_{ic} = \zeta_{ic}{}^{y}{}_{ia}\,,$$

$$\zeta_{ij} = \zeta_{ia}{}^{z}{}_{jb} = \zeta_{ib}{}^{x}{}_{jc} = \zeta_{ic}{}^{y}{}_{ja} = \zeta_{ja}{}^{z}{}_{ib} = \zeta_{jb}{}^{x}{}_{ic} = \zeta_{jc}{}^{y}{}_{ia}\,. \tag{20.8}$$

20.5. Survey of Models

Table I tends to list most of the important models from triatomic through those with seven atoms. Two additional seven-atomic models are treated in Chapter 16, Part I.

The roman numerals in the column headed 'S' indicate the respective parts of the article series on tentatively standardized symmetry coordinates. In the same column 'o' refers to Chapter 8 of Ref. 1. The bibliographies of these references are believed to give a clue to the vast number of works on this topic, which are too numerous to be listed here. For the models which have not been subjected to the tentative standardization some key references are included in Table I. No attempt was made to provide a complete survey of the literature in this field.

Table 20-I. Survey of molecular models.

No.*	Type	Symmetry	Description	Example	*S*	Ref.
	Triatomic					
301	Z_3	D_{3h}	Regular trigonal	-	o	1
302	XY_2	$D_{\infty h}$	Linear symmetrical	CO_2	o	1
303	XY_2	C_{2v}	Bent symmetrical	H_2O	i	2
304	XYZ	$C_{\infty v}$	Linear	HCN,COS	i	2
305	XYZ	C_s	Bent	NOCl	i	2
	Four-atomic					
401	Z_4	T_d	Tetrahedral	P_4	o	1
402	Z_4	D_{4h}	Planar square	-	o	1
403	Z_4	D_{2h}	Planar rectangular	-	i	2
404	Z_4	D_{2d}	Puckered ring	-	i	2
405	XY_3	D_{3h}	Planar symmetrical	BF_3,SO_3	o,i	1,2
406	XY_3	C_{3v}	Regular pyramidal	NH_3	o,i	1,2
407	X_2Y_2	$D_{\infty h}$	Linear symmetrical	C_2H_2	o,i	1,2
408	X_2Y_2	D_{2h}	Planar rhombic	Li_2F_2	i	2
409	X_2Y_2	C_{2v}	cis-X_2Y_2	cis-N_2F_2	i	2
410	X_2Y_2	C_{2h}	trans-X_2Y_2	tr-N_2F_2	i	2
411	X_2Y_2	C_2	Nonplanar symmetrical	H_2O_2	i	2
412	X_2YZ	C_{2v}	Planar cyclic	$LiNaF_2$	Chapter 15	
413	XY_2Z	C_{2v}	Planar symmetrical	CH_2O	i	2
414	XY_2Z	C_s	Pyramidal	NHF_2,$SOCl_2$	i	2
415	WXYZ	$C_{\infty v}$	Linear	C_2HCl	i	2
416	WXYZ	C_s	With linear XYZ chain	HNCO	i	2
417	WXYZ	C_s	Planar	HCOF	i	2
418	WXYZ	C_s	cis-WXYZ	cis-HNO_2	xv	16
419	WXYZ	C_s	trans-WXYZ	tr-HNO_2	xv	16
	Five-atomic					
501	Z_5	D_{5h}	Regular pentagonal	-	-	-
502	XY_4	T_d	Tetrahedral	CH_4	o,ii	1,3
503	XY_4	D_{4h}	Planar symmetrical	XeF_4	o,ii	1,3
504	XY_4	C_{4v}	Regular pyramidal	-	ii	3
505	XY_4	D_{2h}	Planar dihedral	-	ii	3

Table 20-I (Continued)

No.*	Type	Symmetry	Description	Example	*S*	Ref.
506	XY_4	D_{2d}	Twisted dihedral	-	ii	3
507	X_3Y_2	D_{3h}	Bipyramidal	-	vii	8
508	XY_3Z	C_{3v}	Pyramidal-axial	CH_3Cl	ii	3
509	$X(YZ)_2$	$D_{\infty h}$	Linear symmetrical	C_3O_2	ii	3
510	XY_2Z_2	D_{2h}	Planar right-angled	-	ii	3
511	XY_2Z_2	C_{2v}	Planar cyclic	-	xii	13
512	$X(YZ)_2$	C_{2v}	Bent with linear XYZ chains	$S(CN)_2$	ii	3
513	XY_2Z_2	C_{2v}	Planar symmetrical	-	ii	3
514	XY_2Z_2	C_{2v}	Twisted symmetrical	CH_2Cl_2, SO_2Cl_2	ii	3
515	XY_2Z_2	C_2	Generalized XY_2Z_2	-	xv	16
516	XY_2Z_2	C_{2v}	cis-XY_2Z_2 with linear XY_2	-	ii	3
517	XY_2Z_2	C_{2h}	trans-XY_2Z_2 with linear XY_2	-	ii	3
518	XY_2Z_2	C_2	With linear XY_2 part	-	ii	3
519	XY_2ZW	C_{2v}	Planar with linear XZW	CH_2CO	xv	16
520	XY_2WZ	C_{2v}	Planar with linear WXZ	-	xv	16
521	XY_2ZW	C_s	Twisted	CH_2FCl, SO_2FCl	xv	16
522	XY_2ZW	C_s	With planar XY_2Z part	-	xv	16
523	XY_2ZW	C_s	With linear XZW	-	xv	16
524	XY_2ZW	C_s	cis-XY_2ZW	cis-NH_2OH	xv	16
525	XY_2ZW	C_s	trans-XY_2ZW	tr-NH_2OH	xv	16
526	UVXYZ	$C_{\infty v}$	Linear	HCCCN	xv	16
527	XYZUV	C_s	Planar	HCOOH, HNO_3	xv	16, Chapter 14
	Six-atomic					
601	Z_6	O_h	Octahedral	-	o	1
602	Z_6	D_{6h}	Regular hexagonal	-	o	1
603	Z_6	D_{3d}	Puckered ring	S_6	v	6
604	XY_5	D_{5h}	Planar symmetrical	-	-	-
605	X_3Y_3	D_{3h}	Planar trigonal	-	v	6
606	X_3Y_3	C_{3v}	Cyclic	-	xvi	17
607	X_4Y_2	C_{2h}	Cyclic	-	xvi	17

Table 20-I (Continued)

No.*	Type	Symmetry	Description	Example	*S*	Ref.
608	X_2Y_4	D_{2h}	Planar symmetrical	C_2H_4, N_2O_4	iii	4
609	X_2Y_4	D_{2d}	Twisted symmetrical	B_2Cl_4	iii	4
610	X_2Y_4	D_2	Generalized X_2Y_4	-	iii	4
611	X_2Y_4	C_{2v}	cis-X_2Y_4	-	xvi	17
612	X_2Y_4	C_{2h}	trans-X_2Y_4	P_2Cl_4	xvi	17
613	X_2Y_4	C_2	Generalized X_2Y_4	N_2H_4	xvi	17
614	XY_4Z	C_{4v}	Pyramidal-axial	$XeOF_4$	x	11
615	XY_3Z_2	D_{3h}	Bipyramidal	PCl_5, PF_2Cl_3	ix	10
616	$X_2(YZ)_2$	$D_{\infty h}$	Linear	C_4H_2	xv	16
617	$X_2Y_2Z_2$	D_{2h}	Planar bridged	-	v	6
618	$X_2Y_2Z_2$	C_{2v}	Planar cis-$X_2Y_2Z_2$	cis-$C_2H_2Cl_2$	xvi	17
619	$X_2Y_2Z_2$	C_{2h}	Planar trans-$X_2Y_2Z_2$	tr-$C_2H_2Cl_2$	xvi	17
620	XY_3ZW	C_{3v}	Pyramidal-axial with linear XZW	CH_3CN	x	11
621	XY_3WZ	C_{3v}	Pyramidal-axial with linear WXZ	-	-	-
622	WXY_2Z_2	C_{2v}	Planar	CH_2CCl_2	xvi	17
623	WXY_2Z_2	C_{2v}	Twisted	-	xvi	17
624	XY_2Z_2W	C_{2v}	With central atom X	SOF_4	xiii	14
625	$WX(YZ)_2$	C_{2v}	Planar with linear XYZ	$CO(CN)_2$	Chapter 14	
626	XY_2ZUV	C_{2v}	Planar with linear XZUV	-	xvi	17
627	XY_2ZUV	C_s	With central atom X	$SO_2F(OH)$	xvi	17
628	XY_2ZUV	C_s	Planar XY_2ZU, linear XZU	-	xvi	17
629	UVWXYZ	$C_{\infty v}$	Linear	HCCCCCl	xv	16
630	XWYZUV	C_s	Planar with linear XZUV	CHOCCH	-	-
	Seven-atomic					
701	XY_6	O_h	Octahedral	SF_6	o	1
702	XY_6	D_{6h}	Planar symmetrical	-	xi	12
703	XY_4Z_2	D_{4h}	Bipyramidal	-	-	-
704	$X(YZ)_3$	D_{3h}	Planar symmetrical (linear XYZ)	-	-	25,26
705	$X(YZ)_3$	C_{3v}	Pyramidal with linear XYZ	$P(CN)_3$	-	27,28
706	$X(YZ)_3$	C_{3h}	Planar with nonlinear XYZ	$B(OH)_3$	-	29-33

Table 20-I (Continued)

No.*	Type	Symmetry	Description	Example	S	Ref.
707	$W(XY_2)_2$	D_{2h}	Planar symmetrical (linear XWX)	-	-	-
708	$W(XY_2)_2$	D_{2d}	Twisted with linear XWX	allene	iii	4
709	$W(XY_2)_2$	C_{2v}	With bent XWX	ethylene oxide	-	34,35
710	$W(XY_2)_2$	C_{2v}	cis-WX_2Y_4 with linear XWX	-	-	-
711	$W(XY_2)_2$	C_{2h}	trans-WX_2Y_4 with linear XWX	$H_5O_2^+$	-	36
712	XY_4ZW	C_{4v}	Pyramidal-axial with linear XZW	SF_5Cl	-	37,38
713	XY_4WZ	C_{4v}	Pyramidal-axial with linear WXZ	-	-	-
714	$W(XYZ)_2$	$D_{\infty h}$	Linear	C_6N_2	xv	16
715	$X(YZ)_2W_2$	C_{2v}	With linear XYZ	$CH_2(CN)_2$	-	39
716	$X(YZ)_2W_2$	C_{2v}	With nonlinear XYZ	cyclopropene	-	40,41
717	$X(YV)_2Z_2$	C_{2v}	Planar five-membered ring	1,3,4-thiadiazole	xii	13
718	$XY_2(ZW)_2$	C_{2v}	Planar five-membered ring	1,2,5-thiadiazole	xii	13
719	$XY_2(UV)_2$	C_{2v}	With central atom X	$SO_2(OH)_2$	xiii	14
720	XY_3ZUV	C_{3v}	Pyramidal-axial with linear XZUV	SiH_3NCS	x	11
721	XY_2ZUV_2	C_s	With symmetry plane through ZXU	CF_3NO_2	-	42

* The hundreds indicate number of atoms.

REFERENCES

1 S.J.Cyvin: Molecular Vibrations and Mean Square Amplitudes, Universitetsforlaget, Oslo, and Elsevier, Amsterdam, 1968.

2 S.J.Cyvin, J.Brunvoll, B.N.Cyvin, I.Elvebredd, and G.Hagen, Mol.Phys. 14, 43 (1968).

3 S.J.Cyvin, B.N.Cyvin, I.Elvebredd, G.Hagen, and J.Brunvoll, Kgl. Norske Videnskab. Selskabs Skrifter, No. 22 (1972).

4 B.N.Cyvin, S.J.Cyvin, G.Hagen, I.Elvebredd, and J.Brunvoll,

Z.Naturforschg. 24a, 643 (1969).
5 S.J.Cyvin, I.Elvebredd, J.Brunvoll, and G.Hagen, Acta Chem. Scand. 22, 1491 (1968).
6 B.Vizi and S.J.Cyvin, Acta Chem.Scand. 22, 2012 (1968).
7 G.Hagen and S.J.Cyvin, J.Phys.Chem. 72, 1451 (1968).
8 I.Elvebredd and S.J.Cyvin, J.Mol.Structure 2, 321 (1968).
9 B.Vizi and S.J.Cyvin, Acta Chim.Hung. 59, 91 (1969).
10 J.Brunvoll and S.J.Cyvin, J.Mol.Structure 3, 155 (1969).
11 J.Brunvoll and S.J.Cyvin, J.Mol.Structure 6, 289 (1970).
12 B.N.Cyvin, S.J.Cyvin, and A.Müller, Acta Chem.Scand. 23, 1352 (1969).
13 B.N.Cyvin, S.J.Cyvin, and R.Stølevik, Acta Chem.Scand. 22, 3034 (1968).
14 S.J.Cyvin, Acta Chem.Scand. 24, 1499 (1970).
15 B.Vizi and S.J.Cyvin, Acta Chim.Hung. 64, 351 (1970).
16 B.Vizi, S.J.Cyvin, and B.N.Cyvin, Acta Chim.Hung.
17 S.J.Cyvin, G.Hagen, and B.N.Cyvin, Z.Naturforschg. 25a, 363 (1970).
18 R.S.Mulliken, J.Chem.Phys. 23, 1997 (1955).
19 G.Herzberg: Molecular Spectra and Molecular Structure. Vol. II. Infrared and Raman Spectra of Polyatomic Molecules, D. van Nostrand, New York 1945.
20 D.R.J.Boyd and H.C.Longuet-Higgins, Proc.Roy.Soc.(London) A213, 55 (1952).
21 R.S. McDowell, J.Chem.Phys. 43, 319 (1965).
22 C. di Lauro and I.M.Mills, J.Mol.Spectry. 21, 386 (1966).
23 S.G.W.Ginn and S.Reichman, J.Mol.Spectry. 25, 406 (1968).
24 J.Duinker and I.M.Mills, Spectrochim.Acta 24A, 417 (1968).
25 C.W.F.T.Pistorius, Bull.Soc.Chim.Belg. 67, 566 (1968).
26 R.W.Mooney, S.J.Cyvin, and J.Brunvoll, Acta Chem.Scand. 19, 983 (1965).
27 F.A.Miller, S.G.Frankiss, and O.Sala, Spectrochim.Acta 21, 775 (1965).
28 G.Nagarajan, Acta Phys.Hung. 20, 323 (1966).
29 C.W.F.T.Pistorius, J.Chem.Phys. 31, 1454 (1959).
30 R.W.Mooney, S.J.Cyvin, J.Brunvoll, and L.A.Kristiansen, J.Chem. Phys. 42, 3741 (1965).
31 S.J.Cyvin, R.W.Mooney, J.Brunvoll, and L.A.Kristiansen, Acta

Chem.Scand. 19, 1031 (1965).

32 L.A.Kristiansen, R.W.Mooney, S.J.Cyvin, and J.Brunvoll, Acta Chem.Scand. 19, 1749 (1965).

33 R.Ottinger, S.J.Cyvin, R.W.Mooney, L.A.Kristiansen, and J. Brunvoll, Acta Chem.Scand. 20, 1389 (1966).

34 K.Venkateswarlu and G.Thyagarajan, Proc.Indian Acad.Sci. A52, 101 (1960).

35 J.M.Freeman and T.Henshall, Can.J.Chem. 46, 2135 (1968).

36 E.Chemouni, M.Fournier, J.Rozière, and J.Potier, J.Chim.Phys. 67, 517 (1970).

37 K.Venkateswarlu and K.Sathianandan, Opt.Spektroskopiya 11, 24 (1961).

38 W.A.Yeranos, Mol.Phys. 9, 455 (1965).

39 E.Hirota and Y.Morino, Bull.Chem.Soc.Japan 33, 158 (1960).

40 R.W.Mitchell, E.Dorko, and J.A.Merritt, J.Mol.Spectry. 26, 197 (1968).

41 G.Hagen, Acta Chem.Scand. 23, 2311 (1962).

42 A.Castelli, A.Palm, and Ch.Alexander,Jr., J.Chem.Phys. 44, 1577 (1966).

CHAPTER 21

Characteristic Mean Amplitudes of Vibration

A. MÜLLER, E. J. BARAN, and K. H. SCHMIDT

For the most important bonds the mean amplitudes of vibration l_{X-Y}, which have been calculated from spectroscopic data, are summarized and compared. The l_{X-Y}-values are characteristic if the corresponding stretching frequencies $\nu(X-Y)$ also are characteristic. Some rules for mean amplitudes are given.

Für die am häufigsten in Molekülen auftretenden Bindungen werden mittlere Schwingungsamplituden l_{X-Y} aus spektroskopischen Daten tabellarisch aufgeführt und verglichen. Es zeigt sich, daß die l_{X-Y}-Werte für gebundene Atome unter der Voraussetzung, daß die zugehörigen $\nu(X-Y)$-Valenzschwingungen charakteristi sind, sich ebenfalls als charakteristisch erweisen. Es werden verschiedene Regeln für mittlere Schwingungsamplituden angegeben.

Se tabulan y comparan las amplitudes medias de vibración l_{X-Y}, calculadas a partir de datos espectroscópicos, para los enlaces que se presentan más frecuentemente en diversos tipos de moléculas. Se muestra que los valores de l_{X-Y} son característicos para un enlace X-Y, siempre que la frecuencia de estiramiento $\nu(X-Y)$ también lo sea. Asimismo, se presentan diversas reglas par las amplitudes medias de vibración.

In recent years many mean amplitudes of vibration have been calculated from spectroscopic data by several authors. If one compares the values which have been reported, it is clearly seen that the mean amplitudes vary very much from one author to another. In earlier papers we have mentioned these facts and have given some reasons for the discrepancies [536, 545].* We have also published some approximation methods in order to eliminate some of the difficulties; see [536, 545, 660, 559, 534, 535, 661, 556], Ref. 1, as well as [523]. In one of the papers we mentioned that characteristic values exist not only for frequencies and force constants, but also for mean amplitudes of vibration for bonded and nonbonded atoms [559]. (We will consider here only those for bonded atoms.) This means: if the stretching force constant and the corresponding frequency for a definite bond in different molecules are nearly the same, this will also be true for mean amplitudes of vibration.

It is the main aim of this paper to demonstrate these facts for several molecules and to summarize the mean amplitudes, l, of bonded atoms for the most important bonds in several molecules. In addition some rules for mean amplitudes of vibration are given.

In Table I** l values for boron-halogen compounds are reported. It is seen that the ^{10}B—^{11}B isotope effect is not very large, but that l in the ^{10}B compounds are always larger than those in the ^{11}B compounds. On the other hand, in the ions BCl_4^- and BBr_4^- l for the boron-halogen bonds are larger than l in the corresponding BX_3 compounds. This is due to the fact that the BX_3 force constants in the planar molecules are larger than those in the tetrahedral ions because of the existence of $(p\pi-p\pi)$ bonds in the planar molecules.

In Table II the l values for the carbon-hydrogen bond in several organic and inorganic compounds are summarized. The l values for the C-H bond are about 0.075 Å and for the C-D bond about 0.065 Å. The ν_{C-H} frequency is highly characteristic, and therefore the corresponding l are similar in all molecules. They are very large compared with other l because of the low mass of the H atom and the relatively high force constant of the C-H bond. It is interesting to

* In some papers of Nagarajan and one of the present authors (A.M.) the published values are not reliable [547-552, 640-644]; for corrected values, see [538, 553, 554, 560, 563, 663, 881].

** *Note to tables:* All mean amplitude values are in Å and pertain to 298 K. Both ordinary molecules and ions are listed under the heading 'Molecule'.

Table 21-I. Boron-halogen and boron-oxygen bonds: l_{B-X} in BX_3.

Molecule	l	Ref.	Molecule	l	Ref.
$^{10}BF_3$	0.043_1	[232]	$^{11}BF_3$	0.042_5	[232]
$^{10}BCl_3$	0.050_0	[232]	$^{11}BCl_3$	0.049_3	[232]
$^{10}BBr_3$	0.0518	[232]	$^{11}BBr_3$	0.0510	[232]
$^{10}BI_3$	0.0561	[229]	$^{11}BI_3$	0.0554	[229]
BF_4^-	0.050	[538]	$B(OH)_3$	0.0435	[663]
BCl_4^-	0.060	[538]	$B(OH)_4^-$	0.055	[538]
BBr_4^-	0.062	[538]	BO_3^{---}	0.0470	[663]

Table 21-II. Carbon-hydrogen and carbon-deuterium bonds: $l_{C-H(C-D)}$.

Molecule	l	Ref.	Molecule	l	Ref.
CH_4	0.07750	[229]	CD_4	0.06643	[229]
C_6H_6 (benzene)	0.07705	[229]	C_6D_6	0.06589	[229]
$H_2C{=}CH_2$	0.07673	[229]	$D_2C{=}CD_2$	0.06574	[229]
C_3H_6 (cyclopropane)	0.0751	[229]	C_3D_6	0.0643	[229]
HC≡CH	0.0743	[229]	DC≡CD	0.0637	[229]
HC≡CCN	0.074	[229]			
$HCOO^-$	0.0802	[543]	$DCOO^-$	0.0683	[543]
CHF_3	0.0777	[229]	CH_2FCl	0.0776	[229]
$CHCl_3$	0.0777	[229]	CH_2FBr	0.0776	[229]
$CHBr_3$	0.0777	[229]	CH_2ClI	0.0782	[229]
CH_3Cl	0.0761	[229]	CH_2BrI	0.0783	[229]
CH_3Br	0.0759	[229]	CHFClBr	0.0775	[229]
CH_3I	0.0758	[229]			

compare l_{C-H} of CH_4 (sp^3 hybrid), C_2H_4 (sp^2 hybrid), and C_2H_2 (sp hybrid). With increasing force constant (due to a larger s-character) the l_{C-H} value is found to decrease.

In Table III the values for carbon-chalcogen bonds are given.

Table 21-III. Carbon-oxygen, -sulphur, -selenium, and -tellurium bonds: l_{C-X} (X = O, S, Se, Te).

Molecule	l_{C-O}	Ref.
CO	0.0337	[229]
CO_2	0.03459	[229]
CO_3^{--}	0.0428	[663]
CH_2O	0.0377	[232,543]
CD_2O	0.0375	[232,543]
HCO_2^-	0.0409	[543]
DCO_2^-	0.0408	[543]
CF_2O	0.0362	[232,543]
CCl_2O	0.0369	[232,543]
CBr_2O	0.0368	[232,543]
COSe	0.0353	[232]
$Ni(CO)_4$	0.0348	[229]

Molecule	l_{C-Te}	Ref.
CSTe	0.0428	[232]

Molecule	l_{C-S}	Ref.
CS	0.0389	[229]
CS_2	0.0387	[229]
CSSe	0.0389	[232]
CSTe	0.0391	[232]
CS_3^{--}	0.0488	[663]
CF_2S	0.0385	[232,543]
CCl_2S	0.0420	[232,543]

Molecule	l_{C-Se}	Ref.
CSe	0.0398	[229]
CSe_2	0.0396	[229]
COSe	0.0392	[232]
CSSe	0.0399	[232]
CSe_3^{--}	0.0486	[663]

Table 21-IV. Carbon-carbon and carbon-nitrogen bonds: l_{CC}, $l_{C\equiv N}$.

Molecule	l_{CC}	Ref.	Molecule	l_{CC}	Ref.
$H_2C{=}CH_2$	0.04280	[229]	$D_2C{=}CD_2$	0.04273	[229]
HC≡CH	0.0357	[229]	DC≡CD	0.0357	[229]
HC≡C-CN	0.036(C≡C)	[229]	HC≡C-CN	0.042(C-C)	[229]
C_2F_4	0.0392	[229]	C_2Br_4	0.0430	[229]
C_2Cl_4	0.0425	[229]			

Molecule	$l_{C\equiv N}$	Ref.	Molecule	$l_{C\equiv N}$	Ref.
FCN	0.0344	[232]	HCN	0.03417	[229]
ClCN	0.0345	[232]	DCN	0.03410	[229]
BrCN	0.0345	[232]	C_2N_2	0.0349	[229]
ICN	0.0348	[232]	$S(CN)_2$	0.0349	[229]
HC≡C-CN	0.0345	[229]			

Table 21-V. Carbon-halogen bonds: l_{C-X} (X = F, Cl, Br, I).

Molecule	l_{C-F}	Ref.	Molecule	l_{C-Br}	Ref.
CF_4	0.044	[538]	CBr_4	0.055	[538]
FCN	0.0392	[229]	BrCN	0.0436	[229]
CH_2FCl	0.0464	[229]	CH_2FBr	0.0508	[229]
CH_2FBr	0.0463	[229]	CH_2BrI	0.0510	[229]
CHFClBr	0.0464	[229]	CHFClBr	0.0516	[229]
CHF_3	0.0451	[229]	CH_3Br	0.0519	[229]
C_2F_4	0.0429	[229]	$CHBr_3$	0.0532	[229]
CF_2O	0.0436	[232,543]	C_2Br_4	0.0499	[229]
CF_2S	0.0457	[232,543]	CBr_2O	0.0506	[232,543]

Molecule	l_{C-Cl}	Ref.	Molecule	l_{C-I}	Ref.
CCl_4	0.054	[538]	CI_4	0.063	[538]
ClCN	0.0427	[232]	ICN	0.0480	[229]
CH_2FCl	0.0498	[229]	CH_2ClI	0.0528	[229]
CHFClBr	0.0491	[229]	CH_2BrI	0.0527	[229]
CH_3Cl	0.0485	[229]	CH_3I	0.0549	[229]
C_2Cl_4	0.0481	[229]	C_2I_2	0.0472	[229]
CCl_2O	0.0498	[232,543]			
CCl_2S	0.0509	[232,543]			

Also here characteristic values exist. For example l_{C-Se} in CSe, CSe_2, COSe, and CSSe is nearly the same. Contrary to this l_{C-Se} in CSe_3^{--} is larger because of the low bond order (or force constant) in this ion.

In the case of l_{C-C} (Table IV) one can see that the mean amplitudes of vibration of the C=C bonds are larger than those of the C≡C bonds. It is well known than the C≡N frequency is highly characteristic. Therefore we should expect that the corresponding force constant and $l_{C\equiv N}$ would also be characteristic. Indeed, Table IV shows that all values are nearly identical. It is quite certain that $l_{C\equiv N}$ in different molecules can be transferred without making a significant error.

l_{C-F} in different molecules is also very similar (Table V).

Table 21-VI. Nitrogen-oxygen, -sulphur, and -halogen bonds: $l_{N-O(N-S,N-X)}$ (X = F, Cl, Br).

Molecule	l_{N-O}	Ref.
NO	0.0344	[229]
NO_2	0.0382	[229]
NOF	0.04226	[229]
NOCl	0.04226	[229]
NOBr	0.04229	[229]
FNO_2	0.0380	[543]
$ClNO_2$	0.0388	[543]
$(OH)NO_2$	0.039	[543]
$(OD)NO_2$	0.0391	[543]
NO_3^-	0.0418	[663]
$(CH_3)_2NO_2$	0.0394	[543]

Molecule	l_{N-S}	Ref.
NSCl	0.0392	[554]
NSF_3	0.0346	[229]

Molecule	l_{N-X}	Ref.
NOF	0.05701	[229]
FNO_2	0.0481	[543]
NF_3	0.048_7	[736]
$ClNO_2$	0.0434	[543]
ONCl	0.05754	[229]
ONBr	0.05899	[229]

An exception is the C-F bond in F-C≡N. This is due to the fact that the C-F force constant in this molecule is higher than in the other molecules mentioned in this table. The same is true for the C-Cl and C-Br bonds. The variation of the l_{C-I} values is larger because the stretching vibration of C-I is not very characteristic, since the C mass is small compared with the I mass.

In the case of different nitrogen-oxygen compounds the l_{N-O} value becomes lower as the stretching force constant increases. NO has the lowest l value because it has the highest bond order among the molecules in Table VI. We find the highest l_{N-O} value in the nitrate ion because of the delocalization of π bonds over three bonds.

Table VII shows some l values for bonds between halogens and third row metals: Al, Ga, In, Tl.

From Table VIII one can see that the Si-H and Si-D mean amplitudes of vibration are on the one hand very large and on the other hand very characteristic, and therefore transferable from one molecule to the other. Similar characteristic values may be found for l_{Si-F} and l_{Si-Cl}.

The mean amplitudes of vibration for several germanium and tin hydrogen and halogen bonds are given in Table IX. In the series

Table 21-VII. Aluminum-, gallium-, indium-, and thallium-halogen bonds: l_{Me-X} (Me = metal, X = halogen).

Molecule	l	Ref.	Molecule	l	Ref.
AlF_3	0.0429	[229]	$InCl_4^-$	0.053	[538]
$AlCl_4^-$	0.055	[538]	$InBr_4^-$	0.056	[538]
$AlCl_3$	0.0494	[229]	InI_4^-	0.063	[538]
$GaCl_4^-$	0.052	[538]	$TlBr_4^-$	0.056	[538]
$GaBr_4^-$	0.056	[538]			

Table 21-VIII. Silicon-hydrogen, -deuterium, and -halogen bonds: $l_{Si-H(Si-D)}$, l_{Si-X} (X = F, Cl, Br, I).

Molecule	l_{Si-H}	Ref.
SiH_4	0.0888	[229]
SiH_3F	0.0888	[543]
SiH_3Cl	0.0888	[543]
SiH_3Br	0.0888	[543]
SiH_3I	0.0888	[543]
$SiHF_3$	0.0865	[543]
$SiHCl_3$	0.0876	[543]
$SiHBr_3$	0.0880	[543]
SiH_2Cl_2	0.08776	[229]
SiH_2Br_2	0.08806	[229]
SiH_2D_2	0.08784	[229]
Molecule	l_{Si-D}	**Ref.**
SiD_4	0.0753	[229]
$SiDF_3$	0.0735	[229]
SiH_2D_2	0.07886	[229]

Molecule	l_{Si-F}	Ref.
SiF_4	0.0394	[229]
SiH_3F	0.0409	[543]
$SiHF_3$	0.0402	[543]
$SiDF_3$	0.03994	[229]
$SiFCl_3$	0.0413	[543]

Molecule	l_{Si-Cl}	Ref.
$SiCl_4$	0.0459	[662]
SiH_3Cl	0.0461	[543]
$SiFCl_3$	0.0416	[543]
$SiCl_3Br$	0.0460	[543]
$SiClBr_3$	0.0471	[543]
$SiCl_3I$	0.0467	[543]
$SiClI_3$	0.0480	[543]

Molecule	l_{Si-Br}	Ref.
$SiBr_4$	0.0488	[229]
SiH_3Br	0.0476	[543]
$SiHBr_3$	0.0476	[543]
$SiCl_3Br$	0.0502	[543]
$SiClBr_3$	0.0477	[543]
$SiBr_3I$	0.0463	[543]

Molecule	l_{Si-I}	Ref.
SiI_4	0.0546	[229]
SiH_3I	0.0516	[543]
$SiCl_3I$	0.0510	[543]
$SiClI_3$	0.0528	[543]

Table 21-IX. Germanium, tin, and lead bonds with hydrogen, deuterium, and halogen: $l_{Me-H(Me-D)}$, l_{Me-X} (Me = Ge, Sn, Pb; X = F, Cl, Br, I).

Molecule	$l_{Ge-H(Ge-D)}$	Ref.	Molecule	l_{Ge-X}	Ref.
GeH_4	0.0895	[229]	GeF_4	0.039	[538]
GeD_4	0.0755	[229]	$GeCl_4$	0.045	[538,421]
$GeHCl_3$	0.0886	[543]	$GeBr_4$	0.048	[538]
$GeHBr_3$	0.0895	[543]	GeI_4	0.054	[538]
GeH_3F	0.0893	[543]	GeH_3F	0.0415	[543]
GeH_3Cl	0.0893	[543]	GeH_3Cl	0.0465	[543]
GeH_3Br	0.0894	[543]	GeH_3Br	0.0477	[543]
GeH_3I	0.0895	[543]	GeH_3I	0.0517	[543]
Molecule	l_{Sn-H}	**Ref.**	**Molecule**	l_{Sn-D}	**Ref.**
SnH_4	0.0942	[229]	SnD_4	0.0794	[229]
Molecule	l_{Sn-Cl}	**Ref.**	**Molecule**	l_{Sn-Br}	**Ref.**
$SnCl_4$	0.0472	[421]	$SnBr_4$	0.048	[538]
$SnCl_3Br$	0.0468	[543]	$SnCl_3Br$	0.0468	[543]
$SnClBr_3$	0.0480	[543]	$SnClBr_3$	0.0481	[543]
Molecule	l_{Sn-I}	**Ref.**	**Molecule**	l_{Pb-Cl}	**Ref.**
SnI_4	0.055	[538]	$PbCl_4$	0.049	[538]

Table 21-X. Phosphorus-hydrogen, -deuterium, -oxygen, and -sulphur bonds: l_{P-H}, l_{P-D}, l_{P-O}, l_{P-S}.

Molecule	$l_{P-H(P-D)}$	Ref.
PH_3	0.0869	[229]
PD_3	0.0737	[229]
Molecule	l_{P-S}	**Ref.**
PSF_3	0.0386	[229]
$PSCl_3$	0.0391	[229]
$PSBr_3$	0.0402	[229]
PSO_3^{---}	0.0479	[543]

Molecule	l_{P-O}	Ref.
PO_4^{---}	0.041	[538]
HPO_4^{--}	0.0398	[543]
PSO_3^{---}	0.0405	[543]
POF_3	0.0340	[543]
$POCl_3$	0.0354	[543]
$POBr_3$	0.0359	[543]

Table 21-XI. Phosphorus-halogen bonds: l_{P-X} (X = F, Cl, Br, I).

Molecule	l_{P-F}	Ref.
PF_3	0.041	[229]
PF_5	0.0448(ax)	[244]
PF_5	0.0406(eq)	[244]
POF_3	0.0390	[543]
PSF_3	0.0396	[280]
PF_2Cl	0.0418	[312]
PF_2Br	0.0419	[312]
PF_2I	0.0420	[312]
$PFCl_2$	0.0424	[312]
$PFBr_2$	0.0425	[312]

Molecule	l_{P-Cl}	Ref.
PCl_3	0.050	[229]
PCl_4^+	0.043	[538]
PCl_5	0.0577(ax)	[229]
PCl_5	0.0502(eq)	[229]
$POCl_3$	0.0464	[543]
$PSCl_3$	0.0484	[280]
PF_2Cl	0.0470	[312]
$PFCl_2$	0.0485	[312]
PCl_2Br	0.0507	[312]
PCl_2I	0.0513	[312]
PBr_2Cl	0.0520	[312]
$PClI_2$	0.0524	[312]

Molecule	l_{P-Br}	Ref.
PBr_3	0.051	[229]
PBr_4^+	0.049	[538]
$POBr_3$	0.0464	[543]
$PSBr_3$	0.0503	[280]
PF_2Br	0.0459	[312]
PCl_2Br	0.0502	[312]
$PFBr_2$	0.0491	[312]
$PClBr_2$	0.0519	[312]
PBr_2I	0.0539	[312]
$PBrI_2$	0.0539	[312]

Molecule	l_{P-I}	Ref.
PI_3	0.057	[229]
PF_2I	0.0471	[312]
PCl_2I	0.0543	[312]
PBr_2I	0.0546	[312]
$PClI_2$	0.0573	[312]
$PBrI_2$	0.0574	[312]

GeF_4-$GeCl_4$-$GeBr_4$-GeI_4 the mean amplitudes increase. The same featur is observed for Ge-halogen in the series GeH_3F-GeH_3Cl-GeH_3Br-GeH_3I, and for $SnCl_4$-$SnBr_4$-SnI_4. The reason for this is that the influence of decreasing force constants on l is larger than the mass influenc

In accordance with the well known fact that the P-O bond orde decreases in the series POF_3-$POCl_3$-$POBr_3$ we find the inverse sequer for l_{P-O} (Table X). The bond order in PO_4^{---} is lower than that in the other three molecules mentioned above. Consequently one finds a higher mean amplitude. The same is valid for the thiophosphoryl

Table 21-XII. Arsenic and antimony bonds with hydrogen, deuterium, halogen, and sulphur: $l_{Me-H(Me-D)}$, l_{Me-X}, l_{Me-S} (Me = As, Sb; X = F, Cl, Br, I).

Molecule	$l_{Me-H(Me-D)}$	Ref.
AsH_3	0.0909	[229]
AsD_3	0.0769	[229]
SbH_3	0.0970	[229]
SbD_3	0.0822	[229]

Molecule	l_{Me-F}	Ref.
AsF_3	0.0425	[229]
AsF_5	0.0410(ax)	[244]
AsF_5	0.0386(eq)	[244]
SbF_3Cl_2	0.0422(eq)	[229]

Molecule	l_{As-Cl}	Ref.
$AsCl_3$	0.051	[229]
$AsCl_2Br$	0.0510	[312]
$AsClBr_2$	0.0510	[312]
$AsCl_4^+$	0.042	[538]

Molecule	l_{Sb-Cl}	Ref.
$SbCl_3$	0.053	[229]
SbF_3Cl_2	0.0461	[229]

Molecule	l_{Me-Br}	Ref.
$AsBr_3$	0.053	[229]
$AsCl_2Br$	0.0515	[312]
$AsClBr_2$	0.0522	[312]
$SbBr_3$	0.052	[229]

Molecule	l_{Me-I}	Ref.
AsI_3	0.058	[229]
SbI_3	0.0706	[229]

Molecule	l_{Me-S}	Ref.
AsS_4^{---}	0.048	[545]
SbS_4^{---}	0.049	[545]

halides and for SPO_3^{---}. In phosphorus-halogen molecules (Table XI) regularities between the l_{P-Hal} values and the corresponding force constants are found. Highly characteristic values exist for AsH_3 and SbH_3 (Table XII). l_{S-O} values for molecules with S-O bonds having strong π bonds are rather characteristic (see Table XIII). The differences in the l's of S-F bonds should be mainly due to differences in the corresponding bond strengths. Table XIV shows l values for bonds involving Se or Te.

In Table XV the l values for different halogen hydrides are given. With increasing mass of the halogen atom the mean amplitude increases. From Table XVI one can easily see that in the series MnO_4^--MnO_4^{--}-MnO_4^{---} the l_{Mn-O} values increase because of decreasing force constants. For similar molecules like ReO_3Cl and ReO_3Br the l_{Re-O} values are precisely the same. This is also true for l_{Cr-O} in CrO_3F^- and CrO_3Cl^-. The comparison of the pseudo-isoelectronic ions

Table 21-XIII. Sulphur-hydrogen, -deuterium, -oxygen, and -halogen bonds: $l_{S-H(S-D)}$, l_{S-O}, l_{S-X} (X = F, Cl).

Molecule	l_{S-H}	Ref.	Molecule	l_{S-D}	Ref.
H_2S	0.07956	[229]	D_2S	0.06743	[229]
Molecule	l_{S-O}	**Ref.**	**Molecule**	l_{S-F}	**Ref.**
SO_2	0.03546	[229]	SF_4	0.046(ax)	[230]
SO_3	0.0349	[232,663,250]	SF_4	0.043(eq)	[230]
SO_3^{--}	0.041	[545]	SF_6	0.0419	[226,247]
SO_4^{--}	0.039	[538]	SO_3F^-	0.0426	[543]
$S_2O_3^{--}$	0.0390	[543]	SO_2F_2	0.04282	[229]
HSO_3^-	0.0375	[543]	SOF_2	0.04354	[229]
SO_3F^-	0.0364	[543]	SOF_4	0.04060(ax)	[229]
SO_3Cl^-	0.0375	[543]	SOF_4	0.04545(eq)	[229]
SOF_2	0.03945	[229]	NSF_3	0.0440	[229]
$SOCl_2$	0.03957	[229]	**Molecule**	l_{S-Cl}	**Ref.**
$SOBr_2$	0.03972	[229]	SCl_2	0.0487	[229]
SOF_4	0.03617	[229]	SO_3Cl^-	0.0485	[543]
SO_2F_2	0.03969	[229]	SO_2Cl_2	0.05136	[229]
SO_2Cl_2	0.03976	[229]	$SOCl_2$	0.05392	[229]

Table 21-XIV. Selenium and tellurium bonds with hydrogen, deuterium, oxygen, and halogen: $l_{Se-H(Se-D)}$, l_{Me-O}, l_{Me-X} (Me = Se, Te; X = F, Cl).

Molecule	l	Ref.	Molecule	l	Ref.
H_2Se	0.0848	[229]	SeF_6	0.0399	[247]
D_2Se	0.0715	[229]	$SeCl_3^+$	0.047	[545]
SeO_3^{--}	0.041	[545]	TeF_6	0.0376	[247]
SeO_4^{--}	0.039	[538]	$TeCl_3^+$	0.047	[545]
TeO_3^{--}	0.042	[545]			

MoO_4^{--}, WO_4^{--}, TcO_4^-, and ReO_4^- shows a great similarity of the mea amplitude values because of similar bond properties though the masses vary very much from one ion to the other. This is again due

Table 21-XV. Halogen-hydrogen and halogen-oxygen bonds: l_{H-X}, l_{X-O} (X = F, Cl, Br, I).

Molecule	l_{H-X}	Ref.
HF	0.0652	[229]
HCl	0.0759	[229]
HBr	0.0800	[229]
HI	0.0854	[229]
Molecule	l_{X-O}	**Ref.**
F_2O	0.0510	[229]
$HClO_4$	0.0367	[543]
ClO_4^-	0.038	[538, 69]
ClO_3^-	0.040	[545]
ClO_2	0.0385	[229]
Cl_2O	0.0437	[229]
BrO_4^-	0.039	[69]
BrO_3^-	0.040	[545]
IO_4^-	0.039	[69]
IO_3^-	0.039	[545]

Table 21-XVI. Transition metal-oxygen, -sulphur, and -selenium bonds: l_{Me-X} (Me = transition metal; X = O, S, Se).

Molecule	l_{Me-O}	Ref.
VO_4^{---}	0.043	[538]
VOF_3	0.0366	[543]
$VOCl_3$	0.0369	[543]
$VOBr_3$	0.0371	[543]
CrO_4^{--}	0.040	[538]
CrO_3F^-	0.0389	[543]
CrO_3Cl^-	0.0388	[543]
MoO_4^{--}	0.039	[538]
WO_4^{--}	0.037	[538]
MnO_4^-	0.040	[538]
MnO_4^{--}	0.041	[538]
MnO_4^{---}	0.043	[538]
TcO_4^-	0.037	[538]
ReO_4^-	0.036	[538]
ReO_3S^-	0.0360	[543]
ReO_3Cl	0.0346	[543]
ReO_3Br	0.0346	[543]
FeO_4^{--}	0.042	[538]
RuO_4	0.037	[538]
RuO_4^-	0.039	[538]
RuO_4^{--}	0.039	[538]
OsO_4	0.035	[538]
OsO_3N^-	0.0366	[537]
Molecule	l_{Me-S}	**Ref.**
VS_4^{---}	0.049	2
NbS_4^{---}	0.047	2
TaS_4^{---}	0.045	2
MoS_4^{--}	0.043	[538]
WS_4^{--}	0.040	[538]
ReS_4^-	0.0391	[540]
ReO_3S^-	0.0384	[543]
Molecule	l_{Me-Se}	**Ref.**
VSe_4^{---}	0.048	2
$NbSe_4^{---}$	0.046	2
$TaSe_4^{---}$	0.044	2
$MoSe_4^{--}$	0.044	2
WSe_4^{--}	0.040	2

Table 21-XVII. Transition metal-halogen and -carbon bonds: l_{Me-X}, l_{Me-C} (Me = transition metal; X = F, Cl, Br, I).

Molecule	l_{Me-X}	Ref.
$TiCl_4$	0.047	[538]
$TiBr_4$	0.048	[538]
$ZrCl_4$	0.046	[538]
$HfCl_4$	0.0464	[229]
$HfBr_4$	0.0509	[229]
HfI_4	0.0471	[229]
VF_5	0.0415(ax)	[881]
VF_5	0.0405(eq)	[881]
VCl_4	0.0471	[229]
VOF_3	0.0404	[543]
$VOCl_3$	0.0459	[543]
$VOBr_3$	0.0465	[543]
		cont.

cont.		
CrO_3Cl^-	0.0485	[543]
WF_6	0.054	[538]
ReO_3Cl	0.0550	[543]
ReO_3Br	0.0571	[543]
$FeCl_4^{--}$	0.054	[538]
OsF_6	0.0385	[229]
RhF_6	0.0413	[247]
IrF_6	0.0389	[247]
PtF_6	0.0400	[247]

Molecule	l_{Me-C}	Ref.
$Mo(CO)_6$	0.0664	[229]
$Cr(CO)_6$	0.0556	[229]
$Ni(CO)_4$	0.0522	[229]

Table 21-XVIII. Zinc-, cadmium, and mercury-halogen bonds: l_{Me-X} (Me = Zn, Cd, Hg; X = F, Cl, Br, I).

Molecule	l	Ref.	Molecule	l	Ref.
$ZnCl_4^{--}$	0.065	[538]	$HgCl_4^{--}$	0.060	[538]
$ZnBr_4^{--}$	0.072	[538]	HgF_2	0.0427	[229]
ZnI_4^{--}	0.083	[538]	$HgCl_2$	0.0444	[229]
$CdBr_4^{--}$	0.071	[538]	$HgBr_2$	0.0448	[229]
CdI_4^{--}	0.079	[538]	HgI_2	0.0495	[229]

to the fact that the influence of the f_{ij} values on l_{ij} is greater than the influence of the different masses. Finally some l values for bonds involving transition metals and halogens are shown in Tables XVII and XVIII.

We wish to summarize the main points below. The l values which we have given in the tables can help to get an idea of the magnitude of the mean amplitudes of vibration for bonded atoms in different molecules and ions. We have mentioned already that the l's are almost characteristic for definite bonds if the corresponding force constants or stretching vibration frequencies are characteristic. In these cases the following formulae are valid [534, 523, 545, 559].

$$l_{X-Y}^2 = kT(f_{X-Y})^{-1} + (h^2/64\pi^2 kT)(\mu_X+\mu_Y), \qquad (21.1)$$

$$(l_{X-Y}^a)^2 - (l_{X-Y}^b)^2 = kT[(f_{X-Y}^a)^{-1} - (f_{X-Y}^b)^{-1}]; \qquad (21.2)$$

a and b indicate two different molecules. It is seen that l for the same bond (X-Y) should be the same under this approximation if the force constant is the same in different molecules. If the force constant increases l should decrease. In isoelectronic rows (e.g. $ZnCl_4^{--}$-$GaCl_4^-$-$GeCl_4$-$AsCl_4^+$) the l values decrease with increasing positive charge. This can be explained with the above formulae since the masses are nearly the same, but the force constants increase with increasing positive charge. It is possible to calculate approximately the mean amplitudes of vibration with the above simple formulae [545].

The l values in the row MnO_4^--MnO_4^{--}-MnO_4^{---} (in general: in isostructural ions with the same $\mathbf{G}$ matrix) increase with decreasing force constants f. In an earlier paper Müller [535] has given a theory of molecular vibrations of isostructural ions. In the case of all vibrational frequencies below 1200 cm^{-1} the following approximate relation between the $\mathbf{\Sigma}$ matrices for the isostructural ions a and b can be derived.

$$\mathbf{\Sigma}^a - \mathbf{\Sigma}^b = kT[(\mathbf{F}^a)^{-1} - (\mathbf{F}^b)^{-1}]. \qquad (21.3)$$

It is seen that the $\mathbf{\Sigma}$ matrix of the ion b can be calculated if the force constants of the ions a and b and the $\mathbf{\Sigma}$ matrix of a are known. From Eq. (3) it appears that a bond with a low force constant should have a high l value. A comparison of l's for the manganates(VII,VI,V) calculated with the formula (3) and those with the complete $\mathbf{\Sigma}$ matrix is found elsewhere [535].

For the case of characteristic vibrations we have derived an approximate method for the calculation of l in complex molecules.

For a molecule in which a frequency assignment can be made for at least parts of the molecule it is easy to calculate the mean amplitudes of vibration, say for ZXY- or XY_2- fragments as examples. With the **L**-matrix approximation method of Müller [536] we get the following expressions for the $\mathbf{\Sigma}$ matrix elements and l values of bonded atoms; [660, 661, 556, 561] for the l values of nonbonded atoms and further details, see Ref. [561].

XY_2-fragment ($\nu_1 = \nu_s$, $\nu_2 = \delta_s$, $\nu_3 = \nu_{as}$)*

$$\Sigma_{11} = G_{11}\delta_1, \quad \Sigma_{12} = G_{12}\delta_1, \quad \Sigma_{22} = (\delta_2 \det \mathbf{G} + \delta_1 G_{12}^2)/G_{11}, \quad \Sigma_{33} = G_{33}\delta_3,$$

$$l_{X-Y}^2 = \tfrac{1}{2}(\Sigma_{11} + \Sigma_{33}).$$

ZXY-fragment ($\nu_1 = \nu_{XY}$, $\nu_2 = \nu_{XZ}$, $\nu_3 = \delta_{XYZ}$)

$$\Sigma_{11} = G_{11}\delta_1, \quad \Sigma_{12} = G_{12}\delta_1, \quad \Sigma_{13} = G_{13}\delta_1, \quad \Sigma_{22} = (\delta_1 G_{12}^2 + \delta_2 A)/G_{11},$$

$$\Sigma_{23} = (\delta_1 G_{12} G_{13} + \delta_2 B)/G_{11}, \quad \Sigma_{33} = (\delta_1 G_{13}^2/G_{11}) + (\delta_2 B^2/G_{11} A) + (\delta_3 \det \mathbf{G}/A),$$

$$l_{X-Y}^2 = \Sigma_{11}, \quad l_{X-Z}^2 = \Sigma_{22}.$$

Here $A = G_{11}G_{22} - G_{12}^2$ and $B = G_{11}G_{23} - G_{12}G_{13}$. If $C = G_{12}G_{23} - G_{13}G_{22}$, then $\det \mathbf{G} = AG_{33} - BG_{23} + CG_{13}$.

It has been shown [561, 560, 683, 68] that, for the case of characteristic vibrations, l values calculated with the method of Müller et al. are nearly the same as those calculated with more rigorous methods.

The method described above includes vibrational coupling in the fragments, but it neglects the coupling with the rest of the molecule. The results improve as the mass of the X atom increases. It seems certain that the results obtained by this method are bett than those from Eqs. (1) and (2), which completely neglect the mix of vibrations in the fragments.

---

*The formulae for Σ_{11}, Σ_{12}, and Σ_{22} are valid for all $n = 2$ problems (**L**-matrix approximation method) [536, 661, 556].

REFERENCES*

* See also Bibliography after Chapter 22.

1 A.Müller, R.Kebabcioglu, and S.J.Cyvin, J.Mol.Structure 3, 507 (1969).

2 A.Müller, K.H.Schmidt, K.H.Tytko, J.Bouwma, and F.Jellinek, Spectrochim.Acta 28A, 381 (1972).

CHAPTER 22

Supplements to Review of Previous Work on Mean Amplitudes of Vibration

S. J. CYVIN

This is a supplement to Chapter 2 of Ref. 1, which reviews previous work on mean amplitudes of vibration. Tables I and II are supplements to Tables 2-II and 2-III of the cited monograph.[1] The majority of these supplements are included in the Russian translation of Ref. 1.[2] The monograph[1] is referred to as "Volume I" in a series of supplements[3-6] to Chapter 12 (and Chapter 15) containing numerical values of mean amplitudes (and shrinkage effects). This designation is also used in the present chapter. In Tables I and II '(*)' indicates that numerical values are reported for the appropriate molecules in Chapter 12 of Volume I. The molecules from the supplements,[3-6] which contain quoted mean amplitudes along with original calculations, are not reviewed in the present chapter. The reader should also refer to other parts of this book for original data of mean amplitudes.

Precise references to the works cited in Tables I and II are found in the next chapter. The bibliography therein also includes some works not cited in these tables because they do not contain original numerical results. Among these works are the theoretical papers from the Soviet electron-diffraction group:

EZHOV and RAMBIDI (1967), VILKOV and SADOVA (1967a), AKISHIN, RAMBIDI, and SPIRIDONOV (1967), VILKOV and MASTRYUKOV (1968), VILKOV, SPIRIDONOV, and SADOVA (1968), RAMBIDI and EZHOV (1968), VILKOV (1968).

An excellent review article on electron diffraction studies of structures of molecules in the gas phase during 50 years' research, containing 204 references, has been furnished by VILKOV, RAMBIDI, and SPIRIDONOV (1967). BASTIANSEN (1968) has contributed with a chapter to the cited book, published on the occation of Linus Pauling's 70 years' anniversary. A theoretical contribution to the interpretation of electron diffraction measurements is also due to ANDERSEN, SEIP, STRAND, and STØLEVIK (1969). A review of the activities of the Oslo electron diffraction group was presented by HARGITTAI (1969c). The article contains 70 references.

The studies of isotope effects, including the secondary isotope effects, on molecular structure and mean amplitudes as advanced mainly by L. S. Bartell, have not been fully reviewed in Volume I. Some supplementary references are given below.

BARTELL (1960c); BARTELL (1961b), Bartell (1961a) [Papers I, II]; BARTELL (1962), BARTELL and ROSKOS (1966).

J. Rundgren's dissertation consists of four papers with valuable contributions to the theory of electron diffraction and to spectroscopic analyses of diatomic molecules. Three of these papers are not found in the present tables, viz. RUNDGREN (1965), RUNDGREN (1967a), and RUNDGREN (1967b).

The following papers contain various theoretical aspects in spectroscopical studies of mean amplitudes and related quantities.

MAYANTS (1963), MAYANTS (1964), MAYANTS, GALPERN, and AVERBUKH (1965).

An approach to mean amplitude calculations by SVERDLOV and KUKINA (1967) is reported in the very useful book by Sverdlov et al.[7] NAGARAJAN and PERUMAL (1968a) have treated the cubic XY_8 model theoretically [cf. also CYVIN (1968)]. Additional references:

MÜLLER (1967), MÜLLER and PEACOCK (1968a), MÜLLER and PEACOCK (1968b), PEACOCK and MÜLLER (1968), MÜLLER and PEACOCK (1969a), MÜLLER and PEACOCK (1969b); YERANOS (1968a); VENKATESWARLU and BABU JOSEPH (1968a); PULAY, BOROSSAY, and TÖRÖK (1968); KUKINA (1969); STØLEVIK, CYVIN, and FADINI (1970).

REFERENCES

1 S.J.Cyvin: Molecular Vibrations and Mean Square Amplitudes, Universitetsforlaget, Oslo, and Elsevier, Amsterdam, 1968. [In this chapter called "Volume I"].

2 S.Cyvin: Kolebaniya molekul i srednekvadratichnye amplitudy (Molecular vibrations and mean square amplitudes; Russian transl.), Izd. "MIR", Moscow 1971.

3 S.J.Cyvin, Kgl.Norske Videnskab.Selskabs Skrifter, No. 1 (1969).

4 S.J.Cyvin, Kgl.Norske Videnskab.Selskabs Skrifter, No. 7 (1971).

5 S.J.Cyvin, Z.anorg.allgem.Chem. 378, 117 (1970).

6 S.J.Cyvin and B.Vizi, Acta Chim.Hung. 70, 55 (1971).

7 L.M.Sverdlov, M.A.Kovner, and E.P.Krainov: Kolebatelnye spektry mnogoatomnykh molekul (Vibrational spectra of polyatomic molecules), Izd. Nauka, Moscow 1970.

Table 22-I. References to mean amplitudes of vibration from electron diffractic None of the references in Table 2-II of Volume I are duplicated, but are occasionally referred to as *previous literature:* Chapt. I-2. (*) indicates that numerical values are quoted in Chapt. I-12 (Volume I).

Compound (Number of atoms in parentheses)	Reference
(2) N_2, O_2, NO, Cl_2, Br_2, I_2	(*) *previous literature:* Chapt. I-2
$(HF)_n$	Janzen and Bartell (1969)
(3) CO_2	(*) *previous literature:* Chapt. I-2; Murata, Kuchitsu, and Kimura (1970)
CS_2	(*) *previous literature:* Chapt. I-2; Kato, Konaka, Iijima, and Kimura (1969), Konaka and Kimura (1970)
$(H;D)_2O$, NO_2, Cl_2S	(*) *previous literature:* Chapt. I-2
Cl_2O	Beagley, Clark, and Hewitt (1968)
SO_2	Haase and Winnewisser (1968)
NF_2 radical	Bohn and Bauer (1967a)
(4) As_4	(*) *previous literature:* Chapt. I-2
$AlCl_3$	Zasorin and Rambidi (1967a)

Table 22-I (Continued) -- electron diffraction --

CH_2O	Kato, Konaka, Iijima, and Kimura (1969)
Li_2F_2	Akishin and Rambidi (1960a), Akishin and Rambidi (1960b)
(tr;cis)-N_2F_2	Bohn and Bauer (1967b)
$N(H;D)_3$	(*) *previous literature:* Chapt. I-2; Kuchitsu, Guillory, and Bartell (1968)
PH_3, PCl_3, AsI_3	(*) *previous literature:* Chapt. I-2
PF_3	Morino, Kuchitsu, and Moritani (1969)
AsF_3	Clippard and Bartell (1970a), Konaka and Kimura (1970)
$AsCl_3$	Konaka and Kimura (1970)
NF_2Cl	Vilkov and Nazarenko (1967)
$SOCl_2$	Hargittai (1968c), Hargittai and Cyvin (1969), Hargittai (1969a), Hargittai (1969b)
H_2O_2 (H_2O_2-H_2O systems)	Maltsev and Nekrasov (1970)
H_2S_2	Winnewisser and Haase (1968)
S_2Cl_2	(*) *previous literature:* Chapt. I-2; Beagley, Eckersley, Brown, and Tomlinson (1969)
(5) $C(H;D;F;Cl)_4$, $GeCl_4$, $SnBr_4$	(*) *previous literature:* Chapt. I-2
$SiCl_4$	(*) *previous literature:* Chapt. I-2; Ryan and Hedberg (1969)
$SnCl_4$	(*) *previous literature:* Chapt. I-2; Fujii and Kimura (1970)
OsO_4	(*) *previous literature:* Chapt. I-2
XeO_4	Gundersen, Hedberg, and Huston (1970)
$C(H;F)_3Cl$	(*) *previous literature:* Chapt. I-2
NOF_3	Plato, Hartford, and Hedberg (1970)
$POCl_3$	(*) *previous literature:* Chapt. I-2; Vilkov, Khaykin, Vasilev, and Tulyakova (1968)
$PSCl_3$	Vilkov, Khaykin, Vasilev, and Tulyakova (1968)
$FClO_3$	Clark, Beagley, Cruickshank, and Hewitt (1970a)
SF_4	(*) *previous literature:* Chapt. I-2
$(U;Th)F_4$	Ezhov, Akishin, and Rambidi (1969a), Ezhov, Akishin, and Rambidi (1969b)

Table 22-I (Continued) -- electron diffraction --

	(U;Th)(Cl;Br)$_4$	Ezhov, Akishin, and Rambidi (1969b)
	SO_2Cl_2	Hargittai (1968a), Hargittai (1968b), Hargittai and Cyvin (1969), Hargittai (1969b)
	XeF_4	Bohn, Katada, Martinez, and Bauer (1963)
	C_3O_2	(*) *previous literature:* Chapt. I-2; Almenningen, Arnesen, Seip, and Seip (1968), Tanimoto,Kuchitsu, and Morino (1970), Clark and Seip (1970)
	HCOOH	*previous literature:* Chapt. I-2; Almenningen, Bastiansen, and Motzfeldt (1969)
	F_2CN_2 (cyclic)	Hencher and Bauer (1967)
(6)	Se_6	Barzdain and Alekseev (1968)
	C_2(H;D;F;Cl)$_4$, CH_2CF_2, cis-$C_2H_2Br_2$	(*) *previous literature:* Chapt. I-2
	$C_2H_2O_2$ (glyoksal)	Kuchitsu, Fukuyama, and Morino (1967-68)
	$C_2O_2Cl_2$	Hjortaas (1967a)
	B_2Cl_4	*previous literature:* Chapt. I-2; Ryan and Hedberg (1969)
	IF_5	Cyvin, Brunvoll, and Robiette (1969)
	PF_5	Hansen and Bartell (1965), Holmes and Deiters (1969), Bartell (1970)
	PCl_5	Bartell (1970)
	AsF_5	Clippard and Bartell (1970a), Bartell (1970)
	VF_5	Bartell (1970)
	N_2F_4	Bohn and Bauer (1967a)
	SOF_4	(*) *previous literature:* Chapt. I-2; Hencher, Cruickshank, and Bauer (1968), Gundersen and Hedberg (1969)
	HCCCCBr	Seip, Strand, and Stølevik (1969), Almenningen, Hargittai, Kloster-Jensen, and Stølevik (1970)
	NH_4Cl(g)	Shibata (1970)
	$HClO_4$	Clark, Beagley, Cruickshank, and Hewitt (1970b)
(7)	(S;Se)F_6	(*) *previous literature:* Chapt. I-2
	(W;U)F_6	*previous literature:* Chapt. I-2; Kimura, Schomaker, Smith, and Weinstock (1968)
	(Np;Pu;Os;Ir)F_6	Kimura, Schomaker, Smith, and Weinstock (1968)

Table 22-I (Continued) -- electron diffraction --

ReF_6	Jacob and Bartell (1970a)
XeF_6	Gavin and Bartell (1968), Bartell and Gavin (1968)
P_4S_3	Akishin, Rambidi, and Ezhov (1960)
C_3H_4 (allene), $C(F;Br)_3NO_2$	(*) *previous literature:* Chapt. I-2
CCl_3NO_2	(*) *previous literature:* Chapt. I-2; Karle (1966b)
$SiCl_3NCO$	Hilderbrandt and Bauer (1969)
CH_3CHO	Kato, Konaka, Iijima, and Kimura (1969)
CH_2CHCN (acrylonitrile)	Fukuyama and Kuchitsu (1970)
$C_2N_2SH_2$ (1,3,4-thiadiazole)	Seip, Strand, and Stølevik (1969), Cyvin, Cyvin, Hagen, and Markov (1969), Markov and Stølevik (1970)
(8) Ge_2H_6	Beagley and Monaghan (1970)
ReF_7	Jacob and Bartell (1970b)
$(COOH)_2$	Náhlovská, Náhlovský, and Strand (1970)
C_4H_4 (1,3-butatriene)	(*) *previous literature:* Chapt. I-2
C_3H_4O (acrolein)	Kuchitsu, Fukuyama, and Morino (1967-68), Trætteberg (1970a)
(9) C_3H_6 (cyclopropane), $(CH_3)_2O$, $(SiH_3)_2(O;S)$	(*) *previous literature:* Chapt. I-2
$(CH_3)_2Be$	Almenningen, Haaland, and Morgan (1969)
$(SiF_3)_2O$	Airey, Glidewell, Rankin, Robiette, Sheldrick, and Cruickshank (1970)
CH_3OSiH_3	Glidewell, Rankin, Robiette, Sheldrick, Beagley, and Freeman (1970)
$C_2O_3H_4$ (ethylene ozonide)	Almenningen, Kolsaker, Seip, and Willadsen (1969)
CH_3PF_4	Bartell and Hansen (1965), Bartell (1970)
$SiCl_2(NCO)_2$	Hilderbrandt and Bauer (1969)
(10) P_4O_6	Beagley, Cruickshank, Hewitt, and Jost (1969)
C_4H_6 (1,3-butadiene)	*previous literature:* Chapt. I-2; Kuchitsu, Fukuyama, and Morino (1967-68)
CH_3CCCH_3	Tanimoto, Kuchitsu, and Morino (1969)
C_4F_6 (perfluoro-cyclobutene)	Chang, Porter, and Bauer (1971)

Table 22-I (Continued) -- electron diffraction --

	$C_4H_4Cl_2$ (cis-3,4-dichlorobutyne-1)	(*) *previous literature:* Chapt. I-2
	$C_2(CN)_4$	Hope (1968)
	$(CH_3)_2N_2$	Almenningen, Anfinsen, and Haaland (1970b), Chang, Porter, and Bauer (1970a)
	$(CF_3)_2N_2$, CH_3NNCF_3	Chang, Porter, and Bauer (1970a)
	$(CH_3)_2NH$	Beagley and Hewitt (1968)
	$(CH_3)_2CO$	Kato, Konaka, Iijima, and Kimura (1969), Iijima (197
	$(CF_3)_2CO$	Hilderbrandt, Andreassen, and Bauer (1970)
	$(HCOOH)_2$	Almenningen, Bastiansen, and Motzfeldt (1969)
	$CH_3OC_2H_3$	Owen and Seip (1970)
	Si_2BF_7	Chang, Porter, and Bauer (1970b)
(11)	$Fe(CO)_5$	(*) *previous literature:* Chapt. I-2; Almenningen, Haaland, and Wahl (1969b), Beagley, Cruickshank, Pinder, Robiette, and Sheldrick (1969)
	BeB_2H_8	Almenningen, Gundersen, and Haaland (1968b)
	$(SiH_3)_2CH_2$	Almenningen, Seip, and Seip (1970)
	$(CH_3)_2SO_2$	Oberhammer and Zeil (1970a)
	$(CH_3)_2NNO$	Vilkov and Nazarenko (1968), Rademacher and Stølevik (1969)
	$(CF_3)_2CNH$	Hilderbrandt, Andreassen, and Bauer (1970)
	$SiCl(NCO)_3$	Hilderbrandt and Bauer (1969)
(12)	C_6H_6 (benzene), C_6F_6, C_6H_6 (dimethyldiacetylene)	(*) *previous literature:* Chapt. I-2
	$(CH_2)_3C_3$ (trimethylenecyclopropane)	Dorko, Hencher, and Bauer (1968)
	$(CH_2)_2C_4H_2$ (3,4-dimethylenecyclobutene)	Skancke (1968)
	(tr;cis)-$(CH_3)_2C_2H_2$	Almenningen, Anfinsen, and Haaland (1970a)
	C_4F_8 (perfluorocyclobutane)	Chang, Porter, and Bauer (1971)
	$(CH_3)_2PF_3$	Bartell and Hansen (1965)
	$(CH_3)_2NNO_2$	Stølevik and Rademacher (1969)
	$(CF_3)_2CCH_2$	Hilderbrandt, Andreassen, and Bauer (1970)

Table 22-I (Continued) -- electron diffraction --

	$(CH_3)_2SONH$	Oberhammer and Zeil (1970b)
	$(CH_3)_2NBCl_2$	Clippard and Bartell (1970b)
	$(SiH_3)_2NBF_2$	Robiette, Sheldrick, and Sheldrick (1970)
	$C_2Si_2H_4Cl_4$ (1,1-3,3-tetrachloro-1,3-disilacyclobutane)	Vilkov, Kusakov, Nametkin, and Oppenheim (1968)
	$HMn(CO)_5$	Robiette, Sheldrick, and Simpson (1969)
(13)	C_5H_8 (bicyclo-[1.1.1]pentane)	Cyvin, Elvebredd, Hagen, and Andersen (1968)
	C_5H_8 (bicyclo-[2.1.0]pentane)	Bohn and Tai (1970)
	C_5H_8 (spiropentane)	Dallinga, van der Draai, and Toneman (1968)
	C_5H_8 (cyclopentene)	Davis and Muecke (1970)
	$CH_2CHCCH_3CH_2$	Vilkov and Sadova (1967b)
	(tr;gauche)-$CH_3COC_2H_5$	Abe, Kuchitsu, and Shimanouchi (1969)
	C_4H_8O (tetrahydrofuran)	Almenningen, Seip, and Willadsen (1969)
	C_4H_8S (tetrahydrothiophene)	Náhlovská, Náhlovský, and Seip (1969)
	C_4H_8Se (tetrahydroselenophene)	Náhlovská, Náhlovský, and Seip (1970)
	$(CH_3)_3N$	Beagley and Hewitt (1968)
	$(SiH_3)_3N$	*previous literature:* Chapt. I-2; Rankin, Robiette, Sheldrick, Aylett, Ellis, and Monaghan (1969), Beagley and Conrad (1970)
	$(GeH_3)_3N$	Glidewell, Rankin, and Robiette (1970)
	$(SiH_3)_3P$	Beagley, Robiette, and Sheldrick (1968a)
	$(SiH_3)_3As$	Beagley, Robiette, and Sheldrick (1968b)
	$(SiH_3)_2NCH_3$	Glidewell, Rankin, Robiette, and Sheldrick (1969)
	$CH_3CON(CH_2)_2$	Vilkov, Nazarenko, and Kostyanovskii (1968)
	C_6H_5HgBr	Vilkov and Anashkin (1968)
	$Si(NCO)_4$	Hjortaas (1967b)
	$(CH_3)_2NSO_2Cl$	Vilkov and Hargittai (1967)
(14)	C_6H_8 (1,3,5-trhexatriene)	*previous literature:* Chapt. I-2; Trætteberg (1968a)

Table 22-I (Continued) -- electron diffraction --

	C_6H_8 (1,3,5-cis-hexatriene)	Trætteberg (1968b)
	C_6H_8 (1,3-cyclohexadiene)	Dallinga and Toneman (1967-68a), Trætteberg (1968c), Oberhammer and Bauer (1969)
	C_6H_8 (1,4-cyclohexadiene)	Dallinga and Toneman (1967-68b), Oberhammer and Bauer (1969)
	$(CH_3)_3CF$	Haas, Haase, and Zeil (1967)
	$(CH_3)_3CCl$	*previous literature:* Chapt. I-2; Haase, Kamphusmann and Zeil (1967)
	$(CH_3)_2CCH_2NH$	Naumov and Semashko (1969)
(15)	C_5H_{10} (cyclopentane)	Adams, Geise, and Bartell (1970)
	C_7H_8 (norbornadiene)	Dallinga and Toneman (1968b)
	$CH_3Mn(CO)_5$	Seip and Seip (1970)
	$C_6H_5POCl_2$	Vilkov, Sadova, and Zilber (1967)
(16)	$Al(BH_4)_3$	Almenningen, Gundersen, and Haaland (1968a)
	C_8H_8 (cyclooctatetraene)	(*) *previous literature:* Chapt. I-2
	C_6H_{10} (cyclohexene)	Chiang and Bauer (1969)
	$(CH_3)_2C_4H_4$ (2,3-dimethylbutadiene)	Aten, Hedberg, and Hedberg (1968)
	C_9H_7 (indenyl radical)	Schäfer (1968)
(17)	$(CH_3)_4C$	Livingston, Lurie, and R.Rao (1960), Beagley,Brown, and Monaghan (1969)
	C_7H_{10} (1,3-cycloheptadiene), C_7H_{10} (bicycloheptene)	Chiang and Bauer (1966)
	C_7H_9Cl (4-chloronortricyclene)	Chiang, Wilcox, and Bauer (1969)
	$Ni(PF_3)_4$	Marriott, Salthouse, Ware, and Freeman (1970), Almenningen, Andersen, and Astrup (1970)
	$(CH_2)_3CFe(CO)_3$	Almenningen, Haaland, and Wahl (1969a)
(18)	C_6F_{12} (dodecafluorocyclohexane)	Hjortaas and Strømme (1968)

Table 22-I (Continued) -- electron diffraction --

	$(CF_3)_4N_2$	Bartell and Higginbotham (1965a)
	$(SiH_3)_4N_2$	Glidewell, Rankin, Robiette, and Sheldrick (1970)
	$(CH_3)_4P_2$	McAdam, Beagley, and Hewitt (1970)
(19)	C_7H_{12} (norbornane)	Dallinga and Toneman (1968a), Chiang, Wilcox, and Bauer (1968)
	$C_7H_{10}Cl_2$ (1,4-dichloronorbornane)	Chiang, Wilcox, and Bauer (1968)
	C_7H_{12} (bicyclo[3.1.1]heptane)	Dallinga and Toneman (1969)
	$(CH_3)_4N_2S$	Hargittai, Hargittai, and Hernádi (1970)
(20)	C_8H_{12} (1,3-cyclooctadiene)	Trætteberg (1970b)
	C_8H_{12} (syn;anti)-tricyclo[4.2.0.0^{2,5}]-octane	Andersen and Fernholt (1970)
	$B_2H_2(CH_3)_4$	Carroll and Bartell (1968)
	$(CH_3)_4NSON$	Hargittai and Vilkov (1970)
(21)	$(C_5H_5)_2Be$	*previous literature:* Chapt. I-2; Haaland (1968)
	$(C_5H_5)_2(Sn;Pb)$	Almenningen, Haaland, and Motzfeldt (1967a)
	$(C_5H_5)_2Fe$	*previous literature:* Chapt. I-2; Haaland and Nilsson (1968)
	$(C_5H_5)_2Ni$	Hedberg and Hedberg (1970)
	$(C_5H_5)_2Ru$	Haaland and Nilsson (1968)
	$(CH_3)_2CHC_6H_5$(cumene)	Vilkov, Sadova, and Mochalov (1968)
(22)	$Mn_2(CO)_{10}$	Gapotchenko, Alekseev, Anisimov, Kolobova, and Ronova (1968), Almenningen, Jacobsen, and Seip (1969a)
	C_8H_{14} (cis,cis;tr,tr;cis,tr)-3,4-dimethyl-2,4-hexadiene	Trætteberg (1970c)
	$C_{12}F_{10}$ (decafluorobiphenyl)	Almenningen, Hartmann, and Seip (1968)
	$C_4H_7C_6H_5$ (phenylcyclobutane)	Vilkov, Sadova, and Mochalov (1968)

Table 22-I (Continued) -- electron diffraction --

(23) $(C_6H_5)_2Hg$	Vilkov, Anashkin, and Mamaeva (1968)
(24) neo-$B_{10}C_2H_{12}$	Vilkov, Mastryukov, Zhigach, and Siryatskaya (1966), Mastryukov, Vilkov, Zhigach, and Siryatskaya (1969)
neo-$B_{10}C_2H_{10}I_2$	Vilkov, Khaykin, Zhigach, and Siryatskaya (1968)
(26) $C_{10}H_{16}$ (cyclo-decadiene)	Almenningen, Jacobsen, and Seip (1969b)
(27) $(C_4H_9)_2Be$	Almenningen, Haaland, and Nilsson (1968)
(29) $ClSi[N(CH_3)_2]_3$	Vilkov and Tarasenko (1969)
(37) $Sn[N(CH_3)_2]_4$	Vilkov, Tarasenko, and Prokofev (1970)
(42) $C_6H_{10}(C_4H_9)_2$(dibutyl cyclohexane)	*previous literature:* Chapt. I-2; van Bekkum, Hoefnagel, de Lavieter, van Veen, Verkade, Wemmers, Wepster, Palm, Schäfer, Dekker, Mosselman, and Somsen (1968)
(53) $Si[Si(CH_3)_3]_4$	Bartell, Clippard, and Boates (1970)
(55) $BeN_2[Si(CH_3)_3]_4$	Clark and Haaland (1970)

Table 22-II. References to mean amplitudes of vibration and related quantities from spectroscopic data. None of the references in Table 2-III of Volume I are duplicated, but are occasionally referred to as *previous literature:* Chapt.I-2. (*) indicates that numerical values (original calculations, or more frequently quotations) are reported in Chapter I-12 (Volume I). In some cases shrinkage effects for the same molecules are reported in Chapter I-15.

Compound (Number of atoms in parentheses)	Reference
(2) $^1H_2$, D_2, $^{14}N_2$, $^{16}O_2$, $^{32}S_2$, $^{80}Se_2$, $^{35}Cl_2$, $^{79}Br^{81}Br$, Br_2, $^{127}I_2$, $^1H(^{19}F;^{35}Cl)$, $^{12}C^{16}O$, $^{14}N^{16}O$	(*) *previous literature:* Chapt. I-2

Table 22-II (Continued) -- spectroscopic --

T_2, $^{31}P_2$, Te_2, $^{1}H(Br;^{127}I)$, $^{12}C(^{32}S;Se)$	(*)
H_2, I_2	(*) *previous literature:* Chapt. I-2; *incl. anharmonicity:* Rundgren (1967c)
$^{27}Al(^{19}F;Cl;^{79}Br;^{127}I)$, Li(F;Cl;Br;I), $(^{23}Na;K;^{85}Rb;^{133}Cs)(^{19}F;$ $^{35}Cl;^{37}Cl;^{79}Br;^{81}Br;^{127}I)$	Cyvin (1971a)
Cs(Cl;Br;I)	*previous literature:* Chapt. I-2
(3) Linear triatomic	
$(^{12}C;^{13}C;^{14}C)O_2$, $C(S;Se)_2$	(*) *previous literature:* Chapt. I-2; $(^{12}C;^{13}C;^{14}C)$ $^{16}O_2$, $(^{12}C;^{13}C)^{18}O_2$, CS_2, CSe_2: Ramaswamy and Srinivasan (1969c)
COS	(*) *previous literature:* Chapt. I-2; Shanmugasundaram (1969), Ramaswamy and Ranganathan (1970)
COSe	(*) *previous literature:* Chapt. I-2
CS(Se;Te)	(*) *previous literature:* Chapt. I-2; Ramaswamy and Ranganathan (1970)
$(^{12}C;^{13}C)^{14}N_2$	Ramaswamy and Srinivasan (1969c)
XeF_2	*previous literature:* Chapt. I-2; Yeranos (1967)
$(Be;Mg;Zn;Cd;Ti;V;Cr;$ $Mn;Fe;Co;Ni;Cu)(F;Cl;$ $Br;I)_2$	*previous literature:* Chapt. I-2; *incl. shrinkage effect:* Cyvin and Vizi (1968)
$(Ca;Sr;Ba)(F;Cl;Br;I)_2$	*incl.perpendicular amplitudes and shrinkage effect:* Nagarajan (1965c)
$Hg(F;Cl;Br;I)_2$	(*) *previous literature:* Chapt. I-2; *incl.shrinkage effect:* Cyvin and Vizi (1968)
$(^{14}N;^{15}N)(^{14}N;^{15}N)O$	(*) *previous literature:* Chapt. I-2; N_2O: Ramaswamy and Ranganathan (1970)
$(^{1}H;D;T)(^{12}C;^{13}C)N$	(*)
FCN	(*) *previous literature:* Chapt. I-2
(H;D;Cl;Br;I)CN	(*) *previous literature:* Chapt. I-2; Ramaswamy and Ranganathan (1970)
(H;D)NC	Ramaswamy and Ranganathan (1970)
$(H;D)F_2^-$	*previous literature:* Chapt. I-2; Shanmugasundaram (1968)

Table 22-II (Continued) -- spectroscopic --

$I(Cl;Br;I)_2^-$, $(Np;Pu;Am)O_2^{++}$	Sanyal, Singh, and Pandey (1969b)
(3) Bent triatomic	
$(H;D;T)(H;D;T)(^{16}O;^{18}O)$, (*) $(H;D)(H;D)(^{32}S;^{34}S)$, $(^{32}S;^{34}S)(^{16}O;^{18}O)(^{16}O;^{18}O)$, $(^{16}O;^{18}O)(^{16}O;^{18}O)_2$, $(^{35}Cl;^{37}Cl)O_2$, $(^{35}Cl;^{37}Cl)_2(O;S)$	
$(H;D)_2(O;Se)$, $(^{14}N;^{15}N)O_2$, F_2O	(*) *previous literature:* Chapt. I-2
SO_2	*previous literature:* Chapt. I-2; Török and Hun (1969)
Cl_2O	*previous literature:* Chapt. I-2; Beagley, Clark, and Hewitt (1960)
$(Th;Zr)O_2$	*incl. generalized amplitudes:* Nagarajan (1967)
$(H;D)OCl$	(*) *previous literature:* Chapt. I-2
NOF	(*) *previous literature:* Chapt. I-2; Török and Pulay (1969), Ramaswamy and Namasivayam (1970)
$(^{14}N;^{15}N)^{16}OF$	Peacock, Heidborn, and Müller (1969)
NOCl	*previous literature:* Chapt. I-2; Jones, Ryan, and Asprey (1968)
$(^{14}N;^{15}N)OCl$, NOBr	(*) *previous literature:* Chapt. I-2; Peacock, Heidborn, and Müller (1969)
$^{14}N^{16}O^{79}Br$	Laane, Jones, Ryan, and Asprey (1969)
NSF	*previous literature:* Chapt. I-2; Peacock, Heidborn, and Müller (1969)
NSCl	Müller, Nagarajan, Glemser, Cyvin, and Wegener (1967), Peacock, Heidborn, and Müller (1969)
(4) P_4	(*) *previous literature:* Chapt. I-2
(4) Planar symmetrical XY_3	
BH_3	*incl.perpendicular amplitudes and shrinkage effect:* Nagarajan (1966n')
$B(F;Cl;Br;I)_3$	(*) *previous literature:* Chapt. I-2; $^{11}BX_3$: Peacock, Müller, and Kebabcioglu (1968b), Müller (1968b)

Table 22-II (Continued) -- spectroscopic --

$^{11}B(OH)_3$	Peacock, Müller, and Kebabcioglu (1968b)
AlF_3	(*) *incl.perpendicular amplitudes and shrinkage effect:* Nagarajan (1966n'), Shanmugasundaram and Nagarajan (1969a)
$AlCl_3$	(*) *previous literature:* Chapt. I-2; *incl.perpendicular amplitudes and shrinkage effect:* Nagarajan (1966n'); *incl. shrinkage effect:* Zasorin and Rambidi (1967b), Cyvin and Brunvoll (1969b)
$^{12}C(^{16}O;^{18}O)_3$, $^{13}C^{16}O_3$	Nagarajan and Durig (1968b)
GaC_3, $(C;Si)(H;F)_3$, GeH_3, NO_3	*incl.perpendicular amplitudes and shrinkage effect:* Nagarajan (1966n')
SO_3	(*) *previous literature:* Chapt. I-2; Peacock,Müller, and Kebabcioglu (1968b); *incl. shrinkage effect:* Cyvin, Brunvoll, and Stølevik (1969)
$(Mo;W;U)O_3$	(*) *previous literature:* Chapt. I-2
$HgCl_3^-$	*incl.perpendicular amplitudes and shrinkage effect:* Nagarajan (1966n')
BO_3^{---}	*previous literature:* Chapt. I-2; *incl. perpendicular amplitudes and shrinkage effect:* Nagarajan (1966n'); $^{10}BO_3^{---}$: Peacock, Müller, and Kebabcioglu (1968b)
CO_3^{--}, NO_3^-	*previous literature:* Chapt. I-2; *incl. perpendicular amplitudes and shrinkage effect:* Nagarajan (1966n'); Peacock, Müller, and Kebabcioglu (1968b)
$C(S;Se)_3^{--}$	*previous literature:* Chapt. I-2; Peacock, Müller, and Kebabcioglu (1968b)

(4) Planar XY_2Z

$BF_2(Cl;Br)$, $BF(Cl;Br)_2$, BCl_2Br, $BClBr_2$	Nagarajan and Müller (1967b); *generalized amplitudes for* ^{10}B *and* ^{11}B: Venkateswarlu and Bhamambal (1968a)
CH_2O	(*) *previous literature:* Chapt. I-2; Müller and Nagarajan (1967d); *generalized amplitudes:* Venkateswarlu, Babu Joseph, and Malathy Devi (1967); Müller, Krebs, Fadini, Glemser, Cyvin, Brunvoll, Cyvin, Elvebredd, Hagen, and Vizi (1968), Müller, Peacock, Schulze, and Heidborn (1969), Kato, Konaka, Iijima, and Kimura (1969); *approximate:* Baran (1970)

Table 22-II (Continued) -- spectroscopic --

CD_2O	(*) *previous literature:* Chapt. I-2; *generalized amplitudes:* Venkateswarlu, Babu Joseph, and Malathy Devi (1967); Müller, Krebs, Fadini, Glemser, Cyvin, Brunvoll, Cyvin, Elvebredd, Hagen, and Vizi (1968); *approximate:* Baran (1970)
CF_2O	(*) *previous literature:* Chapt. I-2; Müller and Nagarajan (1967d), Müller, Krebs, Fadini, Glemser, Cyvin, Brunvoll, Cyvin, Elvebredd, Hagen, and Vizi (1968), Müller, Peacock, Schulze, and Heidborn (1969 *approximate:* Baran (1970)
$C(Cl;Br)_2O$	(*) *previous literature:* Chapt. I-2; Müller and Nagarajan (1967d), Müller, Krebs, Fadini, Glemser, Cyvin, Brunvoll, Cyvin, Elvebredd, Hagen, and Vizi (1968); *approximate:* Baran (1970)
$[N(H;D)_2]_2CO$	Puranik and Sirdeshmukh (1968)
CF_2S	Müller and Nagarajan (1967d), Müller, Krebs, Fadini, Glemser, Cyvin, Brunvoll, Cyvin, Elvebredd, Hagen, ar Vizi (1968), Müller, Peacock, Schulze, and Heidborn (1969); *approximate:* Baran (1970)
CCl_2S, NO_2F	Müller and Nagarajan (1967d); *generalized amplitudes:* Venkateswarlu, Babu Joseph, and Malathy Devi (1967); Müller, Krebs, Fadini, Glemser, Cyvin, Brunvoll, Cyvin, Elvebredd, Hagen, and Vizi (1968); *approximate* Baran (1970)
NO_2Cl	Müller and Nagarajan (1967d); *generalized amplitudes:* Venkateswarlu, Babu Joseph, and Malathy Devi (1967); Müller, Krebs, Fadini, Glemser, Cyvin, Brunvoll, Cyvin, Elvebredd, Hagen, and Vizi (1968), Müller, Peacock, Schulze, and Heidborn (1969); *approximate:* Baran (1970)
$(OH)NO_2$	Müller and Nagarajan (1967d), Müller, Krebs, Fadini, Glemser, Cyvin, Brunvoll, Cyvin, Elvebredd, Hagen, ar Vizi (1968), Müller, Peacock, Schulze, and Heidborn (1969); *approximate:* Baran (1970)
$(OD)NO_2$	Müller, Krebs, Fadini, Glemser, Cyvin, Brunvoll, Cyvin, Elvebredd, Hagen, and Vizi (1968);*approximate:* Baran (1970)

Table 22-II (Continued) -- spectroscopic --

$(CH_3)NO_2$	Müller and Nagarajan (1967d), Müller, Krebs, Fadini, Glemser, Cyvin, Brunvoll, Cyvin, Elvebredd, Hagen, and Vizi (1968)
ClF_3	(*) *previous literature:* Chapt. I-2; Müller and Nagarajan (1967d), Müller, Krebs, Fadini, Glemser, Cyvin, Brunvoll, Cyvin, Elvebredd, Hagen, and Vizi (1968); *approximate:* Baran (1970)
BrF_3	Müller and Nagarajan (1967d), Müller, Krebs, Fadini, Glemser, Cyvin, Brunvoll, Cyvin, Elvebredd, Hagen, and Vizi (1968); *approximate:* Baran (1970)
HCO_2^-	Müller and Nagarajan (1967d), Müller, Krebs, Fadini, Glemser, Cyvin, Brunvoll, Cyvin, Elvebredd, Hagen, and Vizi (1968); *approximate:* Baran (1970)
DCO_2^-	Müller, Krebs, Fadini, Glemser, Cyvin, Brunvoll, Cyvin, Elvebredd, Hagen, and Vizi (1968);*approximate:* Baran (1970)
$[-OCO_2]^{--}$, $[(-O)_2CO]^{--}$	Müller and Nagarajan (1967d), Müller, Krebs, Fadini, Glemser, Cyvin, Brunvoll, Cyvin, Elvebredd, Hagen, and Vizi (1968)

(4) Regular pyramidal XY_3

$N(H;D)_3$	(*) *previous literature:* Chapt. I-2; Kuchitsu, Guillory, and Bartell (1968), Ramaswamy and Swaminathan (1969)
NT_3	(*) *previous literature:* Chapt. I-2; Ramaswamy and Swaminathan (1969)
NF_3	(*) *previous literature:* Chapt. I-2; Müller, Krebs, and Peacock (1968), Sawodny, Ruoff, Peacock, and Müller (1968); *incl. perpendicular amplitude corrections:* Cyvin, Cyvin, and Müller (1969); Ramaswamy and Mohan (1970)
NCl_3	Ramaswamy and Mohan (1970)
$P(H;D;T)_3$	(*) *previous literature:* Chapt. I-2; Ramaswamy and Swaminathan (1969)
PF_3	(*) *previous literature:* Chapt. I-2; Müller, Krebs, and Peacock (1968), Morino, Kuchitsu, and Moritani (1969), Cyvin, Cyvin, and Müller (1969), Ramaswamy and Krishna Rao (1969a)

Table 22-II (Continued) -- spectroscopic --

PCl_3	(*) *previous literature:* Chapt. I-2; Levin (1967), Müller, Krebs, and Peacock (1968); *incl.perpendicular amplitude corrections:* Cyvin, Cyvin, and Müller (196[illegible] Ramaswamy and Krishna Rao (1969a)
PBr_3	(*) *previous literature:* Chapt. I-2; Müller, Krebs, [illegible] Peacock (1968), Cyvin, Cyvin, and Müller (1969), Ramaswamy and Krishna Rao (1969c)
PI_3	(*) *previous literature:* Chapt. I-2; Müller, Krebs, [illegible] Peacock (1968), Cyvin, Cyvin, and Müller (1969)
$As(H;D;T)_3$	(*) *previous literature:* Chapt. I-2; Ramaswamy and Swaminathan (1969)
$As(F;Cl)_3$	(*) *previous literature:* Chapt. I-2; Müller, Krebs, Peacock (1968), Cyvin, Cyvin, and Müller (1969); *inc. perpendicular amplitude corrections:* Konaka (1970)
$AsBr_3$	(*) *previous literature:* Chapt. I-2; Müller, Krebs, Peacock (1968), Cyvin, Cyvin, and Müller (1969)
AsI_3	(*) *previous literature:* Chapt. I-2; Müller, Krebs, Peacock (1968); *incl. perpendicular amplitude corrections:* Cyvin, Cyvin, and Müller (1969)
$Sb(H;D;T)_3$	(*) *previous literature:* Chapt. I-2; Ramaswamy and Swaminathan (1969)
$Sb(Cl;Br)_3$	(*) *previous literature:* Chapt. I-2; Müller, Krebs, Peacock (1968), Cyvin, Cyvin, and Müller (1969), Rai and Thakur (1970)
SbI_3, $Bi(Br;I)_3$	(*) *previous literature:* Chapt. I-2; Cyvin, Cyvin, a[illegible] Müller (1969), Rai and Thakur (1970)
$BiCl_3$	(*) *previous literature:* Chapt. I-2; Rai and Thakur (1970)
XeO_3	(*) *previous literature:* Chapt. I-2; Müller, Krebs, [illegible] Peacock (1968), Cyvin, Cyvin, and Müller (1969)
$(P;As;Sb)(CCH)_3$, $(N;P;As;Sb;Bi)(CH_3)_3$	Sanyal, Pandey, and Singh (1969a)
$GeCl_3^-$, $Sn(Cl;Br)_3^-$, $(S;Se;Te)O_3^{--}$, $(Cl;Br;I)O_3^-$	*previous literature:* Chapt. I-2; Müller, Krebs, and Peacock (1968)

Table 22-II (Continued) -- spectroscopic --

$(Se;Te)Cl_3^+$	Müller, Krebs, and Peacock (1968)
(4) Pyramidal XY_2Z	
NH_2D, NHD_2	(*) *previous literature:* Chapt. I-2; Purnachandra Rao and Rama Murthy (1968)
$N(H;D)_2T$, $N(H;D)T_2$	(*) Purnachandra Rao and Rama Murthy (1968)
$(P;As)H_2D$, $(P;As)HD_2$	
$(P;As)(H;D)_2T$, $(P;As)(H;D)T_2$	Purnachandra Rao and Rama Murthy (1968)
SbH_2D, $SbHD_2$	(*) *previous literature:* Chapt. I-2
NHF_2	*previous literature:* Chapt. I-2; *generalized amplitudes:* Venkateswarlu, Rajalakshmi, and Babu Joseph (1968)
NDF_2	*previous literature:* Chapt. I-2; *generalized amplitudes:* Venkateswarlu, Rajalakshmi, and Babu Joseph (1968); Ramaswamy and Mohan (1970)
NH_2Cl, $NHCl_2$	*generalized amplitudes:* Venkateswarlu, Rajalakshmi, and Babu Joseph (1968)
NF_2Cl, $NFCl_2$	Shanmugasundaram and Nagarajan (1969b), Ramaswamy and Mohan (1970)
PF_2Cl	Nagarajan, Durig, and Müller (1967), Elvebredd, Vizi, Cyvin, Müller, and Krebs (1968)
$PFCl_2$	*incl. generalized amplitudes:* Venkateswarlu and Rajalakshmi (1964); Nagarajan, Durig, and Müller (1967), Elvebredd, Vizi, Cyvin, Müller, and Krebs (1968)
$P(F;Cl)_2Br$, $P(F;Cl)Br_2$	Nagarajan, Durig, and Müller (1967), Elvebredd, Vizi, Cyvin, Müller, and Krebs (1968)
$P(F;Cl;Br)_2I$, $P(Cl;Br)I_2$, $AsCl_2Br$, $AsClBr_2$	Elvebredd, Vizi, Cyvin, Müller, and Krebs (1968)
$(CH_3)PF_2$	Ramaswamy and Krishna Rao (1969d)
SOF_2	(*) *previous literature:* Chapt. I-2; Pichai, Krishna Pillai, and Ramaswamy (1967), Müller and Nagarajan (1967c), Müller, Peacock, and Heidborn (1968)

Table 22-II (Continued) -- spectroscopic --

$SOCl_2$	(*) *previous literature:* Chapt. I-2; Müller and Nagarajan (1967c), Müller, Peacock, and Heidborn (1968), Hargittai and Cyvin (1969)
$SOBr_2$	(*) *previous literature:* Chapt. I-2; Müller and Nagarajan (1967c), Müller, Peacock, and Heidborn (1968)
$SeOF_2$	*incl. generalized amplitudes:* Venkateswarlu and Rajalakshmi (1964); Müller and Nagarajan (1967c), Müller, Peacock, and Heidborn (1968), Ramaswamy and Jayaraman (1970b)
$SeOCl_2$	*incl. generalized amplitudes:* Venkateswarlu and Rajalakshmi (1964); Müller and Nagarajan (1967c), Müller, Peacock, and Heidborn (1968)
(4) Linear four-atomic (incl. symmetrical X_2Y_2)	
$C_2(H;D)_2$, C_2I_2, $^{12}C_2{}^{14}N_2$	(*) *previous literature:* Chapt. I-2
$C_2H(F;Cl;Br)$	*previous literature:* Chapt. I-2; *incl. perpendicular amplitudes and shrinkage effects:* Venkateswarlu and Mathew (1968b)
$C_2D(F;Cl;Br)$	*incl. perpendicular amplitudes and shrinkage effects:* Venkateswarlu and Mathew (1968b)
(4) Planar X_2Y_2	
cis-N_2F_2	*generalized amplitudes:* Venkateswarlu and Babu Joseph (1967b); Hagen (1968), Nagarajan and Durig (1968a)
tr-N_2F_2	*generalized amplitudes:* Venkateswarlu and Babu Joseph (1967b); Hagen (1968)
$N_2O_2^{--}$	Nagarajan and Sivaprakasam (1970b)
(4) Nonplanar X_2Y_2	
$(H;D)_2O_2$, H_2S_2	Elvebredd and Cyvin (1969)
S_2Cl_2	*previous literature:* Chapt. I-2; Cyvin, Elvebredd, Hagen, and Brunvoll (1968), Elvebredd and Cyvin (1969); *incl. generalized amplitudes:* Unnikrishnan Nayar and Aruldhas (1970)

Table 22-II (Continued) -- spectroscopic --

S_2Br_2	*previous literature:* Chapt. I-2; Elvebredd and Cyvin (1969); *incl. generalized amplitudes:* Unnikrishnan Nayar and Aruldhas (1970)
(4) Other planar four-atomic	
$(H;D)N_3$, HNC(O;S)	(*) *previous literature:* Chapt. I-2; *generalized amplitudes and shrinkage effect:* Venkateswarlu and Malathy Devi (1967)
(H;D)COF	(*) *previous literature:* Chapt. I-2; *generalized amplitudes:* Venkateswarlu and Malathy Devi (1967)
trans-HNO_2	*generalized amplitudes:* Unnikrishnan Nayar and Aruldhas (1968b)
(5) Tetrahedral XY_4	
$C(H;D)_4$	(*) *previous literature:* Chapt. I-2; Ramaswamy and Ranganathan (1968)
CT_4	(*) *previous literature:* Chapt. I-2
CF_4	(*) *previous literature:* Chapt. I-2; Müller and Cyvin (1968); *approximate:* Müller, Krebs, and Peacock (1968); Müller (1968b)
$(^{12}C;^{13}C)F_4$	(*)
$C(Cl;Br;I)_4$	(*) *previous literature:* Chapt. I-2; Müller and Cyvin (1968); *approximate:* Müller, Krebs, and Peacock (1968)
$Si(H;D)_4$	(*) *previous literature:* Chapt. I-2; Ramaswamy and Ranganathan (1968)
SiT_4	(*) *previous literature:* Chapt. I-2
SiF_4	(*) *previous literature:* Chapt. I-2; Müller and Cyvin (1968); *approximate:* Müller, Krebs, and Peacock (1968); Müller (1968b)
$(^{28}Si;^{30}Si)F_4$	(*)
$SiCl_4$	(*) *previous literature:* Chapt. I-2; Müller and Cyvin (1968); *incl. shrinkage effect:* Kebabcioglu, Müller, Peacock, and Lange (1968); *approximate:* Müller, Krebs, and Peacock (1968); Müller (1968b); *incl. shrinkage effect:* Peacock, Müller, and Kebabcioglu (1968a)

Table 22-II (Continued) -- spectroscopic --

$Si(Br;I)_4$	(*) *previous literature:* Chapt. I-2; Müller and Cyvin (1968); *approximate:* Müller, Krebs, and Peacock (1968)
$Ge(H;D)_4$	(*) *previous literature:* Chapt. I-2; Ramaswamy and Ranganathan (1968)
GeT_4	(*)
GeF_4	(*) *previous literature:* Chapt. I-2; Müller and Cyvin (1968); *approximate:* Müller, Krebs, and Peacock (1968)
$GeCl_4$	(*) *previous literature:* Chapt. I-2; Müller and Cyvin (1968); *incl. shrinkage effect:* Kebabcioglu, Müller, Peacock, and Lange (1968); *approximate:* Müller, Krebs, and Peacock (1968); Müller (1968b)
$Ge(Br;I)_4$	(*) *previous literature:* Chapt. I-2; Müller and Cyvin (1968); *approximate:* Müller, Krebs, and Peacock (1968)
SnH_4	(*) *previous literature:* Chapt. I-2
$Sn(D;T)_4$	(*)
$SnCl_4$	(*) *previous literature:* Chapt. I-2; Müller and Cyvin (1968); *incl. shrinkage effect:* Kebabcioglu, Müller, Peacock, and Lange (1968); *approximate:* Müller, Krebs, and Peacock (1968); *perpendicular amplitudes:* Fujii and Kimura (1970)
$Sn(Br;I)_4$	(*) *previous literature:* Chapt. I-2; Müller and Cyvin (1968); *approximate:* Müller, Krebs, and Peacock (1968)
PbF_4	(*) *previous literature:* Chapt. I-2
$PbCl_4$	(*) *previous literature:* Chapt. I-2; Müller and Cyvin (1968); *approximate:* Müller, Krebs, and Peacock (1968)
$Pb(Br;I)_4$, TiF_4	(*) *previous literature:* Chapt. I-2
$Ti(Cl;Br)_4$	(*) *previous literature:* Chapt. I-2; Müller and Cyvin (1968); *approximate:* Müller, Krebs, and Peacock (1968)
ZrF_4	(*) *previous literature:* Chapt. I-2
$ZrCl_4$	(*) *previous literature:* Chapt. I-2; Müller and Cyvin (1968); *approximate:* Müller, Krebs, and Peacock (1968)

Table 22-II (Continued) -- spectroscopic --

$Zr(Br;I)_4$, $Hf(Cl;Br;I)_4$	(*) *previous literature:* Chapt. I-2
VCl_4	(*) *previous literature:* Chapt. I-2; Müller and Cyvin (1968); *approximate:* Müller, Krebs, and Peacock (1968)
RuO_4	(*) *previous literature:* Chapt. I-2; Krebs, Müller, and Roesky (1967), Müller and Cyvin (1968); *incl. perpendicular amplitudes and shrinkage effect:* Müller, Krebs, Cyvin, and Diemann (1968);*approximate:* Müller, Krebs, and Peacock (1968); Müller (1968b), Levin and Abramowitz (1969)
OsO_4	(*) *previous literature:* Chapt. I-2; Müller and Cyvin (1968); *approximate:* Müller, Krebs, and Peacock (1968); Müller (1968b)
XeO_4	Yeranos (1968b)
$Ni(PF_3)_4$	Marriott, Salthouse, Ware, and Freeman (1970)
$ZnCl_4^{--}$, $CdBr_4^{--}$	*previous literature:* Chapt. I-2; Müller and Cyvin (1968); *approximate:* Müller, Krebs, and Peacock (1968); *parallel and perpendicular amplitudes:* Sanyal, Pandey, and Singh (1969b)
$Zn(Br;I)_4^{--}$, CdI_4^{--}	*previous literature:* Chapt. I-2; Müller and Cyvin (1968); *approximate:* Müller, Krebs, and Peacock (1968)
$HgCl_4^{--}$	Müller and Cyvin (1968); *approximate:* Müller, Krebs, and Peacock (1968); *parallel and perpendicular amplitudes:* Sanyal, Pandey, and Singh (1969b)
HgI_4^{--}	*parallel and perpendicular amplitudes:* Sanyal, Pandey, and Singh (1969b)
BF_4^-	*previous literature:* Chapt. I-2; $^{11}BF_4^-$: Müller and Cyvin (1968); *approximate* (^{11}B): Müller, Krebs, and Peacock (1968); $^{11}BF_4^-$: Müller (1968b)
BCl_4^-	Müller, Nagarajan, and Fadini (1966), *erratum:* Müller, Nagarajan, and Fadini (1967); Müller and Cyvin (1968); *approximate:* Müller, Krebs, and Peacock (1968)

Table 22-II (Continued) -- spectroscopic --

BBr_4^-	Müller, Nagarajan, and Fadini (1966), *erratum:* Müller, Nagarajan, and Fadini (1967); Müller and Cyvin (1968); *approximate:* Müller, Krebs, and Peacock (1968); Müller (1968b)
$B(OH)_4^-$	Müller and Cyvin (1968); *approximate:* Müller, Krebs, and Peacock (1968)
$AlCl_4^-$, $Ga(Cl;Br)_4^-$, $In(Cl;Br;I)_4^-$, $TlBr_4^-$, $P(Cl;Br)_4^+$	*previous literature:* Chapt. I-2; Müller and Cyvin (1968); *approximate:* Müller, Krebs, and Peacock (1968
NH_4^+	*previous literature:* Chapt. I-2; Ramaswamy and Ranganathan (1968)
ND_4^+	Ramaswamy and Ranganathan (1968)
$AsCl_4^+$	*previous literature:* Chapt. I-2; Nagarajan and Müller (1967c), Müller and Cyvin (1968); *approximate:* Müller Krebs, and Peacock (1968)
$SbCl_4^+$	Nagarajan and Müller (1967c), Müller and Cyvin (1968) *approximate:* Müller, Krebs, and Peacock (1968)
$FeCl_4^{--}$	Müller and Cyvin (1968); *approximate:* Müller, Krebs, and Peacock (1968)
SiO_4^{----}, PO_4^{---}	*previous literature:* Chapt. I-2; Müller and Cyvin (1968); *approximate:* Müller, Krebs, and Peacock (1968 *parallel and perpendicular amplitudes:* Sanyal, Pandey and Singh (1969b)
AsO_4^{---}, $(As;Sb)S_4^{---}$	*previous literature:* Chapt. I-2; Müller and Cyvin (1968); *approximate:* Müller, Krebs, and Peacock (1968
VO_4^{---}	*previous literature:* Chapt. I-2; Müller, Krebs, Rittner, and Stockburger (1967), Müller and Cyvin (1968); *approximate:* Müller, Krebs, and Peacock (1968
VS_4^{---}	Müller, Krebs, Rittner, and Stockburger (1967)
$(Mo;W)O_4^{--}$	*previous literature:* Chapt. I-2; Müller and Cyvin (1968); *approximate:* Müller, Krebs, and Peacock (1968 *parallel and perpendicular amplitudes:* Sanyal, Pandey and Singh (1969b)
$(S;Se;Cr)O_4^{--}$	*previous literature:* Chapt. I-2; Müller and Cyvin (1968); *approximate:* Müller, Krebs, and Peacock (1968
FeO_4^{--}	*parallel and perpendicular amplitudes:* Sanyal, Pandey and Singh (1969b)

Table 22-II (Continued) -- spectroscopic --

$(Mo;W)S_4^{--}$	*previous literature:* Chapt. I-2; Müller, Krebs, Rittner, and Stockburger (1967), Müller and Cyvin (1968); *approximate:* Müller, Krebs, and Peacock (1968); Müller and Diemann (1969)
ClO_4^-	*previous literature:* Chapt. I-2; Müller and Cyvin (1968): *approximate:* Müller, Krebs, and Peacock (1968); Baran, Aymonino, and Müller (1970)
$(Br;I)O_4^-$	Baran, Aymonino, and Müller (1970)
MnO_4^-	*previous literature:* Chapt. I-2; Krebs, Müller, and Roesky (1967), Müller and Cyvin (1968); *approximate:* Müller, Krebs, and Peacock (1968)
MnO_4^{--}, MnO_4^{---}	Krebs, Müller, and Roesky (1967), Müller and Cyvin (1968); *approximate:* Müller, Krebs, and Peacock (1968), Müller (1968a)
TcO_4^-	*previous literature:* Chapt. I-2; Müller and Cyvin (1968); *approximate:* Müller, Krebs, and Peacock (1968); *parallel and perpendicular amplitudes:* Sanyal, Pandey, and Singh (1969b)
ReO_4^-	*previous literature:* Chapt. I-2; Müller and Cyvin (1968); *approximate:* Müller, Krebs, and Peacock (1968)
ReS_4^-	Müller, Diemann, and Krishna Rao (1970)
FeO_4^{--}	Müller and Cyvin (1968); *approximate:* Müller, Krebs, and Peacock (1968); *parallel and perpendicular amplitudes:* Sanyal, Pandey, and Singh (1969b)
RuO_4^-, RuO_4^{--}	Krebs, Müller, and Roesky (1967), Müller and Cyvin (1968); *approximate:* Müller, Krebs, and Peacock (1968), Müller (1968a)

(5) Pyramidal-axial XY_3Z

$(C;Si;Ge)H_3D$, $C(H;D)_3T$, (*)	
$(C;Si;Ge)HD_3$, $C(H;D)T_3$	
$C(H;D)_3(Cl;Br;I)$,	(*) *previous literature:* Chapt. I-2
$C(H;D)(F;Cl;Br)_3$	
$C(H;D)I_3$	Anantarama Sarma and Tiruvenganna Rao (1969)

Table 22-II (Continued) -- spectroscopic --

$SiHD_3$	(*) *previous literature:* Chapt. I-2; *generalized amplitudes:* Venkateswarlu and Malathy Devi (1968b)
SiH_3F	(*) *previous literature:* Chapt. I-2; Nagarajan and Müller (1967a); *generalized amplitudes:* Venkateswarlu and Malathy Devi (1968b); Müller, Krebs, Fadini, Glemser, Cyvin, Brunvoll, Cyvin, Elvebredd, Hagen, an Vizi (1968)
SiH_3(Cl;Br)	Nagarajan and Müller (1967a), Müller, Krebs, Fadini, Glemser, Cyvin, Brunvoll, Cyvin, Elvebredd, Hagen, an Vizi (1968); *approximate:* Baran (1970)
SiH_3I	(*) *previous literature:* Chapt. I-2; Nagarajan and Müller (1967a); *generalized amplitudes:* Venkateswarlu and Malathy Devi (1968b); Müller, Krebs, Fadini, Glemser, Cyvin, Brunvoll, Cyvin, Elvebredd, Hagen, an Vizi (1968), Müller, Peacock, Schulze, and Heidborn (1969); *approximate:* Baran (1970)
$SiHF_3$	(*) *previous literature:* Chapt. I-2; Nagarajan and Müller (1967a); *generalized amplitudes:* Venkateswarlu and Malathy Devi (1968b); Müller, Krebs, Fadini, Glemser, Cyvin, Brunvoll, Cyvin, Elvebredd, Hagen, an Vizi (1968); *approximate:* Baran (1970)
$SiDF_3$	(*) *previous literature:* Chapt. I-2; *generalized amplitudes:* Venkateswarlu and Malathy Devi (1968b)
$SiHCl_3$	(*) *previous literature:* Chapt. I-2; Nagarajan and Müller (1967a); *generalized amplitudes:* Venkateswarlu and Malathy Devi (1968b); Müller, Krebs, Fadini, Glemser, Cyvin, Brunvoll, Cyvin, Elvebredd, Hagen, ar Vizi (1968); *approximate:* Baran (1970)
$SiHBr_3$	(*) *previous literature:* Chapt. I-2; Nagarajan and Müller (1967a); *generalized amplitudes:* Venkateswarlu and Malathy Devi (1968b); Müller, Krebs, Fadini, Glemser, Cyvin, Brunvoll, Cyvin, Elvebredd, Hagen, ar Vizi (1968), Müller, Peacock, Schulze, and Heidborn (1969); *approximate:* Baran (1970)
$SiFCl_3$	Nagarajan and Müller (1967a), Müller, Krebs, Fadini, Glemser, Cyvin, Brunvoll, Cyvin, Elvebredd, Hagen, ar Vizi (1968); *approximate:* Baran (1970)

Table 22-II (Continued) -- spectroscopic --

$SiCl_3(Br;I)$, $SiClBr_3$	(*) *previous literature:* Chapt. I-2; Nagarajan and Müller (1967a); *generalized amplitudes:* Venkateswarlu and Malathy Devi (1968b); Müller, Krebs, Fadini, Glemser, Cyvin, Brunvoll, Cyvin, Elvebredd, Hagen, and Vizi (1968); *approximate:* Baran (1970)
$SiClI_3$	(*) *previous literature:* Chapt. I-2; Nagarajan and Müller (1967a); *generalized amplitudes:* Venkateswarlu and Malathy Devi (1968b); Müller, Krebs, Fadini, Glemser, Cyvin, Brunvoll, Cyvin, Elvebredd, Hagen, and Vizi (1968)
$SiBr_3I$	Nagarajan and Müller (1967a), Müller, Krebs, Fadini, Glemser, Cyvin, Brunvoll, Cyvin, Elvebredd, Hagen, and Vizi (1968); *approximate:* Baran (1970)
GeH_3F	(*) *previous literature:* Chapt. I-2; Nagarajan and Müller (1967a), Müller, Krebs, Fadini, Glemser, Cyvin, Brunvoll, Cyvin, Elvebredd, Hagen, and Vizi (1968), Müller, Peacock, Schulze, and Heidborn (1969); *approximate:* Baran (1970)
$GeH_3(Cl;Br)$	(*) *previous literature:* Chapt. I-2; Nagarajan and Müller (1967a), Müller, Krebs, Fadini, Glemser, Cyvin, Brunvoll, Cyvin, Elvebredd, Hagen, and Vizi (1968); *approximate:* Baran (1970)
GeD_3Cl	(*) *previous literature:* Chapt. I-2
GeH_3I, $GeH(Cl;Br)_3$, $GeClBr_3$	Nagarajan and Müller (1967a), Müller, Krebs, Fadini, Glemser, Cyvin, Brunvoll, Cyvin, Elvebredd, Hagen, and Vizi (1968); *approximate:* Baran (1970)
$SnCl_3Br$	Nagarajan and Müller (1967a), Müller, Krebs, Fadini, Glemser, Cyvin, Brunvoll, Cyvin, Elvebredd, Hagen, and Vizi (1968); *generalized amplitudes:* Venkateswarlu, Joseph, Rudrawarrier, and Mathew (1969)
$SnClBr_3$	Nagarajan and Müller (1967a), Müller, Krebs, Fadini, Glemser, Cyvin, Brunvoll, Cyvin, Elvebredd, Hagen, and Vizi (1968), Müller, Peacock, Schulze, and Heidborn (1969); *generalized amplitudes:* Venkateswarlu, Joseph, Rudrawarrier, and Mathew (1969)

Table 22-II (Continued) -- spectroscopic --

$TiCl_3Br$, $TiClBr_3$	Nagarajan and Müller (1967a), Müller, Krebs, Fadini, Glemser, Cyvin, Brunvoll, Cyvin, Elvebredd, Hagen, a Vizi (1968)
NOF_3	Müller, Nagarajan, and Krebs (1967); *generalized amplitudes:* Venkateswarlu, Joseph, Rudrawarrier, and Mathew (1969); Ramaswamy and Mohan (1970)
POF_3	Nagarajan (1968), Müller, Krebs, Fadini, Glemser, Cyvin, Brunvoll, Cyvin, Elvebredd, Hagen, and Vizi (1968), Ramaswamy and Krishna Rao (1969a)
$POCl_3$	Nagarajan (1968), Müller, Krebs, Fadini, Glemser, Cyvin, Brunvoll, Cyvin, Elvebredd, Hagen, and Vizi (1968), Vilkov, Khaykin, Vasilev, and Tulyakova (196 Ramaswamy and Krishna Rao (1969a); *approximate:* Baran (1970)
$POBr_3$	Nagarajan (1968), Müller, Krebs, Fadini, Glemser, Cyvin, Brunvoll, Cyvin, Elvebredd, Hagen, and Vizi (1968), Ramaswamy and Krishna Rao (1969c);*approximate* Baran (1970)
$VO(F;Cl;Br)_3$	Nagarajan (1968), Müller, Krebs, Fadini, Glemser, Cyvin, Brunvoll, Cyvin, Elvebredd, Hagen, and Vizi (1968); *approximate:* Baran (1970)
PSF_3	Nagarajan (1968), Cyvin, Vizi, Müller, and Krebs (196 Ramaswamy and Krishna Rao (1969a); *generalized amplitudes:* Venkateswarlu, Joseph, Rudrawarrier, and Mathe (1969)
$PSCl_3$	Nagarajan (1968), Vilkov, Khaykin, Vasilev, and Tulyakova (1968), Cyvin, Vizi, Müller, and Krebs (196 Ramaswamy and Krishna Rao (1969a); *generalized amplitudes:* Venkateswarlu, Joseph, Rudrawarrier, and Mathe (1969)
$PSBr_3$	Nagarajan, Müller, and Horn (1966), Nagarajan (1968), Cyvin, Vizi, Müller, and Krebs (1969), Ramaswamy and Krishna Rao (1969c)
NSF_3	(*) *previous literature:* Chapt. I-2
$Re(Cl;Br)O_3$	Nagarajan (1968), Müller, Krebs, Fadini, Glemser, Cyvin, Brunvoll, Cyvin, Elvebredd, Hagen, and Vizi (1968); *approximate:* Baran (1970)

Table 22-II (Continued) -- spectroscopic --

$FClO_3$	Müller and Nagarajan (1967a), Müller, Krebs, Fadini, Glemser, Cyvin, Brunvoll, Cyvin, Elvebredd, Hagen, and Vizi (1968), Müller, Peacock, Schulze, and Heidborn (1969), Clark, Beagley, Cruickshank, and Hewitt (1970a)
$(OH)ClO_3$, SPO_3^{---}, FPO_3^{--}, $ClSO_3^-$, $(NO)SO_3^-$	Müller and Nagarajan (1967a), Müller, Krebs, Fadini, Glemser, Cyvin, Brunvoll, Cyvin, Elvebredd, Hagen, and Vizi (1968)
$(OH)PO_3^{--}$, HPO_3^{--}, $(H;F)SO_3^-$, $(OH)(S;Se)O_3^-$, $(F;Cl)CrO_3^-$	Müller and Nagarajan (1967a), Müller, Krebs, Fadini, Glemser, Cyvin, Brunvoll, Cyvin, Elvebredd, Hagen, and Vizi (1968); *approximate:* Baran (1970)
$S_2O_3^{--}$	(*) *previous literature:* Chapt. I-2; Müller and Nagarajan (1967a), Müller, Krebs, Fadini, Glemser, Cyvin, Brunvoll, Cyvin, Elvebredd, Hagen, and Vizi (1968); *approximate:* Baran (1970)
ReO_3S^-	Müller and Nagarajan (1967a), Müller, Krebs, Rittner, and Stockburger (1967), Müller, Krebs, Fadini, Glemser, Cyvin, Brunvoll, Cyvin, Elvebredd, Hagen, and Vizi (1968); *approximate:* Baran (1970)
$OsNO_3^-$	Müller, Baran, Bollmann, and Aymonino (1969); *approximate:* Baran (1970)

(5) Twisted XY_2Z_2

$(C;Ge)H_2D_2$, $C(H;D)_2T_2$	(*)
$CH_2(F;Cl;Br;I)_2$	*generalized amplitudes:* Venkateswarlu, Joseph, and Malathy Devi (1969)
$SiH_2(D;Cl;Br)_2$	(*) *previous literature:* Chapt. I-2; *generalized amplitudes:* Venkateswarlu and Malathy Devi (1968a)
CH_2N_2 (diazirine)	Nagarajan and Sivaprakasam (1970a)
CF_2N_2	*generalized amplitudes:* Venkateswarlu and Rudra Warrier (1970); Nagarajan and Sivaprakasam (1970a)
SF_4	(*) *previous literature:* Chapt. I-2; Cyvin (1969a), Christie and Sawodny (1970)
SeF_4	Ramaswamy and Jayaraman (1970a)
SO_2F_2	(*) *previous literature:* Chapt. I-2; Pichai, Krishna Pillai, and Ramaswamy (1967); *generalized amplitudes:* Venkateswarlu and Malathy Devi (1968a); Cyvin and Hargittai (1969)

Table 22-II (Continued) -- spectroscopic --

SO_2Cl_2	(*) *previous literature:* Chapt. I-2; *generalized amplitudes:* Venkateswarlu and Malathy Devi (1968a); Hargittai and Cyvin (1969)
SeO_2F_2	Ramaswamy and Jayaraman (1970b)
(5) Planar XY_4	
XeF_4	(*) *previous literature:* Chapt. I-2; Yeranos (1967), Cyvin, Cyvin, Müller, and Krebs (1968); *generalized amplitudes and shrinkage effects:* Venkateswarlu and Babu Joseph (1968b)
$Au(Cl;Br)_4^-$, $PtCl_4^{--}$	*previous literature:* Chapt. I-2; Cyvin, Cyvin, Müller, and Krebs (1968)
$PtBr_4^{--}$, $Pt(NH_3)_4^{++}$, $Pd(Cl;Br)_4^{--}$	Cyvin, Cyvin, Müller, and Krebs (1968)
(5) Linear five-atomic	
$Hg(CN)_2$	Nagarajan (1966h')
C_3O_2	(*) *previous literature:* Chapt. I-2; *incl. shrinkage effects:* Brunvoll, Cyvin, Elvebredd, and Hagen (1968) Almenningen, Arnesen, Bastiansen, Seip, and Seip (1968); Ramaswamy and Srinivasan (1969b); *incl.shrinkage effects and higher-order approximation terms:* Tanimoto, Kuchitsu, and Morino (1970); *higher-order approximations:* Clark and Seip (1970)
C_3S_2	*incl. shrinkage effects:* Brunvoll (1968); *incl. perpendicular amplitudes and shrinkage effects:* Diallo (1968); Ramaswamy and Srinivasan (1969b)
HCCCN	(*) *previous literature:* Chapt. I-2; Ramaswamy and Srinivasan (1968)
$(Ag;Au)(CN)_2^-$	Nagarajan (1966h')
(5) Planar XY_2ZW	
$C(H;D)_2CO$ (ketenes)	*incl. generalized amplitudes and shrinkage effect:* Venkateswarlu, Rajalakshmi, and Malathy Devi (1966); Cyvin and Alfheim (1970b)
$C(H;D)_2N_2$	*incl. generalized amplitudes and shrinkage effect:* Venkateswarlu, Rajalakshmi, and Malathy Devi (1966)
$N(H;D)_2CN$	(*) *previous literature:* Chapt. I-2

Table 22-II (Continued) -- spectroscopic --

(5) Other five-atomic	
$S(CN)_2$	(*) *previous literature:* Chapt. I-2; *generalized amplitudes:* Unnikrishnan Nayar and Aruldhas (1968a)
$CH_2(F;I)(Cl;Br)$, CHFClBr	(*)
HCOOH	*incl. perpendicular amplitude corrections:* Cyvin, Alfheim, and Hagen (1970)
$(OH)NO_2$ (HNO_3)	*see also:* planar XY_2Z (four-atomic); Rajeswara Rao and Sakku (1968), Kumar, Padma, and Rajeswara Rao (1969)
$(OD)NO_2$	*see also:* planar XY_2Z (four-atomic); Kumar, Padma, and Rajeswara Rao (1969)
PO_4^{---}	Nagarajan and Perumal (1968a)
(6) Planar X_2Y_4	
C_2H_4	(*) *previous literature:* Chapt. I-2; *generalized amplitudes:* Venkateswarlu, Mariam, and Girijavallabhan (1967); Cyvin and Hagen (1969), Ramaswamy and Devarajan (1968), Fogarasi and Mezey (1970)
C_2D_4	(*) *previous literature:* Chapt. I-2; *generalized amplitudes:* Venkateswarlu, Mariam, and Girijavallabhan (1967); Fogarasi and Mezey (1970)
C_2T_4	(*) *previous literature:* Chapt. I-2
$C_2(F;Cl;Br)_4$	(*) *previous literature:* Chapt. I-2; *generalized amplitudes:* Venkateswarlu, Mariam, and Girijavallabhan (1967); Fogarasi (1970); *incl. generalized amplitudes:* Cyvin (1970b)
C_2I_4	*incl. generalized amplitudes:* Cyvin (1970b)
N_2O_4, $C_2O_4^{----}$	*previous literature:* Chapt. I-2;*generalized amplitudes:* Venkateswarlu, Mariam, and Girijavallabhan (1967)
(6) Other planar six-atomic	
(tr;cis;a)-$C_2H_2D_2$, C_2H_3D, C_2HD_3, $C_2H_2F_2$	(*) *previous literature:* Chapt. I-2
$C_2H_2O_2$ (glyoksal)	Jensen, Hagen, and Cyvin (1969), Fukuyama, Kuchitsu, and Morino (1968a)
$C_2O_2(F;Cl)_2$	*generalized amplitudes:* Venkateswarlu and Bhamambal (1970b)

Table 22-II (Continued) -- spectroscopic --

C_4H_2 (diacetylene)	*previous literature:* Chapt. I-2; *incl. perpendicular amplitudes and shrinkage effects:* Venkateswarlu, Mariam, and Anantarama Sarma (1967); Ramaswamy and Srinivasan (1968), Ramaswamy and Srinivasan (1969a)
C_4N_2 (dicyano-acetylene)	(*) *previous literature:* Chapt. I-2; *incl. perpendicular amplitudes and shrinkage effects:* Venkateswar Mathew, and Malathy Devi (1969); *incl. shrinkage effects:* Cyvin (1969b); Ramaswamy and Srinivasan (1969a)
HCCCC(Cl;Br;I)	*previous literature:* Chapt. I-2; Minasso and Zerbi (1971)

(6) Pyramidal-axial XY_4Z

$(Cl;Br)F_5$	*incl. perpendicular amplitude corrections:* Cyvin, Brunvoll, and Robiette (1969); Ramaswamy and Muthusubramanian (1971)
IF_5	*incl. perpendicular amplitude corrections:* Cyvin, Brunvoll, and Robiette (1969); Ramaswamy and Muthusubramanian (1970)
$XeOF_4$	*generalized amplitudes and shrinkage effects:* Venkateswarlu and Babu Joseph (1968b); *incl. perpendicular amplitude corrections:* Cyvin, Brunvoll, and Robiette (1969)

(6) Pyramidal-axial XY_3ZW

$(^{10}B;^{11}B)(H;D)_3CO$, CH_3CN	(*) *previous literature:* Chapt. I-2; *generalized amplitudes, perpendicual amplitude corrections, and shrinkage effect:* Venkateswarlu, Rajalakshmi, and Purushothaman (1967)
CH_3NC	*previous literature:* Chapt. I-2; *generalized amplitudes, perpendicular amplitude corrections, and shrinkage effect:* Venkateswarlu, Rajalakshmi, and Purushothaman (1967)
CD_3CN, CD_3NC	*generalized amplitudes, perpendicular amplitude corrections, and shrinkage effect:* Venkateswarlu, Rajalakshmi, and Purushothaman (1967)
CF_3CN	(*) *previous literature:* Chapt. I-2

Table 22-II (Continued) -- spectroscopic --

Ge(H;D)$_3$CN	*generalized amplitudes, perpendicular amplitude corrections, and shrinkage effect:* Venkateswarlu and Bhamambal (1970a)
(6) Trigonal bipyramidal	
PH_5(hypothetical)	*incl. atomic amplitudes:* Holmes and Deiters (1969)
PF_5	(*) *previous literature:* Chapt. I-2; Nagarajan and Durig (1967b); *incl. shrinkage effects:* Cyvin and Brunvoll (1969a); *incl. atomic amplitudes:* Holmes and Deiters (1969); Ramaswamy and Krishna Rao (1969b), Ramaswamy and Krishna Rao (1969d), Levin (1970), Bartell (1970)
PCl_5	(*) *previous literature:* Chapt. I-2; *incl. atomic amplitudes:* Holmes and Deiters (1969); Ramaswamy and Krishna Rao (1969b), Bartell (1970)
PF_2Cl_3	(*) *previous literature:* Chapt. I-2; *incl. atomic amplitudes:* Holmes and Deiters (1969); Ramaswamy and Krishna Rao (1969b)
$(CH_3)_2PF_3$	Ramaswamy and Krishna Rao (1969d)
AsF_5	Nagarajan and Durig (1967b); *incl. shrinkage effects:* Cyvin and Brunvoll (1969a); *incl. atomic amplitudes:* Holmes and Deiters (1969); Bartell (1970)
$SbCl_5$	(*) *previous literature:* Chapt. I-2; *incl. atomic amplitudes:* Holmes and Deiters (1969)
SbF_3Cl_2, (Nb;Ta)Cl_5	(*) *previous literature:* Chapt. I-2
$Sb(CH_3)_5$	*incl. atomic amplitudes:* Holmes and Deiters (1969)
VF_5	Müller and Nagarajan (1967b); *incl. shrinkage effects:* Vizi, Brunvoll, and Müller (1968); *incl. atomic amplitudes:* Holmes and Deiters (1969); Bartell (1970)
(6) Other six-atomic	
B_2F_4	Cyvin and Elvebredd (1969)
P_2I_4	*generalized amplitudes:* Venkateswarlu and Girijavallabhan (1968)
PF_4Cl	*incl. atomic amplitudes:* Holmes and Deiters (1969); *generalized amplitudes and shrinkage effect:* Venkateswarlu and Girijavallabhan (1969)

Table 22-II (Continued) -- spectroscopic --

PF_3Cl_2 ($PFCl_2F_2$)	*incl. atomic amplitudes:* Holmes and Deiters (1969)
$PF_4(CH_3)$	*incl. atomic amplitudes:* Holmes and Deiters (1969); Ramaswamy and Krishna Rao (1969d), Bartell (1970)
SOF_4	(*) *previous literature:* Chapt. I-2
$(OH)SO_2(F;Cl)$	Cyvin and Hargittai (1969)
$SC(SH)_2$	*approximate:* Müller, Krebs, and Gattow (1967)
$(OH)ClO_3$ ($HClO_4$)	*see also:* pyramidal-axial XY_3Z (five-atomic); Clark, Beagley, Cruickshank, and Hewitt (1970b)
$(OH)PO_3^{--}$, $(NO)SO_3^-$, $(OH)(S;Se)O_3^-$	*see:* pyramidal-axial XY_3Z (five-atomic)

(7) Octahedral XY_6

SF_6	(*) *previous literature:* Chapt. I-2; Thakur, Rao, an Rai (1970), Ramaswamy and Mohan (1971)
SeF_6	(*) *previous literature:* Chapt. I-2; *incl. shrinkage effects:* Cyvin, Brunvoll, and Müller (1968); Thakur, Rao, and Rai (1970)
TeF_6	(*) *previous literature:* Chapt. I-2; Thakur, Rao, an Rai (1970)
CrF_6	(*)
MoF_6	(*) *previous literature:* Chapt. I-2; Thakur, Rao, an Rai (1970)
$(W;U;Np;Pu)F_6$	(*) *previous literature:* Chapt. I-2; Kimura, Schomak Smith, and Weinstock (1968); *incl. shrinkage effects* Cyvin, Brunvoll, and Müller (1968); Thakur, Rao, and Rai (1970)
WCl_6	*previous literature:* Chapt. I-2; *incl. perpendicular amplitudes and shrinkage effects:* Sanyal, Singh, and Pandey (1969a)
$(Tc;Re;Ru)F_6$	(*) *previous literature:* Chapt. I-2; Thakur, Rao, an Rai (1970)
RhF_6	(*) *previous literature:* Chapt. I-2; *incl. shrinkage effects:* Cyvin, Brunvoll, and Müller (1968); Thakur, Rao, and Rai (1970)
OsF_6	(*) *previous literature:* Chapt. I-2; Kimura, Schomak Smith, and Weinstock (1968), Thakur, Rao, and Rai (1970)

Table 22-II (Continued) -- spectroscopic --

IrF_6	(*) *previous literature:* Chapt. I-2; Kimura, Schomaker, Smith, and Weinstock (1968); *incl. shrinkage effects:* Cyvin, Brunvoll, and Müller (1968); Thakur, Rao, and Rai (1970)
PtF_6	(*) *previous literature:* Chapt. I-2; *incl. shrinkage effects:* Cyvin, Brunvoll, and Müller (1968); Thakur, Rao, and Rai (1970)
$(Si;Ge;Sn)F_6^{--}$, $(P;As;Sb)F_6^{-}$	Rao, Thakur, and Rai (1970); *incl. perpendicular amplitudes and shrinkage effects:* Singh, Pandey, and Singh (1970)
$SeCl_6^{--}$	*previous literature:* Chapt. I-2; *incl. shrinkage effects:* Avasthi and Mehta (1970)
$SeBr_6^{--}$, $Te(Cl;Br)_6^{--}$	*incl. shrinkage effects:* Avasthi and Mehta (1970)
$Pt(Cl;Br)_6^{--}$	*previous literature:* Chapt. I-2; Yeranos (1970)
PtI_6^{--}	Yeranos (1970)
UCl_6^{--}	*incl. perpendicular amplitudes:* Pandey, Singh, and Sanyal (1969); Rao, Thakur, and Rai (1970)

(7) Pyramidal-axial XY_3ZUV

$CH_3CC(H;Cl;Br;I)$, CD_3CCCl	*generalized amplitudes and shrinkage effects:* Venkateswarlu, Mariam, and Mathew (1966); Venkateswarlu and Mathew (1967)
$CF_3CC(H;D)$	*previous literature:* Chapt. I-2; *generalized amplitudes and shrinkage effects:* Venkateswarlu and Mathew (1966)
$Ge(H;D)_3CCH$	*generalized amplitudes and shrinkage effects:* Venkateswarlu, Malathy Devi, and Natarajan (1969)

(7) Other seven-atomic

IOF_5	Ramaswamy and Muthusubramanian (1970)
SF_5Cl	*previous literature:* Chapt. I-2; *incl. generalized amplitudes:* Venkateswarlu and Mariam (1967); Ramaswamy and Mohan (1969)
C_3H_4 (allene)	(*) *previous literature:* Chapt. I-2; Cyvin and Hagen (1969)
C_3D_4	(*) *previous literature:* Chapt. I-2

Table 22-II (Continued) -- spectroscopic --

$C_3(H;D)_4$,(1,2;3,3)-$C_3H_2D_2$ (cyclopropenes)	*incl. perpendicular amplitude corrections:* Cyvin and Hagen (1970)
$C_2(H;D)_4O$, C_2H_4S (ethylene oxide and sulphide)	*previous literature:* Chapt. I-2; *incl. generalized amplitudes and perpendicular amplitude corrections:* Venkateswarlu and Joseph (1970)
CH_3CHO	Kato, Konaka, Iijima, and Kimura (1969); *incl. perpendicular amplitude corrections:* Iijima and Kimura (1969)
CH_2CHCN (acrylonitrile)	Fukuyama and Kuchitsu (1970)
$C_2(Br;I)_3CN$ (trihaloacrylonitrile)	Lie, Klaboe, Kloster-Jensen, Hagen, and Christensen (1970)
$C_2N_2(O;S;Se)H_2$ (1,2,5;1,3,4)-oxadiazole, (1,2,5;1,3,4)-thiadiazole, 1,2,5-selenadiazole	*incl. perpendicular amplitude corrections and atomic amplitudes:* Cyvin, Cyvin, Hagen, and Markov (1969)
$B(OH)_3$	*see:* planar XY_3 (four-atomic)
$(CH_3)NO_2$	*see:* planar XY_2Z (four-atomic)
CH_3PF_2	*see:* pyramidal XY_2Z (four-atomic)
(8) S_8	*incl. generalized amplitudes:* Venkateswarlu and Babu Joseph (1967a); Babu J. (1967); *incl. generalized amplitudes:* Cyvin (1970a)
IF_7	Ramaswamy and Muthusubramanian (1970)
(8) Trigonal X_2Y_6	
$C_2(H;D)_6$	*incl. perpendicular amplitude corrections:* Kuchitsu (1968)
$C_2(F;Cl;Br)_6$	*previous literature:* Chapt. I-2; *framework amplitudes:* Cyvin and Brunvoll (1968)
(8) Bridged $X_2Y_2Z_4$	
$B_2(H;D)_6$	*incl. generalized amplitudes:* Venkateswarlu, Mariam, and Natarajan (1967); *incl. perpendicular amplitude corrections:* Kuchitsu (1968); ^{10}B *and* ^{11}B: Cyvin and Vizi (1969)
Al_2Cl_6, $Al_2(CH_3)_6$, $Al_2(CH_3)_4Cl_2$	*incl. generalized amplitudes:* Venkateswarlu, Mariam, and Natarajan (1967)

Table 22-II (Continued) -- spectroscopic --

(8) Linear eight-atomic	
C_6N_2 (dicyanodiacetylene)	Nagarajan and Durig (1967c)
(8) C_4H_4 (1,3-butatriene)	Nagarajan and Durig (1967a); *incl. generalized amplitudes and shrinkage effects:* Venkateswarlu, Rajalakshmi, and Natarajan (1967); *incl. perpendicular amplitudes and shrinkage effects:* Cyvin and Hagen (1969); Ramaswamy and Srinivasan (1969b)
C_4D_4	*incl. perpendicular amplitudes and shrinkage effects:* Cyvin and Hagen (1969)
C_4T_4	*incl. shrinkage effects:* Cyvin and Hagen (1969)
(tr;cis;a)-$C_4H_2D_2$, C_4H_3D, C_4HD_3	*incl. perpendicular amplitudes:* Cyvin and Hagen (1969)
C_2(Cl;Br;I)$_2$(CN)$_2$ (dihalofumaronitrile)	Lie, Klaboe, Christensen, and Hagen (1970)
$(COOH)_2$ (oxalic acid)	*incl. perpendicular amplitude corrections:* Náhlovská, Náhlovský, and Strand (1970), Cyvin and Alfheim (1970a)
CH_3COOH	*incl. perpendicular amplitude corrections:* Alfheim and Cyvin (1970); *incl. framework amplitudes:* Derissen (1971)
CH_3COOD, CD_3COO(H;D)	Alfheim and Cyvin (1970)
C_3H_4O (acrolein)	Fukuyama, Kuchitsu, and Morino (1968b), Jensen, Hagen, and Cyvin (1969)
CH_3CHCO (methylketene)	*incl.* ^{13}C *and* ^{18}O *isotopic molecules:* Cyvin, Christensen, and Nicolaisen (1970); Cyvin and Alfheim (1970b)
CD_3CDCO	Cyvin, Christensen, and Nicolaisen (1970), Cyvin and Alfheim (1970b)
CH_3CDCO, CD_3CHCO, (tr;gauche)-CH_2DCHCO	Cyvin, Christensen, and Nicolaisen (1970)
[N(H;D)$_2$]$_2$CO	*see:* planar XY_2Z (four-atomic)
(9) $Ni(CO)_4$	(*) *previous literature:* Chapt. I-2; *incl. shrinkage effects:* Jones, McDowell, and Goldblatt (1968)
C_3(H;D)$_6$ (cyclopropanes)	(*) *previous literature:* Chapt. I-2; *incl. generalized amplitudes:* Venkateswarlu and Bhamambal (1967); *incl. perpendicular amplitude corrections:* Cyvin and Hagen (1970)

Table 22-II (Continued) -- spectroscopic --

C_3T_6	(*)
$C_4O_3(H;D)_2$ (maleic anhydride)	Cyvin and Cyvin (1971a)
$C_4(H;D)_4(O;S)$ (furan and thiophene)	*incl. perpendicular amplitude corrections and atomic amplitudes:* Cyvin, Cyvin, Hagen, and Markov (1969)
$CONH_2CH_3$ (acetamide)	Puranik and Sirdeshmukh (1967)
$B(OH)_4^-$	*see:* tetrahedral XY_4 (five-atomic)
PF_4CH_3	*see:* six-atomic
(10) P_4O_6	*incl. atomic amplitudes:* Cyvin and Cyvin (1971b)
CH_3CCCH_3	*generalized amplitudes and shrinkage effects:* Venkateswarlu and Mathew (1968a); Tanimoto, Kuchitsu, and Morino (1969)
CF_3CCCF_3	*previous literature:* Chapt. I-2; Elvebredd (1968); *generalized amplitudes and shrinkage effects:* Venkateswarlu and Mathew (1968a)
C_4H_6 (1,3-butadiene)	Trætteberg, Hagen, and Cyvin (1969); *perpendicular amplitude corrections:* Cyvin, Trætteberg, and Hagen (1969)
C_4D_6, 2-C_4H_5D	Trætteberg, Hagen, and Cyvin (1969)
$(CH_3)_2CO$	Kato, Konaka, Iijima, and Kimura (1969); *incl. perpendicular amplitude corrections:* Iijima (1970)
$(HCOOH)_2$	*incl. perpendicular amplitude corrections:* Alfheim, Hagen, and Cyvin (1971)
$(HCOOD)_2$, $DCOO(H;D)_2$	Alfheim, Hagen, and Cyvin (1971)
$[PCl_4][FeCl_4]$	*atomic amplitudes:* Kistenmacher and Stucky (1968)
$P(CCH)_3$	*see also:* pyramidal XY_3 (four-atomic); Smit and Dijkstra (1971)
$(As;Sb)(CCH)_3$	*see:* pyramidal XY_3 (four-atomic)
$C_5H_5^-$	Brunvoll, Cyvin, and Schäfer (1971b)
(11) $Fe(CO)_5$	(*) *previous literature:* Chapt. I-2
$C_3(H;D)_8$ (propanes)	*incl. perpendicular amplitude corrections and atomic amplitudes:* Cyvin and Vizi (1970)
$(CH_3)_2CCO$	Cyvin and Alfheim (1970b)
(12) C_6H_6 (benzene)	(*) *previous literature:* Chapt. I-2; Schäfer, Southern, Cyvin, and Brunvoll (1970)

Table 22-II (Continued) -- spectroscopic --

C_6D_6	(*) *previous literature:* Chapt. I-2
C_6T_6	(*)
C_6F_6 (hexafluorobenzene)	(*) *previous literature:* Chapt. I-2
sym-$C_6H_3(D;T)_3$, sym-$C_6D_3T_3$, C_6H_5D, (o;m;p)-$C_6H_4D_2$, (vic;asym)-$C_6H_3D_3$, (1,2,3,4;1,3,4,5; 1,2,4,5)-$C_6H_2D_4$, C_6HD_5	(*)
C_6H_6 (dimethyl-diacetylene)	(*) *previous literature:* Chapt. I-2
$C_4(H;D)_8$ (cyclobutanes), C_4F_8	*generalized amplitudes:* Venkateswarlu and Bhamambal (1968b)
$C_6H_4O_2$ (quinone)	Jensen, Hagen, and Cyvin (1969)
$(CH_3)_2PF_3$	*see:* trigonal bipyramidal (six-atomic)
(13) $(Cr;Mo)(CO)_6$	(*) *previous literature:* Chapt. I-2; *incl. shrinkage effects:* Jones, McDowell, and Goldblatt (1969)
$W(CO)_6$	*previous literature:* Chapt. I-2; *incl. shrinkage effects:* Jones, McDowell, and Goldblatt (1969)
C_5H_8 (bicyclo-1,1,1 pentane)	Cyvin, Elvebredd, Hagen, and Andersen (1968)
$[C(H;D)_3]_3N$	Dellepiane and Zerbi (1968)
(tr;gauche)-$CH_3COC_2H_5$	Abe, Kuchitsu, and Shimanouchi (1969)
$(N;P;As;Sb;Bi)(CH_3)_3$	*see:* pyramidal XY_3 (four-atomic)
(14) P_4O_{10}	*incl. atomic amplitudes:* Cyvin and Cyvin (1971b)
(cis;tr)-C_6H_8 (1,3,5-hexatrienes)	*incl. perpendicular amplitude corrections:* Cyvin, Hagen, and Trætteberg (1969)
(16) $C_8(H;D)_8$ (1,3,5,7-cyclooctatetraene)	*incl. perpendicular amplitude corrections:* Trætteberg, Hagen, and Cyvin (1970)
(17) $Ni(PF_3)_4$	*see:* tetrahedral XY_4 (five-atomic)
$Pt(NH_3)_4^{++}$	*see:* planar XY_4 (five-atomic)
(18) C_6H_{12} (cyclohexane)	Cyvin, Vizi, and Hagen (1968)

Table 22-II (Continued) -- spectroscopic --

$C_{10}H_8$ (naphthalene)	*previous literature:* Chapt. I-2; Hagen and Cyvin (1968), Cyvin, Cyvin, and Hagen (1968), Cyvin and Cyvin (1969)
$C_{10}D_8$	*previous literature:* Chapt. I-2; Hagen and Cyvin (1968), Cyvin and Cyvin (1969)
(α;β;1,2,3,4)-$C_{10}H_4D_4$, (amphi;peri)-$C_{10}H_6D_2$	Hagen and Cyvin (1968)
(19) $C_6H_6Cr(CO)_3$	*ligand only:* Schäfer, Southern, Cyvin, and Brunvoll (1970)
(20) $Al_2(CH_3)_4Cl_2$	*see:* bridged $X_2Y_2Z_4$ (eight-atomic)
(21) $(C_5H_5)_2Fe$	Brunvoll, Cyvin, and Schäfer (1971b)
$Sb(CH_3)_5$	*see:* trigonal bipyramidal XY_3Z_2 (six-atomic)
(22) $Mn_2(CO)_{10}$	Brunvoll and Cyvin (1968)
(24) $C_{14}(H;D)_{10}$ (anthracenes),(α;β)-$C_{14}H_6D_4$, meso-$C_{14}H_8D_2$	Cyvin and Cyvin (1969)
(25) $(C_6H_6)_2Cr$	*ligand only:* Schäfer, Southern, Cyvin, and Brunvoll (1970); Brunvoll, Cyvin, and Schäfer (1971a)
$(C_6H_6)_2Cr^+$	*ligand only:* Schäfer, Southern, Cyvin, and Brunvoll (1970)
(26) $Al_2(CH_3)_6$	*see:* bridged $X_2Y_2Z_4$ (eight-atomic)

Bibliography to Works on Mean Amplitudes

S. J. CYVIN

The present list* incorporates the 476 articles with special bearing upon mean-square amplitudes of vibration, as given in CYVIN (1968). The numbers in brackets [ ] from that bibliography are included here in order to facilitate the review work. The bibliography has been supplemented with some omitted articles and numerous later works. Most of them are included in the Russian translation of CYVIN (1968); cf. Ref. 2 of Chapter 22.

*

[1] ABE (MIKAMI), M., KUCHITSU,K.,and SHIMANOUCHI,T. (1969): J.Mol.Structure 4, 245.

[2] ADAMS,W.J., GEISE,H.J., and BARTELL,L.S. (1970): J.Am.Chem. Soc. 92, 5013.

[3] AINSWORTH,J. and KARLE,J. (1952): J.Chem.Phys. 20, 425. [1]

[4] AIRBY,W., GLIDEWELL,C., RANKIN,D.W.H., ROBIETTE,A.G., SHELDRICK,G.M., and CRUICKSHANK,D.W.J. (1970): Trans.Faraday Soc. 66, 551.

[5] AKISHIN,P.A., NAUMOV,V.A., and TATEVSKII,V.M. (1959): Kristallografiya 4, 194. [1.1]

[6] AKISHIN (AKISCHIN), P.A. and RAMBIDI,N.G.(1960a): Z.phys. Chem. 213, 111.

[7] AKISHIN,P.A. and RAMBIDI,N.G.(1960b): Zh.neorg.Khim. 5, 23.

---

* References to the present list are found in several parts of this book. They are given in either of the two ways of giving (a) the front number in brackets, or (b) author(s) and year of publication.

[8] AKISHIN,P.A., RAMBIDI,N.G., and EZHOV,YU.S.(1960): Zh.neorg. Khim. 5, 747.

[9] AKISHIN,P.A., RAMBIDI,N.G., and SPIRIDONOV,V.P.(1967): The Characterization of High Temperature Vapors (Ed. J.L.Margrave) p. 300, Wiley, New York 1967.

[10] AKISHIN,P.A., RAMBIDI,N.G., and ZASORIN,E.Z.(1959a): Kristallografiya 4, 186. [1.2]

[11] AKISHIN,P.A., RAMBIDI,N.G., and ZASORIN,E.Z.(1959b): Kristallografiya 4, 360. [1.3]

[12] ALEKSEEV,N.V. and KITAIGORODSKII,A.I.(1963): Zh.Strukt.Khim. 4, 163. [2]

[13] ALFHEIM,I. and CYVIN,S.J.(1970): Acta Chem.Scand. 24, 3043.

[14] ALFHEIM,I., HAGEN,G., and CYVIN,S.J.(1971): J.Mol.Structure 8, 159.

[15] ALMENNINGEN,A., ANDERSEN,B., and ASTRUP,E.E.(1970): Acta Chem. Scand. 24, 1579.

[16] ALMENNINGEN,A., ANDERSEN,B., and TRÆTTEBERG,M.(1964): Acta Chem.Scand. 18, 603. [3]

[17] ALMENNINGEN,A., ANFINSEN,I.M., and HAALAND,A.(1970a): Acta Chem.Scand. 24, 43.

[18] ALMENNINGEN,A., ANFINSEN,I.M., and HAALAND,A.(1970b): Acta Chem.Scand. 24, 1230.

[19] ALMENNINGEN,A., ARNESEN,S.P., BASTIANSEN,O., SEIP,H.M., and SEIP,R.(1968): Chem.Phys.Letters 1, 569.

[20] ALMENNINGEN,A. and BASTIANSEN,O.(1958): Kgl. Norske Videnskab. Selskabs Skrifter, No. 4. [4]

[21] ALMENNINGEN,A., BASTIANSEN,O., CYVIN,S.J., and SKANCKE,P.N. (1960): Acta Chem.Scand. 14, 959. [5]

[22] ALMENNINGEN,A., BASTIANSEN,O., EWING,V., HEDBERG,K., and TRÆTTEBERG,M.(1963): Acta Chem.Scand. 17, 2455. [6]

[23] ALMENNINGEN,A., BASTIANSEN,O., and FERNHOLT,L.(1958): Kgl. Norske Videnskab. Selskabs Skrifter, No. 3. [7]

[24] ALMENNINGEN,A., BASTIANSEN,O., FERNHOLT,L., and TRÆTTEBERG,M. (1960): J.Chem.Phys. 32, 616. [8]

[25] ALMENNINGEN,A., BASTIANSEN,O., and HAALAND,A.(1964): J.Chem. Phys. 40, 3434. [9]

[26] ALMENNINGEN,A., BASTIANSEN,O., HAALAND,A., and SEIP,H.M.(1965) Angew.Chem. 77, 877 (internat. Edit. 4, 819). [10]

[27] ALMENNINGEN,A., BASTIANSEN,O., and JENSEN,H.(1966): Acta Chem.Scand. 20, 2689. [10.1]

[28] ALMENNINGEN,A., BASTIANSEN,O., and MOTZFELDT,T.(1969): Acta Chem.Scand. 23, 2848.

[29] ALMENNINGEN,A., BASTIANSEN,O., and MUNTHE-KAAS,T.(1956): Acta Chem.Scand. 10, 261. [11]

[30] ALMENNINGEN,A., BASTIANSEN,O., SEIP,R., and SEIP,H.M.(1964): Acta Chem.Scand. 18, 2115. [12]

[31] ALMENNINGEN,A., BASTIANSEN,O., and TRÆTTEBERG,M.(1959): Acta Chem.Scand. 13, 1699. [13]

[32] ALMENNINGEN,A., BASTIANSEN,O., and TRÆTTEBERG,M.(1961): Acta Chem.Scand. 15, 1557. [14]

[33] ALMENNINGEN,A., BASTIANSEN,O., and WALLØE,L.(1967): Selected Topics in Structure Chemistry (Ed. P.Andersen, O.Bastiansen, and S.Furberg), p. 91, Universitetsforlaget, Oslo 1967.[14.1]

[34] ALMENNINGEN,A. and BJORVATTEN,T.(1963): Acta Chem.Scand. 17, 2573. [14.2]

[35] ALMENNINGEN,A., FERNHOLT,L., and SEIP,H.M.(1968): Acta Chem. Scand. 22, 51. [14.3]

[36] ALMENNINGEN,A., GUNDERSEN,G., and HAALAND,A.(1968a): Acta Chem.Scand. 22, 328.

[37] ALMENNINGEN,A., GUNDERSEN,G., and HAALAND,A.(1968b): Acta Chem.Scand. 22, 859.

[38] ALMENNINGEN,A., HAALAND,A., and MORGAN,G.L.(1969): Acta Chem. Scand. 23, 2921.

[39] ALMENNINGEN,A., HAALAND,A., and MOTZFELDT,T.(1967a): J.Organometal.Chem. 7, 97.

[40] ALMENNINGEN,A., HAALAND,A., and MOTZFELDT,T.(1967b): Selected Topics in Structure Chemistry (Ed. P.Andersen, O.Bastiansen, and S.Furberg), p. 105, Universitetsforlaget, Oslo 1967.[14.4]

[41] ALMENNINGEN,A., HAALAND,A., and NILSSON,J.E.(1968): Acta Chem.Scand. 22, 972.

[42] ALMENNINGEN,A., HAALAND,A., and WAHL,K.(1969a): Acta Chem. Scand. 23, 1145.

[43] ALMENNINGEN,A., HAALAND,A., and WAHL,K.(1969b): Acta Chem. Scand. 23, 2245.

[44] ALMENNINGEN,A., HARGITTAI,I., KLOSTER-JENSEN,E., and STØLEVIK,R.(1970): Acta Chem.Scand. 24, 3463.

[45] ALMENNINGEN,A., HARTMANN,A.O., and SEIP,H.M.(1968): Acta Chem.Scand. 22, 1013.

[46] ALMENNINGEN,A., HEDBERG,K., and SEIP,R.(1963): Acta Chem. Scand. 17, 2264. [15]

[47] ALMENNINGEN,A., JACOBSEN,G.G., and SEIP,H.M.(1969a): Acta Chem.Scand. 23, 685.

[48] ALMENNINGEN,A., JACOBSEN,G.G., and SEIP,H.M.(1969b): Acta Chem.Scand. 23, 1495.

[49] ALMENNINGEN,A., KOLSAKER,P., SEIP,H.M., and WILLADSEN,T. (1969): Acta Chem.Scand. 23, 3398.

[50] ALMENNINGEN,A., SEIP,H.M., and SEIP,R.(1970): Acta Chem.Scand. 24, 1697.

[51] ALMENNINGEN,A., SEIP,H.M., and WILLADSEN,T.(1969): Acta Chem. Scand. 23, 2748.

[52] ANANTARAMA SARMA, Y., SUNDARAM,S., and CLEVELAND,F.F.(1964): J.Mol.Spectry. 13, 67. [16]

[53] ANANTARAMA SARMA, Y. and TIRUVENGANNA RAO, P. (1969): Spectroscopia Molecular 18, 26.

[54] ANDERSEN,B. and ANDERSEN,P.(1966a): Acta Chem.Scand. 20, 2728. [16.1]

[55] ANDERSEN,B. and ANDERSEN,P.(1966b): Trans.Am.Cryst.Assoc. 2, 193. [16.2]

[56] ANDERSEN,B. and FERNHOLT,L.(1970): Acta Chem.Scand. 24, 445.

[57] ANDERSEN,B. and MARSTRANDER,A.(1967): Acta Chem.Scand. 21, 1676. [16.3]

[58] ANDERSEN,B., SEIP,H.M., STRAND,T.G., and STØLEVIK,R.(1969): Acta Chem.Scand. 23, 3224.

[59] ANDERSEN,B., STØLEVIK,R., BRUNVOLL,J., CYVIN,S.J., and HAGEN,G (1967): Acta Chem.Scand. 21, 1759. [16.4]

[60] ANDERSEN,P.(1965a): Acta Chem.Scand. 19, 622. [17]

[61] ANDERSEN,P.(1965b): Acta Chem.Scand. 19, 629. [18]

[62] ARNESEN,S.P. and SEIP,H.M.(1966): Acta Chem.Scand. 20, 2711 [18.1]

[63] ATEN,C.F., HEDBERG,L., and HEDBERG,K.(1968): J.Am.Chem.Soc. 90. 2463.

[64] AVASTHI,M.N. and MEHTA,M.L.(1970): Z.Naturforschg. 25a, 566.

[65] BABU J., K. (1967): Opt.Spektroskopiya 22, 340.

[66] BADGLEY,G.R. and LIVINGSTON,R.L.(1954): J.Am.Chem.Soc. 76, 261. [19]

[67] BAKKEN,J.(1958): Acta Chem.Scand. 12, 594. [20]

[68] BARAN,E.J.(1970): Z.Naturforschg. 25a, 1292.

[69] BARAN,E.J., AYMONINO,P.J., and MÜLLER,A.(1970): Anales Asoc. Quim. Argentina 58, 71.

[70] BARTELL,L.S.(1955): J.Chem.Phys. 23, 1219. [21]

[71] BARTELL,L.S.(1959): J.Am.Chem.Soc. 81, 3497. [22]

[72] BARTELL,L.S.(1960a): J.Chem.Phys. 32, 827. [23]

[73] BARTELL,L.S.(1960b): J.Chem.Phys. 32, 832. [24]

[74] BARTELL,L.S.(1960c): Tetrahedron Letters No. 6, 13.

[75] BARTELL,L.S.(1961a): Iowa State J. Science 36, 137.

[76] BARTELL,L.S.(1961b): J.Am.Chem.Soc. 83, 3567.

[77] BARTELL,L.S.(1962): J.Chem.Phys. 36, 3495.

[78] BARTELL,L.S.(1963): J.Chem.Phys. 38, 1827. [25]

[79] BARTELL,L.S.(1965): J.Chem.Phys. 42, 1681. [26]

[80] BARTELL,L.S.(1970): Inorg.Chem. 9, 1594.

[81] BARTELL,L.S. and BONHAM,R.A.(1959): J.Chem.Phys. 31, 400.[27]

[82] BARTELL,L.S. and BONHAM,R.A.(1960): J.Chem.Phys. 32, 824.[28]

[83] BARTELL,L.S. and BROCKWAY,L.O.(1953): Nature 171, 978. [29]

[84] BARTELL,L.S. and BROCKWAY,L.O.(1955): J.Chem.Phys. 23, 1860. [30]

[85] BARTELL,L.S. and BROCKWAY,L.O.(1960): J.Chem.Phys. 32, 512. [31]

[86] BARTELL,L.S., BROCKWAY,L.O., and SCHWENDEMAN,R.H.(1955): J.Chem.Phys. 23, 1854. [32]

[87] BARTELL,L.S. and CARROLL,B.L.(1965a): J.Chem.Phys. 42, 1135. [33]

[88] BARTELL,L.S. and CARROLL,B.L.(1965b): J.Chem.Phys. 42, 3076. [34]

[89] BARTELL,L.S., CLIPPARD,F.B.JR., and BOATES,T.L.(1970): Inorg. Chem. 9, 2436.

[90] BARTELL,L.S. and GAVIN,R.M.JR.(1968): J.Chem.Phys. 48, 2466.

[91] BARTELL,L.S. and GUILLORY,J.P.(1965): J.Chem.Phys. 43, 647. [35]

[92] BARTELL,L.S., GUILLORY,J.P., and PARKS,A.T.(1965): J.Phys. Chem. 69, 3043. [36]

[93] BARTELL,L.S. and HANSEN,K.W.(1965): Inorg.Chem. 4, 1777.

[94] BARTELL,L.S. and HIGGINBOTHAM,H.K.(1965a): Inorg.Chem. 4,1346.

[95] BARTELL,L.S. and HIGGINBOTHAM,H.K.(1965b): J.Chem.Phys. 42, 851. [37]

[96] BARTELL,L.S. and HIRST,R.C.(1959): J.Chem.Phys. 31, 449.[38]

[97] BARTELL,L.S. and KOHL,D.A.(1963): J.Chem.Phys. 39, 3097.[39]

[98] BARTELL,L.S., KOHL,D.A., CARROLL,B.L., and GAVIN,R.M.JR. (1965): J.Chem.Phys. 42, 3079. [40]

[99] BARTELL,L.S. and KUCHITSU,K.(1962a): J.Chem.Phys. 37, 691. [41]

[100] BARTELL,L.S. and KUCHITSU,K.(1962b): J.Phys.Soc.Japan 17, Suppl.B-II, 20. [42]

[101] BARTELL,L.S., KUCHITSU,K., and DE NEUI, R.J. (1960): J.Chem. Phys. 33, 1254. [43]

[102] BARTELL,L.S., KUCHITSU,K., and DE NEUI, R.J. (1961): J.Chem. Phys. 35, 1211. [44]

[103] BARTELL,L.S. and ROSKOS,R.R.(1966); J.Chem.Phys. 44, 457.

[104] BARTELL,L.S., ROTH,E.A., HOLLOWELL,C.D., KUCHITSU,K., and YOUNG,J.E.JR.(1965): J.Chem.Phys. 42, 2683. [45]

[105] BARZDAIN,P.P. and ALEKSEEV,N.V.(1968): Zh.Strukt.Khim. 9,520.

[106] BASTIANSEN,O.(1962): 11. Pohjoismainen Kemistikokous - 11. Nordiska Kemistmötet, Åbo, 65. [46]

[107] BASTIANSEN,O.(1968): Structural Chemistry and Molecular Biology (Ed. A.Rich and N.Davidson), p. 640, Freeman, San Francisco 1968.

[108] BASTIANSEN,O. and BEAGLEY,B.(1964): Acta Chem.Scand. 18, 2077. [47]

[109] BASTIANSEN,O. and CYVIN,S.J.(1957a): Acta Chem.Scand. 11, 1789. [48]

[110] BASTIANSEN,O. and CYVIN,S.J.(1957b): Nature 180, 980. [49]

[111] BASTIANSEN,O. and DE MEIJERE,A. (1966a): Acta Chem.Scand. 20, 516. [50]

[112] BASTIANSEN,O. and DE MEIJERE,A. (1966b): Angew.Chem. 78, 142 (internat. Edit. 5, 124). [51]

[113] BASTIANSEN,O. and DERISSEN,J.L.(1966a): Acta Chem.Scand. 20, 1089. [52]

[114] BASTIANSEN,O. and DERISSEN,J.L.(1966b): Acta Chem.Scand. 20, 1319. [52.1]

[115] BASTIANSEN,O., FRITSCH,F.N., and HEDBERG,K.(1964): Acta

Cryst. 17, 538. [53]
[116] BASTIANSEN,O., HEDBERG,L., and HEDBERG,K.(1957): J.Chem. Phys. 27, 1311; erratum: ibid. 28, 512 (1958). [54]
[117] BASTIANSEN,O. and SKANCKE,A.(1967): Acta Chem.Scand. 21, 587. [54.1]
[118] BASTIANSEN,O. and SKANCKE,P.N.(1960): Advances Chem. Phys. 3, 323. [55]
[119] BASTIANSEN,O. and TRÆTTEBERG,M.(1960): Acta Cryst. 13, 1108. [56]
[120] BASTIANSEN,O. and TRÆTTEBERG,M.(1962): Tetrahedron 17, 147. [57]
[121] BEAGLEY,B.(1965): Trans. Faraday Soc. 61, 1821. [58]
[122] BEAGLEY,B., BROWN,D.P., and MONAGHAN,J.J.(1969): J.Mol. Structure 4, 233.
[123] BEAGLEY,B., CLARK,A.H., and CRUICKSHANK,D.W.J.(1966): Chem. Commun., 458. [58.1]
[124] BEAGLEY,B., CLARK,A.H., and HEWITT,T.G.(1968): J.Chem.Soc. A, 658.
[125] BEAGLEY,B. and CONRAD,A.R.(1970): Trans. Faraday Soc. 66,2740.
[126] BEAGLEY,B., CRUICKSHANK,D.W.J., HEWITT,T.G., and HAALAND,A. (1967): Trans. Faraday Soc. 63, 836. [58.2]
[127] BEAGLEY,B., CRUICKSHANK,D.W.J., HEWITT,T.G., and JOST,K.H. (1969): Trans. Faraday Soc. 65, 1219.
[128] BEAGLEY,B., CRUICKSHANK,D.W.J., PINDER,P.M., ROBIETTE,A.G., and SHELDRICK,G.M.(1969): Acta Cryst. B25, 737.
[129] BEAGLEY,B., ECKERSLEY,G.H., BROWN,D.P., and TOMLINSON,D. (1969): Trans. Faraday Soc. 65, 2300.
[130] BEAGLEY,B. and HEWITT,T.G.(1968): Trans. Faraday Soc. 64,2561.
[131] BEAGLEY,B. and MONAGHAN,J.J.(1970): Trans. Faraday Soc. 66, 2745.
[132] BEAGLEY,B., ROBIETTE,A.G., and SHELDRICK,G.M.(1968a): J.Chem. Soc. A,3002.
[133] BEAGLEY,B., ROBIETTE,A.G., and SHELDRICK,G.M.(1968b): J.Chem. Soc. A, 3006.
[134] BECKA,L.N. and CRUICKSHANK,D.W.J.(1963): Proc.Roy.Soc. (London) A273, 455. [59]
[135] BOHN,R.K. and BAUER,S.H.(1967a): Inorg.Chem. 6, 304.
[136] BOHN,R.K. and BAUER,S.H.(1967b): Inorg.Chem. 6, 309.

[137] BOHN,R.K. and HAALAND,A.(1966): J.Organometal.Chem. 5, 470. [59.1]

[138] BOHN,R.K., KATADA,K., MARTINEZ,J.V., and BAUER,S.H.(1963): Noble-Gas Compounds, p. 238, University of Chicago Press, Chicago 1963.

[139] BOHN,R.K. and TAI,Y.-H.(1970): J.Am.Chem.Soc. 92, 6447.

[140] BOND,A.C. and BROCKWAY,L.O.(1954): J.Am.Chem.Soc. 76, 3312. [60]

[141] BONHAM,R.A. and BARTELL,L.S.(1959a): J.Am.Chem.Soc. 81, 3491. [61]

[142] BONHAM,R.A. and BARTELL,L.S.(1959b): J.Chem.Phys. 31, 702. [62]

[143] BONHAM,R.A., BARTELL,L.S., and KOHL,D.A.(1959): J.Am.Chem.Soc. 81, 4765. [63]

[144] BONHAM,R.A. and MOMANY,F.A.(1961): J.Am.Chem.Soc. 83, 4475. [64]

[145] BONHAM,R.A. and MOMANY,F.A.(1963): J.Phys.Chem. 67, 2474.[65]

[146] BONHAM,R.A. and PEACHER,J.L.(1963): J.Chem.Phys. 38, 2319. [66]

[147] BONHAM,R.A. and UKAJI,T.(1962): J.Chem.Phys. 36, 72. [67]

[148] BRANDT,J.L. and LIVINGSTON,R.L.(1954): J.Am.Chem.Soc. 76, 2096. [68]

[149] BRANDT,J.L. and LIVINGSTON,R.L.(1956): J.Am.Chem.Soc. 78, 3573. [69]

[150] BREARLEY,N. and SUNDARAM,S.(1962): Z.phys.Chem. Neue Folge 32, 219. [70]

[151] BREED,H., BASTIANSEN,O., and ALMENNINGEN,A.(1960): Acta Cryst. 13, 1108. [71]

[152] BREGMAN,J. and BAUER,S.H.(1955): J.Am.Chem.Soc. 77, 1955.[72]

[153] BROOKS,W.V.F., CYVIN,B.N., CYVIN,S.J., KVANDE,P.C., and MEISINGSETH,E.(1963): Acta Chem.Scand. 17, 345. [73]

[154] BROOKS,W.V.F. and CYVIN,S.J.(1962a): Acta Chem.Scand. 16, 820. [74]

[155] BROOKS,W.V.F. and CYVIN,S.J.(1962b): Spectrochim.Acta 18, 397. [75]

[156] BROOKS,W.V.F., CYVIN,S.J., and KVANDE,P.C.(1965): J.Phys. Chem. 69, 1489. [76]

[157] BROUN,T.T. and LIVINGSTON,R.L.(1952): J.Am.Chem.Soc. 74,

6084. [77]

[158] BRUNVOLL,J.(1965): J.Mol.Spectry. 15, 386. [78]

[159] BRUNVOLL,J.(1967a): Acta Chem.Scand. 21, 473. [79]

[160] BRUNVOLL,J.(1967b): Acta Chem.Scand. 21, 820. [80]

[161] BRUNVOLL,J.(1967c): Acta Chem.Scand. 21, 1390. [81]

[162] BRUNVOLL,J.(1968): Chem.Phys.Letters 2, 116.

[163] BRUNVOLL,J. and CYVIN,S.J.(1963a): Acta Chem.Scand. 17, 1405. [82]

[164] BRUNVOLL,J. and CYVIN,S.J.(1963b): Acta Chem.Scand. 17, 1412. [83]

[165] BRUNVOLL,J. and CYVIN,S.J.(1968): Acta Chem.Scand. 22, 2709.

[166] BRUNVOLL,J., CYVIN,S.J., ELVEBREDD,I., and HAGEN,G.(1968): Chem.Phys.Letters 1, 566.

[167] BRUNVOLL,J., CYVIN,S.J., and SCHÄFER,L.(1971a): J.Organometal.Chem. 27, 69.

[168] BRUNVOLL,J., CYVIN,S.J., and SCHÄFER,L.(1971b): J.Organometal.Chem. 27, 107.

[169] BYE,B.H. and CYVIN,S.J.(1963): Acta Chem.Scand. 17, 1804.[84]

[170] CARROLL,B.L. and BARTELL,L.S.(1968): Inorg.Chem. 7, 219.

[171] CHANG,C.H., PORTER,R.F., and BAUER,S.H.(1970a): J.Am.Chem.Soc. 92, 5313.

[172] CHANG,C.H., PORTER,R.F., and BAUER,S.H.(1970b): J.Phys.Chem. 74, 1363.

[173] CHANG,C.H., PORTER,R.F., and BAUER,S.H.(1971): J.Mol.Structure 7, 89.

[174] CHANTRY,G.W. and EWING,V.C.(1962): Mol.Phys. 5, 209. [85]

[175] CHANTRY,G.W. and WOODWARD,L.A.(1960): Trans. Faraday Soc. 56, 1110. [86]

[176] CHIANG,J.F. and BAUER,S.H.(1966): J.Am.Chem.Soc. 88, 420.

[177] CHIANG,J.F. and BAUER,S.H.(1969): J.Am.Chem.Soc. 91, 1898.

[178] CHIANG,J.F., WILCOX,C.F.JR., and BAUER,S.H.(1968): J.Am.Chem.Soc. 90, 3149.

[179] CHIANG,J.F., WILCOX,C.F.JR., and BAUER,S.H.(1969): Tetrahedron 25, 369.

[180] CHRISTIE,K.O. and SAWODNY,W.(1970): J.Chem.Phys. 52, 6320.

[181] CLARK,A.H., BEAGLEY,B., CRUICKSHANK,D.W.J., and HEWITT,T.G. (1970a): J.Chem.Soc. A, 872.

[182] CLARK,A.H., BEAGLEY,B., CRUICKSHANK,D.W.J., and HEWITT,T.G. (1970b): J.Chem.Soc. A, 1613.

[183] CLARK,A.H. and HAALAND,A. (1970): Acta Chem.Scand. 24, 3024.

[184] CLARK,A. and SEIP,H.M.(1970): Chem.Phys.Letters 6, 452.

[185] CLIPPARD,F.B.JR. and BARTELL,L.S.(1970a): Inorg.Chem. 9, 805.

[186] CLIPPARD,F.B.JR. and BARTELL,L.S.(1970b): Inorg.Chem. 9, 2439

[187] CORBET,H.C., DALLINGA,G., OLTMANS,F., and TONEMAN,L.H. (1964): Rec.trav.chim. 83, 789. [87]

[188] COUTTS,J.W. and LIVINGSTON,R.L.(1953): J.Am.Chem.Soc. 75, 1542. [88]

[189] CURTIS,E.C.(1964): J.Mol.Spectry. 14, 279. [89]

[190] CYVIN,B.N. and CYVIN,S.J.(1964): Acta Chem.Scand. 18, 1690. [90]

[191] CYVIN,B.N. and CYVIN,S.J.(1965): Kgl. Norske Videnskab. Selskabs Skrifter, No.5. [91]

[192] CYVIN,B.N. and CYVIN,S.J.(1969): J.Phys.Chem. 73, 1430.

[193] CYVIN,B.N., CYVIN,S.J., and HAGEN,G.(1967): Chem.Phys.Letters 1, 211. [91.1]

[194] CYVIN,S.J.(1957): Acta Chem.Scand. 11, 1499. [92]

[195] CYVIN,S.J.(1958a): Acta Chem.Scand. 12, 233. [93]

[196] CYVIN,S.J.(1958b): Acta Chem.Scand. 12, 1697. [94]

[197] CYVIN,S.J.(1958c): J.Chem.Phys. 29, 583. [95]

[198] CYVIN,S.J.(1959a): Acta Chem.Scand. 13, 334. [96]

[199] CYVIN,S.J.(1959b): Acta Chem.Scand. 13, 1397. [97]

[200] CYVIN,S.J.(1959c): Acta Chem.Scand. 13, 1400. [98]

[201] CYVIN,S.J.(1959d): Acta Chem.Scand. 13, 1809. [99]

[202] CYVIN,S.J.(1959e): Acta Chem.Scand. 13, 2135. [100]

[203] CYVIN,S.J.(1959f): J.Chem.Phys. 30, 337. [101]

[204] CYVIN,S.J.(1959g): J.Mol.Spectry. 3, 467. [102]

[205] CYVIN,S.J.(1959h): Kgl. Norske Videnskab. Selskabs Skrifter, No. 2 [103]

[206] CYVIN,S.J.(1959i): Spectrochim.Acta 15, 56. [104]

[207] CYVIN,S.J.(1959j): Spectrochim.Acta 15, 341. [105]

[208] CYVIN,S.J.(1959k): Spectrochim.Acta 15, 828. [106]

[209] CYVIN,S.J.(1959l): Spectrochim.Acta 15, 835. [107]

[210] CYVIN,S.J.(1959m): Spectrochim.Acta 15, 958. [108]

[211] CYVIN,S.J.(1959n): Tidsskr. Kjemi, Bergvesen, Metallurgi 19, 83. [109]

[212] CYVIN,S.J.(1960a): Acta Polytechn.Scand., Ph. 6. [110]
[213] CYVIN,S.J.(1960b): J.Mol.Spectry. 5, 38. [111]
[214] CYVIN,S.J.(1960c): Spectrochim.Acta 16, 1421. [112]
[215] CYVIN,S.J.(1960d): Spectrochim.Acta 16, 1432. [113]
[216] CYVIN,S.J.(1960e): Spectroscopia Molecular 9, 56. [114]
[217] CYVIN,S.J.(1960f): Tidsskr. Kjemi, Bergvesen, Metallurgi 20, 246. [115]
[218] CYVIN,S.J.(1960g): Z.phys.Chem. Neue Folge 23, 402. [116]
[219] CYVIN,S.J.(1961a): J.Mol.Spectry. 6, 333. [117]
[220] CYVIN,S.J.(1961b): J.Mol.Spectry. 6, 338. [118]
[221] CYVIN,S.J.(1961c): Spectrochim.Acta 17, 1219. [119]
[222] CYVIN,S.J.(1961d): Tidsskr. Kjemi, Bergvesen, Metallurgi 21, 236. [120]
[223] CYVIN,S.J.(1962a): Acta Chem.Scand. 16, 1528. [121]
[224] CYVIN,S.J.(1962b): Tidsskr. Kjemi, Bergvesen, Metallurgi 22, 44. [122]
[225] CYVIN,S.J.(1962c): Tidsskr. Kjemi, Bergvesen, Metallurgi 22, 73. [123]
[226] CYVIN,S.J.(1963a): Acta Chem.Scand. 17, 296. [124]
[227] CYVIN,S.J.(1963b): J.Mol.Spectry. 11, 195. [125]
[228] CYVIN,S.J.(1963c): Z.phys.Chem. Neue Folge 38, 405. [126]
[229] CYVIN,S.J.(1968): Molecular Vibrations and Mean Square Amplitudes, Universitetsforlaget, Oslo, and Elsevier, Amsterdam, 1968.
[230] CYVIN,S.J.(1969a): Acta Chem.Scand. 23, 576.
[231] CYVIN,S.J.(1969b): J.Mol.Structure 3, 520.
[232] CYVIN,S.J.(1969c): Kgl. Norske Videnskab. Selskabs Skrifter, No. 1.
[233] CYVIN,S.J.(1970a): Acta Chem.Scand. 24, 3259.
[234] CYVIN,S.J.(1970b): Czechoslov.J.Phys. B20, 464.
[235] CYVIN,S.J.(1970c): Z.anorg.allgem.Chem. 378, 117.
[236] CYVIN,S.J.(1971a): J.Mol.Structure 8, 43.
[237] CYVIN,S.J.(1971b): Kgl. Norske Videnskab. Selskabs Skrifter, No. 7.
[238] CYVIN,S.J. and ALFHEIM,I.(1970a): Acta Chem.Scand. 24, 2648.
[239] CYVIN,S.J. and ALFHEIM,I.(1970b): Indian J. Pure Appl. Phys. 8, 629.
[240] CYVIN,S.J., ALFHEIM,I., and HAGEN,G.(1970): Acta Chem.Scand.

24, 3038.

[241] CYVIN,S.J. and BAKKEN,J.(1958): Acta Chem.Scand. 12, 1759. [127]

[242] CYVIN,S.J. and BRUNVOLL,J.(1964): Acta Chem.Scand. 18, 1023. [128]

[243] CYVIN,S.J. and BRUNVOLL,J.(1968): Acta Chem.Scand. 22, 2718.

[244] CYVIN,S.J. and BRUNVOLL,J.(1969a): J.Mol.Structure 3, 151.

[245] CYVIN,S.J. and BRUNVOLL,J.(1969b): J.Mol.Structure 3, 453.

[246] CYVIN,S.J., BRUNVOLL,J., CYVIN,B.N., and MEISINGSETH,E. (1964): Bull.Soc.Chim.Belg. 73, 5. [129]

[247] CYVIN,S.J., BRUNVOLL,J., and MÜLLER,A.(1968): Acta Chem.Scand. 22, 2739.

[248] CYVIN,S.J., BRUNVOLL,J., and RAJALAKSHMI,K.V.(1966): Acta Chem.Scand. 20, 1991. [130]

[249] CYVIN,S.J., BRUNVOLL,J., and ROBIETTE,A.G.(1969): J.Mol. Structure 3, 259.

[250] CYVIN,S.J., BRUNVOLL,J., and STØLEVIK,R.(1969): Acta Chem. Scand. 23, 333.

[251] CYVIN,S.J., CHRISTENSEN,D.H., and NICOLAISEN,F.(1970): Chem. Phys.Letters 5, 597; erratum: ibid. 6, 552.

[252] CYVIN,S.J. and CYVIN,B.N.(1971a): J.Mol.Structure 8, 167.

[253] CYVIN,S.J. and CYVIN,B.N.(1971b): Z.Naturforschg. 26a, 901.

[254] CYVIN,S.J., CYVIN,B.N., BRUNVOLL,J., ANDERSEN,B., and STØLEVIK,R.(1967): Selected Topics in Structure Chemistry (Ed. P.Andersen, O.Bastiansen, and S.Furberg), p. 69, Universitetsforlaget, Oslo 1967. [130.1]

[255] CYVIN,S.J., CYVIN,B.N., and HAGEN,G.(1968): Chem.Phys.Letters 2, 341.

[256] CYVIN,S.J., CYVIN,B.N., HAGEN,G., and MARKOV,P.(1969): Acta Chem.Scand. 23, 3407.

[257] CYVIN,S.J., CYVIN,B.N., and MÜLLER,A.(1969): J.Mol.Structure 4, 341.

[258] CYVIN,S.J., CYVIN,B.N., MÜLLER,A., and KREBS,B.(1968): Z.Naturforschg. 23a, 479.

[259] CYVIN,S.J. and ELVEBREDD,I.(1969): Z.anorg.allgem.Chem. 371, 220.

[260] CYVIN,S.J., ELVEBREDD,I., CYVIN,B.N., BRUNVOLL,J., and HAGEN,G.(1967): Acta Chem.Scand. 21, 2405. [130.2]

[261] CYVIN,S.J., ELVEBREDD,I., HAGEN,G., and ANDERSEN,B.(1968): Chem.Phys.Letters 2, 556.

[262] CYVIN,S.J., ELVEBREDD,I., HAGEN,G., and BRUNVOLL,J.(1968): J.Chem.Phys. 49, 3561.

[263] CYVIN,S.J. and HAGEN,G.(1969): Acta Chem.Scand. 23, 2037.

[264] CYVIN,S.J. and HAGEN,G.(1970): Z.Naturforschg. 25b, 350.

[265] CYVIN,S.J., HAGEN,G., and TRÆTTEBERG,M.(1969): Acta Chem. Scand. 23, 3285.

[266] CYVIN,S.J. and HARGITTAI,I.(1969): Acta Chim.Hung. 61, 159.

[267] CYVIN,S.J. and KLÆBOE,P.(1965): Acta Chem.Scand. 19, 697. [131]

[268] CYVIN,S.J. and KRISTIANSEN,L.(1962): Acta Chem.Scand. 16, 2453. [132]

[269] CYVIN,S.J. and KRISTIANSEN,L.A.(1963): Spectroscopia Molecular 12, 15. [133]

[270] CYVIN,S.J. and MEISINGSETH,E.(1961): Acta Chem.Scand. 15, 1289. [134]

[271] CYVIN,S.J. and MEISINGSETH,E.(1962a): J.Phys.Soc.Japan 17, Suppl. B-II, 41. [135]

[272] CYVIN,S.J. and MEISINGSETH,E.(1962b): Kgl. Norske Videnskab. Selskabs Skrifter, No. 2. [136]

[273] CYVIN,S.J. and MEISINGSETH,E.(1963): Acta Chem.Scand. 17, 1805. [137]

[274] CYVIN,S.J., TRÆTTEBERG,M., and HAGEN,G.(1969): Acta Chem. Scand. 23, 1456.

[275] CYVIN,S.J. and VIZI,B.(1968): Veszprémi Vegyipari Egyetem Közleményei 11, 83.

[276] CYVIN,S.J. and VIZI,B.(1969): Acta Chim.Hung. 59, 85.

[277] CYVIN,S.J. and VIZI,B.(1970): Acta Chim.Hung. 64, 357.

[278] CYVIN,S.J. and VIZI,B.(1971): Acta Chim.Hung. 70, 55.

[279] CYVIN,S.J., VIZI,B., and HAGEN,G.(1968): Veszprémi Vegyipari Egyetem Közleményei 11, 91.

[280] CYVIN,S.J., VIZI,B., MÜLLER,A., and KREBS,B.(1969): J.Mol. Structure 3, 173.

[281] DALLINGA,G. and TONEMAN,L.H.(1967): Rec.trav.chim. 86, 171. [137.1]

[282] DALLINGA,G. and TONEMAN,L.H.(1967-68a): J.Mol.Structure 1,11.

[283] DALLINGA,G. and TONEMAN,L.H.(1967-68b): J.Mol.Structure 1, 117.

[284] DALLINGA,G. and TONEMAN,L.H.(1968a): Rec.trav.chim. 87, 795.

[285] DALLINGA,G. and TONEMAN,L.H.(1968b): Rec.trav.chim. 87, 805.

[286] DALLINGA,G. and TONEMAN,L.H.(1969): Rec.trav.chim. 88, 185.

[287] DALLINGA,G., VAN DER DRAAI, R.K., and TONEMAN,L.H.(1968): Rec.trav.chim. 87, 897.

[288] DANFORD,M.D. and LIVINGSTON,R.L.(1955): J.Am.Chem.Soc. 77, 2944. [138]

[289] DANFORD,M.D. and LIVINGSTON,R.L.(1959): J.Am.Chem.Soc. 81, 4157. [139]

[290] DAVIS,M.I., BOGGS,J.E., COEFFEY,D.JR., and HANSON,H.P.(1965) J.Phys.Chem. 69, 3727. [140]

[291] DAVIS,M.I. and HANSON,H.P.(1965a): J.Phys.Chem. 69, 3405. [141]

[292] DAVIS,M.I. and HANSON,H.P.(1965b): J.Phys.Chem. 69, 4091. [141.1]

[293] DAVIS,M.I., KAPPLER,H.A., and COWAN,D.J.(1964): J.Phys.Chem. 68, 2005. [142]

[294] DAVIS,M.I. and MUECKE,T.W.(1970): J.Phys.Chem. 74, 1104.

[295] DAYKIN,P.N. and SUNDARAM,S.(1962): Z.phys.Chem. Neue Folge 32, 222. [143]

[296] DE ALTI, G., GALASSO,V., and COSTA,G.(1965): Spectrochim.Act 21, 649. [144]

[297] DEBYE,P.(1941): J.Chem.Phys. 9, 55. [145]

[298] DECIUS,J.C.(1953): J.Chem.Phys. 21, 1121. [146]

[299] DECIUS,J.C.(1963): J.Chem.Phys. 38, 241. [147]

[300] DELLEPIANE,G. and ZERBI,G.(1968): Spectrochim.Acta 24A, 2151

[301] DE MEIJERE, A. (1966): Acta Chem.Scand. 20, 1093. [148]

[302] DERISSEN,J.L.(1971): J.Mol.Structure 7, 67.

[303] DIALLO,A.O.(1968): Can.J.Chem. 46, 2641.

[304] DITZEL,E.F., MEISTER,A.G., PIOTROWSKI,E.A., CLEVELAND,F.F., ANANTARAMA SARMA,Y., and SUNDARAM,S.(1964): Can.J.Chem. 42, 2841. [149]

[305] DONOHUE,J. and CARON,A.(1966): J.Phys.Chem. 70, 603. [150]

[306] DORKO,E.A., HENCHER,J.L., and BAUER,S.H.(1968): Tetrahedron 24, 2425.

[307] EL-SABBAN,M.Z., DANTI,A., and ZWOLINSKI,B.J.(1966): J.Chem. Phys. 44, 1770. [151]

[308] EL-SABBAN,M.Z. and ZWOLINSKI,B.J.(1966): J.Mol.Spectry. 19, 231. [152]

[309] EL-SABBAN,M.Z. and ZWOLINSKI,B.J.(1967): J.Mol.Spectry. 22, 23. [152.1]

[310] ELVEBREDD,I.(1968): Acta Chem.Scand. 22, 1606.

[311] ELVEBREDD,I. and CYVIN,S.J.(1969): Z.anorg.allgem.Chem. 370, 310.

[312] ELVEBREDD,I., VIZI,B., CYVIN,S.J., MÜLLER,A., and KREBS,B. (1968): J.Mol.Structure 2, 158.

[313] EWING,V.C. and SUTTON,L.E.(1963): Trans. Faraday Soc. 59, 1241. [153]

[314] EZHOV,YU.S., AKISHIN,P.A., and RAMBIDI,N.G.(1969a): Zh.Strukt. Khim. 10, 571.

[315] EZHOV,YU.S., AKISHIN,P.A., and RAMBIDI,N.G.(1969b): Zh.Strukt. Khim. 10, 763.

[316] EZHOV,YU.S. and RAMBIDI,N.G.(1967): Zh.Strukt.Khim. 8, 342.

[317] FOGARASI,G.(1970): Acta Chim.Hung. 66, 87.

[318] FOGARASI,G. and MEZEY,P.(1970): Acta Chim.Hung. 63, 167.

[319] FREMSTAD,D., BRUNVOLL,J., and CYVIN,S.J.(1964): Acta Chem. Scand. 18, 2184. [154]

[320] FUJII,H. and KIMURA,M.(1970): Bull.Chem.Soc.Japan 43, 1933.

[321] FUKUYAMA,T. and KUCHITSU,K.(1970): J.Mol.Structure 5, 131.

[322] FUKUYAMA,T., KUCHITSU,K., and MORINO,Y.(1968a): Bull.Chem.Soc. Japan 41, 3019.

[323] FUKUYAMA,T., KUCHITSU,K., and MORINO,Y.(1968b): Bull.Chem.Soc. Japan 41, 3021.

[324] GALASSO,V. and BIGOTTO,A.(1965): Spectrochim.Acta 21, 2085. [155]

[325] GALASSO,V., DE ALTI, G., and COSTA,G.(1965): Spectrochim.Acta 21, 669. [156]

[326] GAPOTCHENKO,N.I., ALEKSEEV,N.V., ANISIMOV,K.N., KOLOBOVA,N.E., and RONOVA,I.A.(1968): Zh.Strukt.Khim. 9, 892.

[327] GAVIN,R.M.JR. and BARTELL,L.S.(1968): J.Chem.Phys. 48, 2460.

[328] GLIDEWELL,C., RANKIN,D.W.H., and ROBIETTE,A.G.(1970): J.Chem.

Soc. A, 2935.

[329] GLIDEWELL,C., RANKIN,D.W.H., ROBIETTE,A.G., and SHELDRICK,G.M (1969): J.Mol.Structure 4, 215.

[330] GLIDEWELL,C., RANKIN,D.W.H., ROBIETTE,A.G., and SHELDRICK,G.M (1970): J.Chem.Soc. A, 318.

[331] GLIDEWELL,C., RANKIN,D.W.H., ROBIETTE,A.G., SHELDRICK,G.M., BEAGLEY,B., and FREEMAN,J.M.(1970): J.Mol.Structure 5, 417.

[332] GOLDISH,E., HEDBERG,K., MARSH,R.E., and SCHOMAKER,V.(1955): J.Am.Chem.Soc. 77, 2948. [157]

[333] GOLDISH,E., HEDBERG,K., and SCHOMAKER,V.(1956): J.Am.Chem.Soc 78, 2714; erratum: ibid. 79, 6577 (1957). [158]

[334] GUILLORY,J.P. and BARTELL,L.S.(1965): J.Chem.Phys. 43, 654. [159]

[335] GUNDERSEN,G. and HEDBERG,K.(1969): J.Chem.Phys. 51, 2500.

[336] GUNDERSEN,G., HEDBERG,K., and HUSTON,J.L.(1970): J.Chem.Phys. 52, 812.

[337] HAALAND,A.(1965): Acta Chem.Scand. 19, 41. [160]

[338] HAALAND,A.(1968): Acta Chem.Scand. 22, 3030.

[339] HAALAND,A. and NILSSON,J.E.(1968): Acta Chem.Scand. 22, 2653.

[340] HAALAND,A. and SCHÄFER,L.(1967): Acta Chem.Scand. 21, 2474. [160.1]

[341] HAAS,B., HAASE,J., and ZEIL,W.(1967): Z.Naturforschg. 22a, 1646.

[342] HAASE,J., KAMPHUSMANN,H.D., and ZEIL,W.(1967): Z.phys.Chem. Neue Folge 55, 225.

[343] HAASE,J., STEINGROSS,W., and ZEIL,W.(1967): Z.Naturforschg. 22a, 195. [160.2]

[344] HAASE,J. and WINNEWISSER,M.(1968): Z.Naturforschg. 23a, 61.

[345] HAASE,J. and ZEIL,W.(1965): Z.phys.Chem. Neue Folge 45, 202. [161]

[346] HAGEN,G.(1967): Acta Chem.Scand. 21, 465. [162]

[347] HAGEN,G.(1968): J.Mol.Structure 2, 160.

[348] HAGEN,G. and CYVIN,S.J.(1968): J.Phys.Chem. 72, 1446.

[349] HAMILTON,W.C. and HEDBERG,K.(1952): J.Am.Chem.Soc. 74, 5529. [163]

[350] HANSEN,K.W. and BARTELL,L.S.(1965): Inorg.Chem. 4, 1775.

[351] HARGITTAI,I.(1968a): Acta Chim.Hung. 57, 403.

[352] HARGITTAI,I.(1968b): Magyar Kémiai Folyóirat 74, 362.
[353] HARGITTAI,I.(1968c): Magyar Kémiai Folyóirat 74, 596.
[354] HARGITTAI,I.(1969a): Acta Chim.Hung. 59, 351.
[355] HARGITTAI,I.(1969b): Acta Chim.Hung. 60, 231.
[356] HARGITTAI,I.(1969c): Kémiai Közlemények 32, 253.
[357] HARGITTAI,I. and CYVIN,S.J.(1969): Acta Chim.Hung. 61, 51.
[358] HARGITTAI,I., HARGITTAI,M., and HERNADI,J.(1970): Magyar Kémiai Folyóirat 76, 63.
[359] HARGITTAI,I. and VILKOV,L.V.(1970): Acta Chim.Hung. 63, 143.
[360] HARVEY,R.B. and BAUER,S.H.(1953): J.Am.Chem.Soc. 75, 2840. [164]
[361] HARVEY,R.B. and BAUER,S.H.(1954): J.Am.Chem.Soc. 76, 859.[165]
[362] HASTINGS,J.M. and BAUER,S.H.(1950): J.Chem.Phys. 18, 13. [166]
[363] HAUGEN,W. and TRÆTTEBERG,M.(1966): Acta Chem.Scand. 20, 1726. [166.1]
[364] HAUGEN,W. and TRÆTTEBERG,M.(1967): Selected Topics in Structure Chemistry (Ed. P.Andersen, O.Bastiansen, and S.Furberg), p. 113, Universitetsforlaget, Oslo 1967. [166.2]
[365] HEDBERG,K.(1955): J.Am.Chem.Soc. 77, 6491. [167]
[366] HEDBERG,K.(1966): Trans.Am.Cryst.Assoc. 2, 79. [167.1]
[367] HEDBERG,K. and IWASAKI,M.(1960): Acta Cryst. 13, 1108. [168]
[368] HEDBERG,K. and IWASAKI,M.(1962a): J.Chem.Phys. 36, 589. [169]
[369] HEDBERG,K. and IWASAKI,M.(1962b): J.Phys.Soc.Japan 17, Suppl. B-II, 32. [170]
[370] HEDBERG,K. and IWASAKI,M.(1964): Acta Cryst. 17, 529. [171]
[371] HEDBERG,K. and RYAN,R.(1964): J.Chem.Phys. 41, 2214. [172]
[372] HEDBERG,K. and SCHOMAKER,V.(1951): J.Am.Chem.Soc. 73, 1482; erratum: ibid. 73, 5929 (1951). [173]
[373] HEDBERG,K. and STOSICK,J.(1952): J.Am.Chem.Soc. 74, 954. [174]
[374] HEDBERG,L. and HEDBERG,K.(1970): J.Chem.Phys. 53, 1228.
[375] HENCHER,J.L. and BAUER,S.H.(1967): J.Am.Chem.Soc. 89, 5527.
[376] HENCHER,J.L., CRUICKSHANK,D.W.J., and BAUER,S.H.(1968): J.Chem.Phys. 48, 518.
[377] HIGGINBOTHAM,H.K. and BARTELL,L.S.(1965): J.Chem.Phys. 42, 1131. [175]
[378] HILDERBRANDT,R.L., ANDREASSEN,A.L., and BAUER,S.H.(1970): J.Phys.Chem. 74, 1586.

[379] HILDERBRANDT,R.L. and BAUER,S.H.(1969): J.Mol.Structure 3, 325.

[380] HIROTA,E.(1958): Bull.Chem.Soc.Japan 31, 130. [176]

[381] HJORTAAS,K.E.(1967a): Acta Chem.Scand. 21, 1379.

[382] HJORTAAS,K.E.(1967b): Acta Chem.Scand. 21, 1381.

[383] HJORTAAS,K.E. and STRØMME,K.O.(1968): Acta Chem.Scand. 22, 2965.

[384] HOLMES,R.R. and DEITERS,R.M.(1969): J.Chem.Phys. 51, 4043.

[385] HOPE,H.(1968): Acta Chem.Scand. 22, 1057.

[386] IBERS,J.A.(1959): Acta Cryst. 12, 251. [177]

[387] IBERS,J.A. and STEVENSON,D.P.(1958): J.Chem.Phys. 28, 929. [178]

[388] IIJIMA,T.(1970): Bull.Chem.Soc.Japan 43, 1049.

[389] IIJIMA,T. and KIMURA,M.(1969): Bull.Chem.Soc.Japan 42, 2159.

[390] IWASAKI,M.(1958): Bull.Chem.Soc.Japan 31, 1071. [179]

[391] IWASAKI,M.(1959): Bull.Chem.Soc.Japan 32, 194. [180]

[392] IWASAKI,M., FRITSCH,F.N., and HEDBERG,K.(1964): Acta Cryst. 17, 533. [181]

[393] IWASAKI,M. and HEDBERG,K.(1960): Acta Cryst. 13, 1108. [182]

[394] IWASAKI,M. and HEDBERG,K.(1962a): J.Chem.Phys. 36, 594. [183]

[395] IWASAKI,M. and HEDBERG,K.(1962b): J.Chem.Phys. 36, 2961. [184

[396] IWASAKI,M., NAGASE,S., and KOJIMA,R.(1957): Bull.Chem.Soc. Japan 30, 230. [185]

[397] JACOB,E.J. and BARTELL,L.S.(1970a): J.Chem.Phys. 53, 2231.

[398] JACOB,E.J. and BARTELL,L.S.(1970b): J.Chem.Phys. 53, 2235.

[399] JAMES,R.W.(1932): Phys.Z. 33, 737. [186]

[400] JANZEN,J. and BARTELL,L.S.(1969): J.Chem.Phys. 50, 3611.

[401] JENSEN,H.H., HAGEN,G., and CYVIN,S.J.(1969): J.Mol.Structure 4, 51.

[402] JONES,L.H., MC DOWELL, R.S., and GOLDBLATT,M.(1968): J.Chem. Phys. 48, 2663.

[403] JONES,L.H., MC DOWELL, R.S., and GOLDBLATT,M.(1969): Inorg. Chem. 8, 2349.

[404] JONES,L.H., RYAN,R.R., and ASPREY,L.B.(1968): J.Chem.Phys. 49, 581.

[405] JONES,M.E., HEDBERG,K., and SCHOMAKER,V.(1955): J.Am.Chem.Soc 77, 5278; erratum: ibid. 78, 6421 (1956). [187]

[406] KAMESWARA RAO, P. and BABU RAO, P. (1964): Z.phys.Chem. Neue Folge 42, 166. [188]
[407] KARLE,I.L.(1952): J.Chem.Phys. 20, 65. [189]
[408] KARLE,I.L.(1955): J.Chem.Phys. 23, 1739. [190]
[409] KARLE,I.L. and KARLE,J.(1949): J.Chem.Phys. 17, 1052. [191]
[410] KARLE,I.L. and KARLE,J.(1950): J.Chem.Phys. 18, 963. [192]
[411] KARLE,I.L. and KARLE,J.(1952): J.Chem.Phys. 20, 63. [193]
[412] KARLE,I.L. and KARLE,J.(1954): J.Chem.Phys. 22, 43. [194]
[413] KARLE,I.L. and KARLE,J.(1962): J.Chem.Phys. 36, 1969. [195]
[414] KARLE,J.(1947): J.Chem.Phys. 15, 202. [196]
[415] KARLE,J.(1954): J.Chem.Phys. 22, 1246. [197]
[416] KARLE,J.(1966a): J.Chem.Phys. 45, 4149. [197.1]
[417] KARLE,J.(1966b): Trans.Am.Cryst.Assoc. 2, 117.
[418] KARLE,J. and HAUPTMAN,H.(1950): J.Chem.Phys. 18, 875. [198]
[419] KARLE,J. and KARLE,I.L.(1950): J.Chem.Phys. 18, 957. [199]
[420] KATO,C., KONAKA,S., IIJIMA,T., and KIMURA,M.(1969): Bull.Chem. Soc.Japan 42, 2148.
[421] KEBABÇIOGLU,R., MÜLLER,A., PEACOCK,C.J., and LANGE,L.(1968): Z.Naturforschg. 23a, 703.
[422] KEIDEL,F.A. and BAUER,S.H.(1956): J.Chem.Phys. 25, 1218.[200]
[423] KIMURA,K.(1957): J.Chem.Phys. 27, 1213. [201]
[424] KIMURA,K. and BAUER,S.H.(1963): J.Chem.Phys. 39, 3172. [202]
[425] KIMURA,K., KATADA,K., and BAUER,S.H.(1966): J.Am.Chem.Soc. 88, 416. [203]
[426] KIMURA,K. and KIMURA,M.(1956): J.Chem.Phys. 25, 362. [204]
[427] KIMURA,K. and KIMURA,M.(1960): J.Chem.Phys. 32, 1398. [205]
[428] KIMURA,K. and KUBO,M.(1959): J.Chem.Phys. 30, 151. [206]
[429] KIMURA,K. and KUBO,M.(1960): J.Chem.Phys. 32, 1776. [207]
[430] KIMURA,K., SCHOMAKER,V., SMITH,D.W., and WEINSTOCK,B.(1968): J.Chem.Phys. 48, 4001.
[431] KIMURA,K., SUZUKI,S., KIMURA,M., and KUBO,M.(1957): J.Chem. Phys. 27, 320. [208]
[432] KIMURA,K., SUZUKI,S., KIMURA,M., and KUBO,M.(1958): Bull.Chem. Soc.Japan 31, 1051. [209]
[433] KIMURA,M. and KIMURA,K.(1963): J.Mol.Spectry. 11, 368. [210]
[434] KIMURA,M., KIMURA,K., AOKI,M., and SHIBATA,S.(1956): Bull. Chem.Soc.Japan 29, 95. [211]
[435] KIMURA,M., KIMURA,K., and SHIBATA,S.(1956): J.Chem.Phys. 24,

622. [212]

[436] KISTENMACHER,T.J. and STUCKY,G.D.(1968): Inorg.Chem. 7, 2150.

[437] KLABOE (KLÆBOE), P., KLOSTER-JENSEN,E., and CYVIN,S.J.(1967): Spectrochim.Acta 23A, 2733. [212.1]

[438] KNUDSEN,R.E., GEORGE,C.F., and KARLE,J.(1966): J.Chem.Phys. 44, 2334. [213]

[439] KONAKA,S.(1970): Bull.Chem.Soc.Japan 43, 3107.

[440] KONAKA,S., ITO,T., and MORINO,Y.(1966): Bull.Chem.Soc.Japan 39, 1146. [213.1]

[441] KONAKA,S. and KIMURA,M.(1970): Bull.Chem.Soc.Japan 43, 1693.

[442] KONAKA,S., MURATA,Y., KUCHITSU,K., and MORINO,Y.(1966): Bull. Chem.Soc.Japan 39, 1134. [213.2]

[443] KREBS,B. and MÜLLER,A.(1966): Spectrochim.Acta 22, 1532. [213.3]

[444] KREBS,B., MÜLLER,A., and ROESKY,H.W.(1967): Mol.Phys. 12,469.

[445] KRISHNA PILLAI, M.G. and GNANADESIKAN,S.G.(1965): Indian J. Pure Appl. Phys. 3, 273. [214]

[446] KRISHNA PILLAI, M.G. and PERUMAL,A.(1964): Bull.Soc.Chim.Belg. 73, 641. [215]

[447] KRISHNA PILLAI, M.G., RAMASWAMY,K., and GNANADESIKAN,S.G. (1966): Czechoslov.J.Phys. B16, 150. [216]

[448] KRISHNA PILLAI, M.G., RAMASWAMY,K., and PERUMAL,A.(1965a): Indian J. Pure Appl. Phys. 3, 180. [217]

[449] KRISHNA PILLAI, M.G., RAMASWAMY,K., and PERUMAL,A.(1965b): Indian J. Pure Appl. Phys. 3, 276. [218]

[450] KRISHNA PILLAI, M.G., RAMASWAMY,K., and PICHAI,R.(1965): Austral.J.Chem. 18, 1575. [219]

[451] KRISTIANSEN,L. and CYVIN,S.J.(1963): J.Mol.Spectry. 11, 185. [220]

[452] KUCHITSU,K.(1957): Bull.Chem.Soc.Japan 30, 391. [221]

[453] KUCHITSU,K.(1959): Bull.Chem.Soc.Japan 32, 748. [222]

[454] KUCHITSU,K.(1961): J.Mol.Spectry. 7, 399. [223]

[455] KUCHITSU,K.(1966): J.Chem.Phys. 44, 906. [224]

[456] KUCHITSU,K.(1967a): Bull.Chem.Soc.Japan 40, 498. [224.1]

[457] KUCHITSU,K.(1967b): Bull.Chem.Soc.Japan 40, 505. [224.2]

[458] KUCHITSU,K.(1968): J.Chem.Phys. 49, 4456.

[459] KUCHITSU,K. and BARTELL,L.S.(1961): J.Chem.Phys. 35, 1945. [225]

[460] KUCHITSU,K. and BARTELL,L.S.(1962a): J.Chem.Phys. 36, 2460. [226]

[461] KUCHITSU,K. and BARTELL,L.S.(1962b): J.Chem.Phys. 36, 2470. [227]

[462] KUCHITSU,K. and BARTELL,L.S.(1962c): J.Phys.Soc.Japan 17, Suppl. B-II, 23. [228]

[463] KUCHITSU,K., FUKUYAMA,T., and MORINO,Y.(1967-68): J.Mol. Structure 1, 463.

[464] KUCHITSU,K., GUILLORY,J.P., and BARTELL,L.S.(1968): J.Chem. Phys. 49, 2488.

[465] KUCHITSU,K. and KONAKA,S.(1966): J.Chem.Phys. 45, 4342.[228.1]

[466] KUKINA,V.S.(1969): Opt.Spektroskopiya 26, 111.

[467] KUMAR,S.P., PADMA,V.A., and RAJESWARA RAO, N. (1969): Indian J. Pure Appl. Phys. 7, 373.

[468] LAANE,J., JONES,L.H., RYAN,R.R., and ASPREY,L.B.(1969): J.Mol.Spectry. 30, 489.

[469] LEMAIRE,H.P. and LIVINGSTON,R.L.(1952): J.Am.Chem.Soc. 74, 5732. [229]

[470] LEVIN,I.W.(1967): J.Chem.Phys. 47, 4685.

[471] LEVIN,I.W.(1970): J.Mol.Spectry. 33, 61.

[472] LEVIN,I.W. and ABRAMOWITZ,S.(1969): J.Chem.Phys. 50, 4860.

[473] LEVIN,I.W. and BERNEY,CH.V.(1966): J.Chem.Phys. 44, 2557. [230]

[474] LIE,S.B., KLABOE,P., CHRISTENSEN,D.H., and HAGEN,G.(1970): Spectrochim.Acta 26A, 1861.

[475] LIE,S.B., KLABOE,P., KLOSTER-JENSEN,E., HAGEN,G., and CHRISTENSEN,D.H.(1970): Spectrochim.Acta 26A, 2077.

[476] LIVINGSTON,R.L., LURIE,C., and R. RAO, C.N.(1960): Nature 185, 458.

[477] LIVINGSTON,R.L., PAGE,W.L., and RAMACHANDRA RAO, C.N.(1960): J.Am.Chem.Soc. 82, 5048. [231]

[478] LIVINGSTON,R.L. and RAMACHANDRA RAO, C.N. (1959a): J.Am.Chem. Soc. 81, 285. [232]

[479] LIVINGSTON,R.L. and RAMACHANDRA RAO, C.N. (1959b): J.Am.Chem. Soc. 81, 3584. [233]

[480] LIVINGSTON,R.L. and RAMACHANDRA RAO, C.N. (1959c): J.Chem. Phys. 30, 339. [234]

[481] LIVINGSTON,R.L. and RAMACHANDRA RAO, C.N. (1960): J.Phys.Chem 64, 756. [235]

[482] LIVINGSTON,R.L., RAMACHANDRA RAO,C.N., KAPLAN,L.H., and ROCKS,L.(1958): J.Am.Chem.Soc. 80, 5368. [236]

[483] LIVINGSTON,R.L. and VAUGHAN,G.(1956a): J.Am.Chem.Soc. 78, 2711. [237]

[484] LIVINGSTON,R.L. and VAUGHAN,G.(1959b): J.Am.Chem.Soc. 78, 4866. [238]

[485] LONG,D.A. and CHAU,J.Y.H.(1962): Trans. Faraday Soc. 58, 2328. [239]

[486] LONG,D.A. and SEIBOLD,E.A.(1960): Trans. Faraday Soc. 56, 1105. [240]

[487] MALTSEV,YU.A. and NEKRASOV,L.I.(1970): Zh.Strukt.Khim. 10,769

[488] MANLEY,T.R. and WILLIAMS,D.A.(1965): Spectrochim.Acta 21, 1467. [241]

[489] MARIAM,S. and MALATHY DEVI, V. (1966): Indian J. Pure Appl. Phys. 4, 344. [241.1]

[490] MARKOV,P. and STØLEVIK,R.(1970): Acta Chem.Scand. 24, 2525.

[491] MARRIOTT,J.C., SALTHOUSE,J.A., WARE,M.J., and FREEMAN,J.M. (1970): Chem.Commun., 595.

[492] MASTRYUKOV,V.S., VILKOV,L.V., ZHIGACH,A.F., and SIRYATSKAYA, V.N.(1969): Zh.Strukt.Khim. 10, 136.

[493] MAYANTS,L.S.(1963): Doklady Akad. Nauk SSSR 151, 624.

[494] MAYANTS,L.S.(1964): Zh.Fiz.Khim. 38, 623.

[495] MAYANTS,L.S., GALPERN,E.G., and AVERBUKH,B.S.(1965): Opt. Spektroskopiya 18, 933.

[496] MC ADAM, A., BEAGLEY,B., and HEWITT,T.G.(1970): Trans.Faraday Soc. 66, 2732.

[497] MEISINGSETH,E.(1962): Acta Chem.Scand. 16, 778. [242]

[498] MEISINGSETH,E., BRUNVOLL,J., and CYVIN,S.J.(1964): Kgl. Norske Videnskab. Selskabs Skrifter, No. 7. [243]

[499] MEISINGSETH,E. and CYVIN,S.J.(1961): Acta Chem.Scand. 15, 2021. [244]

[500] MEISINGSETH,E. and CYVIN,S.J.(1962a): Acta Chem.Scand. 16, 1321. [245]

[501] MEISINGSETH,E. and CYVIN,S.J.(1962b): Acta Chem.Scand. 16, 2452. [246]

[502] MEISINGSETH,E. and CYVIN,S.J.(1962c): J.Mol.Spectry. 8, 464. [247]

[503] MIJLHOFF,F.C.(1965): Rec.trav.chim. 84, 74. [248]

[504] MINASSO,B. and ZERBI,G.(1971): J.Mol.Structure 7, 59.

[505] MOMANY,F.A. and BONHAM,R.A.(1964): J.Am.Chem.Soc. 86, 162. [249]

[506] MOMANY,F.A., BONHAM,R.A., and DRUELINGER,M.L.(1963): J.Am.Chem.Soc. 85, 3075. [250]

[507] MOMANY,F.A., BONHAM,R.A., and MC COY, W.H. (1963): J.Am.Chem.Soc. 85, 3077. [251]

[508] MORINO,Y.(1950): J.Chem.Phys. 18, 395. [252]

[509] MORINO,Y.(1960): Acta Cryst. 13, 1107. [253]

[510] MORINO,Y. and CYVIN,S.J.(1961): Acta Chem.Scand. 15, 483.[254]

[511] MORINO,Y., CYVIN,S.J., KUCHITSU,K., and IIJIMA,T.(1962): J.Chem.Phys. 36, 1109. [255]

[512] MORINO,Y. and HIROTA,E.(1955): J.Chem.Phys. 23, 737. [256]

[513] MORINO,Y. and HIROTA,E.(1958): J.Chem.Phys. 28, 185. [257]

[514] MORINO,Y. and IIJIMA,T.(1962a): J.Phys.Soc.Japan 17, Suppl. B-II, 27. [258]

[515] MORINO,Y. and IIJIMA,T.(1962b): Bull.Chem.Soc.Japan 35, 1661. [259]

[516] MORINO,Y. and IIJIMA,T.(1963): Bull.Chem.Soc.Japan 36, 412. [260]

[517] MORINO,Y., IIJIMA,T., and MURATA,Y.(1960): Bull.Chem.Soc. Japan 33, 46. [261]

[518] MORINO,Y. and KUCHITSU,K.(1958): J.Chem.Phys. 28, 175. [262]

[519] MORINO,Y., KUCHITSU,K., and MORITANI,T.(1969): Inorg.Chem. 8, 867.

[520] MORINO,Y., KUCHITSU,K., and MURATA,Y.(1965): Acta Cryst. 18, 549. [263]

[521] MORINO,Y., KUCHITSU,K., and OKA,T.(1962): J.Chem.Phys. 36, 1108. [264]

[522] MORINO,Y., KUCHITSU,K., and SHIMANOUCHI,T.(1952): J.Chem.Phys. 20, 726. [265]

[523] MORINO,Y., KUCHITSU,K., TAKAHASHI,A., and MAEDA,K.(1953): J.Chem.Phys. 21, 1927. [266]

[524] MORINO,Y. and MURATA,Y.(1965a): Bull.Chem.Soc.Japan 38, 104. [267]

[525] MORINO,Y. and MURATA,Y.(1965b): Bull.Chem.Soc.Japan 38, 114. [268]

[526] MORINO,Y., MURATA,Y., ITO,T., and NAKAMURA,J.(1962): J.Phys. Soc.Japan 17, Suppl. B-II, 37. [269]

[527] MORINO,Y., NAKAMURA,J., and MOORE,P.W.(1962): J.Chem.Phys. 36, 1050. [270]

[528] MORINO,Y., NAKAMURA,Y., and IIJIMA,T.(1960): J.Chem.Phys. 32, 643. [271]

[529] MORINO,Y. and UEHARA,H.(1966): J.Chem.Phys. 45, 4543. [271.1]

[530] MORINO,Y., UKAJI,T., and ITO,T.(1966a): Bull.Chem.Soc.Japan 39, 64. [272]

[531] MORINO,Y., UKAJI,T., and ITO,T.(1966b): Bull.Chem.Soc.Japan 39, 71. [273]

[532] MORINO,Y., UKAJI,T., and ITO,T.(1966c): Bull.Chem.Soc.Japan 39, 191. [274]

[533] MÜLLER,A.(1966): Naturwissensch. 53, 701. [274.1]

[534] MÜLLER,A.(1967): Z.phys.Chem. 236, 305.

[535] MÜLLER,A.(1968a): Z.phys.Chem. 238, 107.

[536] MÜLLER,A.(1968b): Z.phys.Chem. 238, 116.

[537] MÜLLER,A., BARAN,E.J., BOLLMANN,F., and AYMONINO,P.J.(1969): Z.Naturforschg. 24b, 960.

[538] MÜLLER,A. and CYVIN,S.J.(1968): J.Mol.Spectry. 26, 315.

[539] MÜLLER,A. and DIEMANN,E.(1969): Chem.Ber. 102, 945.

[540] MÜLLER,A., DIEMANN,E., and KRISHNA RAO, V.V. (1970): Chem. Ber. 103, 2961.

[541] MÜLLER,A., KREBS,B., and CYVIN,S.J.(1967): Acta Chem.Scand. 21, 2399. [274.2]

[542] MÜLLER,A., KREBS,B., CYVIN,S.J., and DIEMANN,E.(1968): Z.anorg.allgem.Chem. 359, 194.

[543] MÜLLER,A., KREBS,B., FADINI,A., GLEMSER,O., CYVIN,S.J., BRUNVOLL,J., CYVIN,B.N., ELVEBREDD,I., HAGEN,G., and VIZI,B. (1968): Z.Naturforschg. 23a, 1656.

[544] MÜLLER,A., KREBS,B., and GATTOW,G.(1967): Z.anorg.allgem. Chem. 349, 74.

[545] MÜLLER,A., KREBS,B., and PEACOCK,C.J.(1968): Z.Naturforschg. 23a, 1024.

[546] MÜLLER,A., KREBS,B., RITTNER,W., and STOCKBURGER,M.(1967): Ber.Bunsenges.Physik.Chem. 71, 182.

[547] MÜLLER,A. and NAGARAJAN,G.(1966): Z.Naturforschg. 21b, 508. [275]
[548] MÜLLER,A. and NAGARAJAN,G.(1967a): Z.anorg.allgem.Chem. 349, 87.
[549] MÜLLER,A. and NAGARAJAN,G.(1967b): Z.Chem. 7, 35.
[550] MÜLLER,A. and NAGARAJAN,G.(1967c): Z.phys.Chem. 235, 57.
[551] MÜLLER,A. and NAGARAJAN,G.(1967d): Z.phys.Chem. 235, 113.
[552] MÜLLER,A., NAGARAJAN,G., and FADINI,A.(1966): Z.anorg.allgem. Chem. 347, 269.
[553] MÜLLER,A., NAGARAJAN,G., and FADINI,A.(1967): Z.anorg.allgem. Chem. 353, 223.
[554] MÜLLER,A., NAGARAJAN,G., GLEMSER,O., CYVIN,S.J., and WEGENER,J.(1967): Spectrochim.Acta 23A, 2683.
[555] MÜLLER,A., NAGARAJAN,G., and KREBS,B.(1967): Spectroscopia Molecular 16, 31.
[556] MÜLLER,A. and PEACOCK,C.J.(1968a): Mol.Phys. 14, 393.
[557] MÜLLER,A. and PEACOCK,C.J.(1968b): Z.Chem. 8, 69.
[558] MÜLLER,A. and PEACOCK,C.J.(1969a): Ber.Bunsengesellsch. 72, 1057.
[559] MÜLLER,A. and PEACOCK,C.J.(1969b): J.Mol.Spectry. 30, 345.
[560] MÜLLER,A., PEACOCK,C.J., and HEIDBORN,U.(1968): Z.Naturforschg. 23a, 1687.
[561] MÜLLER,A., PEACOCK,C.J., SCHULZE,H., and HEIDBORN,U.(1969): J.Mol.Structure 3, 252.
[562] MURATA,Y., KUCHITSU,K., and KIMURA,M.(1970): Japanese J. Appl. Phys. 9, 591.

[563] NAGARAJAN,G.(1962a): Bull.Soc.Chim.Belg. 71, 329. [276]
[564] NAGARAJAN,G.(1962b): Bull.Soc.Chim.Belg. 71, 337. [277]
[565] NAGARAJAN,G.(1962c): Bull.Soc.Chim.Belg. 71, 347. [278]
[566] NAGARAJAN,G.(1962d): Bull.Soc.Chim.Belg. 71, 361. [279]
[567] NAGARAJAN,G.(1963a): Austral.J.Chem. 16, 908. [280]
[568] NAGARAJAN,G.(1963b): Bull.Soc.Chim.Belg. 72, 16. [281]
[569] NAGARAJAN,G.(1963c): Bull.Soc.Chim.Belg. 72, 351. [282]
[570] NAGARAJAN,G.(1963d): Bull.Soc.Chim.Belg. 72, 524. [283]
[571] NAGARAJAN,G.(1963e): Bull.Soc.Chim.Belg. 72, 537. [284]
[572] NAGARAJAN,G.(1963f): Bull.Soc.Chim.Belg. 72, 647. [285]
[573] NAGARAJAN,G.(1963g): Bull.Soc.Chim.Belg. 72, 657. [286]

[574] NAGARAJAN,G.(1963h): Current Science 32, 64. [287]
[575] NAGARAJAN,G.(1963i): Current Science 32, 448. [288]
[576] NAGARAJAN,G.(1963j): Indian J. Pure Appl. Phys. 1, 232. [289
[577] NAGARAJAN,G.(1963k): Indian J. Pure Appl. Phys. 1, 322. [290
[578] NAGARAJAN,G.(1963l): Indian J. Pure Appl. Phys. 1, 324. [291
[579] NAGARAJAN,G.(1963m): Z.phys.Chem. 224, 256. [292]
[580] NAGARAJAN,G.(1964a): Acta Phys.Austriaca 17, 240. [293]
[581] NAGARAJAN,G.(1964b): Acta Phys.Austriaca 17, 246. [294]
[582] NAGARAJAN,G.(1964c): Acta Phys.Austriaca 18, 1. [295]
[583] NAGARAJAN,G.(1964d): Acta Phys.Austriaca 18, 11. [296]
[584] NAGARAJAN,G.(1964e): Acta Phys.Austriaca 18, 23. [297]
[585] NAGARAJAN,G.(1964f): Bull.Soc.Chim.Belg. 73, 665. [298]
[586] NAGARAJAN,G.(1964g): Bull.Soc.Chim.Belg. 73, 768. [299]
[587] NAGARAJAN,G.(1964h): Bull.Soc.Chim.Belg. 73, 799. [300]
[588] NAGARAJAN,G.(1964i): Bull.Soc.Chim.Belg. 73, 811. [301]
[589] NAGARAJAN,G.(1964j): Bull.Soc.Chim.Belg. 73, 850. [302]
[590] NAGARAJAN,G.(1964k): Indian J.Phys. 38, 289. [303]
[591] NAGARAJAN,G.(1964l): Indian J. Pure Appl.Phys. 2, 17. [304]
[592] NAGARAJAN,G.(1964m): Indian J. Pure Appl.Phys. 2, 86. [305]
[593] NAGARAJAN,G.(1964n): Indian J. Pure Appl.Phys. 2, 145. [306]
[594] NAGARAJAN,G.(1964o): Indian J. Pure Appl.Phys. 2, 179. [307]
[595] NAGARAJAN,G.(1964p): Indian J. Pure Appl.Phys. 2, 205. [308]
[596] NAGARAJAN,G.(1964q): Indian J. Pure Appl.Phys. 2, 237. [309]
[597] NAGARAJAN,G.(1964r): Indian J. Pure Appl.Phys. 2, 278. [310]
[598] NAGARAJAN,G.(1964s): Indian J. Pure Appl.Phys. 2, 341. [311]
[599] NAGARAJAN,G.(1964t): Indian J. Pure Appl.Phys. 2, 343. [312]
[600] NAGARAJAN,G.(1964u): J.Chim.Phys. 61, 335. [313]
[601] NAGARAJAN,G.(1964v): J.Chim.Phys. 61, 338. [314]
[602] NAGARAJAN,G.(1964w): J.Mol.Spectry. 12, 198. [315]
[603] NAGARAJAN,G.(1964x): J.Mol.Spectry. 13, 361. [316]
[604] NAGARAJAN,G.(1965a): Acta Phys.Polon. 28, 875. [316.1]
[605] NAGARAJAN,G.(1965b): Bull.Soc.Chim.Belg. 74, 187. [317]
[606] NAGARAJAN,G.(1965c): Indian J. Phys. 39, 405.
[607] NAGARAJAN,G.(1966a): Acta Phys.Austriaca 21, 225. [317.1]
[608] NAGARAJAN,G.(1966b): Acta Phys.Austriaca 21, 355. [317.2]
[609] NAGARAJAN,G.(1966c): Acta Phys.Austriaca 24, 20. [317.3]
[610] NAGARAJAN,G.(1966d): Acta Phys.Hung. 20, 331. [318]
[611] NAGARAJAN,G.(1966e): Acta Phys.Polon. 29, 831. [318.1]

[612] NAGARAJAN,G.(1966f): Acta Phys.Polon. 29, 841. [318.2]
[613] NAGARAJAN,G.(1966g): Acta Phys.Polon. 30, 743. [318.3]
[614] NAGARAJAN,G.(1966h): Czechoslov.J.Phys. B16, 157. [319]
[615] NAGARAJAN,G.(1966h'): Indian J.Phys. 40, 319.
[616] NAGARAJAN,G.(1966i): Indian J. Pure Appl. Phys. 4, 151. [320]
[617] NAGARAJAN,G.(1966j): Indian J. Pure Appl. Phys. 4, 158. [321]
[618] NAGARAJAN,G.(1966k): Indian J. Pure Appl. Phys. 4, 237.[321.1]
[619] NAGARAJAN,G.(1966l): Indian J. Pure Appl. Phys. 4, 244.[321.2]
[620] NAGARAJAN,G.(1966m): Indian J. Pure Appl. Phys. 4, 347.[321.3]
[621] NAGARAJAN,G.(1966n): Indian J. Pure Appl. Phys. 4, 351.[321.4]
[622] NAGARAJAN,G.(1966n'): Indian J. Pure Appl. Phys. 4, 423.
[623] NAGARAJAN,G.(1966o): Indian J. Pure Appl. Phys. 4, 456.[321.5]
[624] NAGARAJAN,G.(1966p): Z.Naturforschg. 21a, 244. [321.6]
[625] NAGARAJAN,G.(1967): Z.phys.Chem. 234, 406.
[626] NAGARAJAN,G.(1968): Z.phys.Chem. 237, 297.
[627] NAGARAJAN,G. and DURIG,J.R.(1967a): Bull.Soc.Roy.Sci.Liège 36, 111.
[628] NAGARAJAN,G. and DURIG,J.R.(1967b): Bull.Soc.Roy.Sci.Liège 36, 334.
[629] NAGARAJAN,G. and DURIG,J.R.(1967c): Bull.Soc.Roy.Sci.Liège 36, 552.
[630] NAGARAJAN,G. and DURIG,J.R.(1968a): Acta Phys.Polon. 34, 319.
[631] NAGARAJAN,G. and DURIG,J.R.(1968b): Monatsh.Chem. 99, 473.
[632] NAGARAJAN,G., DURIG,J.R., and MÜLLER,A.(1967): Monatsh.Chem. 98, 1545.
[633] NAGARAJAN,G. and HARIHARAN,T.A.(1965a): Acta Phys.Austriaca 19, 349. [322]
[634] NAGARAJAN,G. and HARIHARAN,T.A.(1965b): Bull.Soc.Chim.Belg. 74, 201. [323]
[635] NAGARAJAN,G. and HARIHARAN,T.A.(1966): Acta Phys.Austriaca 21, 366. [323.1]
[636] NAGARAJAN,G. and LIPPINCOTT,E.R.(1965): J.Chem.Phys. 42, 1809. [324]
[637] NAGARAJAN,G. and LIPPINCOTT,E.R.(1966): Bull.Soc.Chim.Belg. 75, 555. [324.1]
[638] NAGARAJAN,G., LIPPINCOTT,E.R., and STUTMAN,J.M.(1965a): J.Phys.Chem. 69, 2017. [325]
[639] NAGARAJAN,G., LIPPINCOTT,E.R., and STUTMAN,J.M.(1965b):

Z.Naturforschg. 20a, 786. [326]
[640] NAGARAJAN,G. and MÜLLER,A.(1966): Z.Naturforschg. 21b, 393. [327]
[641] NAGARAJAN,G. and MÜLLER,A.(1967a): Monatsh.Chem. 98, 68.
[642] NAGARAJAN,G. and MÜLLER,A.(1967b): Monatsh.Chem. 98, 73.
[643] NAGARAJAN,G. and MÜLLER,A.(1967c): Z.anorg.allgem.Chem. 349, 82.
[644] NAGARAJAN,G., MÜLLER,A., and HORN,H.G.(1966): Z.Chem. 6,319.
[645] NAGARAJAN,G. and PERUMAL,A.(1968a): Acta Phys.Hung. 25, 119.
[646] NAGARAJAN,G. and PERUMAL,A.(1968b): Acta Phys.Polon. 34, 41.
[647] NAGARAJAN,G. and SIVAPRAKASAM,R.(1970a): Acta Phys.Polon. A37, 327.
[648] NAGARAJAN,G. and SIVAPRAKASAM,R.(1970b): Z.phys.Chem. 244, 117.
[649] NÁHLOVSKÁ,Z., NÁHLOVSKÝ,B., and SEIP,H.M.(1969): Acta Chem. Scand. 23, 3534.
[650] NÁHLOVSKÁ,Z., NÁHLOVSKÝ,B., and SEIP,H.M.(1970): Acta Chem. Scand. 24, 1903.
[651] NÁHLOVSKÁ,Z., NÁHLOVSKÝ,B., and STRAND,T.G.(1970): Acta Chem.Scand. 24, 2617.
[652] NAUMOV,V.A. and SEMASHKO,V.N.(1969): Zh.Strukt.Khim. 10,542.

[653] OBERHAMMER,H. and BAUER,S.H.(1969): J.Am.Chem.Soc. 91, 10.
[654] OBERHAMMER,H. and ZEIL,W.(1970a): J.Mol.Structure 6, 399.
[655] OBERHAMMER,H. and ZEIL,W.(1970b): Z.Naturforschg. 25a, 845.
[656] OWEN,N.L. and SEIP,H.M.(1970): Chem.Phys.Letters 5, 162.

[657] PANDEY,A.N., SINGH,H.S., and SANYAL,N.K.(1969): Current Science 38, 108.
[658] PAPOUŠEK,D. and PLÍVA,J.(1965): Spectrochim.Acta 21, 1147. [328]
[659] PAULI,G.H., MOMANY,F.A., and BONHAM,R.A.(1964): J.Am.Chem. Soc. 86, 1286. [329]
[660] PEACOCK,C.J., HEIDBORN,U., and MÜLLER,A.(1969): J.Mol.Spectry 30, 338.
[661] PEACOCK,C.J. and MÜLLER,A.(1968): J.Mol.Spectry. 26, 454.
[662] PEACOCK,C.J., MÜLLER,A., and KEBABCIOGLU,R.(1968a): J.Mol. Spectry. 27, 351.

[663] PEACOCK,C.J., MÜLLER,A., and KEBABCIOGLU,R.(1968b): J.Mol. Structure 2, 163.

[664] PICHAI,R., KRISHNA PILLAI, M.G., and RAMASWAMY,K.(1967): Austral.J.Chem. 20, 1055.

[665] PLANJE,M.C., TONEMAN,L.H., and DALLINGA,G.(1965): Rec.trav. chim. 84, 232. [330]

[666] PLATO,V., HARTFORD,W.D., and HEDBERG,K.(1970): J.Chem.Phys. 53, 3488.

[667] PULAY,P., BOROSSAY,GY., and TÖRÖK,F.(1968): J.Mol.Structure 2, 336.

[668] PURANIK,P.G. and RAO,E.V.(1963): Proc. Indian Acad. Sci. A58, 368. [331]

[669] PURANIK,P.G. and RAO,E.V.(1966): Indian J. Pure Appl. Phys. 4, 229. [331.1]

[670] PURANIK,P.G. and SIRDESMUKH,L.(1967): Indian J. Pure Appl. Phys. 5, 334.

[671] PURANIK,P.G. and SIRDESHMUKH,L.(1968): Proc. Indian Acad. Sci. A67, 99.

[672] PURNACHANDRA RAO, B. and RAMA MURTHY, V. (1968): Indian J. Pure Appl. Phys. 6, 339.

[673] PURUSHOTHAMAN,C.(1964): Proc. Indian Acad. Sci. A60, 431.[332]

[674] RADEMACHER,P. and STØLEVIK,R.(1969): Acta Chem.Scand. 23,660.

[675] RADHAKRISHNAN,M.(1964):Z.phys.Chem. Neue Folge 40, 189. [333]

[676] RAI,S.N. and THAKUR,S.N.(1970): Indian J. Pure Appl. Phys. 8, 367.

[677] RAJALAKSHMI,K.V.(1964): Proc. Indian Acad. Sci. A60, 51. [334]

[678] RAJALAKSHMI,K.V. and CYVIN,S.J.(1966): Acta Chem.Scand. 20, 2611. [335]

[679] RAJESWARA RAO, N. and SAKKU,S.(1968): Indian J. Pure Appl. Phys. 6, 4.

[680] RAMACHANDRA RAO, C.N. and LIVINGSTON,R.L.(1958): Current Science 27, 330. [336]

[681] RAMASWAMY,K. and DEVARAJAN,V.(1968): Acta Phys.Polon. 34,985.

[682] RAMASWAMY,K. and JAYARAMAN,S.(1970a): Indian J. Pure Appl. Phys. 8, 625.

[683] RAMASWAMY,K. and JAYARAMAN,S.(1970b): J.Mol.Structure 5, 325.

[684] RAMASWAMY,K. and KRISHNA RAO, B.(1969a): Z.phys.Chem. 240,127.

[685] RAMASWAMY,K. and KRISHNA RAO,B.(1969b): Z.phys.Chem. 242,18.

[686] RAMASWAMY,K. and KRISHNA RAO,B.(1969c): Z.phys.Chem. 242,155.

[687] RAMASWAMY,K. and KRISHNA RAO,B.(1969d): Z.phys.Chem. 242,215.

[688] RAMASWAMY,K. and MOHAN,N.(1969): Indian J. Pure Appl. Phys. 7, 459.

[689] RAMASWAMY,K. and MOHAN,N.(1970): Z.Naturforschg. 25b, 169.

[690] RAMASWAMY,K. and MOHAN,N.(1971): J.Mol.Structure 7, 51.

[691] RAMASWAMY,K. and MUTHUSUBRAMANIAN,P.(1970): J.Mol.Structure 6, 205.

[692] RAMASWAMY,K. and MUTHUSUBRAMANIAN,P.(1971): J.Mol.Structure 7, 45.

[693] RAMASWAMY,K. and NAMASIVAYAM,R.(1970): Z.Naturforschg. 25b, 465.

[694] RAMASWAMY,K. and RANGANATHAN,V.(1968): Indian J. Pure Appl. Phys. 6, 651.

[695] RAMASWAMY,K. and RANGANATHAN,V.(1970): Z.Naturforschg. 25b, 657.

[696] RAMASWAMY,K., SATHIANANDAN,K., and CLEVELAND,F.F.(1962): J.Mol.Spectry. 9, 107. [337]

[697] RAMASWAMY,K. and SRINIVASAN,K.(1968): Austral.J.Chem. 21,575.

[698] RAMASWAMY,K. and SRINIVASAN,K.(1969a): Austral.J.Chem. 22, 1123.

[699] RAMASWAMY,K. and SRINIVASAN,K.(1969b): J.Mol.Structure 3,473.

[700] RAMASWAMY,K. and SRINIVASAN,K.(1969c): J.Mol.Structure 4,135.

[701] RAMASWAMY,K. and SWAMINATHAN,S.(1969): Austral.J.Chem. 22, 291.

[702] RAMBIDI,N.G.(1962): Zh.Strukt.Khim. 3, 131. [338]

[703] RAMBIDI,N.G.(1963): Zh.Strukt.Khim. 4, 167. [339]

[704] RAMBIDI,N.G.(1964): Zh.Strukt.Khim. 5, 179. [340]

[705] RAMBIDI,N.G. and AKISHIN,P.A.(1961): Zh.Strukt.Khim. 2, 251. [341]

[706] RAMBIDI,N.G. and EZHOV,YU.S.(1967): Zh.Strukt.Khim. 8, 12. [341.1]

[707] RAMBIDI,N.G. and EZHOV,YU.S.(1968): Zh.Strukt.Khim. 9, 363.

[708] RAMBIDI,N.G. and SHCHEDRIN,B.M.(1964): Zh.Strukt.Khim. 5, 663. [342]

[709] RAMBIDI,N.G. and SPIRIDONOV,V.P.(1964a): Teplofiz.Vysokikh Temper. 2, 280. [343]

[710] RAMBIDI,N.G. and SPIRIDONOV,V.P.(1964b): Teplofiz.Vysokikh Temper. 2, 464. [344]

[711] RAMBIDI,N.G., SPIRIDONOV,V.P., and ALEKSEEV,N.V.(1962): Zh.Strukt.Khim. 3, 347. [345]

[712] RAMBIDI,N.G. and ZASORIN,E.Z.(1966): Zh.Strukt.Khim. 7,483. [345.1]

[713] RAMBIDI,N.G., ZASORIN,E.Z., and SHCHEDRIN,B.M.(1964): Zh. Strukt.Khim. 5, 503. [346]

[714] RANKIN,D.W.H., ROBIETTE,A.G., SHELDRICK,G.M., SHELDRICK,W.S., AYLETT,B.J., ELLIS,I.A., and MONAGHAN,J.J.(1969): J.Chem.Soc. A, 1224.

[715] RAO,D.V.R.A., THAKUR,S.N., and RAI,D.K.(1970): Proc. Indian Acad.Sci. A71, 42.

[716] RAW,C.J.G.(1962): J.Chem.Phys. 36, 1397. [347]

[717] RAW,C.J.G. and KUENZ,C.(1961): J.Chem.Phys. 35, 1529. [348]

[718] REITAN,A.(1958a): Acta Chem.Scand. 12, 131. [349]

[719] REITAN,A.(1958b): Acta Chem.Scand. 12, 785. [350]

[720] REITAN,A.(1958c): Kgl. Norske Videnskab. Selskabs Skrifter, No. 2. [351]

[721] ROBIETTE,A.G., SHELDRICK,G.M., and SHELDRICK,W.S.(1970): J.Mol.Structure 5, 423.

[722] ROBIETTE,A.G., SHELDRICK,G.M., and SIMPSON,R.N.F.(1969): J.Mol.Structure 4, 221.

[723] RUNDGREN,J.(1965): Arkiv Fysik 30, 61.

[724] RUNDGREN,J.(1967a): Arkiv Fysik 35, 31.

[725] RUNDGREN,J.(1967b): Arkiv Fysik 35, 269.

[726] RUNDGREN,J.(1967c): Arkiv Fysik 35, 361.

[727] RYAN,R.R. and HEDBERG,K.(1969): J.Chem.Phys. 50, 4986.

[728] SANYAL,N.K., PANDEY,A.N., and SINGH,H.S.(1969a): Indian J. Pure Appl. Phys. 7, 526.

[729] SANYAL,N.K., PANDEY,A.N., and SINGH,H.S.(1969b): J.Quant. Spectrosc.Radiat.Transfer. 9, 1035.

[730] SANYAL,N.K., SINGH,H.S., and PANDEY,A.N.(1969a): Indian J. Phys. 43, 361.

[731] SANYAL,N.K., SINGH,H.S., and PANDEY,A.N.(1969b): J.Quant. Spectrosc.Radiat.Transfer. 9, 1647.

[732] SASAKI,Y., KIMURA,K., and KUBO,M.(1959): J.Chem.Phys. 31,

477. [352]

[733] SATHIANANDAN,K. and MARGRAVE,J.L.(1963): J.Mol.Spectry. 10, 442. [353]

[734] SATHIANANDAN,K., RAMASWAMY,K., and CLEVELAND,F.F.(1962): J.Mol.Spectry. 8, 470. [354]

[735] SATHIANANDAN,K., RAMASWAMY,K., SUNDARAM,S., and CLEVELAND, F.F.(1964): J.Mol.Spectry. 13, 214. [355]

[736] SAWODNY,W., RUOFF,A., PEACOCK,C.J., and MÜLLER,A.(1968): Mol.Phys. 14, 433.

[737] SCHÄFER,L.(1968): J.Am.Chem.Soc. 90, 3919.

[738] SCHÄFER,L. and SEIP,H.M.(1967): Acta Chem.Scand. 21, 737. [355.1]

[739] SCHÄFER,L., SOUTHERN,J.F., CYVIN,S.J., and BRUNVOLL,J.(1970): J.Organometal.Chem. 24, C13.

[740] SEIP,H.M.(1965): Acta Chem.Scand. 19, 1955. [356]

[741] SEIP,H.M.(1967): Selected Topics in Structure Chemistry (Ed. P.Andersen, O.Bastiansen, and S.Furberg), p. 25, Universitetsforlaget, Oslo 1967. [356.1]

[742] SEIP,H.M. and SEIP,R.(1966): Acta Chem.Scand. 20, 2698. [356.2]

[743] SEIP,H.M. and SEIP,R.(1970): Acta Chem.Scand. 24, 3431.

[744] SEIP,H.M. and STØLEVIK,R.(1966a): Acta Chem.Scand. 20, 385. [357]

[745] SEIP,H.M. and STØLEVIK,R.(1966b): Acta Chem.Scand. 20, 1535. [357.1]

[746] SEIP,H.M., STRAND,T.G., and STØLEVIK,R.(1969): Chem.Phys. Letters 3, 617.

[747] SHANMUGASUNDARAM,G.(1968): Acta Phys.Polon. 34, 547.

[748] SHANMUGASUNDARAM,G.(1969): Acta Phys.Polon. 35, 483.

[749] SHANMUGASUNDARAM,G. and NAGARAJAN,G.(1969a): Z.phys.Chem. 240, 363.

[750] SHANMUGASUNDARAM,G. and NAGARAJAN,G.(1969b): Z.phys.Chem. 242, 312.

[751] SHIBATA,S.(1962): J.Phys.Soc.Japan 17, Suppl. B-II, 34.[358]

[752] SHIBATA,S.(1963): J.Phys.Chem. 67, 2256. [359]

[753] SHIBATA,S.(1970): Acta Chem.Scand. 24, 705.

[754] SHIBATA,S. and BARTELL,L.S.(1965): J.Chem.Phys. 42, 1147. [360]

[755] SHIBATA,S., BARTELL,L.S., and GAVIN,R.M.JR.(1964): J.Chem. Phys. 41, 717. [361]

[756] SINGH,B.P., PANDEY,A.N., and SINGH,H.S.(1970): Indian J. Pure Appl. Phys. 8, 193.

[757] SINGH,O.N. and RAI,D.K.(1965): Can.J.Phys. 43, 378. [362]

[758] SKANCKE,A.(1968): Acta Chem.Scand. 22, 3239.

[759] SMIT,W.M.A. and DIJKSTRA,G.(1971): J.Mol.Structure 7, 223.

[760] SMITH,D.W. and HEDBERG,K.(1956): J.Chem.Phys. 25, 1282.[363]

[761] SPIRIDONOV,V.P.(1964a): Zh.Strukt.Khim. 5, 3. [364]

[762] SPIRIDONOV,V.P.(1964b): Zh.Strukt.Khim. 5, 359. [365]

[763] SPIRIDONOV,V.P., RAMBIDI,N.G., and ALEKSEEV,N.V.(1963): Zh. Strukt.Khim. 4, 779. [366]

[764] SPIRIDONOV,V.P., RAMBIDI,N.G., and ALEKSEEV,N.V.(1965): Zh. Strukt.Khim. 6, 481. [367]

[765] SPITZER,R., HOWELL,W.J.JR., and SCHOMAKER,V.(1942): J.Am. Chem.Soc. 64, 62. [368]

[766] SRINIVASACHARYA,K.G. and SANTHAMMA,C.(1965): J.Mol.Spectry. 15, 435. [369]

[767] STEVENSON,D.P. and IBERS,J.A.(1960): J.Chem.Phys. 33, 762. [370]

[768] STØLEVIK,R., ANDERSEN,B., CYVIN,S.J., and BRUNVOLL,J.(1967): Acta Chem.Scand. 21, 1581. [370.1]

[769] STØLEVIK,R., CYVIN,S.J., and FADINI,A.(1970): Acta Chem. Scand. 24, 746.

[770] STØLEVIK,R. and RADEMACHER,P.(1969): Acta Chem.Scand. 23,672.

[771] STRAND,T.G.(1966): J.Chem.Phys. 44, 1611. [371]

[772] STRAND,T.G.(1967a): Acta Chem.Scand. 21, 1033. [371.1]

[773] STRAND,T.G.(1967b): Acta Chem.Scand. 21, 2111. [371.2]

[774] STRAND,T.G. and COX,H.L.JR.(1966): J.Chem.Phys. 44, 2426. [372]

[775] SUNDARAM,S.(1961): J.Mol.Spectry. 7, 53. [373]

[776] SUNDARAM,S.(1962a): Z.phys.Chem. Neue Folge 34, 225. [374]

[777] SUNDARAM,S.(1962b): Z.phys.Chem. Neue Folge 34, 233. [375]

[778] SVERDLOV,L.M. and KUKINA,V.S.(1967): Opt.Spektroskopiya 23, 172.

[779] SWICK,D.A. and KARLE,I.L.(1955): J.Chem.Phys. 23, 1499.[376]

[780] SWICK,D.A., KARLE,I.L., and KARLE,J.(1954): J.Chem.Phys. 22, 1242. [377]

[781] TANIMOTO,M., KUCHITSU,K., and MORINO,Y.(1969): Bull.Chem.Soc. Japan 42, 2519.

[782] TANIMOTO,M., KUCHITSU,K., and MORINO,Y.(1970): Bull.Chem.Soc. Japan 43, 2776.

[783] THAKUR,S.N., RAO,D.V.R.A., and RAI,D.K.(1970): Indian J. Pure Appl. Phys. 8, 196.

[784] TÖRÖK,F. and HUN,GY.B.(1969): Acta Chim.Hung. 59, 303.

[785] TÖRÖK,F. and PULAY,P.(1969): J.Mol.Structure 3, 283.

[786] TOYAMA,M. OKA,T., and MORINO,Y.(1964): J.Mol.Spectry. 13, 193. [378]

[787] TRÆTTEBERG,M.(1964): J.Am.Chem.Soc. 86, 4265. [379]

[788] TRÆTTEBERG,M.(1966): Acta Chem.Scand. 20, 1724. [379.1]

[789] TRÆTTEBERG,M.(1968a): Acta Chem.Scand. 22, 628.

[790] TRÆTTEBERG,M.(1968b): Acta Chem.Scand. 22, 2294.

[791] TRÆTTEBERG,M.(1968c): Acta Chem.Scand. 22, 2305.

[792] TRÆTTEBERG,M.(1970a): Acta Chem.Scand. 24, 373.

[793] TRÆTTEBERG,M.(1970b): Acta Chem.Scand. 24, 2285.

[794] TRÆTTEBERG,M.(1970c): Acta Chem.Scand. 24, 2295.

[795] TRÆTTEBERG,M., HAGEN,G., and CYVIN,S.J.(1969): Acta Chem. Scand. 23, 74.

[796] TRÆTTEBERG,M., HAGEN,G., and CYVIN,S.J.(1970): Z.Naturforschg 25b, 134.

[797] UKAJI,T. and BONHAM,R.A.(1962a): J.Am.Chem.Soc. 84, 3627. [380]

[798] UKAJI,T. and BONHAM,R.A.(1962b): J.Am.Chem.Soc. 84, 3631. [381]

[799] UKAJI,T. and KUCHITSU,K.(1966): Bull.Chem.Soc.Japan 39,2153. [381.1]

[800] UNNIKRISHNAN NAYAR, V. and ARULDHAS,G.(1968a): Indian J. Pure Appl. Phys. 6, 126.

[801] UNNIKRISHNAN NAYAR, V. and ARULDHAS,G.(1968b): Indian J. Pure Appl. Phys. 6, 521.

[802] UNNIKRISHNAN NAYAR, V. and ARULDHAS,G.(1970): Current Science 39, 204.

[803] VAN BEKKUM, H., HOEFNAGEL,M.A., DE LAVIETER, L., VAN VEEN, A., VERKADE.P.E., WEMMERS,A., WEPSTER,B.M., PALM,J.H., SCHÄFER,L., DEKKER,H., MOSSELMAN,C., and SOMSEN,G.(1968):

Rec.trav.chim. 87, 1363.

[804] VENKATA CHALAPATHI, V. and VENKATA RAMIAH, K. (1966): *Proc. Indian Acad. Sci*. A64, 148. [381.2]

[805] VENKATESWARLU,K.(1965): *J.Scient.Industr.Res*. 24, 10.[382]

[806] VENKATESWARLU,K. and BABU JOSEPH, K. (1967a): *Bull.Roy.Soc. Liège* 36, 173.

[807] VENKATESWARLU,K. and BABU JOSEPH, K. (1967b): *Current Science* 36, 605.

[808] VENKATESWARLU,K. and BABU JOSEPH, K. (1968a): *Acta Chim. Hung*. 55, 351.

[809] VENKATESWARLU,K. and BABU JOSEPH, K. (1968b): *Acta Phys. Hung*. 24, 139.

[810] VENKATESWARLU,K., BABU JOSEPH, K., and MALATHY DEVI, V. (1967): *Indian J. Pure Appl. Phys*. 5, 14.

[811] VENKATESWARLU,K. and BHAMAMBAL,P.(1967): *Z.Naturforschg*. 22b, 946.

[812] VENKATESWARLU,K. and BHAMAMBAL,P.(1968a): *Indian J. Pure Appl.Phys*. 6, 467.

[813] VENKATESWARLU,K. and BHAMAMBAL,P.(1968b): *Indian J. Pure Appl.Phys*. 6, 530.

[814] VENKATESWARLU,K. and BHAMAMBAL,P.(1970a): *Acta Phys.Polon*. A37, 661.

[815] VENKATESWARLU,K. and BHAMAMBAL,P.(1970b): *Indian J. Pure Appl.Phys*. 8, 180.

[816] VENKATESWARLU,K. and GIRIJAVALLABHAN,C.P.(1968): *Z.Naturforschg*. 23b, 1300.

[817] VENKATESWARLU,K. and GIRIJAVALLABHAN,C.P.(1969): *Z.Naturforschg*. 24b, 1256.

[818] VENKATESWARLU,K. and JOSEPH,P.A.(1970): *J.Mol.Structure* 6, 145.

[819] VENKATESWARLU,K., JOSEPH,P.A., and MALATHY DEVI, V. (1969): *Acta Phys.Polon*. 35, 875.

[820] VENKATESWARLU,K., JOSEPH,P.A., RUDRAWARRIER,M.K., and MATHEW,M.P.(1969): *Indian J. Pure Appl. Phys*. 7, 279.

[821] VENKATESWARLU,K. and MALATHY DEVI, V. (1965a): *Current Science* 34, 144. [383]

[822] VENKATESWARLU,K. and MALATHY DEVI, V. (1965b): *Current Science* 34, 373. [384]

[823] VENKATESWARLU,K. and MALATHY DEVI, V. (1965c): Indian J. Pure Appl. Phys. 3, 195. [385]

[824] VENKATESWARLU,K. and MALATHY DEVI, V. (1965d): Proc.Indian Acad.Sci. A61, 272. [386]

[825] VENKATESWARLU,K. and MALATHY DEVI, V. (1967): Current Science 36, 118.

[826] VENKATESWARLU,K. and MALATHY DEVI, V. (1968a): Current Science 37, 401.

[827] VENKATESWARLU,K. and MALATHY DEVI, V. (1968b): Proc.Indian Acad.Sci. A67, 71.

[828] VENKATESWARLU,K., MALATHY DEVI, V., and NATARAJAN,A.(1969): Proc. Indian Acad. Sci. A70, 126.

[829] VENKATESWARLU,K. and MARIAM,S.(1965a): Indian J. Pure Appl. Phys. 3, 472. [386.1]

[830] VENKATESWARLU,K. and MARIAM,S.(1965b): Proc. Indian Acad. Sci. A61, 260. [387]

[831] VENKATESWARLU,K. and MARIAM,S.(1966): Czechoslov.J.Phys. B16, 290. [388]

[832] VENKATESWARLU,K. and MARIAM,S.(1967): Bull.Roy.Soc.Liège 36, 178.

[833] VENKATESWARLU,K., MARIAM,S., and ANANTARAMA SARMA, Y. (1966): Acta Phys.Polon. 30, 913. [388.1]

[834] VENKATESWARLU,K., MARIAM,S., and ANANTARAMA SARMA, Y. (1967): Opt.Spektroskopiya 22, 210.

[835] VENKATESWARLU,K., MARIAM,S., and GIRIJAVALLABHAN,C.P. (1967): Bull.Roy.Soc.Liège 36, 576.

[836] VENKATESWARLU,K., MARIAM,S., and MATHEW,M.P.(1965): Proc. Indian Acad. Sci. A62, 159. [389]

[837] VENKATESWARLU,K., MARIAM,S., and MATHEW,M.P.(1966): Bull. Roy.Soc.Liege 35, 337.

[838] VENKATESWARLU,K., MARIAM,S., and NATARAJAN,A.(1967): Acta Phys.Polon. 32, 213.

[839] VENKATESWARLU,K., MARIAM,S., and RAJALAKSHMI,K.(1965): Bull.Acad.Roy.Belg. (5.Ser.) 51, 359. [390]

[840] VENKATESWARLU,K. and MATHEW,M.P.(1966): Current Science 35, 562.

[841] VENKATESWARLU,K. and MATHEW,M.P.(1967): Indian J. Pure Appl. Phys. 5, 17.

[842] VENKATESWARLU,K. and MATHEW,M.P.(1968a): Acta Chim.Hung. 55, 203.

[843] VENKATESWARLU,K. and MATHEW,M.P.(1968b): Z.Naturforschg. 23b, 1296.

[844] VENKATESWARLU,K., MATHEW,M.P., and MALATHY DEVI, V. (1969): J.Mol.Structure 3, 119.

[845] VENKATESWARLU,K. and PURUSHOTHAMAN,C.(1965): Indian J. Pure Appl. Phys. 3, 377. [391]

[846] VENKATESWARLU,K. and PURUSHOTHAMAN,C.(1966a): Acta Phys. Polon. 30, 801. [391.1]

[847] VENKATESWARLU,K. and PURUSHOTHAMAN,C.(1966b): Czechoslov. J.Phys. B16, 144. [392]

[848] VENKATESWARLU,K., PURUSHOTHAMAN,C., and BABU JOSEPH, K. (1966): Acta Phys.Polon. 30, 807. [392.1]

[849] VENKATESWARLU,K. and RAJALAKSHMI,K.V.(1964): Proc.Symposium Raman and Infrared Spectroscopy, Ernakulam, 185.

[850] VENKATESWARLU,K. and RAJALAKSHMI,K.V.(1965): Proc. Indian Acad. Sci. A61, 255. [393]

[851] VENKATESWARLU,K., RAJALAKSHMI,K.V., and BABU JOSEPH, K. (1968): Acta Chim.Hung. 55, 209.

[852] VENKATESWARLU,K., RAJALAKSHMI,K.V., and MALATHY DEVI, V. (1966): Bull.Soc.Roy.Sci.Liège 35, 350.

[853] VENKATESWARLU,K., RAJALAKSHMI,K.V., and NATARAJAN,A.(1967): Bull.Roy.Soc.Sci.Liège 36, 347.

[854] VENKATESWARLU,K., RAJALAKSHMI,K.V., and PURUSHOTHAMAN,C. (1967): Bull.Roy.Soc.Liège 36, 583.

[855] VENKATESWARLU,K., RAJALAKSHMI,K.V., and THANALAKSHMI,R. (1963): Proc. Indian Acad. Sci. A58, 290. [394]

[856] VENKATESWARLU,K. and RUDRA WARRIER, M.K. (1970): Indian J. Pure Appl. Phys. 8, 142.

[857] VILKOV,L.V.(1964): Zh.Strukt.Khim. 5, 809. [395]

[858] VILKOV,L.V.(1968): Zh.Strukt.Khim. 9, 518.

[859] VILKOV,L.V., AKISHIN,P.A., and LITOVTSEVA,I.N.(1966): Zh. Strukt.Khim. 7, 3. [395.1]

[860] VILKOV,L.V., AKISHIN,P.A., and SALOVA,G.E.(1965): Zh.Strukt. Khim. 6, 355. [396]

[861] VILKOV,L.V. and ANASHKIN,M.G.(1968): Zh.Strukt.Khim. 9, 690.

[862] VILKOV,L.V., ANASHKIN,M.G., and MAMAEVA,G.I.(1968): Zh.

Strukt.Khim. 9, 372.

[863] VILKOV,L.V. and HARGITTAI,I.(1967): Acta Chim.Hung. 52,423.

[864] VILKOV,L.V., KHAYKIN,L.S., VASILEV,A.F., and TULYAKOVA,T.F. (1968): Zh.Strukt.Khim. 9, 1071.

[865] VILKOV,L.V., KHAYKIN,L.S., ZHIGACH,A.F., and SIRYATSKAYA, V.N.(1968): Zh.Strukt.Khim. 9, 889.

[866] VILKOV,L.V., KUSAKOV,M.M., NAMETKIN,N.S., and OPPENHEIM,V.D. (1968): Dokl.Akad.Nauk.SSSR 183, 830.

[867] VILKOV,L.V. and MASTRYUKOV,V.S.(1968): Zh.Strukt.Khim. 9,587.

[868] VILKOV,L.V., MASTRYUKOV,V.S., and AKISHIN,P.A.(1964): Zh. Strukt.Khim. 5, 183. [397]

[869] VILKOV,L.V., MASTRYUKOV,V.S., ZHIGACH,A.F., and SIRYATSKAYA, V.N.(1966): Zh.Strukt.Khim. 7, 883.

[870] VILKOV,L.V. and NAZARENKO,I.I.(1967): Zh.Strukt.Khim. 8,346.

[871] VILKOV,L.V. and NAZARENKO,I.I.(1968): Zh.Strukt.Khim. 9,887.

[872] VILKOV,L.V., NAZARENKO,I.I., and KOSTYANOVSKII,R.G.(1968): Zh.Strukt.Khim. 9, 1075.

[873] VILKOV,L.V., RAMBIDI,N.G., and SPIRIDONOV,V.P.(1967): Zh. Strukt.Khim. 8, 786.

[874] VILKOV,L.V. and SADOVA,N.I.(1967a): Zh.Strukt.Khim. 8, 344.

[875] VILKOV,L.V. and SADOVA,N.I.(1967b): Zh.Strukt.Khim. 8, 398.

[876] VILKOV,L.V., SADOVA,N.I., and MOCHALOV,S.S.(1968): Dokl. Akad.Nauk SSSR 179, 896.

[877] VILKOV,L.V., SADOVA,N.I., and ZILBER,I.YU.(1967): Zh.Strukt. Khim. 8, 528.

[878] VILKOV,L.V., SPIRIDONOV,V.P., and SADOVA,N.I.(1968): Zh. Strukt.Khim. 9, 187.

[879] VILKOV,L.V. and TARASENKO,N.A.(1969): Chem.Commun., 1176.

[880] VILKOV,L.V., TARASENKO,N.A., and PROKOFEV,A.K.(1970): Zh. Strukt.Khim. 11, 129.

[881] VIZI,B., BRUNVOLL,J., and MÜLLER,A.(1968): Acta Chem.Scand. 22, 1279.

[882] WAIT,S.C.JR.(1962): J.Chem.Phys. 36, 1396. [398]

[883] WASER,J. and SCHOMAKER,V.(1953): Revs. Modern Phys. 25, 671. [399]

[884] WINNEWISSER,M. and HAASE,J.(1968): Z.Naturforschg. 23a, 56.

[885] WONG,CH. and SCHOMAKER,V.(1957): J.Phys.Chem. 61, 358. [400]

[886] WONG,CH. and SCHOMAKER,V.(1958): J.Chem.Phys. 28, 1010.[401]

[887] YERANOS,W.A.(1967): Mol.Phys. 12, 529.
[888] YERANOS,W.A.(1968a): Z.Naturforschg. 23a, 617.
[889] YERANOS,W.A.(1968b): Z.Naturforschg. 23a, 618.
[890] YERANOS,W.A.(1970): J.Mol.Spectry. 34, 533.

[891] ZASORIN,E.Z. and RAMBIDI,N.G.(1967a): Zh.Strukt.Khim. 8,391.
[892] ZASORIN,E.Z. and RAMBIDI,N.G.(1967b): Zh.Strukt.Khim. 8,591.
[893] ZASORIN,E.Z., RAMBIDI,N.G., and AKISHIN,P.A.(1963): Zh. Strukt.Khim. 4, 910.

Supplements and Further Developments

S. J. CYVIN

S.1. Introduction

The chapters of the present volume are apparently scattered over a wide area of topics. Nevertheless they are more or less strongly interconnected. Mean-square amplitude matrices and mean amplitudes of vibration are some of the key words throughout, and some of the chapters are supplements or continuations of the topics within Cyvin's monograph[1,2] 'Molecular Vibrations and Mean Square Amplitudes'.* However, the scope is much wider in the present volum In addition to the calculation of mean amplitudes from spectroscopi data (new results are reported, obtained partly from previous literature data and from new spectroscopical experiments) much emphasis is laid on the deduction of these quantities from modern gas electron diffraction data. New experimental works are reported, in whic the deduction of mean amplitudes represents only a part of the results. Other chapters are still more indirectly (if at all) connected to the concept of mean square amplitude matrices and mean amplitudes of vibration. Many of the different topics have been advanced rapidly during the processing of this book. The present chapter is intended to keep up with these developments by summarizi the relevant recent references. It also contains supplementary surveys of previous literature within certain topics.

A new, practical method of calculating mean amplitudes of vibration and the perpendicular amplitude correction coefficients has been reported.[3] The developed computer program for this purpose utilizes the recently published normal coordinate analysis procedur of Gwinn.[4] It is intended to publish[5] some of the mathematical

---

* The present book is cited erroneously (according to a preliminary title) on p. 27 of the Russian translation.[2]

details involved, in order to clarify the new aspects of the mentioned method[3] in terms of conventional symbolism[1] from the spectroscopic point of view.

New and interesting developments in the theory of mean-square amplitude matrices are due to Mayants.[6]

S.2. Chapter 1

In this section a survey of some basic concepts in the theory of molecular vibrations is given in addition to Chapter 1. It was attempted to develop a suitable notation comprising the most current usage of symbols, although it is not as standardized in the literature as one might wish.

The well-known secular equation of molecular vibrations in the harmonic approximation reads

$$\mathbf{GFL} = \mathbf{L}\boldsymbol{\lambda}. \qquad (S.1)$$

$\boldsymbol{\lambda}$ is the diagonal matrix composed of the eigenvalues of $\mathbf{GF}$. The matrix of eigenvectors, $\mathbf{L}$ [see also Eq. (1.22)*] is normalized in such a way that [cf. also Eqs. (1.25) and (1.26)]

$$\mathbf{G} = \mathbf{LL}', \quad \mathbf{F} = (\mathbf{L}^{-1})'\boldsymbol{\lambda}\mathbf{L}^{-1}. \qquad (S.2)$$

In the standard problem of calculating force constants the $\boldsymbol{\lambda}$ matrix is assumed to be known (its elements are proportional to the squared vibrational frequencies), but not the $\mathbf{L}$ matrix. Let $\bar{\mathbf{L}}$ be given by[7]

$$\bar{\mathbf{L}} = \mathbf{L}\boldsymbol{R}, \qquad (S.3)$$

where $\boldsymbol{R}$ is an arbitrary orthogonal matrix. An arbitrary matrix $\bar{\mathbf{L}}$ defined by (3) clearly satisfies the first one of the relations (2). If the matrix $\mathbf{F}$ is constructed from $\bar{\mathbf{L}}$ according to the second one of the relations (2) also the basic equation (1) will be satisfied. Hence a given $\boldsymbol{\lambda}$ matrix is consistent with an infinite number of $\mathbf{F}$ matrices obtainable in the described way. It is also true

---

* Here we assume that $\mathbf{S}$ in $\mathbf{S} = \mathbf{L}\mathbf{Q}$ represents a complete set of independent internal coordinates.

that all the force-constant matrices consistent with a given set of frequencies may be expressed in terms of an arbitrary orthogonal matrix, which may be restricted to have the determinant equal to +1 (see subsequent relations and Chapter 4).

It is expedient to start an analysis of the problem of calculating force constants with the diagonalization of $\mathbf{G}$. This approach was used in an early work of Taylor[8] and taken up with renewed interest in modern works by several authors. Ref. 1 of Chapter 4 is the first article in a series of papers[9-16] mainly by Pulay and Török; see also Refs. 17,18. The relevant paper of Toman and Plíva[19] is also cited in Chapter 4.

Let the $\mathbf{G}$ matrix be diagonalized by the orthogonal matrix $\mathbf{A}$,* which consequently satisfies the relations

$$\mathbf{A}'\mathbf{G}\mathbf{A} = \Gamma, \quad \mathbf{G} = \mathbf{A}\Gamma\mathbf{A}', \tag{S.4}$$

where Γ is diagonal. Furthermore

$$\mathbf{G}^{\frac{1}{2}} = \mathbf{A}\Gamma^{\frac{1}{2}}\mathbf{A}' \tag{S.5}$$

Many methods of calculating force constants, several of them deduced on the basis of the relations (4),(5), have been proposed.[20-29] It should not be omitted to mention that several mistakes or misinterpretations have been detected[7,30-32] in these deductions. Particularly in the critical review by Averbukh, Mayants, and Shaltuper[7] Fadini's method[20-23] [also A.Fadini, thesis, Technische Hochschule Stuttgart 1967] is stated to be entirely artificial. Fallacies in the other methods[24-28] are pointed out as well. To my opinion this critical analysis[7] has not yet had the impact on the research in this field as it deserves.

In the Billes potential[24] the $\mathbf{L}$ matrix is prescribed to be[7]

$$\mathbf{L}_B = \mathbf{A}\Gamma^{\frac{1}{2}} = \mathbf{G}^{\frac{1}{2}}\mathbf{A}, \tag{S.6}$$

and is assumed to satisfy the basic relations (1),(2). Accordingly[7]

$$\mathbf{F}_B = \mathbf{A}\Gamma^{-\frac{1}{2}}\lambda\Gamma^{-\frac{1}{2}}\mathbf{A}' = \mathbf{A}\Gamma^{-1}\lambda\mathbf{A}'. \tag{S.7}$$

---

* This $\mathbf{A}$ matrix should not be confused with $\mathbf{A}$ of Eq. (1.10).

This means that $\mathbf{A}$ is assumed also to diagonalize $\mathbf{F}_B$, i.e.[24,30]

$$\mathbf{A}'\mathbf{F}_B\mathbf{A} = \mathbf{\Phi}, \quad \mathbf{F}_B = \mathbf{A\Phi A}', \quad \boldsymbol{\lambda} = \mathbf{\Gamma\Phi}. \tag{S.8}$$

With the aid of Eq. (6) the same potential energy matrix may be expressed as

$$\mathbf{F}_B = \mathbf{G}^{-\frac{1}{2}}\mathbf{A}\boldsymbol{\lambda}\mathbf{A}'\mathbf{G}^{-\frac{1}{2}}. \tag{S.9}$$

According to Eq. (6) and the theorem contained in Eq. (3) it is clear that $\mathbf{L} = \mathbf{L}_B\boldsymbol{R} = \mathbf{A\Gamma}^{\frac{1}{2}}\boldsymbol{R} = \mathbf{G}^{\frac{1}{2}}\mathbf{A}\boldsymbol{R}$ gives $\mathbf{L}$ matrices which (with properly constructed $\mathbf{F}$) satisfy the basic relations (1),(2). It appears that an arbitrary $\mathbf{L}$ may be written as

$$\mathbf{L} = \mathbf{G}^{\frac{1}{2}}\mathbf{A}\boldsymbol{R} = \mathbf{G}^{\frac{1}{2}}\boldsymbol{U}, \tag{S.10}$$

where

$$\boldsymbol{U} = \mathbf{A}\boldsymbol{R} \tag{S.11}$$

is an arbitrary orthogonal matrix (with the determinant equal to +1); cf. also Eq. (4.3). The corresponding force-constant matrices read

$$\mathbf{F} = \mathbf{G}^{-\frac{1}{2}}\boldsymbol{U}\boldsymbol{\lambda}\boldsymbol{U}'\mathbf{G}^{-\frac{1}{2}}, \tag{S.12}$$

as is found similarly to Eq. (9); cf. also Eq. (4.4).

Some of the mentioned force fields are easily characterized by specifying the matrix $\boldsymbol{R}$ or $\boldsymbol{U}$ in Eqs.(10)-(12). The Billes potential corresponds to $\boldsymbol{R} = \mathbf{E}$ or $\boldsymbol{U} = \mathbf{A}$. The force field proposed by Herranz and Castaño[25] emerges when $\boldsymbol{R} = \mathbf{A}'$ or $\boldsymbol{U} = \mathbf{E}$. It is consistent with

$$\mathbf{L}_{HC} = \mathbf{A\Gamma}^{\frac{1}{2}}\mathbf{A}' = \mathbf{G}^{\frac{1}{2}}, \tag{S.13}$$

and, accordingly

$$\mathbf{F}_{HC} = \mathbf{A\Gamma}^{-\frac{1}{2}}\mathbf{A}'\boldsymbol{\lambda}\mathbf{A\Gamma}^{-\frac{1}{2}}\mathbf{A}' = \mathbf{G}^{-\frac{1}{2}}\boldsymbol{\lambda}\mathbf{G}^{-\frac{1}{2}}. \tag{S.14}$$

The transformation of the $\mathbf{G}$ matrix as expressed by Eq. (4) corresponds to a coordinate transformation,[30] say

$$\mathbf{S} = \mathbf{A}\boldsymbol{\mathfrak{S}}, \quad \boldsymbol{\mathfrak{S}} = \mathbf{A}'\mathbf{S}. \tag{S.15}$$

A similar coordinate transformation is utilized by Gussoni and Zerbi;[33] cf. also Eq. (7.15) in Chapter 7. The vibrational Hamiltonian in the S basis reads

$$\mathcal{H} = \frac{1}{2}\mathsf{P}_S{}'\mathsf{G}\,\mathsf{P}_S + \frac{1}{2}\mathsf{S}'\mathsf{F}\mathsf{S}\,, \tag{S.16}$$

and transforms into the following form in the $\mathfrak{S}$ basis:

$$\mathcal{H}_{\mathfrak{S}} = \frac{1}{2}\mathsf{P}_{\mathfrak{S}}{}'\Gamma\,\mathsf{P}_{\mathfrak{S}} + \frac{1}{2}\mathfrak{S}'\mathsf{F}_{\mathfrak{S}}\mathfrak{S}\,. \tag{S.17}$$

The transformation between the different energy matrices are easily found to be

$$\mathsf{G} = \mathbf{A}\Gamma\mathbf{A}'\,, \qquad \mathsf{F} = \mathbf{A}\mathsf{F}_{\mathfrak{S}}\mathbf{A}'. \tag{S.18}$$

The first one of the relations (18) is already contained in Eq.(4). On the other hand, the $\mathsf{F}_{\mathfrak{S}}$ matrix is not necessarily a diagonal matrix, as it would be if we restricted it to obey the Billes potential. If one defines $\mathsf{L}_{\mathfrak{S}}$ by $\mathfrak{S} = \mathsf{L}_{\mathfrak{S}}\mathsf{Q}$, one has clearly

$$\mathsf{L}_{\mathfrak{S}} = \mathbf{A}'\mathsf{L}\,. \tag{S.19}$$

It may be expedient to work in another coordinate set basis, χ, which is produced by proper scaling of the $\mathfrak{S}$ coordinates in order to achieve

$$\mathcal{H}_{\chi} = \frac{1}{2}\mathsf{P}_{\chi}{}'\mathsf{P}_{\chi} + \frac{1}{2}\chi'\mathsf{F}_{\chi}\chi\,; \tag{S.20}$$

cf. Eq. (8.1) in Chapter 8. The χ and $\mathfrak{S}$ coordinates are closely inter-related (they are in fact proportional);

$$\mathfrak{S} = \Gamma^{\frac{1}{2}}\chi\,. \tag{S.21}$$

Let the transformation between χ and S be given by

$$\mathsf{S} = \mathbf{V}\chi\,, \qquad \chi = \mathbf{V}^{-1}\mathsf{S}\,. \tag{S.22}$$

The following relations hold in analogy with Eq. (18):

$$\mathsf{G} = \mathbf{V}\mathbf{V}'\,, \qquad \mathsf{F} = (\mathbf{V}^{-1})'\mathsf{F}_{\chi}\mathbf{V}^{-1}\,. \tag{S.23}$$

Furthermore, if $\chi = \mathsf{L}_{\chi}\mathsf{Q}$, one has

$$\mathbf{L}_\chi = \mathbf{V}^{-1}\mathbf{L} \,. \tag{S.24}$$

The χ coordinates are utilized in Chapter 9; see especially Eqs. (9.3)-(9.5).*

Below we give the connection between the transformation matrices $\mathbf{A}$ and $\mathbf{V}$ of Eqs. (15) and (22), respectively. It is easily found

$$\mathbf{V} = \mathbf{A}\Gamma^{\frac{1}{2}}, \quad \mathbf{V}^{-1} = \Gamma^{-\frac{1}{2}}\mathbf{A}'. \tag{S.25}$$

From Eqs. (6) and (25) it follows

$$\mathbf{L}_B = \mathbf{V}. \tag{S.26}$$

In Ref. 34 (which is Ref. 3 of Chapter 3, Part II) Sørensen has primarily introduced the symbol $\mathbf{V}$ in a general sense as the transformation matrix of $\mathbf{S} = \mathbf{V}\mathbf{s}$ which satisfies $\mathbf{V}\mathbf{V}' = \mathbf{G}$; see also Eqs. (3.47)-(3.50). The $\mathbf{V}$ matrix of Eqs. (23)-(25) is one of the possibilities. Sørensen[34] has defined the $\mathbf{V}$ matrix in a special sense by imposing the condition $V_{ij} = 0$ for $i < j$ on its elements. Furthermore, the relations derived with the aid of this $\mathbf{V}$ matrix (including those of Chapter 3, Part II) should retain their validity when redundancies are present among the $\mathbf{S}$ coordinates. The 'method of progressive rigidity'[29,35] for calculations of force constants may be characterized by choosing $\mathbf{L}$ as equal to the trinagular $\mathbf{V}$ matrix with a proper choice of the sequence of the coordinates in $\mathbf{S}$ (the most 'rigid' first when $V_{ij} = 0$ for $i < j$). Triangular $\mathbf{L}$ matrices have also been extensively studied by Müller et al. (the $\mathbf{L}$ -matrix approximation method); see, e.g., Chapter 21 and references cited therein.

S.3. Chapter 4

The methods described in Sections 4.3.2 and 4.3.3 have been applied to formic acid monomer (see also Chapter 14, Part II). The results are going to be published elsewhere.[36]

---

* In Chapter 9 the $\mathbf{L}_\chi$ matrix is designated $\mathbf{W}$.

S.4. Chapter 6

The method of specific imposition of a potential parameter (Section 6.2.4) is applied in Chapter 14, Part II. A detailed description of the method with applications to C_2Cl_4 is communicated elsewhere.[37]

S.5. Chapter 7

A method for cartesian symmetry coordinates is mentioned in Chapter 7. Following the same line (cf. Ref. 20 of Chapt. 7) another method has been suggested.[38]

The work of Zerbi and his co-workers appears to be very timely and natural, since these days there is a general tendency towards using computers also for nonnumerical problems in physics and chemistry. In this connection attention should be drawn to the recent book edited by Loebl.[39] In particular, the chapter written by S. Flodmark and E. Blokker "Symmetry Adaptation of Physical States by Means of Computers" seems relevant in the present context As regards the application of computers in finding irreducible subspaces of the amplitude space of vibrating systems, the work of Zerbi et al. has recently been supplemented by a computer program written by Ra [Ø.Ra, 'SYMVEC', a FORTRAN IV Routine for Generating Irreducible Manifolds in Molecule and Crystal Displacement Space (in) thesis, NTH, Trondheim 1972]. A condensed version of the repor on this program is being prepared for publication.[40] This program, which has been tested and debugged rather carefully, successfully automates the symmetry analysis of a vibrating crystal. Being based on the application of projective representations of wave-vector groups, the routine treats symmorphic and non-symmorphic lattices on and equal footing, and can handle any point in the Brillouin zone. Given the projective representations relevant to line groups, polymers and chain molecules can be analysed without any modification of the program. Description of molecular crystals on the atomi or a molecular level is optional, since means of analysing also rigid-body motion have been incorporated in the routine. By an appropriate arrangement of input data the program covers also vibrating molecules.

S.6. Chapters 8 and 9

In a recent paper[41] many-body techniques involving propagators and Green's functions are used to recover the classical results of Lifshitz[42] on the harmonic vibrations of imperfect systems. This rederivation of some of Lifshitz' key formulae presumably offers a comparatively transparent illustration of the application of field methodology to statistical mechanics. Although Ref. 41 deals with crystal lattices, the theory can be carried over to molecules without further ado, save a trivial change of notation.

The power-series expansion of Σ derived in Chapter 9 has been successfully applied in numerical computations on isotopic molecules of adamantane[43] (see also Chapter 19).

As foreshadowed in the concluding section of Chapter 9 this new and non-spectral method of calculating Σ has been adapted to equal-time amplitude correlation functions in crystals.[44] By conducting comparative numerical assessment studies involving CaF_2 it has been shown that the new method provides a considerable reduction in the computing time normally required in theoretical calculations of Debye-Waller factors.

Recently the molecular version has been extended with the purpose of finding a computational recipe particularly suited to Σ calculations in the low temperature limit.[45] To this end the replacement of the O-shaped circular path of integration with a D-shaped path proved convenient. The final result can also be obtained from the previous expansion in residue matrices by subjecting this series to Poisson or Euler transformations.

S.7. Chapters 10 and 12

The use of rotational constants determined by high-resolution spectroscopy for the improvement of the quality of electron diffraction analyses, mentioned briefly in Section 10.5, is discussed in more detail in Ref. 106 of Chapt. 10. In a recent study of the structure of propynal by gas electron diffraction,[46] it was only possible by electron diffraction alone to determine an average value of the C=O and C≡C distances and that of the C-H distances because they were very nearly equal to each other. However, by a combined analysis of electron diffraction intensities[46] and the rotational constants for the normal species and two mono-deuterated species from microwave spectroscopy,[47] it was possible to determine the structure with no assumption except that the C≡C-H angle is 180°. Since a great many of the molecules of chemical interest have nearly equal internuclear distances, this technique can be applied if precise rotational constants are available.

An application of the principles described in Chapter 12, i.e. mutual conversions of geometrical parameters and their physical significance, has been discussed in a recent review paper (Ref. 106 of Chapt. 10 and Ref. 99 of Chapt. 12). This article starts with a classification of molecules into four categories from the standpoint of the use of electron diffraction and spectroscopic methods for determining molecular geometry and motions. (a) Electron diffraction is the only possible means for determining precise molecular geometry in the gas phase. (b) Spectroscopy is much more effective than electron diffraction. (c, d) The joint use of diffraction and spectroscopic methods is possible and should be profitable, where in (c) electron diffraction alone can determine a unique structure and in (d) neither method alone can provide a unique set of geometrical parameters without making an assumption. Examples of molecules in all categories studied recently from this standpoint are listed with references. In particular, group Vb halides, conjugated aliphatic molecules, strained molecules, and molecules with large amplitudes of vibration are discussed in some detail. Particular emphasis is laid on the advantage of comparing bond distances in analogous molecules in the r_g representation and bond angles in the r_z or r_α representations instead of the r_s and r_0 representations.

S.8. Chapter 14

Part I. The harmonic force field for nitric acid is reported in Ref. 48. That work also contains a normal coordinate analysis with calculated mean amplitudes for trans- and cis nitrous acid.

Part II. Ref. 10 in Part II of Chapter 14 is a part of an article series[49-52] entitled 'Molecular Vibrations and Mean Amplitudes of Carboxylic Acids'. It deals with formic acid monomer[49] and dimer,[52] along with acetic acid[50] and oxalic acid[51] monomers. In addition to the electron diffraction investigation of monomer and dimer formic acid[53] (cited as Ref. 9 in Part II of Chapt. 14), several other carboxylic acids have been investigated by electron diffraction: monomer and dimer acetic acid,[54] monomer and dimer propionic acid,[55] and oxalic acid.[56]

S.9. Chapter 15

A detailed report on the matrix isolation infrared investigation of $LiNaF_2$ is reported elsewhere.[57] That work also contains a summary of the present force constant analysis, but mean amplitudes have not been published previously. Calculated thermodynamic functions for $^7Li_2F_2$, Na_2F_2, and $^7LiNaF_2$ have been given.[58]

Similar investigations have been conducted for $LiAlF_4$[59,60] and $NaAlF_4$.[60] The molecules are supposed to have fluorine bridges and C_{2v} symmetry in the gas phase. Thermodynamic functions are given.[61] Work is currently under way on other fluorine bridge compounds, $LiBeF_3$ and Li_2BeF_4. Calculated entropy values have been compared with those from thermodynamic equilibrium pressure measurements for $LiAlF_4$[62] and $NaAlF_4$.[63]

Other well-known molecules with halogen bridges are the 'borane-like' M_2X_6 compounds, where M = metal and X = halogen. A normal coordinate analysis on Al_2Cl_6 and Fe_2Cl_6 is recently published,[64] and more work in the same direction is in progress.

Al_2Cl_6 is one of the components in alkali chloride - aluminum chloride molten systems, which recently have been investigated by Raman spectroscopy.[65-67] These investigations give evidence for the successive formation of the species $AlCl_4^-$, $Al_2Cl_7^-$, Al_nCl_{3n+1} ($n \geq 3$) and Al_2Cl_6. Similar species formations have also been detected in $GaCl_3$ containing systems.[68,69]

S.10. Chapter 16

After completion of the present work the vibrational analysis of $(C_6H_6)_2Cr$ treated as a whole molecule in terms of alternative symmetry coordinate has been published.[70] The two sets of symmetry coordinates are analogous to the two sets described for eclipsed XY_6 in Part I of Chapter 16. The paper[70] also treats $(C_6D_6)_2Cr$. The effect of the central atom mass changes on some vibrational frequencies have also been calculated and compared with experimental values.[71] Perpendicular amplitude correction coefficients for free and complexed benzene are reported.[72]

Studies parallel to those of benzene sandwich compounds have been conducted for cyclopentadienides.[72-76] Ferrocene, $(C_5H_5)_2Fe$,[74,75] and ruthenocene, $(C_5H_5)_2Ru$,[76] are two of the most important compounds of this type. Recently the $(C_5H_5)_2Ni$ molecule has been investigated by gas electron diffraction;[77] cf. also Table 22-I. It is intended to perform some calculations of mean amplitudes from spectroscopic data to be compared with the electron diffraction results for this molecule. Table 22-I includes references to other cyclopentadienides studied quite recently by electron diffraction.

Extensive studies of an other type of benzene complexes, $C_6H_6Cr(CO)_3$ (benzene chromium tricarbonyl) have been performed,[78-80] and more work in the same direction is in progress. In this case, similarly to the case of dibenzene chromium, the most prominent frequency shift from free to complexed benzene is again nicely explained by kinematic coupling. A general discussion of this phenomenon has been given.[81]

S.11. Chapter 17

At the end of many X-ray structure refinements of molecular crystals the anisotropic temperature factors (b_{ij}) have been interpreted in terms of rigid-body motion, translational and librational. This was first done by Cruickshank.[82-84] In his procedure the temperature factors are converted to the total atomic mean-square amplitudes, U_{ij} . The U_{ij} values form a 3×3 symmetric tensor $\mathbf{U}_a$ for every atom a, and it should be possible to express all of them in terms of the 6 + 6 parameters of the symmetric tensors $\mathbf{T}$ and $\mathbf{L}$ * for translation and libration, respectively. Cruickshank[85] has derived $\mathbf{T}$ and $\mathbf{L}$ tensors for anthracene. In the detailed refinements of the naphthalene[86] and benzene[87,88] structures Cruickshank et al. again have given the $\mathbf{T}$ and $\mathbf{L}$ tensors obtained from data fitting on the basis of rigid-body motions. However, the possible necessity of a correction for internal molecular vibrations is mentioned.[86]

In the naphthalene work[86] theoretical estimates of sphericall[y]

---

* In the original papers denoted by $\boldsymbol{\omega}$.

averaged mean-square amplitudes for the atoms C_1, C_5, and C_9 (cf. Fig. 17-5) at room temperature are quoted from Higgs,[89] viz. 0.0017, 0.0020, and 0.0013 $Å^2$, respectively. The corresponding figures at 298 K from the analysis of Chapter 17, Part II, are 0.0024, 0.0020, and 0.0023 $Å^2$. A survey of the lattice vibrations of benzene, naphthalene, and anthracene has been published.[90]

In the extensive X-ray work on hexamethylene tetramine by Becka and Cruickshank[91,92] refinements of the atomic mean-square amplitudes were made using approximate spectroscopic values for the contributions from internal molecular vibrations. Similar refinements were made in the neutron-diffraction work on benzene,[93,94] cyclobutane,[94] glycolic acid,[95] and other molecules.[96]

Cruickshank's theory[82-84] is appropriate only to centrosymmetric molecules. The general case has been treated by Schomaker and Trueblood,[97] and interpreted in lattice-dynamical terms by Pawley.[98] The screw rotation tensor $\mathbf{S}$ is introduced in addition to the $\mathbf{T}$ and $\mathbf{L}$ tensors. It is suggested that the assumption of rigid-body motion should be included throughout the least-squares structural refinements. This method[99,100] was successfully applied using a simplified model without screw-rotation coefficients.[101] However a screw motion for the sulphate group is required in sodium alum.[102] Recent applications to hexamethylene tetramine[103] and perdeuteronaphthalene[104] have been reported.

Other applications of the Cruickshank, and Schomaker and Trueblood analyses have been given.[105-107] The notation for harmonic vibration tensors, and the representation of $\mathbf{U}$ with respect to different bases have been discussed by Cerrini[108] and Cruickshank.[109] Johnson and Levy[110,111] have given general discussions of thermal motion analyses, including probability and higher-order aspects. In a recent review,[112] a number of examples are given where the three tensors $\mathbf{T}$, $\mathbf{L}$, and $\mathbf{S}$ are determined directly from the diffraction data.

The theory of lattice dynamics has been advanced substantially since the advent of neutron inelastic scattering. In this connection the information on full phonon dispersion curves of various crystals is of great interest. Investigations in this field have been done for hexamethylene tetramine,[113] anthracene,[114] and naphthalene.[115]

In the recent computation on naphthalene[115] the effect of internal molecular vibrations on the external mode dispersion is found to be substantial. Model fitting to measured phonon dispersion curves is discussed by Pawley[116] starting from an analytic model (an advance on Ref. 114). Perdeutero-hexamethylene-tetramine is the only molecular crystal whose phonons have been measured so far with neutron coherent inelastic scattering,[117,118] and for which model fitting has been attempted.

An extensive review[119] and reports on two conferences [M.A. Nusimovici (Editor), 'Phonons', a report on the meeting at Rennes, France, July 1971. Flammarion; Report on the I.A.E.A. conference on neutron scattering, Grenoble, France, March 1972 (in press)] provide much useful material.

S.12. Chapter 18

Part I. Complete sets of calculated mean amplitudes from spectroscopic data for 1-silacyclobutane, 1,1-difluoro-1-silacyclobutane, and 1,1-dichloro-1-silacyclobutane have been published.[120] Two of these compounds were subjected to the electron diffraction work reported in Part I of Chapter 18. Refined interpretations of the electron diffraction data are now in progress.

Part II. The work on 3-chloro-1-propanol was presented by Brunvoll on the scientific session for the 10th anniversary celebra tion of founding of the Center for Studies on Chemical Structures, Budapest, October 1971. A report in Hungarian is published.[121]

S.13. Chapter 19

The electron-diffraction investigation on adamantane has been reported very briefly in a preliminary communication.[122] Quantum-mechanical MINDO/2 calculations have been performed for adamantane.[1 Some computations for isotopic adamantanes[43] are mentioned under Section S.6.

Acknowledgments: The writer wishes to express his gratitude to all authors of this book for wonderful co-operation. In the last instan many of them have kindly contributed with suggestions to the different sections of this supplementary chapter. Special thanks are due to Professors Børge Bak (University of Copenhagen) and Otto Bastiansen (University of Oslo) for helpful referee works and technical advice. Finally the scientific comments on Section S.2 provided by Professor L.S. Mayants (Academy of Sciences of USSR) are gratefully acknowledged.

REFERENCES

1 S.J.Cyvin: Molecular Vibrations and Mean Square Amplitudes, Universitetsforlaget, Oslo, and Elsevier, Amsterdam, 1968.
2 S.Cyvin: Kolebaniya molekul i srednekvadratichnye amplitudy (Molecular vibrations and mean square amplitudes; Russian transl.), Izd. "MIR", Moscow 1971.
3 R.Stølevik, H.M.Seip, and S.J.Cyvin, Chem.Phys.Letters (in press).
4 W.D.Gwinn, J.Chem.Phys. 55, 477 (1971).
5 S.J.Cyvin, to be published.
6 L.S.Mayants, Dokl.Akad.Nauk SSSR 202, 124 (1972).
7 B.S.Averbukh, L.S.Mayants, and G.B.Shaltuper, J.Mol.Spectry. 30, 310 (1969).
8 W.J.Taylor, J.Chem.Phys. 18, 1301 (1950).
9 P.Pulay and F.Török, Acta Chim.Hung. 44, 287 (1965).
10 P.Pulay and F.Török, Acta Chim.Hung. 47, 273 (1966).
11 P.Pulay, Acta Chim.Hung. 52, 49 (1967).
12 F.Török, Acta Chim.Hung. 52, 205 (1967).
13 F.Török and P.Pulay, Acta Chim.Hung. 56, 285 (1968).
14 F.Török, Acta Chim.Hung. 57, 141 (1968).
15 P.Pulay, Acta Chim.Hung. 57, 373 (1968).
16 F.Török and Gy.B.Hun, Acta Chim.Hung. 59, 303 (1969).
17 F.Török and P.Pulay, J.Mol.Structure 3, 1 (1969).
18 F.Török and P.Pulay, J.Mol.Structure 3, 283 (1969).
19 S.Toman and J.Plíva, J.Mol.Spectry. 21, 362 (1966).
20 A.Fadini, Z.ang.Mathem.Mech. 44, 506 (1964).
21 A.Fadini, Z.ang.Mathem.Mech. 45, T29 (1965).
22 W.Sawodny, A.Fadini, and K.Ballein, Spectrochim.Acta 21, 995 (1965).
23 A.Fadini, Z.Naturforschg. 21a, 426 (1966).

24 F.Billes, Acta Chim.Hung. 47, 53 (1966).
25 J.Herranz and G.Castaño, Spectrochim.Acta 22, 1965 (1966).
26 G.S.Koptev and V.M.Tatevskii, Vestn.MGU, Ser.Khim., No.6, 3 (1966).
27 G.S.Koptev, N.F.Stepanov, and V.M.Tatevskii, Vestn.MGU, Ser. Khim., No. 3, 86 (1967).
28 H.J.Becher and R.Mattes, Spectrochim.Acta 23A, 2499 (1967).
29 D.E.Freeman, J.Mol.Spectry. 27, 27 (1968).
30 S.J.Cyvin, L.A.Kristiansen, and J.Brunvoll, Acta Chim.Hung. 51, 217 (1967).
31 D.E.Freeman, J.Mol.Spectry. 22, 305 (1967).
32 L.S.Mayants and B.S.Averbukh, Zh.Strukt.Khim. 8, 565 (1967).
33 M.Gussoni and G.Zerbi, J.Mol.Spectry. 26, 485 (1968).
34 G.O.Sørensen, J.Mol.Spectry. 36, 359 (1970).
35 P.Torkington, J.Chem.Phys. 17, 1026 (1949).
36 S.J.Cyvin, Gy.H.Borossay, G.Kovács, and F.Török, to be published.
37 S.J.Cyvin, J.Chem.Phys. (in press).
38 G.Dellepiane, R.Tubino, and L. Degli Antoni Ferri, J.Chem.Phys. (in press).
39 E.M.Loebl (Editor): Group Theory and its Applications Vol. II, Academic Press, New York 1971.
40 Ø.Ra, to be published.
41 Ø.Ra, Acta Chem.Scand. 25, 2373 (1971).
42 E.M.Lifshitz, Nuovo Cimento Suppl. 10 3, 716 (1956).
43 S.J.Cyvin, Ö.Ra, and J.Brunvoll, Indian J. Pure Appl. Phys. 9, 890 (1971).
44 Ø.Ra, to be published.
45 Ø.Ra, J.Mol.Structure (in press).
46 M.Sugié, T.Fukuyama, and K.Kuchitsu, J.Mol.Structure (in press).
47 C.C.Costain and J.R.Morton, J.Chem.Phys. 31, 389 (1959).
48 S.J.Cyvin, B.N.Cyvin, and B.Vizi, Acta Chim.Hung. (in press).
49 S.J.Cyvin, I.Alfheim, and G.Hagen, Acta Chem.Scand. 24, 3038 (1970).
50 I.Alfheim and S.J.Cyvin, Acta Chem.Scand. 24, 3043 (1970).
51 S.J.Cyvin and I.Alfheim, Acta Chem.Scand. 24, 2648 (1970).
52 I.Alfheim, G.Hagen, and S.J.Cyvin, J.Mol.Structure 8, 159 (1971)
53 A.Almenningen, O.Bastiansen, and T.Motzfeldt, Acta Chem.Scand.

23, 2848 (1969).
54 J.L.Derissen, J.Mol.Structure 7, 67 (1971).
55 J.L.Derissen, J.Mol.Structure 7, 81 (1971).
56 Z.Náhlovská, B.Náhlovský, and T.G.Strand, Acta Chem.Scand. 24, 2617 (1970).
57 S.J.Cyvin, B.N.Cyvin, and A.Snelson, J.Phys.Chem. 74, 4338 (1970).
58 B.N.Cyvin, S.J.Cyvin, D. Bhogeswara Rao, and A.Snelson, Acta Chem.Scand. 25, 470 (1971).
59 S.J.Cyvin, B.N.Cyvin, D. Bhogeswara Rao, and A.Snelson, Z.anorg. allgem.Chem. 380, 212 (1971).
60 S.J.Cyvin, B.N.Cyvin, and A.Snelson, J.Phys.Chem. 75, 2609 (1971).
61 A.Snelson and S.J.Cyvin, Z.anorg.allgem.Chem. (in press).
62 D. Bhogeswara Rao, High Temperature Science 2, 381 (1970).
63 K.Grjotheim, K.Motzfeldt, and D. Bhogeswara Rao, Light Metals 1971 (Ed. T.G.Edgeworth), p. 223, Proceedings of Symposia 100th AIME Annual Meeting New York, 1971.
64 S.J.Cyvin and O.Törset, Rev.Chim.minérale 9, 179 (1972).
65 H.A.Øye, E.Rytter, P.Klæboe, and S.J.Cyvin, Acta Chem.Scand. 25, 559 (1971).
66 E.Rytter, H.A.Øye, S.J.Cyvin, B.N.Cyvin, and P.Klæboe, to be published.
67 S.J.Cyvin, P.Klaboe, E.Rytter, and H.A.Øye, J.Chem.Phys. 52, 2776 (1970).
68 H.A.Øye and W.Bues, Inorg.Nucl.Chem.Letters 8, 31 (1972).
69 H.A.Øye and W.Bues, to be published.
70 S.J.Cyvin, J.Brunvoll, and L.Schäfer, J.Chem.Phys. 54, 1517 (1971).
71 S.J.Cyvin, B.N.Cyvin, J.Brunvoll, and L.Schäfer, Acta Chem. Scand. 24, 3420 (1970).
72 J.Brunvoll, S.J.Cyvin, and L.Schäfer, Acta Chem.Scand. 25, 2357 (1971).
73 J.Brunvoll, S.J.Cyvin, and L.Schäfer, Acta Chem.Scand. 24, 3427 (1970).
74 J.Brunvoll, S.J.Cyvin, and L.Schäfer, J.Organometal.Chem. 27, 107 (1971).

75 L.Schäfer, J.Brunvoll, and S.J.Cyvin, J.Mol.Structure 11, 459 (1972).
76 J.Brunvoll, S.J.Cyvin, and L.Schäfer, Chem.Phys.Letters 13, 286 (1972).
77 L.Hedberg and K.Hedberg, J.Chem.Phys. 53, 1228 (1970).
78 L.Schäfer, G.M.Begun, and S.J.Cyvin, Spectrochim.Acta 28A, 803 (1972).
79 S.Kjelstrup, S.J.Cyvin, J.Brunvoll, and L.Schäfer, J.Organometal.Chem. 36, 137 (1972).
80 J.Brunvoll, S.J.Cyvin, and L.Schäfer, J.Organometal.Chem. 36, 143 (1972).
81 L.Schäfer, S.J.Cyvin, and J.Brunvoll, Tetrahedron 27, 6177 (1971).
82 D.W.J.Cruickshank, Acta Cryst. 9, 747 (1956).
83 D.W.J.Cruickshank, Acta Cryst. 9, 754 (1956).
84 D.W.J.Cruickshank, Acta Cryst. 9, 757 (1956).
85 D.W.J.Cruickshank, Acta Cryst. 9, 915 (1956); errata: ibid. 10, 470 (1957).
86 D.W.J.Cruickshank, Acta Cryst. 10, 504 (1957).
87 E.G.Cox, D.W.J.Cruickshank, and J.A.S.Smith, Proc.Roy.Soc. (London) A247, 1 (1958).
88 D.W.J.Cruickshank, J.Chem.Phys. 52, 5971 (1970).
89 P.W.Higgs, Acta Cryst. 8, 99 (1955).
90 D.W.J.Cruickshank, Revs. Modern Phys. 30, 163 (1958).
91 L.N.Becka and D.W.J.Cruickshank, Proc.Roy.Soc. (London) A273, 435 (1963).
92 L.N.Becka and D.W.J.Cruickshank, Proc.Roy.Soc. (London) A273, 455 (1963).
93 G.E.Bacon, N.A.Curry, and S.A.Wilson, Proc.Roy.Soc. (London) A279, 98 (1964).
94 C.K.Johnson, Thermal Neutron Diffraction (Ed. B.T.M.Willis), p. 132, Oxford University Press, London 1970.
95 R.D.Ellison, C.K.Johnson, and H.A.Levy, Acta Cryst. B27, 333 (1971).
96 G.S.Pawley, Acta Cryst. A27, 80 (1971).
97 V.Schomaker and K.N.Trueblood, Acta Cryst. B24, 63 (1968).
98 G.S.Pawley, Acta Cryst. B24, 485 (1968).

99 G.S.Pawley, Acta Cryst. 16, 1204 (1963).
100 G.S.Pawley, Acta Cryst. 17, 457 (1964).
101 G.S.Pawley, Acta Cryst. 20, 631 (1966).
102 M.I.Kay and D.T.Cromer, Acta Cryst. B26, 1349 (1970).
103 J.A.K.Duckworth, B.T.M.Willis, and G.S.Pawley, Acta Cryst. A26, 263 (1970).
104 G.S.Pawley and E.A.Yeats, Acta Cryst. B25, 2009 (1969).
105 D.M.Burns, W.G.Ferrier, and J.T.McMullan, Acta Cryst. 22, 623 (1967).
106 D.M.Burns, W.G.Ferrier, and J.T.McMullan, Acta Cryst. B24, 734 (1968).
107 J.Baudour and Y.Delugeard, Acta Cryst. A27, 222 (1971).
108 S.Cerrini, Acta Cryst. A27, 130 (1971).
109 D.W.J.Cruickshank, Acta Cryst. A27, 677 (1971).
110 C.K.Johnson, Crystallographic Computing (Ed. F.R.Ahmed), p. 207, Munksgaard, Copenhagen 1970.
111 C.K.Johnson and H.A.Levy (in) International Tables for X-Ray Crystallography, Vol. IV, Kynoch, Birmingham (in press).
112 G.S.Pawley, Advances in Structure Research by Diffraction Methods, Vol. 4 (Ed. W.Hoppe and R.Mason), Pergamon, 1971.
113 W.Cochran and G.S.Pawley, Proc.Roy.Soc. (London) A280, 1 (1964).
114 G.S.Pawley, phys.stat.sol. 20, 347 (1967).
115 G.S.Pawley and S.J.Cyvin, J.Chem.Phys. 52, 4073 (1970).
116 G.S.Pawley, phys.stat.sol. 49, 475 (1972).
117 B.M.Powell and G.Dolling, Proc.Roy.Soc. (London) A319, 209 (1970).
118 G.Dolling, B.M.Powell, and G.S.Pawley, to be published.
119 G.Venkataraman and V.C.Sahni, Revs.Modern Phys. 42, 409 (1970).
120 B.N.Cyvin, S.J.Cyvin, L.V.Vilkov, and V.Mastryukov, Rev.Chim. minérale 8, 877 (1971).
121 O.Bastiansen, J.Brunvoll, and I.Hargittai, Kémiai Közlemények (in press).
122 I.Hargittai and K.Hedberg, Chem.Commun., 1499 (1971).
123 N.Bodor and M.J.S.Dewar, J.Am.Chem.Soc. 92, 4270 (1970).

Author Index

This list contains all author names from the different chapters. The numbere references do not overlap with the Bibliography after Chapter 22. All names fror that bibliography are included here in CAPITAL letters.

Subject Index

Page numbers in *italics* refer to numerical values.